Lecture Notes in Computer Science 9439

Commenced Publication in 1973
Founding and Former Series Editors:
Gerhard Goos, Juris Hartmanis, and Jan van Leeuwen

Editorial Board

More information about this series at http://www.springer.com/series/7407

Christian Scheideler (Ed.)

Structural Information and Communication Complexity

22nd International Colloquium, SIROCCO 2015
Montserrat, Spain, July 14–16, 2015
Post-Proceedings

 Springer

Editor
Christian Scheideler
Department of Computer Science
University of Paderborn
Paderborn
Germany

ISSN 0302-9743 ISSN 1611-3349 (electronic)
Lecture Notes in Computer Science
ISBN 978-3-319-25257-5 ISBN 978-3-319-25258-2 (eBook)
DOI 10.1007/978-3-319-25258-2

Library of Congress Control Number: 2015950867

LNCS Sublibrary: SL1 – Theoretical Computer Science and General Issues

Springer Cham Heidelberg New York Dordrecht London

Printed on acid-free paper

Springer International Publishing AG Switzerland is part of Springer Science+Business Media
(www.springer.com)

Preface

This volume contains the papers presented at SIROCCO 2015, the 22nd International Colloquium on Structural Information and Communication Complexity, held during July 14-16, 2015, in Montserrat. Financial support was provided by the Catalan Society of Mathematics and the Spanish Royal Society of Mathematics.

SIROCCO is devoted to the study of the interplay between communication and knowledge in multi-processor systems from both the qualitative and quantitative viewpoints. Special emphasis is given to innovative approaches and fundamental understanding, in addition to efforts to optimize current designs. SIROCCO has a tradition of interesting and productive scientific meetings in a relaxed and pleasant atmosphere, attracting leading researchers in a variety of fields in which communication and knowledge play a significant role. This time, there were 78 submissions from 26 countries. Each submission was reviewed by at least three Program Committee members with the help of external reviewers, and the committee decided to accept 30 papers after electronic discussions. Of these papers, the papers "Under the Hood of the Bakery Algorithm: Mutual Exclusion as a Matter of Priority" by Katia Patkin and Yoram Moses and "Randomized OBDD-Based Graph Algorithms" by Marc Bury won the Best Student Paper Awards. The program also includes six keynotes from Michel Raynal, Miquel Angel Fiol, Nati Linial, Saket Navlakha, Bernhard Haeupler, and Amos Korman.

As the program chair of SIROCCO 2015, I would very much like to thank the Program Committee for all of their hard work during the paper selection process, which ran on a very tight schedule this time. I am also grateful to the external reviewers for their valuable and insightful comments and to EasyChair for providing a system that was indeed easy to use. Also many thanks to the invited speakers for accepting my invitations and giving very interesting and inspiring talks. Finally, I am very grateful to the chair of the Steering Committee, Shay Kutten, for his valuable advice, and the Organizing Committee headed by Xavier Munoz for their time and effort to ensure a successful meeting. Without all of these people it would not have been possible to come up with such a great event.

August 2015   Christian Scheideler

Organization

Program Committee

James Aspnes	Yale University, USA
Ioannis Chatzigiannakis	Sapienza University of Rome, Italy
Andrea Clementi	University of Rome Tor Vergata, Italy
Colin Cooper	King's College London, UK
Faith Ellen	University of Toronto, Canada
Robert Elsässer	University of Salzburg, Austria
Yuval Emek	Technion, Israel
Sándor Fekete	Technical University of Braunschweig, Germany
Pascal Felber	University of Neuchatel, Switzerland
Pierre Fraigniaud	CNRS and University Paris Diderot, France
Taisuke Izumi	Nagoya Institute of Technology, Japan
Adrian Kosowski	Inria Paris, France
Christoph Lenzen	MPI Saarbrücken, Germany
Boaz Patt-Shamir	Tel Aviv University, Israel
Sriram Pemmaraju	University of Iowa, USA
Seth Pettie	University of Michigan, USA
Sergio Rajsbaum	UNAM, Mexico
Andrea Richa	Arizona State University, USA
Harald Räcke	TU München, Germany
Nicola Santoro	Carleton University, Canada
Christian Scheideler	University of Paderborn, Germany
Christian Schindelhauer	University of Freiburg, Germany
Philippas Tsigas	Chalmers University of Technology, Sweden
Roger Wattenhofer	ETH Zürich, Switzerland
Philipp Woelfel	University of Calgary, Canada

Additional Reviewers

Aghazadeh, Zahra	Brahma, Siddhartha
Avin, Chen	Carmel, Yuval
Bal, Deepak	Casteigts, Arnaud
Bampas, Evangelos	Cord-Landwehr, Andreas
Barenboim, Leonid	Czygrinow, Andrzej
Becchetti, Luca	Das, Shantanu
Berenbrink, Petra	Denysyuk, Oksana
Bonato, Anthony	Di Luna, Giuseppe Antonio

Eren, Tolga
Even, Guy
Fekete, Sándor
Gasieniec, Leszek
Gavoille, Cyril
Georgiou, Chryssis
Georgiou, Konstantinos
Godard, Emmanuel
Hadzilacos, Vassos
Haeupler, Bernhard
Hegeman, James
Ilcinkas, David
Jakoby, Andreas
Jurdzinski, Tomasz
Klasing, Ralf
Kuhn, Fabian
Labourel, Arnaud
Larrea, Mikel
Leucci, Stefano
Lotker, Zvi
Mallmann-Trenn, Frederik
Markou, Euripides
Martin, Russell
Mercier, Hugues
Michail, Othon
Miller, Avery
Monaco, Gianpiero

Natale, Emanuele
Navarra, Alfredo
Ortolf, Christian
Palfrader, Peter
Panagopoulou, Panagiota
Pasquale, Francesco
Pavlogiannis, Andreas
Pelc, Andrzej
Peleg, David
Podlipyan, Pavel
Rivera, Nicolás
Rivière, Etienne
Rossi, Gianluca
Ruppert, Eric
Różański, Michał
Scalosub, Gabriel
Schmidt, Christiane
Stauffer, Alexandre
Sutra, Pierre
Tanigawa, Shin-Ichi
Trinker, Horst
Uznański, Przemysław
Vaccaro, Ugo
Viglietta, Giovanni
Westermann, Matthias
Yamauchi, Yukiko

Contents

Communication Patterns and Input Patterns in Distributed Computing

(Invited Talk)

Michel Raynal

Institut Universitaire de France &
IRISA, Université de Rennes, France &
Department of Computing, Polytechnic University, Hong Kong
raynal@irisa.fr

This paper was written during the week January 5-9, 2014.

"Je suis Charlie..." (January 7, 2014)

Abstract. A *communication pattern* is a pattern on messages exchanged in a distributed computation. An *input pattern* is a vector made up of the input parameters of the processes involved in a distributed computation. This paper investigates three such patterns. The first two, which are related to the causality relation associated with a distributed execution, are on causal message delivery and the capture of consistent global states, respectively. The last one, which concerns the consensus problem, is on vectors defined by the input values proposed by processes (this is also called the "condition-based" approach).

An aim of the paper is to promote the concept of *pattern* in distributed computing, both as a way to provide higher abstraction levels (as it is the case in communication patterns), or a tool to investigate computability or optimality issues (as it is the case with input patterns).

Keywords: Agreement problem, Byzantine failure, Causality, Causal message order, Checkpointing, Consensus, Crash failure, Error-correcting code, Input vector, Message pattern, Zigzag path.

1 Introduction

On Patterns Encountered in Computing. In this paper a *pattern* is seen as a specific arrangement of objects (messages, control flows, processes, input data, etc.) whose aim is to provide either regular structures, or an appropriate abstraction level, or an appropriate setting, which facilitate the design of algorithms solving distributed computing problems[1].

Maybe one of the most famous patterns encountered in computing science is the pattern used by William George Horner (1786-1837) to compute a polynomial, namely,

$$(\cdots(((a_n * x + a_{n-1}) * x + a_{n-2}) * x + a_{n-3}) * x + \cdots + a_1) * x + a_0.$$

---

[1] In the SIROCCO context, a pattern can be seen as a specific type of *structural information*.

© Springer International Publishing Switzerland 2015
C. Scheideler (Ed.): SIROCCO 2015, LNCS 9439, pp. 1–15, 2015.
DOI: 10.1007/978-3-319-25258-2_1

The basic pattern $A * x + a_i$ is iteratively used to obtain a very simple algorithm, which uses n multiplications and n additions (let us notice that this pattern-based method was known and used by Zhu Shijie, 1270-1330, under the name *fan fa* [40]).

More generally, all control structures of sequential computing (such as loops, and predicate-based statements) can be seen as familiar computation patterns. In the domain of parallel computing, where one has to solve problems whose solutions can be based on a regular structure, the pattern-based approach called *systolic programming* has proved to be both easy to use and efficient [10].

When considering the distributed setting, the situation is different. Only a few basic patterns have been abstracted and are now recognized as fundamental. One of them, introduced to help structure distributed computations, is called *round-based* computation. This pattern generalizes the notion of iteration to (both synchronous and asynchronous) distributed computing.

Content of the Paper. This paper is a short scientific essay on patterns in distributed computing. As it is an *essay*, its aim is neither to be exhaustive, nor to give research directions. More precisely, the paper considers two kinds of patterns, one related to the causality created by messages exchanged by computing entities (processes), while an other is related to the input data from which the processes have to agree.

Patterns Related to Message Exchange. In addition to the data they carry, messages create a causality relation (from causes to effects) among the events produced by the processes defining a distributed computation. This relation, expressed for the first time in 1978 by Lamport [21], is a master key to solve causality-related problems [35].

The paper presents two causality-based problems related to message exchange patterns. The first one, called *causal message delivery*, is addressed in Section 2. As indicated by its name, its aim is to reduce the asynchrony (noise) in message delivery, namely, for any process p, the delivery of the messages sent to p has to respect their causal sending order. Hence, the aim is here to provide processes with a higher abstraction level where message delivery is always in agreement with the causality relation on their sending.

The second problem, which is related to the computation of consistent global states (also global checkpoints), is addressed in Section 3. As a global state is made up of a local state per process, its consistency requires that no two of its local states causally depend on one another. An important issue is then to ensure that any local state defined as local checkpoint belongs to a consistent global state. The difficulty here is related to the existence of hidden dependencies (captured with the notion of a *zigzag* path [29]). The paper will show how it is possible to cope with these hidden dependencies by demanding processes to take additional local checkpoints so that all checkpoints patterns are such that any local checkpoint belongs to a consistent global state/checkpoint.

Patterns Related to Input Data. The second type of pattern investigated is related to the consensus problem in asynchronous systems where processes may crash, or even commit Byzantine failures. As consensus is impossible to solve in asynchronous systems where even only one process may crash [12], it remains impossible to solve in the presence of Byzantine processes. The paper presents in Section 4 the *condition-based* approach to solve consensus in such a context [25], which may actually be seen as a pattern-based approach. Let the input data (one per process) define what is usually called an *input vector*. The approach consists in defining the greatest set of input vectors such that, if the present input vector belongs to this set, then consensus can be solved. Hence, such sets define "good" input patterns. Interestingly, the paper shows that, as far as the consensus problem is concerned, these input patterns can be characterized as error-correcting codes. Finally, it also shows that if a set S of input patterns is such that consensus can solved in an asynchronous system for any input vector of S, then it is possible to design a round-based algorithm that solves consensus optimally (with respect to the number of rounds) in a synchronous system, for any input vector of S. Hence, what is a computability issue in asynchronous system translates as an optimality issue in a synchronous system.

2 Ensuring the *Causal Delivery* Message Pattern

Causal message delivery was introduced by Birman and Joseph [5]. This message pattern, which can be be seen as the "triangle inequality" of message-passing distributed computing, reduces the asynchrony of the underlying communication network by ensuring that any two messages sent (by the same or different processes) to the same destination process are delivered in their causal sending order. Hence, the causal message delivery pattern provides the processes with a communication abstraction of higher level than send/receive.

To simplify the presentation, we consider here broadcast communication, and we say that, at the application layer, messages are *broadcast* and *delivered*.

Definition. Let $\xrightarrow{ev}$ denote the causality relation on the events produced by the processes of a distributed execution, as defined by Lamport in [21]. Let us remind that this relation is a partial order. Considering a message m broadcast by a process, let $bc(m)$ denote its broadcast event, and $rec_j(m)$ denotes is delivery event at a process p_j.

The causal message delivery pattern is defined as follows. For any pair of messages m and m' we have: $\left(bc(m) \xrightarrow{ev} bc(m')\right) \Rightarrow \left[\forall j : \left(rec_j(m) \xrightarrow{ev} rec_j(m')\right)\right]$.

It is easy to see that, if causal message delivery is restricted to messages sent by the same sender, we obtain the FIFO delivery order. In that sense, causal delivery extends FIFO delivery to any pair of messages whose sending are causally related. Let also observe that, as $\xrightarrow{ev}$ is a partial order, it is possible that for two messages m and m' not related by $\xrightarrow{ev}$, m is delivered before m' at some process p_j, while m' is delivered before m at another process p_k.

A simple example is given in Figure 1. The execution on the left side does not satisfy causal delivery because, while $m_1 \xrightarrow{ev} m_3$, m_3 is delivered before m_1 at p_3. Differently, the execution on the right side satisfies causal delivery. Let us observe that, as the broadcast of m_1 and m_2 are not causally related, these messages can be delivered in different order at any process.

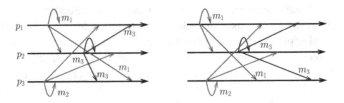

Fig. 1. Illustration of causal message delivery

---

operation broadcast(m) **is**
(1)   **for each** $j \in \{1, ..., n\} \setminus \{i\}$ **do** send($m, broadcast_i[1..n]$) to p_j **end for**;
(2)   delivery of m to the application layer;
(3) $broadcast_i[i] \leftarrow broadcast_i[i] + 1$.

when $(m, broadcast[1..n])$ **is received from** p_j **do**
(4)   **wait** $(\forall k : broadcast_i[k] \geq broadcast[k])$;
(5)   delivery of m to the application layer;
(6) $broadcast_i[j] \leftarrow broadcast_i[j] + 1$.

Fig. 2. An algorithm for causal message delivery (code for p_i)

Ensuring Causal Message Delivery. A very simple algorithm ensuring causal message delivery is presented in Figure 2. This algorithm is due to Raynal, Schiper and Toueg [37].

Each process p_i manages a local array $broadcast_i[1..n]$, initialized to $[0, \ldots, 0]$, such that $broadcast_i[j]$ counts the number of messages broadcast by p_j which have been delivered by p_i. When p_i invokes broadcast(m), it sends to each other process p_j a protocol message including m and the current value of $broadcast_i[1..n]$ (line 1). This indicates to each destination process p_j that m causally depends on, for any x, the $y_x = broadcast_i[x]$ first messages broadcast by p_x. Said differently, the current value of $broadcast_i[1..n]$ captures the causal past of the message m. Then, p_i delivers the message to itself, and consequently increments $broadcast_i[i]$ to take into account this delivery (lines 2-3).

When p_i receives (at the underlying level) a protocol message $(m, broadcast)$ from a process p_j, it waits until an appropriate delivery condition $DC(m)$ is satisfied. When $DC(m)$ becomes true, p_i delivers m and increases consequently $broadcast_i[j]$. $DC(m)$ is a simple requirement that all the messages broadcast in the causal past of m (these messages are "encoded" in the control data $broadcast[1..n]$) must be delivered before m, which translates as $\forall k : broadcast_i[k] \geq broadcast[k]$ (line 4).

A proof of this algorithm, and more efficient algorithms ensuring causal message delivery, can be found in Chapter 12 of [35]. Among several application domains, this message pattern is used in cooperative work and data consistency.

3   Message Pattern: Coping with Hidden Dependencies

3.1   The Concept of a Consistent Global State

Distributed Computation = Partial Order on Local States. An event produced by a process p_i entails its progress from its previous local to its current local state. It follows that, it is possible to deduce from the partial order relation $\xrightarrow{ev}$, defined on events, a partial order relation $\xrightarrow{\sigma}$ on the set of the local states produced by the processes. Let σ_i and σ_j be two local states of processes p_i and p_j, respectively (possibly $i = j$). Intuitively, $\sigma_i \xrightarrow{\sigma} \sigma_j$ if the local state σ_i causally precedes the local state σ_j, namely, there then a causal path starting at σ_i including local states and messages that ends at σ_j (see Chapters 6 and 8 in [35] for formal definitions).

Consistent Global State. We consider here that the communication channels are directed; $c(i, j)$ denotes the channel form p_i to p_j. Moreover, it is assumed that the communication graph connecting the processes is strongly connected.

A full global state of a distributed computation is a pair (Σ, C) where Σ is a vector made up of a local state per process, $[\sigma_1, \ldots, \sigma_n]$, and C is a set including the state of each directed communication channel $c(i, j)$. We use the terminology "global state" when we are interested only in Σ.

A global state $\Sigma = [\sigma_1, \ldots, \sigma_n]$ is *consistent* if for any pair of local states (σ_i, σ_j) we have $\neg(\sigma_i \xrightarrow{\sigma} \sigma_j) \land \neg(\sigma_j \xrightarrow{\sigma} \sigma_i)$ (i.e., none of its local states causally depends on another of its local states).

A full global state Σ, C is *consistent* if (1) Σ is consistent and (2) each (directed) channel state $c(i, j) \in C$ contains all the messages -and only them– sent by p_i to p_j before σ_i and not received by p_j before σ_j (those are the messages that are *in-transit* with respect to the directed pair $\langle \sigma_i, \sigma_j \rangle$).

When computing a full global state, each process has to save one of its local states (sometimes called *local checkpoint*), and each directed pair of processes has to compute the state of the corresponding channel (if any). In order for the resulting full global state to be consistent, the processes have to cooperate in one way or another. The first algorithm to compute a consistent global state of a distributed computation was proposed by Chandy and Lamport [9] (this algorithm assumes FIFO channels).

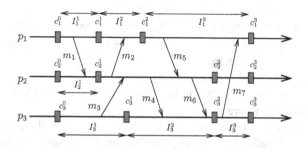

Fig. 3. A checkpoint and communication pattern (CCP)

A distributed computation where processes have saved local states, whose aim is to belong to consistent global states, is represented in Figure 3. Such a space-time diagram is an abstraction of the distributed computation taking into account the causality relation $\xrightarrow{\sigma}$ (or equivalently $\xrightarrow{ev}$) and the subset of local states defined as local checkpoints by the processes. Such an abstraction is called a *checkpoint and communication pattern* (CCP).

The local states saved by the processes (local checkpoints) are identified c_i^x, where i refers to the corresponding process p_i, and x is a sequence number. It is easy to see that, in Figure 3, the global state $[c_1^1, c_2^1, c_3^1]$ is consistent (m_3 is an in-transit message with respect to directed pair $\langle c_3^1, c_1^1 \rangle$), while $[c_1^2, c_2^2, c_3^1]$ is not consistent (the message m_5 is received from the point of view of c_2^2 and not yet sent from the point of view of c_1^2).

3.2   A Few Fundamental Questions and Their Answers

Questions. When considering local checkpoints taken by processes, several questions come to mind.

- Given a set of x ($1 \leq x \leq n$) local checkpoints, each from a distinct process, do they belong to a same consistent global state (i.e., is it possible to add to this set $(n - x)$ local checkpoints, one from each of the missing processes, so that they all together constitute a consistent global state)?
  As an example, considering Figure 3, is it possible to determine on the fly if the singleton $\{c_1^1\}$ can be extended (by adding a local checkpoint from p_2 and one from p_3) to obtain a consistent global state?
- Given a local checkpoint c_i^x, is it possible to determine on the fly, a consistent global state to which it belongs? ("On the fly" means that, if any, the consistent global state must be known by p_i when it locally saves c_i^x).

Definitions. The answer to the previous questions is based on the notion of a *zigzag path* (in short Z-path) introduced in [29], generalized in [18], and used in [15,16,17] to design a generic family of checkpointing/snapshot algorithms which are free from additional control messages (e.g., such as markers).

let an interval I_i^x of a process p_i be the set of events produced by p_i between c_i^{x-1} and c_i^x. A few intervals are indicated in Figure 3. Let $sd(m)$ and $rec(m)$ denote the event "send of m" and "reception of m", respectively.

There is a Z-path connecting a local checkpoint c_i^x to a local checkpoint c_j^y (this path is denoted $c_i^x \xrightarrow{zz} c_j^y$) if $(i = j) \wedge (x < y)$ or $i \neq j$ and there is a sequence of messages $\langle m_1; m_2; \cdots ; m_q \rangle$, $q \geq 1$, such that:
- $sd(m_1)$ occurs after c_i^x, and $rec(m_q)$ occurs before c_j^y,
- for any $\ell \in [1, q)$, if $rec(m_\ell)$ occurs in the interval I_k^z (at process p_k), then $sd(m_{\ell+1})$ occurs in an interval $I_k^{z'}$, where $z' \geq z$.

Let us notice that $m_{\ell+1}$ can be sent by p_k before it has received m_ℓ. Hence, while all causal paths are Z-paths, there are Z-paths that are not causal paths. Such Z-paths characterize hidden dependencies, as captured by Theorem 1. When looking at Figure 3, $\langle m_1; m_4 \rangle$ and $\langle m_5; m_6 \rangle$ are Z-paths which are causal paths, while $\langle m_3; m_2 \rangle$, $\langle m_5; m_4 \rangle$, and $\langle m_7; m_5; m_6 \rangle$ are Z-paths which are not causal paths.

Theorem 1. [18,29] *Let* $C = \{c(1), \ldots, c(x)\}$ *be a set of* x *local checkpoints,* $1 \leq x \leq n$. C *can be extended to obtain a consistent global state iff* $\forall\, k, \ell \in \{1, \ldots, x\}$: $\neg\left(c(k) \xrightarrow{zz} c(\ell)\right)$.

The following corollary is an immediate consequence of the previous theorem.

Corollary 1. *A local checkpoint* c_i^x *such that* $c_i^x \xrightarrow{zz} c_i^x$ *cannot belong to a consistent global state (it is useless).*

When looking at Figure 3, Due to the Z-path $\langle m_5; m_2 \rangle$, we have $c_1^2 \xrightarrow{zz} c_1^2$, from which we conclude that the local checkpoint c_1^2 cannot be part of a consistent global state. We have the same for c_3^2. These local checkpoints are useless.

3.3   Consistent Checkpoint and Communication Pattern

Consistency Definitions. As previously seen, given a distributed execution defined by its causality relation $\xrightarrow{ev}$ (or $\xrightarrow{\sigma}$) and the set S of local checkpoints taken by the processes, we obtain a CCP abstraction denoted $(S, \xrightarrow{zz})$.

A fundamental question is then: "Is a CCP abstraction $(S, \xrightarrow{\sigma})$ consistent?" To answer this question, two consistency conditions suited to CCPs have been defined in the literature.

- Z-cycle-freedom [16,18,29]. This condition states that a CCP is consistent if it has no Z-cycle (i.e., there is no $c \in S$ such that $c \xrightarrow{zz} c$ (such a CCP has no useless local checkpoint).

  From an operational point of view, this consistency condition states that no domino effect [32] can occur when when one wants to compute a consistent global state from local checkpoints.

- RDT-consistency [1,2,39] (RDT stands for Rollback-Dependency Trackability [39]). This is a stronger condition than Z-cycle-freedom. It states that a CCP is consistent if each of its hidden dependency (as captured by $\xrightarrow{zz}$) is "doubled" by a causal dependency. More formally, $\forall c1, c2 \in S : (c1 \xrightarrow{zz} c2) \Rightarrow (c1 \xrightarrow{\sigma} c2)$.

  When considering Figure 3, we have $c_3^0 \xrightarrow{zz} c_1^2$ but we do not have $c_3^0 \xrightarrow{\sigma} c_1^2$, hence the CCP is not RDT-consistent. Differently, we have both $c_1^2 \xrightarrow{zz} c_3^2$ and $c_1^2 \xrightarrow{\sigma} c_3^2$.

  From an operational point of view, if the CCP is RDT-consistent, it is possible to associate with each local checkpoint $c \in S$ a consistent global state including c, this determination being done on the fly and without communicating with other processes. More precisely, given any local checkpoint c, a vector clock value can be associated on the fly with c, which defines a consistent global state.

A Simple Algorithm Ensuring Z-cycle-Freedom. Algorithms computing consistent (full or not) global states of a CCP are described in many papers. A structured presentation of some of them is given in Chapter 8 of [35]. All these algorithms demand the processes to take additional checkpoints (those are called *forced* checkpoints) so that the CCP including both *spontaneous* and forced checkpoints be consistent with respect to the selected consistency condition (a spontaneous local checkpoint is one that is taken by a process on its own initiative).

An important issue in these algorithms is to direct the processes to take as few as possible forced checkpoints, which is a difficult problem (see for example [16] as far as Z-cycle-freedom is concerned, and [2] as far as RDT-consistency is concerned). So, in the following, we consider only Z-cycle-freedom and present a very simple "brute force" algorithm, which is far from being efficient. This algorithm, presented in Figure 4, is based on the following theorem, which assumes that a Lamport's clock date $c.date$ is associated with each local checkpoint $c \in S$.

Theorem 2. [16,35]
$[\forall c1, c2 \in S : (c1 \xrightarrow{zz} c2) \Rightarrow (c1.date < c2.date)] \Leftrightarrow [(S, \xrightarrow{zz})$ is Z-cycle-free$]$.

---

internal operation take_local_checkpoint() **is**
(1) $c \leftarrow$ copy of current local state; $c.date \leftarrow clock_i$;
(2)   save c and its date $c.date$.

when p_i decides to take a spontaneous checkpoint **do**
(3) $clock_i \leftarrow clock_i + 1$; take_local_checkpoint().

when sending MSG(m) **to** p_j **do**
(4)   send MSG($m, clock_i$) to p_j.

when receiving MSG(m, sd) **from** p_j **do**
(5)   **if** $(clock_i < sd)$ **then**
(6) $clock_i \leftarrow sd$; take_local_checkpoint()   % forced local checkpoint
(7)   **end if**;
(8)   Deliver the message m to the application process.

Fig. 4. Building z-cycle-free CCPs (code for p_i)

Each process manages a scalar local clock denoted $clock_i$, which is related only to local checkpoints. When a process p_i defines a local state as a local checkpoint, it timestamps it with its current local date, and saves it (lines 1-2). Each message is required to carry its local date (line 4). Finally, when a message m carrying its sending date sd is received by p_i, it ensures that the first predicate in Theorem 2 is satisfied. Hence, if $clock_i < sd$, p_i updates its clock to sd and takes a forced local checkpoint whose date is then sd (lines 5-7), and consequently logical time increases along all paths including local checkpoints.

Figure 5 presents the previous algorithm is action when the computation with only spontaneous checkpoints is the one of Figure 3. The forced checkpoints are indicated with white rectangles. Clock values are explicitly represented.

To summarize the aim of this section was to show how "bad" patterns can be transformed into "good" patterns when one has to ensure the capture of consistent global states defined from local checkpoints defined on the fly during a distributed computation.

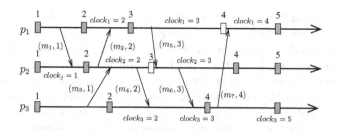

Fig. 5. An example of Z-cycle prevention

4 Input Patterns: The Condition-Based Approach for Agreement

4.1 The Consensus Problem and Its Solvability

The Consensus Problem with Process Crash Failures. The consensus problem is one of the most important problems of fault-tolerant distributed computing. Let us consider the case where up to t processes may unexpectedly crash (premature stop). The consensus problem (a) assumes that each process proposes a value and (b) requires that each non-faulty process decides a value in such a way that the three following properties are satisfied.

- Termination. Any non-faulty process decides a value.
- Validity. A decided value is a value proposed by a process.
- Agreement. No two processes decide different values.

An instance of this problem occurs each time processes have to agree in a "strong" way. Hence, consensus is a basic building block of distributed agreement problems.

Consensus Solvability in Synchronous Systems. The consensus problem was introduced by Lamport, Shostak and Pease more than thirty years ago [22,30] in the context of synchronous systems where processes may commit Byzantine failures (i.e., may behave arbitrarily). In a synchronous system, the processes proceed by executing synchronous rounds. During each round, a process first sends messages, then receives messages, and finally executes local computation. The fundamental property that characterizes a synchronous system is the fact that a message sent during a round is received during the very same round.

It is shown in [22,30] that $t < n/3$ is a necessary and sufficient requirement to solve the consensus problem in synchronous systems prone to Byzantine failures. In synchronous systems with process crash failure model, there is no constraint on t (i.e., $t < n$).

Consensus Solvability in Asynchronous Systems. The situation is different in asynchronous systems, namely, it is impossible to design a deterministic consensus algorithm in the presence of even a single process that may crash, and this is independent of the communication medium, namely, message-passing [12] or atomic read/write registers [23].

Several approaches have been proposed to circumvent the previous impossibility. One of them consists in enriching the underlying systems with eventual synchrony properties [11]. Another consists in enriching the asynchronous system with an oracle called *failure detector* [8]. Such a device is a distributed module that provides each process with (possibly unreliable) information on failures. According to the quality of the information supplied by these modules, several classes of failure detectors can be defined. The aim is then to find the weakest failure detector class that allows consensus to be solved [7]. Failure detector-based algorithms that allows consensus to be solved are described in [34].

Another approach consists in weakening the termination property and looking for a randomized algorithm. In this case, the termination property becomes: any correct process eventually decides with probability 1. Randomized consensus algorithms can be found in [4,24,31,34].

4.2 The Condition-Based Approach

A Pattern-Based Approach: Definitions. Another approach to circumvent consensus impossibility in asynchronous systems was proposed in [25], and investigated in [20,27,26]. This approach, which is a pattern-based approach, is called *condition-based*. Initially designed to address the consensus problem, it was extended to the k-set agreement problem in [6,28].

An input vector is a vector $I[1..n]$ with an entry per process, such that $I[i]$ contains the value proposed by process p_i. A condition is a set of input vectors defined from the same pattern. The idea that underlies the condition-based approach comes from the following question: "Is it possible to characterize sets of input vectors (i.e., conditions) for which consensus can be solved despite asynchrony and up to t faulty processes?"

Due to impossibility results related to consensus in asynchronous systems prone to failures [12,23], not any set C of input vectors allows to solve consensus in such a context. The notion of x-*legality* was introduced in [25] to capture vector patterns that allow the previous question to be answered. The following notations are used.
- $\mathcal{V}$ denotes the set of values that can be proposed.
- $\#(a, I)$ denotes the number of occurrences of the value $a \in \mathcal{V}$ in the vector $I[1..n]$.
- dist$(I1, I2)$ denotes the Hamming distance between the vectors $I1$ and $I2$ (i.e., the number of entries in which they differ).

Definition 1. *A condition C is x-legal if there is a function $h : C \mapsto \mathcal{V}$ such that:*

- $\forall I \in C$: $\#(h(I), I) > x$, *and*
- $\forall I1, I2 \in C$: $\big(h(I1) \neq h(I2)\big) \Rightarrow$ dist$(I1, I2) > x)$.

The parameter x is called the *degree* of the condition. Its value is related on the failure model (crash on Byzantine failure).

The intuition that underlies this definition is the following. For each of its input vectors, a condition C must allows a value to be unambiguously selected in order for it to become the decided value. The function $h()$ is the selection function. The first constraint of x-*legality* states that the decided value has to be present "enough", where "enough" is captured by "more than x times". The aim of the second constraint is to ensure that no two non-faulty processes, which –due to failures– may obtain different views of the actual input vector, do not decide differently. This is captured by the statement that any two input vectors of the condition, from which different values are decided, must be far apart "enough" from each other.

Let $C[x]$ be the set of all x-legal conditions. A simple example of a condition $C \in C[x]$ is the following. Denoted C_{max}^x, this condition is defined as follows, where $\mathrm{max}(I)$ denote the greatest element of the input vector I:

$$C_{max}^x \stackrel{def}{=} \{I : \#(\mathrm{max}(I), I) > x\}.$$

It is easy to show that this condition is both not empty and x-legal. Simpler of more sophisticated conditions can be defined (see [25,27]).

The x-legal conditions, for all $x \geq 0$, define a strict hierarchy on classes of conditions. Considering vectors of size n, we have:

$$C[n-1] \subset \cdots \subset C[x] \subset C[x-1] \subset \cdots \subset C[0],$$

where $C[0]$ is the set including all possible conditions (the largest of them being the condition including all the vectors of V^n).

Asynchronous Systems: At the Limit of Solvability. A main result of the condition-based approach is the following.

Theorem 3. [25] *Let us consider an asynchronous distributed system made up of n processes, where at most t processes may commit crash failures and where $t < n$ if communication is through atomic read/write registers, and $t < n/2$ if communication is by message passing. The consensus problem can be solved using a condition C in such a system iff C is t-legal.*

Hence, from a condition-based point of view, t-legality is the necessary and sufficient requirement to bypass the impossibility of consensus in crash-prone asynchronous systems.

Another important result, published in [27], establishes a complexity hierarchy relating classes of conditions, when processes communicate through atomic read/write registers and at most t of them may crash. The complexity is measured here by the number of shared memory accesses, called step complexity. More precisely, let $C1$ and $C2$ be two conditions, defined from the same function $h()$, such that $C1$ is y-legal (hence, $C1 \in C[y]$) and $C2$ is y'-legal (hence, $C2 \in C[y']$), where $t \leq y' < y \leq \mathrm{min}(n-t, 2t)$. A generic condition-based algorithm is presented in [27], which solves consensus more efficiently when the underlying condition is $C1$ than when it is $C2$. More precisely, when instantiated with $C1 \in C[y]$, this algorithm directs a process to issue at most $O(n \log_2(\lceil \frac{2t-y}{2} \rceil + 1))$ read/write operations to atomic shared read/write registers.

This means that conditions (sets of input vectors) with smaller sizes allow for more and more efficient algorithms. Hence, the following complexity hierarchy in crash-prone asynchronous read/write systems (where condition inclusion is from more efficient algorithms to less efficient algorithms): $C[\mathrm{min}(n-t, 2t)] \subset \cdots C[y+1] \subset C[y] \cdots \subset C[t]$. The degree y, $\mathrm{min}(n-t, 2t) \leq y \leq t$ represents the difficulty of the class $C[y]$ (the greater y, the more efficient the condition-based approach).

A side-effect of this hierarchy lies in the observation that, using more restricted conditions than the one which are $(\min(n-t, 2t))$-legal, does not provide more efficient condition-based read/write consensus algorithms.

Synchronous Systems: At the Limit of Efficiency. The condition-based approach was extended to synchronous message-passing systems in [26]. As consensus can be solved in such systems for any input vector despite up to $t < n$ process crash failures, the aim was not to address computability issues, but complexity issues, measured as the lowest number of synchronous rounds for the processes to decide in worst case scenarios.

This paper shows the following. It is possible to solve consensus more an more efficiently when the degree x of the condition increases from 0 to t. More precisely, the algorithm presented in [26] directs the processes to decide in at most $(t + 1 - x)$ rounds, when instantiated with a condition $C \in C[x]$ (i.e., an x-legal condition) where $0 \leq x \leq t$.

Synchronous Systems vs Asynchronous Systems. It is worth remarking that the class of conditions that allow for the most efficient consensus algorithm in synchronous systems, namely the class $C[t]$ made up of all t-legal conditions, is the largest one that allows to bypass the consensus impossibility in pure asynchronous read/write or message-passing systems. It follows that optimality on the synchronous side, and decidability on the asynchronous side, are the two faces of the very same coin. This discussion is summarized in Figure 6.

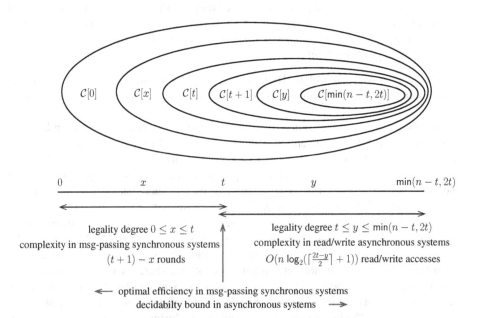

Fig. 6. Hierarchy on the condition classes $C[z]$, $0 \leq z \leq \min(n - t, 2t)$

4.3   Input Patterns vs Error-Correcting Codes

One way to address a distributed agreement problem such as consensus is to consider that an input vector encodes a value, namely the value that has to be decided from this input vector (an input vector is then seen as as a codeword).

Error-Correcting Codes: A Short Reminder. [2] An error-correcting code (ECC) problem arises when one wants to transmit a message m over a channel that can introduce errors. The universe of values V is the alphabet over which messages are constructed. There is a universe of possible messages from which m is selected, then a *coding function* $c()$ is applied to obtain a *codeword* $c = c(m)$, which is transmitted over the channel. The channel can introduce value errors by changing at most f_v symbols of c, or erasure errors (an erasure occurs when the received symbol does not belong to the alphabet) by changing at most f_c symbols of c to some value $\perp$. The resulting word, c', is received at the other end of the channel, where a decoding function $d()$ is applied to c', and the original message is recovered, $m = d(c')$.

We assume all codewords are of the same length, n, over the alphabet V (i.e., a block code). A *code* C is a set of codewords. The problem is then, given the universe of possible messages, to design a coding function $c()$, an associated decoding function $d()$, and a code C that allow the receiver to recover the word sent from the word it receives. It is said that the code is (f_v, f_c)-*error/erasure decoding*. The ECC theory has been widely studied and has applications in many diverse branches of mathematics and engineering (see any textbook, e.g., [3]). A basic theorem is the following:

Theorem 4. [3] *A code C is (f_v, f_c)-error/erasure decoding iff its minimal Hamming distance is $\geq 2f_v + f_c + 1$.*

Error-Correcting Codes vs Consensus. It appears that an erasure error corresponds to a process crash, while a value error corresponds to a Byzantine behavior. As a result, it follows that it is possible to adapt the condition-based approach to Byzantine failures, as shown in [13]. This paper provided also a new proof of the impossibility to build (f_c, f_v)-*perfect* codes. This proof reduces the construction of such codes to a distributed agreement problem in a distributed asynchronous system prone to process failures. The impossibility follows then directly from the consensus impossibility in such systems [12,23].

5   Conclusion

Aim of the Paper. The aim of this paper was to show that the notion of a *pattern* can be a useful notion in the distributed computing area. To this end, two communication patterns and one input pattern have been presented.

Personal Remark. This paper was written to be the companion paper of the SIROCCO invited talk associated with the Prize "Innovation in Distributed Computing". I choose to present three of my works, which were done at different periods, 1991, 1997-2002,

---

[2] The text of this paragraph is from [13].

and 2003-2007. As, we have seen, two of them are related to causality, and one to distributed agreement, but all of them are based on the notion of a pattern.

More generally, I think that *algorithmics* lies at the core of computing science [14], and that, in addition to automata, *synchronization* and *non-determinism* are among its fundamental concepts [19,36,38].

Acknowledgments. I want to thank all my co-authors and my PhD students, with whom (a) I had long discussions on distributed computing, and (b) I enjoyed both the simplicity and the beauty of some distributed algorithms. Among them, a warm thank to A. Mostéfaoui and S. Rajsbaum.

References

1. Baldoni, R., Hélary, J.M., Mostéfaoui, A., Raynal, M.: A communication-induced check-pointing protocol that ensures rollback-dependency trackability. In: Proc. 27th IEEE Symposium on Fault-Tolerant Computing (FTCS-27), pp. 68–77. IEEE Press (1997)
2. Baldoni, R., Hélary, J.M., Raynal, M.: Rollback-dependency trackability: a minimal characterization and its protocol. Information and Computation 165(2), 144–173 (2001)
3. Baylis, J.: Error-Correcting Codes: a Mathematical Introduction, p. 219. Chapman & Hall Mathematics (1998)
4. Ben-Or, M.: Another advantage of free choice: completely asynchronous agreement protocols. In: Proc. 2nd Annual ACM Symposium on Principles of Distributed Computing(PODC 1983), pp. 27–30. ACM Press (1983)
5. Birman, K.P., Joseph, T.A.: Reliable communication in presence of failures. ACM Transactions on Computer Systems 5(1), 47–76 (1987)
6. Bonnet, F., Raynal, M.: Conditions for set agreement with an application to synchronous systems. Springer Journal Computer Science and Technology 24(3), 418–433 (2009)
7. Chandra, T.D., Hadzilacos, V., Toueg, S.: The weakest failure detector for solving consensus. Journal of the ACM 43(4), 685–722 (1996)
8. Chandra, T., Toueg, S.: Unreliable failure detectors for reliable distributed systems. Journal of the ACM 43(2), 225–267 (1996)
9. Chandy, K.M., Lamport, L.: Distributed snapshots: determining global states of distributed systems. ACM Transactions on Computer Systems 3(1), 63–75 (1985)
10. Chandy, K.M., Misra, J.: Systolic algorithms as programs. Distributed Computing 1(3), 177–183 (1986)
11. Dwork, C., Lynch, N., Stockmeyer, L.: Consensus in the presence of partial synchrony. Journal of the ACM 35(2), 288–323 (1988)
12. Fischer, M.J., Lynch, N.A., Paterson, M.S.: Impossibility of distributed consensus with one faulty process. Journal of the ACM 32(2), 374–382 (1985)
13. Friedman, R., Mostéfaoui, A., Rajsbaum, S., Raynal, M.: Distributed agreement problems and their connection with error-correcting codes. IEEE Transactions on Computers 56(7), 865–875 (2007)
14. Harel, D., Feldman, Y.: Algorithmics, the spirit of computing, 572 p. Springer (2012)
15. Hélary, J.-M., Mostéfaoui, A., Raynal, M.: Communication-induced determination of consistent snapshots. IEEE Trans. on Parallel and Distributed Systems 10(9), 865–877 (1999)
16. Hélary, J.-M., Mostéfaoui, A., Netzer, R.H.B., Raynal, M.: Communication-based prevention of useless checkpoints in distributed computations. Distr. Comput. 13(1), 29–43 (2000)

17. Hélary, J.-M., Mostéfaoui, A., Raynal, M.: Interval consistency of asynchronous distributed computations. Journal of Computer and System Sciences 64(2), 329–349 (2002)
18. Hélary, J.-M., Netzer, R.H.B., Raynal, M.: Consistency issues in distributed checkpoints. IEEE Transactions on Software Engineering 25(4), 274–281 (1999)
19. Herlihy, M., Shavit, N.: The art of multiprocessor programming, 508 p. Morgan Kaufmann (2008). ISBN 978-0-12-370591-4
20. Izumi, T., Masuzawa, T.: Condition adaptation in synchronous consensus. IEEE Transactions on Computers 55(7), 843–853 (2006)
21. Lamport, L.: Time, clocks, and the ordering of events in a distributed system. Communications of the ACM 21(7), 558–565 (1978)
22. Lamport, L., Shostak, R., Pease, M.: The Byzantine generals problem. ACM Transactions on Programming Languages and Systems 4(3), 382–401 (1982)
23. Loui, M., Abu-Amara, H.: Memory requirements for for agreement among Unreliable Asynchronous processes. Adv. Computing Research 4, 163–183 (1987)
24. Mostéfaoui, A., Moumen, H., Raynal, M.: Signature-free asynchronous Byzantine consensus with $t < n/3$ and $O(n^2)$ messages. In: Proc. 33th ACM Symposium on Principles of Distributed Computing (PODC 2014), pp. 2–9. ACM Press (2014)
25. Mostéfaoui, A., Rajsbaum, S., Raynal, M.: Conditions on input vectors for consensus solvability in asynchronous distributed systems. Journal of the ACM 50(6), 922–954 (2003)
26. Mostéfaoui, A., Rajsbaum, S., Raynal, M.: Synchronous condition-based Consensus. Distributed Computing 18(5), 325–343 (2006)
27. Mostéfaoui, A., Rajsbaum, S., Raynal, M., Roy, M.: Condition-based consensus solvability: a hierarchy of conditions and efficient protocols. Distr. Computing 17(1), 1–20 (2004)
28. Mostéfaoui, A., Rajsbaum, S., Raynal, M., Travers, C.: The combined power of conditions and information on failures to solve asynchronous set agreement. SIAM Journal of Computing 38(4), 1574–1601 (2008)
29. Netzer, R.H.B., Xu, J.: Necessary and sufficient conditions for consistent global snapshots. IEEE Transactions on Parallel and Distributed Systems 6(2), 165–169 (1995)
30. Pease, M., Shostak, R., Lamport, L.: Reaching agreement in the presence of faults. Journal of the ACM 27, 228–234 (1980)
31. Rabin, M.: Randomized Byzantine generals. In: Proc. 24th IEEE Symposium on Foundations of Computer Science (FOCS 1983), pp. 116–124. IEEE Computer Society Press (1983)
32. Randell, B.: System structure for software fault-tolerance. IEEE Transactions on Software Engineering SE1(2), 220–232 (1975)
33. Raynal, M.: Fault-tolerant agreement in synchronous message-passing systems, 165 p. Morgan & Claypool Publishers (2010). ISBN 978-1-60845-525-6
34. Raynal, M.: Communication and agreement abstractions for fault-tolerant asynchronous distributed systems, 251 p. Morgan & Claypool Publ. (2010). ISBN 978-1-60845-293-4
35. Raynal, M.: Distributed algorithms for message-passing systems, 515 p. Springer (2013). ISBN 978-3-642-38122-5
36. Raynal, M.: Concurrent programming: algorithms, principles, and foundations, p. 530. Springer (2013). ISBN 978-3-642-32026-2
37. Raynal, M., Schiper, A., Toueg, S.: The causal ordering abstraction and a simple way to implement it. Information Processing Letters 39(6), 343–350 (1991)
38. Taubenfeld, G.: Synchronization algorithms and concurrent programming, 423 p. Pearson Education/Prentice Hall (2006). ISBN 0-131-97259-6
39. Wang, Y.-M.: Consistent global checkpoints that contain a given set of local checkpoints. IEEE Transactions on Computers 46(4), 456–468 (1997)
40. Zhu, S.: Jade mirror of the four unknowns (1303), Chinese and English bilingual, vol. 1 & 2. Liaoning Education Press, China (2006). ISBN 7-5382-6923-1

Clock Synchronization and Estimation in Highly Dynamic Networks: An Information Theoretic Approach

Ofer Feinerman[1,*] and Amos Korman[2]

[1] The Shlomo and Michla Tomarin Career Development Chair,
The Weizmann Institute of Science, Rehovot, Israel
`ofer.feinerman@weizmann.ac.il`
[2] CNRS and University Paris Diderot, Paris, 75013, France
`amos.korman@liafa.univ-paris-diderot.fr`

Abstract. We consider the *External Clock Synchronization* problem in dynamic sensor networks. Initially, sensors obtain inaccurate estimations of an external time reference and subsequently collaborate in order to synchronize their internal clocks with the external time. For simplicity, we adopt the *drift-free* assumption, where internal clocks are assumed to tick at the same pace. Hence, the problem is reduced to an estimation problem, in which the sensors need to estimate the initial external time. In this context of distributed estimation, this work is further relevant to the problem of collective approximation of environmental values by biological groups.

Unlike most works on clock synchronization that assume static networks, this paper focuses on an extreme case of highly dynamic networks. We do however impose a restriction on the dynamicity of the network. Specifically, we assume a non-adaptive scheduler adversary that dictates an arbitrary, yet *independent*, meeting pattern. Such meeting patterns fit, for example, with short-time scenarios in highly dynamic settings, where each sensor interacts with only few other arbitrary sensors.

We propose an extremely simple clock synchronization (or an estimation) algorithm that is based on weighted averages, and prove that its performance on any given independent meeting pattern is highly competitive with that of the best possible algorithm, which operates without any resource or computational restrictions, and further knows the whole meeting pattern in advance. In particular, when all distributions involved are Gaussian, the performances of our scheme coincide with the optimal performances. Our proofs rely on an extensive use of the concept of Fisher information. We use the Cramér-Rao bound and our definition of a *Fisher Channel Capacity* to quantify information flows and to obtain lower bounds on collective performance. This opens the door for further rigorous quantifications of information flows within collaborative sensors.

* O.F. has been supported in part by the Clore Foundation, the Israel Science Foundation (FIRST grant no. 1694/10) and the Minerva Foundation. A.K. has been supported in part by the ANR project DISPLEXITY. This work has received funding from the European Research Council (ERC) under the European Unions Horizon 2020 research and innovation programme (grant agreement No 648032).

C. Scheideler (Ed.): SIROCCO 2015, LNCS 9439, pp. 16–30, 2015.
DOI: 10.1007/978-3-319-25258-2_2

1 Introduction

1.1 Background and Motivation

Representing and communicating information is a main interest of theoretical distributed computing. However, such studies often seem disjoint from what may be the largest body of work regarding coding and communication: Information theory [7,33]. Perhaps the main reason for this stems from the fact that distributed computing studies are traditionally concerned with noiseless models of communication, in which the content of a message that passes from one node to another is not distorted. This reliability in transmission relies on an implicit assumption that error-corrections is guaranteed by a lower level protocol that is responsible for implementing communication. Indeed, when bandwidth is sufficiently large, one can encode a message with a large number of error-correcting bits in a way that makes communication noise practically a non-issue.

In some distributed scenarios, however, distortion in communication is unavoidable. One example concerns the classical problem of *clock synchronization*, which has attracted much attention from both theoreticians in distributed computing [2,25,22,30], as well as engineers [10,15,34], see [32,37,24,39] for comprehensive surveys. In this problem, processors need to synchronize their clocks (either among themselves only or with respect to a global time reference) relying on relative time measurements between clocks. Due to unavoidable unknown delays in communication, such measurements are inherently noisy. Furthermore, since the source of the noise is the delays, error-correction does not seem to be of any use for reducing the noise. The situation becomes even more complex when processors are mobile, preventing them from reducing errors by averaging repeated measurements to the same processors, and from contacting reliable processors. Indeed, the clock synchronization problem is particularly challenging in the context of wireless sensor networks and ad hoc networks which are typically formed by autonomous, and often mobile, sensors without central control.

Distributed computing models which include noisy communication call for a rigorous comprehensive study that employs information theoretical tools. Indeed, a recent trend in the engineering community is to view the clock synchronization problem from a signal processing point of view, and adopt tools from information theory (e.g., the Cramér-Rao bound) to bound the affect/impact of inherent noise [6,15], see [39] for a survey. However, this perspective has hardly received any attention by theoreticians in distributed computing that mostly focused on worst case message delays [2,25,22,4], which do not seem to be suitable for information theoretic considerations. In fact, very few works on clock synchronisation consider a system with random delays and analyse it following a rigorous theoretical distributed algorithmic type of analysis. An exception to that is the work of Lenzen el al. [23], but also that work does not involve information theory. In this current paper, we study the clock synchronization problem through the purely theoretical distributed algorithmic perspective while adopting the signal processing and information theoretic point of view. In particular, we adopt tools from Fisher Information theory [35,40].

We consider the *external* version of the problem [8,28,30,37] in which processors (referred to as sensors hereafter) collaborate in order to synchronize their clocks with an external *global clock*. Informally, sensors initially obtain inaccurate estimates of a global (external) time $\tau^* \in \mathcal{R}$ reference, and subsequently collaborate to align their internal clocks to be as close as possible to the external clock. To this end, sensors communicate through uni-directional pairwise interactions that include inherently *noisy measurements* of the relative deviation between their internal clocks and, possibly, some complementary information. To focus on the problems caused by the initial inaccurate estimations of τ^* and the noise in the communication we restrict our attention to *drift-free* settings [2,25], in which all clocks tick at the same rate. This setting essentially reduces the problem to the problem of estimating τ^*. See, e.g., [14,36,38] for works on estimation in the engineering community. In this context of distributed estimation, our model is further relevant to collective approximation of environmental values by biological groups [19,26].

With very few exceptions that effectively deal with dynamic settings [9,20], almost all works on clock synchronization (and distributed estimation) considered static networks. Indeed, the construction of efficient clock synchronization algorithms for dynamic networks is considered as a very important and challenging task[1] [32,37]. This paper addresses this challenge by considering highly dynamic networks in which sensors have little or no control on who they interact with. Specifically, we assume a non-adaptive scheduler adversary that dictates in advance a meeting-pattern for the sensors. However, the adversary we assume is not unlimited. Specifically, in this initial work[2] we restrict the adversary to provide *independent-meeting patterns* only, in which it is guaranteed that whenever a sensor views another sensor, their transitive histories are disjoint[3]. Although they are not very good representatives of communication in static networks, independent meeting patterns fit well with highly stochastic communication patterns during short-time scales, in which each sensor observes only few other arbitrary sensors (see discussion in Section 2). Given such a meeting-pattern, we are concerned with minimizing the deviation of each internal clock from the global time.

As our objective is to model small and simple sensors, we are interested in algorithms that employ elementary computations and economic use of communication. We use competitive analysis to evaluate the performances of algorithms, comparing them to the best possible algorithm that operates under the most

---

[1] For example, dynamic meeting patterns prevent the use of classical external clock synchronization algorithms (e.g., [27,30]) that are based on one or few *source* sensors that obtain accurate estimation of the global time and govern the synchronization of other sensors.

[2] We assume independence for simplicity. As evident by this work, the independent case is already rather complex. We leave it to future work to handle more complex dependent scenarios.

[3] Another informal way to view such patterns is that they guarantee that, given the global time, whenever a sensor views another sensor, their local clocks are independent; see Section 2 for a formal definition.

liberal version of the model that allows for unrestricted resources in terms of memory and communication capacities, and individual computational ability.

Due to space considerations, throughout this paper, most proofs are omitted. These proofs can be found in [12].

1.2   Our Contribution

Lower Bounds on Optimal Performance. We first consider algorithm **Opt**, the best possible algorithm operating on the given independent meeting pattern. We note that specifying **Opt** seems challenging, especially since we do not assume a prior distribution on the starting global time, and hence the use of Bayesian statistics seems difficult. Fortunately, for our purposes, we are merely interested in lower bounding the performances of that algorithm. We achieved that by relating the smallest possible variance of a sensor at a given time to the largest possible *Fisher Information (FI)* of the sensor at that time. This measure quantifies the sensor's current knowledge regarding the relative deviation between its local time and the global time. We provide a recursive formula to calculate J_a, the FI at sensor a, for any sensor a. Specifically, initially, the FI at a sensor is the FI in the distribution family governing its initial deviation from the global time (see Section 2 for the formal definitions). When sensor a observes sensor b, the FI at a after this observation (denoted by J'_a) satisfies:

$$J'_a \leq J_a + \frac{1}{\frac{1}{J_b} + \frac{1}{J_N}}, \tag{1}$$

where J_N is the Fisher Information in the noise distribution related to the observation. To obtain this formula we prove a generalized version of the *Fisher information inequality* [35,40]. Relying on the *Cramér-Rao bound* [7], this formula is then used to bound the corresponding variance under algorithm **Opt**. Specifically, the variance of the internal clock of sensor a is at least $1/J_a$.

Equation 1 provides immediate bounds on the convergence time. Specifically, the inequality sets a bound of J_N for the increase in the *FI* per interaction. In analogy to Channel Capacity as defined by Shannon [7] we term this upper bound as the *Fisher Channel Capacity*. Given small $\epsilon > 0$, we define the convergence time $T(\epsilon)$ as the minimal number of observations required by the typical sensor until its variance drops below ϵ^2 (see Section 2 for the formal definition). Let J_0 denote the median initial Fisher Information of sensors. Based on the Fisher Channel Capacity we prove the following.

Theorem 1. *Let $J_0 \ll 1/\epsilon^2$ for some $\epsilon > 0$. Then $T(\epsilon) \geq (\frac{1}{\epsilon^2} - J_0)/J_N$.*

A Highly Competitive Elementary Algorithm. We propose a simple clock synchronization algorithm and prove that its performance on any given independent meeting pattern is highly competitive with that of the optimal one. That is, estimations of global time at each sensor remain unbiased throughout the execution and the variance at any given time is Δ_0-competitive with the best possible variance, where Δ_0 is initial Fisher-tightness (see definition in Section 2). In contrast to the optimal algorithm that may be based on transmitting complex functions in each interaction, and on performing complex internal computations, our simple algorithm is based on far more basic rules. First, transmission is restricted to a single *accuracy* parameter. Second, using the noisy measurement of deviation from the observed sensor, and the accuracy of that sensor, the observing sensor updates its internal clock and accuracy parameter by careful, yet elementary, weighted-averaging procedures.

Our weighted-average algorithm is designed to maximize the flow of Fisher Information in interactions. This is proved by showing that the accuracy parameter is, at all times, both representative of the reciprocal of the sensor's variance and close to the Fisher Information upper bound. In short, we prove the following.

Theorem 2. *There exists a simple weighted-average based clock synchronization algorithm which is Δ_0-competitive (at any sensor and at any time).*

We note that our algorithm does not require the use of sensor identities and can thus be also employed in *anonymous* networks [1,11], yielding the same performances.

Two important corollaries of Theorem 2 follow directly from the definition of the initial Fisher-tightness Δ_0.

Corollary 1. *If the number of distributions governing the initial clocks is a constant (independent of n), then our algorithm is $O(1)$-competitive, at any sensor and at any time.*

Corollary 2. *If all distributions involved are Gaussians, then the variances of our algorithm coincide with those of the optimal one, for each sensor and at any time.*

2   Preliminaries

We consider a collection of n sensors that collaborate in order to synchronize their internal clocks with an external global clock reference. We consider a set $\mathcal{F}$ of sufficiently smooth (see definition in Section 2), probability density distributions (*pdf*) centered at zero. One specific distribution among the *pdf*s in $\mathcal{F}$ is the *noise* distribution, referred to as $N(\eta)$. Each sensor a is associated with a distribution $\Phi_a(x) \in \mathcal{F}$ which governs the initialization deviation of its internal clock from the global time as described in the next paragraph. Depending on the specific model, we assume that sensor a knows various properties of Φ_a. In the most restricted model, sensor a knows only the variance of Φ_a and in the most liberal model (considered for the sake of lower bounds), a knows the full description of Φ_a. Execution is initiated when the global time is some $\tau^* \in \mathcal{R}$, chosen by an adversary.

Two important cases are (1) when $\mathcal{F}$ contains a constant number of distributions (independent of the number of sensors) and (2) when all distributions in F are Gaussian. Both cases serve as reasonable assumptions for realistic scenarios. For the former case we shall show asymptotically optimal performances and for the latter case we shall show strict optimal (non-asymptotical) performance.

Local Clocks. Each sensor a is initialized with a local clock $\ell_a(0) \in \mathcal{R}$, randomly chosen according to $\Phi_a(x - \tau^*)$, independently of all other sensors. That is, as $\Phi_a(x)$ is centred around zero, the initial local time $\ell_a(0)$ is distributed around τ^*, and this distribution is governed by Φ_a. We stress that sensor a does not know the value τ^* and from its own local perspective the execution started at time $\ell_a(0)$. Sensors rely on both social interactions and further environmental cues[4] to improve their estimates of the global time. In between such events sensors are free to perform "shift" operations to adjust their local clocks. To focus on the problems occurred by the initial inaccurate estimations of τ^* and the noise in the communication we restrict our attention to *drift-free* settings [2,25], in which all clocks tick at the same rate, consistent with the global time.

Opinions. The drift-free assumption reduces the external clock-synchronization problem to the problem of estimating τ^*. Indeed, recall that local clocks are initialized to different values but progress at the same rate. Because sensor a can keep the precise time since the beginning of the execution, its deviation from the global time can be corrected had it known the difference between, $\ell_a(0)$, the initial local clock of a, and τ^*, the global time when the execution started. Hence, one can view the goal of sensor a as estimating τ^*. That is, without loss of generality, we may assume that all shifts performed by sensor a throughout the execution are shifts of its initial position $\ell_a(0)$ aiming to align it to be as close as possible to τ^*. Taking this perspective, we associate with each sensor an *opinion* variable x_a, initialized to $x_a(0) := \ell_a(0)$, and the goal of a is to have its opinion be as close as possible to τ^*. We view the opinion x_a as an *estimator* of τ^*, and note that initially, due to the properties of Φ_a, this estimator is unbiased, i.e., mean$(x_a(0) - \tau^*) = 0$. It is required that at any point in the execution, the opinion x_a remains an unbiased estimator of τ^*, and the goal of a is to minimize its Mean Square Error (MSE).

Due to this simple relation between internal clocks and opinions, in the remaining of this paper, we shall adopt the latter perspective and concern ourselves only with optimizing the opinions of sensors as estimators for τ^*, without discussing further the internal clocks.

Rounds. For simplicity of presentation, we assume that the execution proceeds in discrete rounds. We stress however that the rounds represent the order in which communication events occur (as determined by the meeting-pattern, see

---

[4] In order for the model to include environmental cues, one or more of the sensors can be taken to represent the global clock. The initial times of these sensors are chosen according to highly concentrated distributions, Φ_a, around τ^* and remain fixed thereafter.

below), and do not necessarily correspond to the actual time. Given an algorithm A, the opinion maintained by the algorithm at round t (where t is a non-negative integer) at sensor a is denoted by $x_a(t, A)$. As mentioned, the algorithm aims to keep this value as close as possible to τ^*. When A is clear from the context, we may omit writing it and use the term $x_a(t)$ instead.

In each round $t \geq 1$, a sensor may first choose to shift (or not) its opinion, and then, if specified in the meeting pattern, it observes another specified sensor, thus obtaining some information. To summarize, in each round, a sensor executes the following consecutive actions: (1) Perform internal computation; (2) Perform an opinion-shift: $x_a(t) = x_a(t-1) + \Delta(x)$; and (3) Observe (or not) another sensor. For simplicity, all these three operations are assumed to occur instantaneously, that is, in zero time.

Mobility and Adversarial Independent Meeting Patterns. In cases where sensors are embedded in a Euclidian space, distances between positioning of sensors may impact the possible interactions. To account for physical mobility, and be as general as possible, we assume that an oblivious adversary controls the meeting pattern. That is, the adversary decides (before the execution starts), for each round, which sensor observes which other sensor.

A model that includes an unlimited adversary that controls the meeting pattern appears to be too general. In this preliminary work on the subject, we restrict the adversary to provide only *independent* meeting patterns, in which the set of sensors in the transitive history of each observing sensor is disjoint from the one of the observed sensor.

Formally, given a pattern of meetings $\mathcal{P}$, sensor a and round t, we first define the set of *relevant* sensors of a at time t, denoted by $\mathcal{R}_a(t, \mathcal{P})$. At time zero, we define $\mathcal{R}_a(0, \mathcal{P}) := \{a\}$, and at round t, $\mathcal{R}_a(t, \mathcal{P}) := \mathcal{R}_a(t-1, \mathcal{P}) \cup \mathcal{R}(b, t-1, \mathcal{P})$ if a observes b at time $t-1$ (otherwise $\mathcal{R}_a(t, \mathcal{P}) := \mathcal{R}_a(t-1, \mathcal{P})$). A meeting pattern $\mathcal{P}$ is called *independent* if whenever some sensor a observes a sensor b at some time t, then $\mathcal{R}_a(t-1, \mathcal{P}) \cap \mathcal{R}(b, t-1, \mathcal{P}) = \emptyset$. Note that an independent meeting pattern guarantees that given τ^*, the internal clocks of two interacting sensors are independent. However, given τ^* and the internal clock of a, the internal clock of b and the relative time measurement between them are dependent.

Note that independent-meeting patterns are not very good representatives of communication in static networks[5]. On the other hand, independent meeting patterns fit well with highly stochastic short-time scales communication patterns,

---

[5] Indeed, in such patterns a sensor will not contact the same sensor twice, which contradicts many natural communication schemes in static networks. We note, however, that in some cases, a sequence of multiple consecutive observations between sensors can be compressed into a single observation of higher accuracy thus reducing the dependencies between observations, and possibly converting a dependent meeting pattern into an independent one. For example, if sensors have unique identities and sensor a observes sensor b several times is a row, and it is guaranteed that sensor b did not change its state during these observations, then these observations can be treated by a as a single, more accurate, observation of b.

in which each sensor observes only few other arbitrary sensors. In this sense, such patterns can be considered as representing an extreme case of dynamic systems.

Because sensors have no control of when their next interaction will occur, or if it will occur at all, we require that estimates at each sensor be as accurate as possible at *any* point in time. This requirement is stronger than the liveness property that is typically required from distributed algorithms [21].

Convergence Time. Consider a meeting pattern $\mathcal{P}$. Given small $\epsilon > 0$, the *convergence time* $T(\epsilon)$ of an algorithm A is defined as the minimal number of observations made by the typical sensor until its variance is less than ϵ^2. More formally, let ρ denote the first round when we have more than half of the population satisfying $\mathrm{var}(X_a(t, A)) < \epsilon^2$. For each sensor a, let $R(a)$ denote the number of observations made by a until time ρ. The convergence time $T(\epsilon)$ is defined as the median of $R(a)$ over all sensors a. Note that $T(\epsilon)$ is a lower bound on ρ, since since each sensor observes at most one sensor in a round.

Communication. We assume that sensors are anonymous and hence, in particular, they do not know who they observe. Conversely, for the sake of lower bounds, we allow a much more liberal setting, in which sensors have unique identifiers and know who they interact with.

When a sensor a observes another sensor b at some round t, the information transferred in this interaction contains a *passive* component and, possibly, a complementary *active* one. The passive component is a noisy relative deviation measurement between their opinions:

$$\tilde{d}_{ab}(t) = x_b(t) - x_a(t) + \eta,$$

where the additive noise term, η, is chosen from the noise probability distribution $N(\eta) \in \mathcal{F}$ whose variance is known to the sensors. (Note that this measurement is equivalent to the relative deviation measurement between the sensors' current local times because all clocks tick at the same pace.)

Elementary Algorithms. Our reference for evaluating performances is algorithm **Opt** which operates under the most liberal version of our model, which carries no restrictions on memory, communication capacities or internal computational power, and provides the best possible estimators at any sensor and at any time (we further assume that sensors acting under **Opt** know the meeting pattern in advance). In general, algorithm **Opt** may use complex calculations over very wasteful memories that include detailed distribution density functions, and possibly, accumulated measurements. Our main goal is to identify an algorithm whose performance is highly competitive with that of **Opt** but wherein communication and memory are economically used, and the local computations simple. Indeed, when it comes to applications to tiny and limited processors, simplicity and economic use of communication are crucial restrictions.

An algorithm is called *elementary* if the internal state of each sensor a contains a constant number of real[6] numbers, and the internal computations that a sensor can perform consist of a constant number of basic arithmetic operations, namely: addition, subtraction, multiplication, and division.

Competitive Analysis. Fix a finite family $\mathcal{F}$ of smooth *pdf*'s centered at zero (see the definition for smoothness in the next paragraph), and fix an assignment of a distribution $\Phi_a \in \mathcal{F}$ to each sensor a. For an algorithm A and an independent meeting pattern $\mathcal{P}$, let $X_a(t, A, \mathcal{P})$ denote the random variable indicating the opinion of sensor a at round t. Let $\text{mean}(X_a(t, A, \mathcal{P}))$ and $\text{var}(X_a(t, A, \mathcal{P}))$ denote, respectively, the mean and variance of $X_a(t, A, \mathcal{P})$, where these are taken over all possible random initial opinions, communication errors, and possibly, coins flipped by the algorithm. Note that the unbiased assumption requires that $\text{mean}(X_a(t, A, \mathcal{P})) = \tau^*$. An algorithm A is called λ-competitive, if for *any* independent pattern of meetings $\mathcal{P}$, *any* sensor a, and at *any* time t, we have: $\text{var}(X_a(t, A, \mathcal{P})) \leq \lambda \cdot \text{var}(X_a(t, \mathbf{Opt}, \mathcal{P}))$.

Fisher Information and the Cramér-Rao Bound. The Fisher information is a standard way of evaluating the amount of information that a set of random measurements holds about an unknown parameter τ of the distribution from which these measurements were taken. We provide some definitions for this notion; for more information the reader may refer to [7,40].

A single variable probability distribution function (*pdf*) Φ is called *smooth* if it satisfies the following conditions, as stated by Stam [35]: (1) $\Phi(x) > 0$ for any $x \in \mathcal{R}$, (2) the derivative Φ' exists, and (3) the integral $\int \frac{1}{\Phi(y)}(\Phi'(y))^2 dy$ exists, i.e., $\Phi'(y) \to 0$ rapidly enough for $|y| \to \infty$. Note that, in particular, these conditions hold for natural distributions such as the Gaussian distribution. Recall that we consider a finite set $\mathcal{F}$ of smooth one variable *pdf*s, one of them being the noise distribution $N(\eta)$, and all of which are centered zero.

For a smooth *pdf* Φ, let $J_\Phi^\tau := \int \frac{1}{\Phi(y)}(\Phi'(y))^2 dy$ denote the Fisher information in the parameterized family $\{(\Phi(x, \tau)\}_{\tau \in \mathcal{R}} = \{(\Phi(x - \tau)\}_{\tau \in \mathcal{R}}$ with respect to τ. In particular, let $J_N = J_N^\tau$ denote the Fisher information in the parameterized family $\{N(\eta - \tau)\}_{\tau \in \mathcal{R}}$. More generally, consider a multivariable *pdf* family $\{(\Phi(z_1 - \tau, z_2 \ldots z_k))\}_{\tau \in \mathcal{R}}$ where τ is a translation parameter. The Fisher information in this family with respect to τ is defined as:
$$J_\Phi^\tau = \int \frac{1}{\Phi(z_1 - \tau, z_2 \ldots z_k)} \left[\frac{d\Phi(z_1 - \tau, z_2 \ldots z_k)}{d\tau} \right]^2 dz_1, dz_2 \ldots dz_k \text{ if the integral exists.}$$
As previously noted [40], since τ is a translation parameter, Fisher information is both unique (there is no freedom in choosing the parametrization) and independent of τ.

The Fisher information derives its importance by association with the Cramér-Rao inequality [7]. This inequality lower bounds the variance of the best possible

---

[6] We assume real numbers for simplicity. It seems reasonable to assume that when sufficiently accurate approximation is stored instead of the real numbers similar results could be obtained.

estimator of τ^* by the reciprocal of the Fisher information that corresponds to the random variables on which this estimator is based.

Theorem 3. [**The Cramér-Rao inequality**] *Let $\hat{X}$ be any unbiased estimator of $\tau^* \in \mathcal{R}$ which is based on a multi-variable sample $\bar{z} = (z_1, z_2 \ldots z_k)$ taken from $\Phi(z_1 - \tau^*, z_2 \ldots z_k)$. Then* $\mathrm{var}(\hat{X}) \geq 1/J_\Phi^\tau$.

Initial Fisher-Tightness: To define the initial Fisher-tightness parameter Δ_0, we first define the *Fisher-tightness* of a single variable smooth distribution Φ centered at zero, as $\Delta(\Phi) = \mathrm{var}(\Phi) \cdot J_\Phi^\tau$. Note that, by the Cramér-Rao bound, $\Delta(\Phi) \geq 1$ for any such distribution Φ. Moreover, equality holds if Φ is Gaussian [7]. Recall that $\mathcal{F}$ is the finite collection of the smooth distributions containing the distributions Φ_a governing the initial opinions of sensors. The *initial Fisher-tightness* Δ_0 is the maximum of the Fisher-tightness over all distributions in $\mathcal{F}$ and the noise distribution. Specifically, let $\Delta_0 = \max\{\Delta(\Phi) \mid \Phi \in \mathcal{F}\}$. Two important observations are:

- If $\mathcal{F}$ contains a constant number of distributions then Δ_0 is a constant.
- If the distributions in $\mathcal{F}$ are all Gaussians then $\Delta_0 = 1$.

3   Lower Bounds on the Variance of Opt

In this section we provide lower bounds on the performances of algorithm **Opt** over a fixed independent pattern of meetings $\mathcal{P}$. Note that we are interested in bounding the performances of **Opt** and not in specifying its instructions. Identifying the details of **Opt** may still be of interest, but it is beyond the scope of this paper.

For simplicity of presentation, we assume that the rules of **Opt** are deterministic. We note, however, that our results can easily be extended to the case that **Opt** is probabilistic. For simplicity of notations, since this section deals only with algorithm **Opt** acting over $\mathcal{P}$, we use variables, such as the opinion $X_a(t)$ and the memory $Y_a(t)$ of sensor a, without parametrizing them by neither **Opt** nor by $\mathcal{P}$.

Under algorithm **Opt**, we assume that each sensor holds initially, in addition to the variance of Φ_a, the precise functional form of the distribution Φ_a (recall, Φ_a is centered at zero). In addition, we assume that sensors have unique identifiers and that each sensor knows the whole pattern $\mathcal{P}$ in advance. Moreover, we assume that each sensor a knows for each other sensor b, the *pdf* Φ_b governing b's initial opinion. All this information is stored in one designated part of the memory of a.

Since **Opt** does not have any bandwidth constrains, we may assume, without loss of generality, that whenever some sensor a observes another sensor b, it obtains the whole memory content of b. Since **Opt** is deterministic, its previous opinion-shifts can be extracted from its interaction history, which is, without loss of generality, encoded in its memory[7]. Hence, when sensor a observes sensor

---

[7] In case **Opt** is probabilistic, previous shifts can be extracted from the memory plus the results of coin flips which may be encoded in the memory of the sensor as well.

b at some round t, and receives b's memory together with the noisy measurement $\tilde{d}_{ab}(t) = x_b(t) - x_a(t) + \eta$, sensor a may extract all previous opinion-shifts of both itself and b, treating the measurement $\tilde{d}_{ab}(t)$ as a noisy measurement of the deviation between the initial opinions, i.e., $\tilde{d}_{ab}(0) = x_b(0) - x_a(0) + \eta$. In other words, to understand the behavior of **Opt** at round t, one may assume that sensors never shift their opinions until round t, when they use all memory they gathered to shift their opinion in the best possible manner[8]. It follows that apart from the designated memory part that all sensors share, the memory $M_a(t)$ of sensor a at round t contains the initial opinion $X_a(0)$ and a collection $Y_a(t-1) := \{\tilde{d}_{bc}(0)\}_{bc}$ of relative deviation measurements between initial opinions. That is, $M_a(t) = (X_0(t), Y_a(t-1))$. This multi-valued memory variable $M_a(t)$ contains all the information available to a at round t. In turn, this information is used by the sensor to obtain its opinion $X_a(t)$ which is required to serve as an unbiased estimator of τ^*.

The Fisher Information of Sensors. We now define the notion of the Fisher Information associated with a sensor a at round t. This definition will be used to bound from below the variance of $X_a(t)$ under algorithm **Opt**.

Consider the multi-valued memory variable $M_a(t) = (X_0(t), Y_a(t-1))$ of sensor a that at round t. Note that $Y_a(t-1)$ is independent of τ^*. Indeed, once the adversary decides on the value τ^*, all sensors' initial opinions are chosen with respect to τ^*. Hence, since sensors' memories contains only relative deviations between opinions, the memories by themselves do not contain any information regarding τ^*. In contrast, given τ^*, the random variables $Y_a(t-1)$ and $X_a(0)$ are, in general, dependent. Furthermore, in contrast to $Y_a(t-1)$, the value of $X_a(0)$ depends on τ^*, as it is chosen according to $\Phi_a(x - \tau^*)$. Hence, $M_a(t)$ is distributed according to a *pdf* family $\{(m_a(t), \tau)\}$ parameterized by a translation parameter τ. Based on $M_a(t)$, the sensor produces an unbiased estimation $X_a(t)$ of τ^*, that is, it should hold that: $\text{mean}(X_a(t) - \tau^*) = 0$, where the mean is taken with respect to the distribution of the random multi-variable $M_a(t)$.

Definition: The *Fisher Information (FI)* of sensor a at round t, termed $J_a(t)$, is the the Fisher information in the parameterized family $\{(m_a(t), \tau)\}_{\tau \in \mathcal{R}}$ with respect to τ.

By the Cramér-Rao bound, the variance of any unbiased estimator used by the sensor a at round t is bounded from below by the reciprocal of the *FI* of sensor a at that time. That is, we have:

Lemma 1. $\text{var}(X_a(t)) \geq 1/J_a(t)$.

---

[8] This observation implies, in particular, that previous opinion-shifts of sensors do not affect subsequent estimators in a way that may cause a conflict (a conflict may arise, e.g., when optimizing one sensor at one time necessarily makes estimators at another sensor, at a later time, sub-optimal), hence algorithm **Opt** is well-defined.

3.1   An Upper Bound on the Fisher Information $J_a(t)$

Lemma 1 implies that lower bounds on the variance of the opinion of a sensor can be obtained by bounding from above the corresponding *FI*. To this end, we prove the following recursive inequality. To establish the proof we had to extends the Fisher information inequality [35,40] to our multi-variable (possibly dependent) convolution case.

Theorem 4. *The FI of sensor a under algorithm* **Opt** *satisfies:* $J_a(t + 1) \leq J_a(t) + 1/(\frac{1}{J_b(t)} + \frac{1}{J_N})$.

4   A Highly-Competitive Elementary Algorithm

We define an elementary algorithm, termed **ALG**, and prove that its performances are highly-competitive with those of **Opt**. In this algorithm, each sensor a stores in its memory a single parameter $c_a \in \mathcal{R}$ that represents its *accuracy* regarding the quality of its current opinion with respect to τ^*. The initial accuracy of sensor a is set to $c_a(0) = 1/\text{var}(\Phi_a)$. When sensor a observes sensor b at some round t, it receives $c_b(t)$ and $\tilde{d}_{ab}(t)$, and acts as follows. Sensor a first computes the value $\hat{c}_b(t) = c_b(t)/(1 + c_b(t) \cdot \text{var}(N))$, a reduced accuracy parameter for sensor b that takes measurement noise into account, and then proceeds as follows:

Algorithm ALG
- **Update opinion:** $x_a(t + 1) = x_a(t) + \frac{\tilde{d}_{ab}(t) \cdot \hat{c}_b(t)}{c_a(t) + \hat{c}_b(t)}$.
- **Update accuracy :** $c_a(t + 1) = c_a(t) + \hat{c}_b(t)$.

Fix an independent meeting pattern. First, algorithm **ALG** is designed such that at all times, the opinion is preserved as an unbiased estimator of τ^* and the accuracy, $c_a(t)$, remains equal to the reciprocal of the current variance of the opinion $X_a(t, \textbf{ALG})$. That is, we have:

Lemma 2. *At any round t and for any sensor a: (1) the opinion $X_a(t, \textbf{ALG})$ serves as an unbiased estimator of τ^*, and (2) $c_a(t) = 1/\text{var}(X_a(t, \textbf{ALG}))$.*

We are now ready to analyze the competitiveness of algorithm **ALG**, by relating the variance of a sensor a at round t to the corresponding *FI*, namely, $J_a(t)$. Recall that Lemma 1 gives a lower bound on the variance of algorithm **Opt** at a sensor a, which depends on the corresponding *FI* at the sensor. Specifically, we have: $\text{var}(X_a(t, \textbf{Opt})) \geq 1/J_a(t)$. Initially, $J_a(0)$, the *FI* at a sensor a, equals the Fisher information in the parameterized family $\Phi_a(x - \tau)$ with respect to τ, and hence is at most the initial accuracy $c_a(0)$ times Δ_0. We show that the gain in accuracy following an interaction is always at least as large the corresponding upper bound on the gain in Fisher information as given in Theorem 4, divided by the initial Fisher-tightness. That is: $c_a(t+1) - c_a(t) \geq \left(1/(\frac{1}{J_b(t)} + \frac{1}{J_N})\right)/\Delta_0$. Informally, this property of **ALG** can be interpreted as maximizing the Fisher information flow in each interaction up to an approximation factor of Δ_0. By induction, we obtain the following.

Lemma 3. *At every round t, we have $c_a(t) \geq J_a(t)/\Delta_0$.*

Lemmas 1, 2 and 3 can now be combined to yield the following inequality: $\mathrm{var}(X_a(t, \mathbf{ALG})) \leq \Delta_0 \cdot \mathrm{var}(X_a(t, \mathbf{Opt}))$. This establishes Theorem 2.   □

Note that if $|F| = O(1)$ (i.e., F contains a constant number of distributions, independent of the number of sensors) then initial Fisher-tightness Δ_0 is a constant, and hence Theorem 2 states that **ALG** is constant-competitive at any sensor and at any time. In some other natural cases the performances of **ALG** are even better. One such case is when the distributions in $\mathcal{F}$ as well as the noise distribution $N(\eta)$ are all Gaussians. In this case $\Delta_0 = 1$ and Theorem 2 therefore states that the variance of **ALG** equals that of **Opt**, for any sensor at at any time. Another case is when $|F|$ is a constant, the noise is Gaussian, and both the population size n and the round t go to infinity. In this case, the performances of **ALG** become arbitrarily close to those of **Opt**.

5   The Fisher Channel Capacity and Convergence Times

For a fixed independent meeting pattern, $J_a(t)$, the FI at a sensor a and round t, was defined in Section 3 with respect to algorithm **Opt**. We note that this definition applies to any algorithm A as long as it is sufficiently smooth so that the corresponding Fisher informations are well-defined. This quantity $J_a(t, A)$ would respect the same recursive inequality as state in Theorem 4, that is, we have: $J_a(t+1, A) \leq J_a(t, A) + \frac{1}{\frac{1}{J_b(t, A)} + \frac{1}{J_N}}$. This directly implies the following:

$$J_a(t+1, A) - J_a(t, A) \leq J_N . \tag{2}$$

The inequality above sets a bound of J_N for the increase in *FI* per round. In analogy to Channel Capacity as defined by Shannon [7] we term this upper bound as the *Fisher Channel Capacity*.

The restriction on information flow as given by the Fisher Channel Capacity can be translated into lower bounds for convergence time of algorithm **Opt** (and hence also apply for any algorithm). Recall, ρ is the first round when we have more than half of the population satisfying $\mathrm{var}(X_a(t)) < \epsilon^2$. By Lemma 1, a sensor, a, with variance smaller than ϵ^2 must have a large *FI*, specifically, $J_a(\rho) \geq 1/\epsilon^2$. To get some intuition on the convergence time, assume that the number of sensors is odd, and let J_0 denote the median initial *FI* of sensors (this is the median of the *FI*, J_{Φ_a}, over all sensors a), and assume $J_0 \ll 1/\epsilon^2$. By definition, more than a half of the population have initial Fisher information at most J_0. By the Pigeon-hole principle, at least one sensor has an *FI* of, at most, J_0 at $t = 0$ and, at least, $1/\epsilon^2$ at $t = \rho$. Theorem 1 follows by the fact that, by Equation 2, this sensor could increase its *FI* by, at most, J_N in each observation.

References

1. Angluin, D.: Local and global properties in networks of processors. In: STOC, pp. 82–93 (1980)
2. Attiya, H., Herzberg, A., Rajsbaum, S.: Optimal Clock Synchronization under Different Delay Assumptions. SIAM J. Comput. 25(2), 369–389 (1996)
3. Bar-Yossef, Z., Jayram, T.S., Kumar, R., Sivakumar, D.: Info. Theory Methods in Comm. Complexity. IEEE Conf. on Computational Complexity, 93–102 (2002)
4. Biaz, S., Welch, J.L.: Closed form bounds for clock synchronization under simple uncertainty assumptions. Inf. Process. Lett. 80(3), 151–157 (2001)
5. Blachman, N.M.: The convolution inequality for entropy powers. IEEE Transactions on Information Theory 11(2), 267–271 (1965)
6. Chaudhari, Q., Serpedin, E., Wu, Y.C.: Improved estimation of clock offset in sensor networks. In: ICC (2009)
7. Cover, T.M., Thomas, J.A.: Elements of Information Theory, 2nd edn. John Wiley & Sons (2006)
8. Cristian, F.: Probabilistic Clock Synchronization. Distributed Computing 3(3), 146–158 (1989)
9. Dolev, D., Halpern, J., Simons, B., Strong, R.: Dynamic fault-tolerant clock synchronization. Journal of the ACM 42(1), 143–185 (1995)
10. Elson, J., Girod, L., Estrin, D.: Fine-Grained Network Time Synchronization Using Reference Broadcasts. Operating Systems Review 36, 147–163 (2002)
11. Feinerman, O., Haeupler, B., Korman, A.: Breathe before speaking: efficient information dissemination despite noisy, limited and anonymous communication. In: PODC, pp. 114–123 (2014)
12. Feinerman, O., Korman, A.: Clock Synchronization and Estimation in Highly Dynamic Networks: An Information Theoretic Approach (An Arxiv version). http://arxiv.org/pdf/1504.08247v1.pdf
13. El Gamal, A., Kim, Y.: Network Information Theory, 709 p. Cambridge University Press (2012)
14. Gubner, J.: Distributed Estimation and Quantization. IEEE Tran. on Information Theory 39(4) (1993)
15. Jeske, D.: On the maximum likelihood estimation of clock offset. IEEE Trans. Commun. 53(1) (2005)
16. Kar, S., Moura, J.M.F.: Distributed Consensus Algorithms in Sensor Networks With Imperfect Communication: Link Failures and Channel Noise. IEEE Tran. on SIgnal Processing 57(1), 355–369 (2009)
17. Kempe, D., Dobra, A., Gehrke, J.: Gossip-based computation of aggregate information. In: FOCS 2003, pp. 482–449 (2003)
18. Koetter, R., Kschischang, F.R.: Coding for errors and erasures in random network coding. IEEE Transactions on Info. Theory 54(8), 3579–3591 (2008)
19. Korman, A., Greenwald, E., Feinerman, O.: Confidence Sharing: an Economic Strategy for Efficient Information Flows in Animal Groups. PLOS Computational Biology 10(10) (2014)

20. Kuhn, F., Lenzen, C., Locher, T., Oshman, R.: Optimal gradient clock synchronization in dynamic networks. In: PODC, pp. 430–439 (2010)
21. Lamport, L.: Proving the Correctness of Multiprocess Programs. IEEE Transactions on Software Engineering (2), 125–143 (1977)
22. Lenzen, C., Locher, T., Wattenhofer, R.: Tight Bounds for Clock Synchronization. JACM 57(2) (2010)
23. Lenzen, C., Sommer, P., Wattenhofer, R.: PulseSync: An Efficient and Scalable Clock Synchronization Protocol. ACM/IEEE Transactions on Networking (2014)
24. Lenzen, C., Locher, T., Sommer, P., Wattenhofer, R.: Clock synchronization: Open problems in theory and practice. In: van Leeuwen, J., Muscholl, A., Peleg, D., Pokorný, J., Rumpe, B. (eds.) SOFSEM 2010. LNCS, vol. 5901, pp. 61–70. Springer, Heidelberg (2010)
25. Lundelius, J., Lynch, N.: An Upper and Lower Bound for Clock Synchronization. Information and Control 62, 190–204 (1984)
26. McNamara, J.M., Houston, A.I.: Memory and the efficient use of information. Journal of Theoretical Biology 125(4), 385–395 (1987)
27. Mills, D.L.: Internet time synchronization: the network time protocol. IEEE Transactions of Communications 39(10), 1482–1493 (1991)
28. Mills, D.L.: Improved algorithms for synchronizing computer network clocks. Networks 3, 3 (1995)
29. Ostrovsky, R., Patt-Shamir, B.: Optimal and efficient clock synchronization under drifting clocks. In: PODC 1999, pp. 3–12 (1999)
30. Patt-Shamir, B., Rajsbaum, S.: A theory of clock synchronization. In: STOC 1994, pp. 810–819 (1994)
31. Rioul, O.: Information theoretic proofs of entropy power inequalities. IEEE Transactions on Information Theory 57(1), 33–55 (2011)
32. Sivrikaya, F., Yener, B.: Time synchronization in sensor networks: a survey. IEEE Network 18(4) (2004)
33. Shannon, C.: A Mathematical Theory of Communication. Technical Journal 27(3), 379–423 (1948)
34. Solis, R., Borkar, V., Kumar, P.R.: A new distributed time synchronization protocol for multihop wireless networks. In: Proc. 45th IEEE Conference on Decision and Control (CDC) (2006)
35. Stam, A.J.: Some inequalities satisfied by the quantities of information of Fisher and Shannon. Inform. and Control 2, 101–112 (1959)
36. Xiao, L., Boyd, S., Lall, S.: A scheme for robust distributed sensor fusion based on average consensus. In: Proc. of the 4th International Symposium on Information Processing in Sensor Networks (IPSN) (2005)
37. Sundararaman, B., Buy, U., Kshemkalyani, A.D.: Clock synchronization for wireless sensor networks: a survey. Ad Hoc Networks 3, 281–323 (2005)
38. Viswanathan, R., Varshney, P.K.: Distributed detection with multiple sensors I. Fundamentals. Proceedings of the IEEE (1997)
39. Wu, Y.C., Chaudhari, Q.M., Serpedin, E.: Clock Synchronization of Wireless Sensor Networks. IEEE Signal Process. Mag. 28(1), 124–138 (2011)
40. Zamir, R.: A proof of the Fisher Information inequality via a data processing arguement. IEEE Trans. Inf. Theory, 482–491 (2003)

Node Labels in Local Decision

Pierre Fraigniaud[1], Juho Hirvonen[2], and Jukka Suomela[2]

[1] Theoretical Computer Science Federation
CNRS and University Paris Diderot, France
`pierre.fraigniaud@liafa.univ-paris-diderot.fr`
[2] Helsinki Institute for Information Technology HIIT,
Department of Computer Science, Aalto University, Finland
{`juho.hirvonen,jukka.suomela`}`@aalto.fi`

Abstract. The role of unique node identifiers in network computing is well understood as far as *symmetry breaking* is concerned. However, the unique identifiers also *leak information* about the computing environment—in particular, they provide some nodes with information related to the size of the network. It was recently proved that in the context of *local decision*, there are some decision problems such that (1) they cannot be solved without unique identifiers, and (2) unique node identifiers leak a *sufficient* amount of information such that the problem becomes solvable (PODC 2013).

In this work we study what is the *minimal* amount of information that we need to leak from the environment to the nodes in order to solve local decision problems. Our key results are related to *scalar oracles* f that, for any given n, provide a multiset $f(n)$ of n labels; then the adversary assigns the labels to the n nodes in the network. This is a direct generalisation of the usual assumption of unique node identifiers. We give a complete characterisation of the *weakest oracle* that leaks at least as much information as the unique identifiers.

Our main result is the following dichotomy: we classify scalar oracles as *large* and *small*, depending on their asymptotic behaviour, and show that (1) any large oracle is at least as powerful as the unique identifiers in the context of local decision problems, while (2) for any small oracle there are local decision problems that still benefit from unique identifiers.

1 Introduction

This work studies the role of *unique node identifiers* in the context of *local decision problems* in distributed systems. We generalise the concept of node identifiers by introducing *scalar oracles* that choose the labels of the nodes, depending on the size of the network n—in essence, we let the oracle leak some information on n to the nodes—and ask what is the *weakest* scalar oracle that we could use instead of unique identifiers. We prove the following dichotomy: we classify each scalar oracle as *small* or *large*, depending on its asymptotic behaviour, and we show that the large oracles are precisely those oracles that are at least as strong as unique identifiers.

© Springer International Publishing Switzerland 2015
C. Scheideler (Ed.): SIROCCO 2015, LNCS 9439, pp. 31–45, 2015.
DOI: 10.1007/978-3-319-25258-2_3

1.1 Context and Background

The research trends within the framework of distributed computing are most often pragmatic. Problems closely related to real world applications are tackled under computational assumptions reflecting existing systems, or systems whose future existence is plausible. Unfortunately, small variations in the model settings may lead to huge gaps in terms of computational power. Typically, some problems are unsolvable in one model but may well be efficiently solvable in a slight variant of that model. In the context of *network computing*, this commonly happens depending on whether the model assumes that pairwise distinct identifiers are assigned to the nodes. While the presence of distinct identifiers is inherent to some systems (typically, those composed of artificial devices), the presence of such identifiers is questionable in others (typically, those composed of biological or chemical elements). Even if the identifiers are present, they may not necessarily be directly visible, e.g., for privacy reasons.

The absence of identifiers, or the difficulty of accessing the identifiers, limits the power of computation. Indeed, it is known that the presence of identifiers ensures two crucial properties, which are both used in the design of efficient algorithms. One such property is **symmetry breaking**. The absence of identifiers makes symmetry breaking far more difficult to achieve, or even impossible if asymmetry cannot be extracted from the inputs of the nodes, from the structure of the network, or from some source of random bits. The role of the identifiers in the framework of network computing, as far as symmetry breaking is concerned, has been investigated in depth, and is now well understood [1–8,14,16–24,27–29].

The other crucial property of the identifiers is their ability to **leak global information** about the framework in which the computation takes place. In particular, the presence of pairwise distinct identifiers guarantees that at least one node has an identifier at least n in n-node networks. This apparently very weak property was proven to actually play an important role when one is interested in checking the correctness of a system configuration in a decentralised manner. Indeed, it was shown in prior work [10] that the ability to check the legality of a system configuration with respect to some given Boolean predicate differs significantly according to the ability of the nodes to use their identifiers. This phenomenon is of a nature different from symmetry breaking, and is far less understood than the latter.

More precisely, let us define a *distributed language* as a set of system configurations (e.g., the set of properly coloured networks, or the set of networks each with a unique leader). Then let LD be the class of distributed languages that are *locally decidable*. That is, LD is the set of distributed languages for which there exists a distributed algorithm where every node inspects its neighbourhood at constant distance in the network, and outputs *yes* or *no* according to the following rule: all nodes output *yes* if and only if the instance is legal. Equivalently, the instance is illegal if and only if at least one node outputs *no*. Let LDO be defined as LD with the restriction the local algorithm is required to be *identifier oblivious*, that is, the output of every node is the same regardless of the identifiers assigned to the nodes. By definition, LDO ⊆ LD, but [10] proved that this inclusion is strict: there are languages in LD \ LDO. This strict inclusion was obtained by constructing a distributed language that can be decided by an algorithm whose outputs depend heavily on the identifiers assigned to the nodes, and in particular on the fact that at least one node has an identifier whose value is at least n.

The gap between LD and LDO has little to do with symmetry breaking. Indeed, decision tasks do not require that some nodes act differently from the others: on legal instances, all nodes must output *yes*, while on illegal instances, it is permitted (but not required) that all nodes output *no*. The gap between LD and LDO is entirely due to the fact that the identifiers leak information about the size n of the network. Moreover, it is known that the gap between LD and LDO is strongly related to computability issues: there is an identifier-oblivious *non-computable* simulation A' of every local algorithm A that uses identifiers to decide a distributed language [10]. Informally, for every language in LD \ LDO, the unique identifiers are precisely as helpful as providing the nodes with the capability of solving undecidable problems.

1.2 Objective

One objective of this paper is to measure the *amount of information* provided to a distributed system via the labels given to its nodes. For this purpose, we consider the classes LD and LDO enhanced with *oracles*, where an oracle f is a function that provides every node with information about its environment.

We focus on the class of *scalar* oracles, which are functions over the positive integers. Given an $n \geq 1$, a scalar oracle f returns a list $f(n) = (f_1, \ldots, f_n)$ of n labels (bit strings) that are assigned arbitrarily to the nodes of any n-node network in a one-to-one manner. The class LD^f (resp., LDO^f) is then defined as the class of distributed languages decidable locally by an algorithm (resp., by an identifier-oblivious algorithm) in networks labelled with oracle f.

If, for every $n \geq 1$, the n values in the list $f(n)$ are pairwise distinct, then $\mathsf{LD} \subseteq \mathsf{LDO}^f$ since the nodes can use the values provided to them by the oracle as identifiers. However, as we shall demonstrate in the paper, this pairwise distinctness condition is not necessary.

Our goal is to identify the interplay between the classes LD, LDO, LD^f, and LDO^f, with respect to any scalar oracle f, and to characterise the power of identifiers in distributed systems as far as leaking information about the environment is concerned.

1.3 Our Results

Our first result is a characterisation of the weakest oracles providing the same power as unique node identifiers. We say that a scalar oracle f is *large* if, roughly, f ensures that, for any set of k nodes, the largest value provided by f to the nodes in this set grows with k (see Section 2.3 for the precise definition). We show the following theorem.

Theorem 1. *For any computable scalar oracle f, we have $\mathsf{LDO}^f = \mathsf{LD}^f$ if and only if f is large.*

Theorem 1 is a consequence of the following two lemmas. The first says that small oracles (i.e. non-large oracles) do not capture the power of unique identifiers. Note that the following separation result holds for any small oracle, including uncomputable oracles.

Lemma 1. *For any small oracle f, there exists a language $L \in \mathsf{LD} \setminus \mathsf{LDO}^f$.*

The second is a simulation result, showing that any local decision algorithm using identifiers can be simulated by an identifier-oblivious algorithm with the help of *any* large oracle, as long as the oracle itself is computable. Essentially large oracles capture the power of unique identifiers.

Lemma 2. *For any large computable oracle f, we have $\mathsf{LD} \subseteq \mathsf{LDO}^f = \mathsf{LD}^f$.*

Theorem 1 holds despite the fact that small oracles can still produce some large values, and that there exist small oracles guaranteeing that, in any n-node network, at least one node has a value at least n. Such a small oracle would be sufficient to decide the language $L \in \mathsf{LD} \setminus \mathsf{LDO}$ presented in [10]. However, it is not sufficient to decide all languages in LD.

Our second result is a complete description of the hierarchy of the four classes LD, LDO, LD^f, and LDO^f of local decision, using identifiers or not, with or without oracles. The pictures for small and large oracles are radically different.

- For any large oracle f, the hierarchy yields a *total order*:

$$\mathsf{LDO} \subsetneq \mathsf{LD} \subseteq \mathsf{LDO}^f = \mathsf{LD}^f.$$

  The strict inclusion $\mathsf{LDO} \subsetneq \mathsf{LD}$ follows from [10]. The second inclusion $\mathsf{LD} \subseteq \mathsf{LDO}^f$ may or may not be strict depending on oracle f.
- For any small oracle f, the hierarchy yields a *partial order*. We have $\mathsf{LDO}^f \subsetneq \mathsf{LD}^f$ as a consequence of Lemma 1. However, LD and LDO^f are incomparable, in the sense that there is a language $L \in \mathsf{LD} \setminus \mathsf{LDO}^f$ for any small oracle f, and there is a language $L \in \mathsf{LDO}^f \setminus \mathsf{LD}$ for some small oracles f. Hence, the relationships of the four classes can be represented as the following diagram:

$$\mathsf{LD}^f$$
$$\nearrow \qquad \nwarrow$$
$$\mathsf{LDO}^f \qquad\qquad \mathsf{LD}$$
$$\nwarrow \qquad \nearrow$$
$$\mathsf{LDO}$$

All inclusions (represented by arrows) can be strict.

1.4 Additional Related Work

In the context of network computing, oracles and advice commonly appear in the form of *labelling schemes* [9, 15]. A typical example is a *distance labelling scheme*, which is a labelling of the nodes so that the distance between any pair of nodes can be computed or approximated based on the labels. Other examples are *routing schemes* that label the nodes with information that helps in finding a short path between any given source and destination. For graph problems, one could of course encode the entire solution in the advice string—hence the key question is whether a very small amount of advice helps with solving a given problem.

In prior work, it is commonly assumed that the oracle can give a specific piece of advice for each individual node. The advice is localised, and entirely controlled by the oracle. Moreover, the oracle can see the entire problem instance and it can tailor the advice for any given task.

In the present work, we study a much weaker setting: the oracle is only given n, and it cannot choose which label goes to which node. This is a generalisation of, among others, typical models of *networks with unique identifiers*: one commonly assumes that the unique identifiers are a permutation of $\{1, 2, \ldots, n\}$ [21], which in our case is exactly captured by the large scalar oracle

$$f(n) = (1, 2, \ldots, n),$$

or that the unique identifiers are a subset of $\{1, 2, \ldots, n^c\}$ for some constant c [26], which in our case is captured by a subfamily of large scalar oracles. Our model is also a generalisation of *anonymous networks with a unique leader* [14]—the assumption that there is a unique leader is captured by the small scalar oracle

$$f(n) = (0, 0, \ldots, 0, 1).$$

2   Model and Definitions

In this work, we augment the usual definitions of *locally checkable labellings* [23] and *local distributed decision* [10, 11, 13] with scalar oracles.

2.1   Computational Model

We deal with the standard LOCAL model [26] for distributed graph algorithms. In this model, the network is a simple connected graph $G = (V, E)$. Each node $v \in V$ has an *identifier* $\mathsf{id}(v) \in \mathbb{N}$, and all identifiers of the nodes in the network are pairwise distinct. Computation proceeds in synchronous rounds. During a round, each node communicates with its neighbours in the graph, and performs some local computation. There are no limits to the amount of communication done in a single round. Hence, in r communication rounds, each node can learn the complete topology of its radius-r neighbourhood, including the inputs and the identifiers of the nodes in this neighbourhood. In a distributed algorithm, all nodes start at the same time, and each node must halt after some number of rounds, and produce its individual output. The collection of individual outputs then forms the global output of the computation. The running time of the algorithm is the number of communication rounds until all nodes have halted.

We consider *local* algorithms, i.e., constant-time algorithms [27]. That is, we focus on algorithms with a running time that does not depend on the size n of the graph. Any such algorithm, with running time r, can be seen as a function from the set of all possible radius-r neighbourhoods to the set of all possible outputs. An *identifier-oblivious* algorithm is an algorithm whose outputs are independent of the identifiers assigned to the nodes. Note that, from the perspective of an identifier-oblivious algorithm, the set of all possible radius-r degree-d neighbourhoods is finite. This is not the case for every algorithm since there are infinitely many identifier assignments to the nodes in a radius-r degree-d neighbourhood.

Although the LOCAL model does not put any restriction on the amount of individual computation performed at each node, we only consider algorithms that are *computable*.

2.2   Local Decision Tasks

We are interested in the power of constant-time algorithms for *local decision*. A *labelled graph* is a pair (G, x), where G is a simple connected graph, and $x : V(G) \rightarrow \{0, 1\}^*$ is a function assigning a label to each node of G. A *distributed language* L is a set of labelled graphs. Examples of distributed languages include:

- 2-colouring, the language where G is a bipartite graph and $x(v) \in \{0, 1\}$ for all $v \in V(G)$ such that $x(v) \neq x(u)$ whenever $\{u, v\} \in E(G)$;
- parity, the language of graphs with an even number of nodes;
- planarity, the language that consists of all planar graphs.

We say that algorithm A decides L if and only if the output of A at every node is either *yes* or *no*, and, for every instance (G, x), A satisfies:

$$(G, x) \in L \iff \text{all nodes output } yes.$$

Hence, for an instance $(G, x) \notin L$, the algorithm A must ensure that at least one node outputs *no*. We consider two main distributed complexity classes:

- LD (for *local decision*) is the set of languages decidable by constant-time algorithms in the LOCAL model.
- LDO (for *local decision oblivious*) is the set of languages decidable by constant-time identifier-oblivious algorithms in the LOCAL model.

By definition, LDO $\subseteq$ LD, and it is known [10] that this inclusion is strict: there are languages $L \in$ LD $\setminus$ LDO. The fact that we consider only computable algorithms is crucial here—without this restriction we would have LDO $=$ LD [10].

2.3   Distributed Oracles

We study the relationship of classes LD and LDO with respect to *scalar oracles*. Such an oracle f is a function that assigns a list of n values to every positive integer n, i.e.,

$$f(n) = (f_1, f_2, \ldots, f_n)$$

with $f_i \in \{0,1\}^*$. In essence, oracle f can provide some information related to n to the nodes. In an n-node graph, each of the n nodes will receive a value $f_i \in f(n)$, $i \in [n]$. These values are arbitrarily assigned to the nodes in a one-to-one manner. Two different nodes will thus receive f_i and f_j with $i \neq j$. Note that f_i may or may not be different from f_j for $i \neq j$; this is up to the choice of the oracle. The way the values provided by the oracles are assigned to the nodes is under the control of an adversary. One example of an oracle is $f(n) = (1, 2, \ldots, n)$, which provides the nodes with identifiers. Another example is $f(n) = (0, 0, \ldots, 0)$, which provides no information to the nodes.

W.l.o.g., let us assume that $f_i \leq f_{i+1}$ for every i. We use the shorthand $f_k^{(n)}$ for the kth label provided by f on input n, that is, $f(n) = (f_1^{(n)}, f_2^{(n)}, \ldots, f_n^{(n)})$. For a fixed oracle f, we consider two main distributed complexity classes:

- LD^f is the set of languages decidable by constant-time algorithms in networks that are labelled with oracle f.
- LDO^f is the set of languages decidable by constant-time identifier-oblivious algorithms in networks that are labelled with oracle f.

We will separate oracles in two classes, which play a crucial role in the way the four classes LDO, LD, LDO^f, and LD^f interact.

Definition 1. An oracle f is said to be *large* if

$$\forall c > 0, \exists k \geq 1, \forall n \geq k, f_k^{(n)} \geq c.$$

An oracle is *small* if it is not large.

Hence, a large oracle f satisfies that, for any value $c > 0$, there exists a large enough k, such that, in every graph G of size at least k, for every set of nodes $S \subseteq V(G)$ of size $|S| \geq k$, oracle f is providing at least one node of S with a value at least as large as c. In short: every large set of nodes must include at least one node that receives a large value.

Conversely, a small oracle f satisfies that there exists a value $c > 0$ such that, for every k, we can find $n \geq k$ such that, in every n-node graph G, and for every set of nodes $S \subseteq V(G)$ of size $|S| \geq k$, there is an assignment of the values provided by f such that every node in S receives a value smaller that c. In short: there are arbitrarily large sets of nodes which all receive a small value.

For example, oracles $f(n) = (1, 2, \ldots, n)$ and $f(n) = (n, n, \ldots, n)$ are large, while oracles $f(n) = (0, 0, \ldots, 0, 1)$ and $f(n) = (0, 0, \ldots, 0, 2^n)$ are small. We emphasise that small oracles can output very large values.

3   Proof of the Main Theorem

In this section we give the proof of our main result that characterises the power of weak and large oracles with respect to identifier-oblivious local decision.

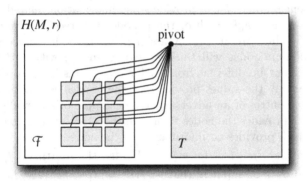

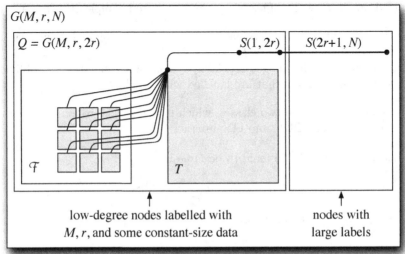

Fig. 1. The construction of Section 3.1.

3.1 Small Oracles Do Not Capture the Power of Unique Identifiers

Fraigniaud et al. [10] showed that there exists a language $L \in \mathsf{LD} \setminus \mathsf{LDO}$. We use a very similar Turing machine construction as in the proof of their Theorem 1. However, we must take into account the additional concern of the values that the oracle assigns to the nodes. We handle this by forcing any small oracle to always give many copies of the same constant label c so that the adversary can cover the interesting parts of the construction with this unhelpful label c. We can then use uncomputability arguments to show that if a certain language were in LDO^f, then we could get a sequential algorithm for uncomputable problems. See Figure 1 for illustrations.

Lemma 1. *For any small oracle f, there exists a language $L \in \mathsf{LD} \setminus \mathsf{LDO}^f$.*

Proof. We assume that for each halting Turing machine M and each locality parameter $r \in \mathbb{N}$, there exists a labelled graph $H(M, r)$ with the following properties:

(P1) There is an identifier-oblivious local checker that verifies that a given labelled graph is a equal to $H(M, r)$ for some M and r.

(P2) The number of nodes in the graph $H(M, r)$ is at least as large as the number of steps M takes on an empty tape.

(P3) Given $H(M, r)$, an identifier-oblivious local checker A with a running time of r cannot decide if M outputs 0 or 1.

(P4) Each label of $H(M, r)$ is a triple $x(v) = (M, r, x'(v))$. The maximum degree of H and the maximum size of $x'(v)$ are constants that only depend on r.

(P5) Graph $H(M, r)$ can be padded with additional nodes without violating properties (P1)–(P4).

The construction of Fraigniaud et al. [10] satisfies these properties. They show how to construct a labelled graph $H(M, r)$ that encodes the execution table of a given Turing machine M such that a local checker with running time r cannot decide if M halts with 0 or 1. The original construction $(H, x) = H(M, r)$ consists of three main parts.

(i) *The execution table T of the Turing machine M.* Let s be the number of steps M takes on an empty tape. Then table T is an $(s + 1) \times (s + 1)$ grid, where node (i, j) holds the contents of the tape at position j after computation step i, and its own coordinates (i, j) modulo 3. Node (i, j) also knows if the head is at position j after step i, and if so, what is the state of M after step i. Node $(0, 0)$ representing the first position of the empty tape is called the *pivot*. The execution table exists essentially to guarantee (P2).

(ii) *The fragment collection $\mathcal{F}$.* This is a collection of subgrids labelled with all syntactically possible ways that are consistent with being in some execution table of M. The dimensions of the fragments are linear in r and independent of M. In each fragment, every 2×2 subgrid is consistent with a state transition of M. It is crucial to observe that there is a finite number of such fragments. Each fragment is connected to the pivot in a way that supports the local verification of the structure. The fragment collection is added to ensure (P3). Informally, if we only had T, then some node (i, s) at the last row of the grid would be able to see the stopping state of M; however, $\mathcal{F}$ will contain some fragments in which M halts with output 0 and some fragments in which M halts with output 1, and the nodes at the last row of T are locally indistinguishable from the nodes in such fragments.

(iii) *Pyramid structure.* This is added to the execution table and to the fragments to ensure (P1). Without any additional structure, a grid with coordinates modulo 3 is locally indistinguishable from, e.g., a grid that is wrapped into a torus. The pyramid structure guarantees that at least one node is able to detect invalid instances.

Finally, since all labellings can be made constant-size, we can ensure (P4). In particular, for any (M, r), there are constantly many syntactically possible r-neighbourhoods of $H(M, r)$. This is a crucial property as it guarantees that there is a sequential algorithm that on all inputs (M, r) halts and, if M halts, outputs all possible labelled r-neighbourhoods of $H(M, r)$.

Let $S(a, b)$ be the labelled path $(s_a, s_{a+1}, \ldots, s_b)$ in which node s_i is labelled with value i. We augment the construction $H(M, r)$ as follows: labelled graph $G(M, r, N)$ consists of $H(M, r)$, plus $S(1, N)$, plus an edge between the pivot of $H(M, r)$ and the first node s_1 of the path $S(1, N)$; we call $S(1, N)$ the *tail* of the construction. The structure of $G(M, r, N)$ is still locally checkable in LDO: any tail must eventually connect to the pivot, and the pivot can detect if there are multiple tails. The key property of the construction is that the nodes in the tail $S(1, N)$ with large labels are far from the nodes of $G(M, r)$ that are aware of M.

We will separate LD and LDO^f using the following language:

$$L = \{G(M, r, N) : r \geq 1, \ N \geq 1, \ \text{and Turing machine } M \text{ outputs } 0\}.$$

We have $L \in$ LD as there will be a node v with $\mathrm{id}(v) \geq s$ which can simulate M for s steps and output *no* if M does not output 0. Next we will argue that L cannot be in LDO^f for any small f.

Let f be a small oracle. For any M and r, we can choose a sufficiently large N as follows. By definition, there exists a c such that for all k oracle f outputs some label $i \in [c]$ at least $\lceil k/c \rceil$ times on some $n \geq k$. Moreover, we can find an infinite sequence of values $k_0, k_1, \ldots$ such that the most common value is some fixed i_0. We select w.l.o.g. the smallest k_j and a suitable n such that $f(n)$ contains at least $k_j/c \geq |H(M, r)| + 2r$ labels equal to i_0. Let $N = n - |H(M, r)|$, and consider $G(M, r, N)$. Now the adversary can construct the following *worst-case labelling*: every node of $G(M, r, 2r) \subseteq G(M, r, N)$ receives the constant input $i_0 \in [c]$; all other labels as assigned to the nodes in $S(2r + 1, N) \subseteq G(M, r, N)$.

It is known that separating the following languages is undecidable (see e.g. [25, p. 65]):

$$L_i = \{M : \text{Turing machine } M \text{ outputs } i\} : i \in \{0, 1\}. \tag{1}$$

For the sake of contradiction, we assume that there is an LDO^f-algorithm A that decides L. We will use algorithm A and constant i_0 defined above to construct a sequential algorithm B that separates L_0 and L_1.

Let r be the running time of A, and consider the execution of A on an instance $G(M, r, N)$ for some M and N. It follows that each node in $S(r + 1, N) \subseteq G(M, r, N)$ must always output *yes*. To see this, note that the claim is trivial if M halts with 0. Otherwise we can always construct another instance $G(M_0, r, N)$ such that M_0 halts with 0 and both $G(M, r, N)$ and $G(M_0, r, N)$ have the same number of nodes. Hence the oracle and the adversary can assign the same labels to $S(r + 1, N)$ in both $G(M, r, N)$ and $G(M_0, r, N)$. If any of these nodes would answer *no* in $G(M, r, N)$, then A would also incorrectly reject the *yes*-instance $G(M_0, r, N) \in L$.

Now given a Turing machine M, algorithm B proceeds as follows. Consider the subgraph $Q = G(M, r, 2r) \subseteq G(M, r, N)$, and assume the worst-case labelling of $G(M, r, N)$ in which all nodes of Q have the constant label i_0. Algorithm B cannot construct Q; indeed, M might not halt, in which case $G(M, r, N)$ would not even exist. However, B can do the following: it can assume that M halts, and then generate a collection $\mathcal{Q}$ that would contain all possible radius-r neighbourhoods of the nodes in $G(M, r, r)$. Collection $\mathcal{Q}$ is finite, its size only depends on r and M, and the key observation is that $\mathcal{Q}$ is computable (in essence, B enumerates all syntactically possible fixed-size fragments of partial execution tables of M).

Then B will simulate A in each neighbourhood of $\mathcal{Q}$. If M halts with 1, then $G(M, r, N) \notin L$, and therefore one of the nodes in $G(M, r, r)$ has to output no; in this case B outputs 1. If M halts with 0, then $G(M, r, N) \in L$, and therefore one of the nodes in $G(M, r, r)$ has to output yes; in this case B outputs 0. The key observation is that B will always halt with some (meaningless) output even if we are given an input $M \notin L_0 \cup L_1$; hence B is a computable function that separates L_0 and L_1. As such a B cannot exist, A cannot exist either.   □

3.2   Large Oracles Capture the Power of Unique Identifiers

In this section we will show that a *computable* large oracle f is sufficient to have $\mathsf{LD} \subseteq \mathsf{LDO}^f = \mathsf{LD}^f$. This result holds even if f only has access to an upper bound $N \geq n$, and the adversary gets to pick an n-subset of labels from $f(N)$. Note that the oracle has to be computable in order for us to invert it locally.

Lemma 2. *For any large computable oracle f, we have $\mathsf{LD} \subseteq \mathsf{LDO}^f = \mathsf{LD}^f$.*

Proof. We begin by showing how to recover an oracle $\hat{f}$ with $\hat{f}_k^{(N)} \geq k$, for all k and $N \geq k$, from a large oracle f. We want to guarantee that each node v receives a label $\ell \geq i$ if in the initial labelling it had the ith smallest label.

By definition, it holds for large oracles that for each natural number ℓ there is a largest index i such that $f_i^{(N)} \leq \ell$; we denote the index by $g(\ell)$. By assumption, a node with label ℓ can locally compute the value $g(\ell)$. We now claim that

$$\hat{f} \colon N \mapsto \{g(f_1), g(f_2), \ldots, g(f_N)\}$$

has the property $\hat{f}_k^{(N)} \geq k$. To see this, assume that we have $f_k^{(N)} = \ell$ for an arbitrary k. Seeing label ℓ, node v knows that, in the worst case, its own label is the $g(\ell)$th smallest. Thus for every k, the node with the kth smallest label will compute a new label at least k.

Now given $\hat{f}$, we can simulate any r-round LD-algorithm A as follows.

1. Each node v with label ℓ_v locally computes the new label $g(\ell_v)$.
2. Each node gathers all labels $g(\ell_u)$ in its r-neighbourhood. Denote by g_v^* the maximum value in the neighbourhood of v.
3. Each node v simulates A on every unique identifier assignment to its local r-neighbourhood from $\{1, 2, \ldots, g_v^*\}$. If for some assignment A outputs no, then v outputs no, and otherwise it outputs yes.

Because of how the decision problem is defined, it is always safe to output *no* when some simulation of A outputs *no*. It remains to be argued that it is safe to say *yes*, if all simulations say *yes*. This requires that *some* subset of simulations of A, one for each node, looks as if there had been a consistent setting of unique identifiers on the graph. Now let id be one identifier assignment with $\mathsf{id}(v) = i$ for the v with ith smallest label, for all i (breaking ties arbitrarily). Since by construction $g(\ell_v) \geq \mathsf{id}(v)$ for all v, there will be a simulation of A for every node v with local identifier assignment id_v such that for all u in the radius-r neighbourhood of v we have $\mathsf{id}_v(u) = \mathsf{id}(u)$.

So far we have seen how to simulate any LD-algorithm A with LDO^f-algorithms. We can apply the same reasoning to simulate any LD^f-algorithm A with LDO^f-algorithms; the only difference is that each node in the simulation has now access to the original oracle labels as well.   □

4   Full Characterisation of LD^f, LDO^f, LD, and LDO

Our goal in this section is to complete the characterisation of the power of scalar oracles with respect to the classes LD and LDO. We aim at giving a robust characterisation that holds also for minor variations in the definition of a scalar oracle. In particular, all of the key results can be adapted to weaker oracles that only receive an upper bound $N \geq n$ on the size of the graph.

4.1   Large Oracles Can Be Stronger than Identifiers

Let us first consider large oracles. By prior work [10] and Lemma 2 we already know that for any computable large oracle f we have a linear order

$$\mathsf{LDO} \subsetneq \mathsf{LD} \subseteq \mathsf{LDO}^f = \mathsf{LD}^f.$$

Trivially, there is a large computable oracle $f(n) = (1, 2, \ldots, n)$ such that

$$\mathsf{LDO} \subsetneq \mathsf{LD} = \mathsf{LDO}^f = \mathsf{LD}^f.$$

We will now show that there is also a large computable oracle f such that

$$\mathsf{LDO} \subsetneq \mathsf{LD} \subsetneq \mathsf{LDO}^f = \mathsf{LD}^f.$$

For a simple proof, we could consider the large oracle $f(n) = (n, n, \ldots, n)$. Now the parity language L that consists of graphs with an even number of nodes is clearly in LDO^f but not in LD. However, this separation is not robust with respect to minor changes in the model of scalar oracles. In particular, if the oracle only knows an upper bound on n, we cannot use the parity language to separate LDO^f from LD.

In what follows, we will show that the *upper bound oracle* f that labels all nodes with some upper bound on $N \geq n$ can be used to separate LDO^f from LD.

Theorem 2. *For the upper bound oracle f there exists a language L such that $L \in \mathsf{LDO}^f \setminus \mathsf{LD}$.*

Proof. The proof uses computability arguments—see the full version [12] for the details.

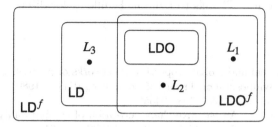

Fig. 2. There is a small oracle f such that each of the languages L_i exists.

4.2   Small Oracles and Identifiers Are Incomparable

In the case of small oracles, we already know that $\mathsf{LDO}^f \subsetneq \mathsf{LD}^f$ for any small oracle f by Lemma 1. Next we characterise the relationship of LDO^f and LD. In essence, we show that these classes are incomparable.

Theorem 3. *There is a single small oracle f so that each of the languages L_1, L_2, and L_3 shown in Figure 2 exist.*

Proof. Let f be the small oracle

$$f(n) = (0, 0, \ldots, 0, b_n),$$

where b_n is an n-bit string such that the ith bit tells whether the ith Turing machine halts. We construct the languages as follows:

L_1: Let $P(n)$ denote the labelled path of length n such that each node has two input labels: n and the distance to a specified leaf node v_0. The correct structure of $P(n)$ is in LDO. Now let

$$L_1 = \{P(M) : \text{Turing machine } M \text{ halts}\}.$$

The node that receives the n-bit oracle label can use it to decide whether the nth Turing machine halts, and therefore $L_1 \in \mathsf{LDO}^f$. Conversely, we have $L_1 \notin \mathsf{LD}$; otherwise we would have a sequential algorithm that solves the halting problem for each Turing machine M by constructing the path $P(M)$ with some fixed identifier assignment and simulating the local verifier.

L_2: We can use the same language

$$L_2 = \{H(M, r) : r \geq 1 \text{ and Turing machine } M \text{ outputs } 0\}$$

that we used in the proof of Lemma 1. It is known that $L_2 \in \mathsf{LD}$ and $L_2 \notin \mathsf{LDO}$ [10]. Since checking the structure of $H(M, r)$ is in LDO, it suffices to note that the node that receives the bit vector b_n of length n can use the *length* of the vector as an upper bound in simulating M. Thus $L_2 \in \mathsf{LDO}^f$.

L_3: Apply Lemma 1.   □

We conclude by noting that Theorem 3 is also robust to minor variations in the definitions. In particular, the oracle does not need to know the exact value of n; it is sufficient that at least one node receives the bit string b_N, where $N \geq n$ is some upper bound on n.

Acknowledgements. Thanks to Laurent Feuilloley for discussions.

References

1. Angluin, D.: Local and global properties in networks of processors. In: Proc. 12th Annual ACM Symposium on Theory of Computing (STOC 1980), pp. 82–93. ACM Press (1980). doi:10.1145/800141.804655
2. Boldi, P., Vigna, S.: An effective characterization of computability in anonymous networks. In: Welch, J.L. (ed.) DISC 2001. LNCS, vol. 2180, pp. 33–47. Springer, Heidelberg (2001)
3. Chalopin, J., Das, S., Santoro, N.: Groupings and pairings in anonymous networks. In: Dolev, S. (ed.) DISC 2006. LNCS, vol. 4167, pp. 105–119. Springer, Heidelberg (2006)
4. Czygrinow, A., Hańćkowiak, M., Wawrzyniak, W.: Fast distributed approximations in planar graphs. In: Taubenfeld, G. (ed.) DISC 2008. LNCS, vol. 5218, pp. 78–92. Springer, Heidelberg (2008)
5. Diks, K., Kranakis, E., Malinowski, A., Pelc, A.: Anonymous wireless rings. Theoretical Computer Science 145(1–2), 95–109 (1995). doi:10.1016/0304-3975(94)00178-L
6. Emek, Y., Pfister, C., Seidel, J., Wattenhofer, R.: Anonymous networks: randomization = 2-hop coloring. In: Proc. 33rd ACM SIGACT-SIGOPS Symposium on Principles of Distributed Computing (PODC 2014), pp. 96–105. ACM Press (2014). doi:10.1145/2611462.2611478
7. Emek, Y., Seidel, J., Wattenhofer, R.: Computability in anonymous networks: Revocable vs. Irrecovable outputs. In: Esparza, J., Fraigniaud, P., Husfeldt, T., Koutsoupias, E. (eds.) ICALP 2014, Part II. LNCS, vol. 8573, pp. 183–195. Springer, Heidelberg (2014)
8. Fich, F., Ruppert, E.: Hundreds of impossibility results for distributed computing. Distributed Computing 16(2–3), 121–163 (2003), doi:10.1007/s00446-003-0091-y
9. Fraigniaud, P., Gavoille, C., Ilcinkas, D., Pelc, A.: Distributed computing with advice: Information sensitivity of graph coloring. In: Arge, L., Cachin, C., Jurdziński, T., Tarlecki, A. (eds.) ICALP 2007. LNCS, vol. 4596, pp. 231–242. Springer, Heidelberg (2007)
10. Fraigniaud, P., Göös, M., Korman, A., Suomela, J.: What can be decided locally without identifiers? In: Proc. 32nd Annual ACM Symposium on Principles of Distributed Computing (PODC 2013), pp. 157–165. ACM Press, New York (2013). doi:10.1145/2484239.2484264
11. Fraigniaud, P., Halldórsson, M.M., Korman, A.: On the impact of identifiers on local decision. In: Baldoni, R., Flocchini, P., Binoy, R. (eds.) OPODIS 2012. LNCS, vol. 7702, pp. 224–238. Springer, Heidelberg (2012)
12. Fraigniaud, P., Hirvonen, J., Suomela, J.: Node Labels in Local Decision (2015). arXiv:1507.00909v1
13. Fraigniaud, P., Korman, A., Peleg, D.: Local distributed decision. In: Proc. 52nd Annual IEEE Symposium on Foundations of Computer Science (FOCS 2011). IEEE Computer Society Press (2011). doi:10.1109/FOCS.2011.17
14. Fraigniaud, P., Pelc, A., Peleg, D., Pérennes, S.: Assigning labels in an unknown anonymous network with a leader. Distributed Computing 14(3), 163–183 (2001). doi:10.1007/PL00008935
15. Gavoille, C., Peleg, D.: Compact and localized distributed data structures. Distributed Computing 16(2–3), 111–120 (2003). doi:10.1007/s00446-002-0073-5

16. Göös, M., Hirvonen, J., Suomela, J.: Lower bounds for local approximation. Journal of the ACM 60(5) 39, 1–23 (2013). doi:10.1145/2528405
17. Hasemann, H., Hirvonen, J., Rybicki, J., Suomela, J.: Deterministic local algorithms, unique identifiers, and fractional graph colouring. Theoretical Computer Science (2014) (to appear). doi:10.1016/j.tcs.2014.06.044
18. Hella, L., Järvisalo, M., Kuusisto, A., Laurinharju, J., Lempiäinen, T., Luosto, K., Suomela, J., Virtema, J.: Weak models of distributed computing, with connections to modal logic. Distributed Computing 28(1), 31–53 (2015). doi:10.1007/s00446-013-0202-3
19. Kranakis, E.: Symmetry and computability in anonymous networks: a brief survey. In: Proc. 3rd Colloquium on Structural Information and Communication Complexity (SIROCCO 1996), pp. 1–16. Carleton University Press (1997)
20. Lenzen, C., Wattenhofer, R.: Leveraging linial's locality limit. In: Taubenfeld, G. (ed.) DISC 2008. LNCS, vol. 5218, pp. 394–407. Springer, Heidelberg (2008)
21. Linial, N.: Locality in distributed graph algorithms. SIAM Journal on Computing 21(1), 193–201 (1992). doi:10.1137/0221015
22. Mayer, A., Naor, M., Stockmeyer, L.: Local computations on static and dynamic graphs. In: Proc. 3rd Israel Symposium on the Theory of Computing and Systems (ISTCS 1995), pp. 268–278. IEEE (1995). doi:10.1109/ISTCS.1995.377023
23. Naor, M., Stockmeyer, L.: What can be computed locally? SIAM Journal on Computing 24(6), 1259–1277 (1995). doi:10.1137/S0097539793254571
24. Norris, N.: Classifying anonymous networks: when can two networks compute the same set of vector-valued functions? In: Proc.1st Colloquium on Structural Information and Communication Complexity (SIROCCO 1994), pp. 83–98. Carleton University Press (1995)
25. Papadimitriou, C.H.: Computational Complexity. Addison-Wesley Publishing Company (1994)
26. Peleg, D.: Distributed Computing: A Locality-Sensitive Approach. SIAM Monographs on Discrete Mathematics and Applications. Society for Industrial and Applied Mathematics, Philadelphia (2000)
27. Suomela, J.: Survey of local algorithms. ACM Computing Surveys 45(2) 24:1–40 (2013). doi:10.1145/2431211.2431223
28. Yamashita, M., Kameda, T.: Computing on anonymous networks: part I—characterizing the solvable cases. IEEE Transactions on Parallel and Distributed Systems 7(1), 69–89 (1996). doi:10.1109/71.481599
29. Yamashita, M., Kameda, T.: Leader election problem on networks in which processor identity numbers are not distinct. IEEE Transactions on Parallel and Distributed Systems 10(9), 878–887 (1999). doi:10.1109/71.798313

Exact Bounds for Distributed Graph Colouring

Joel Rybicki[1,2] and Jukka Suomela[1]

[1] Helsinki Institute for Information Technology HIIT,
Department of Computer Science, Aalto University, Saarbrücken, Germany
[2] Department of Algorithms and Complexity, Max Planck Institute for Informatics,
Saarbrücken, Germany

Abstract We prove exact bounds on the time complexity of distributed graph colouring. If we are given a directed path that is properly coloured with n colours, by prior work it is known that we can find a proper 3-colouring in $\frac{1}{2}\log^*(n) \pm O(1)$ communication rounds. We close the gap between upper and lower bounds: we show that for infinitely many n the time complexity is precisely $\frac{1}{2}\log^* n$ communication rounds.

1 Introduction

One of the key primitives in the area of distributed graph algorithms is *graph colouring in directed paths*. This is a fundamental symmetry-breaking task, widely studied since the 1980s—it is used as a subroutine in numerous efficient distributed algorithms, and it also serves as a convenient starting point in many lower-bound proofs. In the 1990s it was already established that the distributed computational complexity of this problem is $\frac{1}{2}\log^*(n) \pm O(1)$ communication rounds [3,13,20]. We are now able to give *exact* bounds on the distributed time complexity of this problem, and the answer turns out to take a surprisingly elegant form:

Theorem 1. *For infinitely many values of n, it takes exactly $\frac{1}{2}\log^* n$ rounds to compute a 3-colouring of a directed n-coloured path.*

1.1 Problem Setting

Throughout this work we focus on *deterministic* distributed algorithms. As is common in this context, what actually matters is not the number of nodes but the range of their labels. For the sake of concreteness, we study precisely the following problem setting:

> We have a path or a cycle with any number of nodes, and the nodes are properly coloured with colours from $[n] = \{1, 2, \ldots, n\}$.

The techniques that we present in this work can also be used to analyse other variants of the problem—for example, a cycle with n nodes that are labelled with some permutation of $[n]$, or a path with at most n nodes that are labelled with unique identifiers from $[n]$. However, the exact bounds on the time complexity will slightly depend on such details.

© Springer International Publishing Switzerland 2015
C. Scheideler (Ed.): SIROCCO 2015, LNCS 9439, pp. 46–60, 2015.
DOI: 10.1007/978-3-319-25258-2_4

We will assume that there is a globally consistent orientation in the path: each node has at most one predecessor and at most one successor. Our task is to find a proper colouring of the path with c colours, for some number $c \geq 3$. We will call this task *colour reduction from n to c*.

We will use the following model of distributed computing. Each node of the graph is a computational entity. Initially, each node knows the global parameters n and c, its own label from $[n]$, its degree, and the orientations of its incident edges. Computation takes place in synchronous communication rounds. In each round, each node can send a message to each of its neighbours, receive a message from each of its neighbours, update its state, and possibly stop and output its colour. The *running time* of an algorithm is defined to be the number of communication rounds until all nodes have stopped. We will use the following notation:

- $C(n, c)$ is the time complexity of colour reduction from n to c.
- $T(n, c)$ is the time complexity of colour reduction from n to c if we restrict the algorithm so that a node can only send messages to its successor. We call such an algorithm *one-sided*, while unrestricted algorithms are *two-sided*.

We can compose colour reduction algorithms, yielding $C(a, c) \leq C(a, b) + C(b, c)$ and $T(a, c) \leq T(a, b) + T(b, c)$ for any $a \geq b \geq c$. Using a simple simulation argument, it is easy to see that

$$C(n, c) = \lceil T(n, c)/2 \rceil.$$

We will be interested primarily in $C(n, c)$, but function $T(n, c)$ is much more convenient to analyse when we prove upper and lower bounds.

1.2 Prior Work

The asymptotically optimal bounds of

$$\log^*(n) - O(1) \leq T(n, 3) \leq \log^*(n) + O(1)$$

are covered in numerous textbooks and courses on distributed and parallel computing [2, 4, 17, 19, 21]. The proof is almost unanimously based on the following classical results:

Cole–Vishkin Colour Reduction (CV): The upper bound was presented in the modern form by Goldberg, Plotkin, and Shannon [9] and it is based on the technique first introduced by Cole and Vishkin [3]. The key ingredients are a fast colour reduction algorithm that shows that $T(2^k, 2k) \leq 1$ for any $k \geq 3$, and a slow colour reduction algorithm that show that $T(k + 1, k) \leq 2$ for any $k \geq 3$. By iterating the fast colour reduction algorithm, we can reduce the number of colours from n to 6 in $\log^*(n) \pm O(1)$ rounds, and by iterating the slow colour reduction algorithm, we can reduce the number of colours from 6 to 3 in 6 rounds (with one-sided algorithms).

Linial's Lower Bound: The lower bound is the seminal result by Linial [13]. The key ingredient is a speed-up lemma that shows that $T(n, 2^c) \leq T(n,c)-1$ when $T(n,c) \geq 1$. By iterating the speed-up lemma for $\log^*(n) - 3$ times, we have $T(n,4) \geq T(n,k) + \log^*(n) - 3$ for a $k < n$. Clearly $T(n,3) \geq T(n,4)$ and $T(n,k) \geq 1$, and hence $T(n,3) \geq \log^*(n) - 2$.

In the upper bound, many sources—including the original papers by Cole and Vishkin and Goldberg et al.—are happy with the asymptotic bounds of $\log^*(n) + O(1)$ or $O(\log^* n)$. However, there are some sources that provide a more careful analysis. The analysis by Barenboim and Elkin [2] yields $T(n,3) \leq \log^*(n) + 9$, and the analysis in the textbook by Cormen et al. [4] yields $T(n,3) \leq \log^*(n)+7$. In our lecture course [19] we had an exercise that shows how to push it down to

$$T(n,3) \leq \log^*(n) + 6.$$

In the lower bound, there is less variation. Linial's original proof [13] yields $T(n,3) \geq \log^*(n) - 3$, and many sources [2, 11, 19] prove a bound of

$$T(n,3) \geq \log^*(n) - 2.$$

On the side of lower bounds, nothing stronger than Linial's result is known. There are alternative proofs based on Ramsey's theorem [5] that yield the same asymptotic bound of $T(n,3) = \Omega(\log^* n)$, but the constants one gets this way are worse than in Linial's proof.

On the side of upper bounds, however, there is an algorithm that is strictly better than CV: **Naor–Stockmeyer colour reduction (NS)** [15]. While CV yields $T(2^k, 2k) \leq 1$ for any $k \geq 3$, NS yields a strictly stronger claim of $T(\binom{2k}{k}, 2k) \leq 1$ for any $k \geq 2$. However, the exact bounds that we get from NS are apparently not analysed anywhere. Szegedy and Vishwanathan [20] describe a very similar algorithm, but exact bounds for their algorithm are not given either. Hence the state of the art appears to be

$$\log^*(n) - 2 \leq T(n,3) \leq \log^*(n) + 6,$$
$$\frac{1}{2}\log^*(n) - 1 \leq C(n,3) \leq \frac{1}{2}\log^*(n) + 3.$$

Note that we have $\log^* n \leq 5$ for all $n < 10^{19728}$, and hence in practice the constant term 6 dominates the term $\log^* n$ in the upper bound.

1.3   Contributions

In this work we derive *exact bounds* on $C(n,3)$ for infinitely many values of n, and near-tight bounds for all values of n. We prove that for infinitely many values of n

$$C(n,3) = \frac{1}{2}\log^* n,$$

and for all sufficiently large values of n

$$\log^*(n) - 1 \leq T(n, 3) \leq \log^*(n) + 1.$$

With $C(n, 3) = \lceil T(n, 3)/2 \rceil$ this gives a near-complete picture of the exact complexity of colouring directed paths. The key new techniques are as follows:

1. We give a new analysis of NS colour reduction.
2. We give a new lower-bound proof that is strictly stronger than Linial's lower bound.
3. We show that *computational techniques* can be used to prove not only upper bounds but also lower bounds on $T(n, c)$, also for the case of a general n and not just for fixed small values of n and c. We introduce *successor graphs* S_i that are defined so that a graph colouring of S_i with a small number of colours implies an improved bound on $T(n, 3)$.

This work focuses on colour reduction, i.e., the setting in which we are given a proper colouring as an input. Our upper bounds naturally apply directly in more restricted problems (e.g., the input labels are unique identifiers). Our lower bounds results do not hold directly, but the key techniques are still applicable: in particular, the successor graph technique can be used also in the case of unique identifiers.

1.4 Applications

Graph colouring in paths, and the related problems of graph colouring in rooted trees and directed pseudoforests, are key symmetry-breaking primitives that appear as subroutines in numerous distributed algorithms for various graph problems [1, 5, 8, 9, 12, 16].

One of the most direct application of our results is related to colouring *trees*: In essence, colour reduction from n to c in trees with *arbitrary* algorithms is the same problem as colour reduction from n to c in paths with *one-sided* algorithms. Informally, in the worst case the children contain all possible coloured subtrees and hence "looking down" in the tree is unhelpful, and we can equally well restrict ourselves to "looking up" towards the root. Hence our bounds on $T(n, 3)$ can be directly interpreted as bounds on colour reduction from n to 3 in trees.

The bounds have also applications outside distributed computing. A result by Fich and Ramachandran [6] demonstrates that bounds on $C(n, 3)$ have direct implications in the context of *decision trees* and *parallel computing*.

Indeed, the fastest known *parallel* algorithms for colouring linked lists are just adaptations of CV and NS colour reduction algorithms. These algorithms reduce the number of colours very rapidly to a relatively small number (e.g., dozens of colours), and the key bottleneck has been pushing the number of colours down to 3. In particular, reducing the number of colours down to 3 with state-of-the-art algorithms has been much more expensive than reducing it to 4, but this phenomenon has not been understood so far. Prior bounds on $T(n, c)$ have not been able to show that the case of $c = 3$ is necessarily more expensive than $c = 4$. Our improved bounds are strong enough to separate $T(n, 4)$ and $T(n, 3)$.

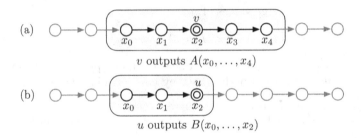

(a) v outputs $A(x_0, \ldots, x_4)$

(b) u outputs $B(x_0, \ldots, x_2)$

Fig. 1. The difference of two-sided and one-sided algorithms. (a) A two-sided algorithm A that runs for 2 rounds. (b) A one-sided algorithm B that runs for 2 rounds.

From the perspective of practical algorithm engineering and programming, this work shows that we should avoid CV colour reduction, but we can be content with NS colour reduction; the former incurs a significant overhead (e.g., in terms of linear scans over the data in parallel computing), but the latter is near-optimal.

2 Preliminaries

Sets and Functions. For any positive integer k, we use $[k]$ to denote the set $\{1, 2, \ldots, k\}$. For any set X, we use $2^X = \{Y \subseteq X\}$ to denote the powerset of X. Define the *iterated logarithm* as

$$\log^{(0)}(x) = x,$$
$$\log^{(i+1)}(x) = \log^{(i)}(\log x) \text{ for all } i \geq 0.$$

In this work, all logarithms are in base 2. Moreover, the *log-star* function is

$$\log^* x = \min\{i : \log^{(i)} x \leq 1\}.$$

Finally, we define the *tetration*, or a power tower, with base 2 as

$$^0 2 = 1,$$
$$^{i+1} 2 = 2^{(^i 2)} \text{ for all } i \geq 0.$$

Algorithms. In this work, we focus on algorithms that run on directed paths. We distinguish between two-sided and one-sided algorithms; see Figure 1. *Two-sided algorithms* correspond to the usual notion of an algorithm in the LOCAL model: an algorithm running for t rounds has to decide on its output using the information available at most t hops away. Formally, a two-sided c-colouring algorithm corresponds to a function

$$A \colon [n]^{2t+1} \to [c].$$

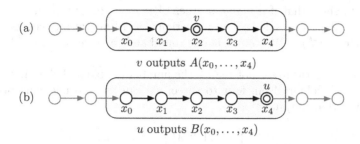

Fig. 2. The correspondence between two-sided and one-sided algorithms. (a) A two-sided algorithm A that runs for 2 rounds. (b) A one-sided algorithm that runs in 4 rounds. Both nodes see the same information, so v can easily simulate B and u can simulate A.

Moreover, as A outputs a proper colouring, the function satisfies $A(x_0, \ldots, x_{2t}) \neq A(x_1, \ldots, x_{2t+1})$ when $x_i \neq x_{i+1}$ for all $i \geq 0$. Note that an end point of the path can always simulate a properly coloured virtual path that extends from it to get exactly $2t + 1$ input values for A.

In contrast to two-sided algorithms, *one-sided algorithms* are algorithms in which nodes can only send messages to successors. Therefore, a one-sided algorithm that runs in t rounds can only gather information from at most t *predecessors*. Formally, a one-sided c-colouring algorithm B that runs for t steps corresponds to a function

$$B \colon [n]^{t+1} \to [c],$$

which satisfies $B(x_0, \ldots, x_t) \neq B(x_1, \ldots, x_{t+1})$ when $x_i \neq x_{i+1}$ for all $i \geq 0$.

It is now easy to see that $C(n, c) = \lceil T(n, c)/2 \rceil$ holds. For example, Figure 2 illustrates how a t-time two-sided algorithm can gather the same information as a $2t$-time one-sided algorithm. We refer to this identity as the following lemma:

Lemma 1. $C(n, c) = \lceil T(n, c)/2 \rceil$.

3 The Upper Bound

In this section, we bound $T(n, c)$ from above. To do this, we analyse the Naor–Stockmeyer (NS) colour reduction algorithm [15]. The NS algorithm is one-sided, thus yielding upper bounds for $T(n, c)$.

Let us first recall the NS colour reduction algorithm. Let $n \leq \binom{2k}{k}$ for some $k \geq 2$ and fix an injection $f \colon [n] \to X$, where $X = \{Y \subseteq [2k] : |Y| = k\}$. That is, we interpret all colours from $[n]$ as distinct k-subsets of $[2k]$.

The algorithm works as follows. First, all nodes send their colour to the successor. Then a node with colour v receiving colour u from its predecessor will output

$$A(u, v) = \min f(u) \setminus f(v).$$

It is easy to show that if $u \neq v \neq w$, then $A(u,v) \in [2k]$ and $A(u,v) \neq A(v,w)$ holds. Thus, A is a one-sided colour reduction algorithm that reduces the number of colours from $\binom{2k}{k}$ to $2k$ colours in one round and we have that $T\big(\binom{2k}{k}, 2k\big) = 1$ for any $k \geq 2$.

The above algorithm cannot reduce the number of colours below 4. To reduce the number of colours from four to three, we can use the following one-sided algorithm B that outputs

$$B(u,v,w) = \begin{cases} \min\{1,2,3\} \setminus \{u,w\} & \text{if } v = 4, \\ v & \text{otherwise.} \end{cases}$$

The algorithm uses two rounds and this is optimal by Lemma 5 in Section 4.

We now show the following upper bounds for $T(n,c)$ using the NS colour reduction algorithm.

Lemma 2. *The function T satisfies the following:*

(a) $T\big(\frac{3}{2} \cdot 2^c, \frac{3}{2} \cdot c\big) = 1$ *for any $c = 4h$, where $h > 1$,*
(b) $T\big(\frac{3}{2} \cdot {}^{r+4}2, \frac{3}{2} \cdot {}^{4}2\big) \leq r$ *for any $r \geq 0$,*
(c) $T\big(\frac{3}{2} \cdot {}^{4}2, 3\big) \leq 5$.

Proof.

(a) As discussed, the NS colour reduction algorithm shows that $T\big(\binom{2k}{k}, 2k\big) = 1$ for $k \geq 2$. Recall the following bound for the central binomial coefficent

$$\binom{2k}{k} \geq \frac{4^k}{\sqrt{4k}}$$

and let $2k = 3c/2$. Since $c \geq 8$ it follows that

$$\binom{2k}{k} \geq \frac{(2 \cdot 2)^{3c/4}}{\sqrt{3c}} = \frac{2^{c/2}}{\sqrt{3c}} \cdot 2^c > \frac{3}{2} \cdot 2^c.$$

(b) To show the claim, it suffices to apply part (a) for r times.
(c) As $\binom{20}{10} > \frac{3}{2} \cdot {}^{4}2$, we can reduce the number of colours to 4 in three rounds as follows: $\binom{20}{10} \rightsquigarrow \binom{6}{3} \rightsquigarrow \binom{4}{2} \rightsquigarrow 4$. By previous discussion, the remaining two rounds can be used to remove the fourth colour.

Theorem 2. $T({}^{h}2, 3) \leq T({}^{h}2 + 1, 3) \leq h + 1$ *holds for any $h > 1$.*

Proof. The cases $2 \leq h \leq 4$ follow from the proof of Lemma 2c. Suppose $h = r+4$ for some $r > 0$. By Lemma 2b and c we can get a 3-colouring in $r + 5 = h + 1$ rounds.

4   The Lower Bound

In this section, we give a new lower bound for the time complexity of one-sided colour reduction algorithms. The proof follows the basic idea of Linial's proof [13] adapted to the case of colour reduction, but we show a new lemma that can be used to tighten the bound.

The proof is structured as follows. First, we show that $T(n, 2^c-2) \leq T(n,c)-1$, that is, given a c-colouring algorithm, we can devise a faster algorithm that uses at most $2^c - 2$ colours; this is just a minor tightening of the usual standard bound, and should be fairly well-known. Second, we prove that a fast 3-colouring algorithm implies a fast 16-colouring algorithm, more precisely, $T(n, 16) \leq T(n, 3) - 2$; this is the key contribution of this section. Together these yield the following new bound:

Theorem 3. *For any $h > 1$, we have $T(^h2, 3) \geq h$.*

4.1   The Speed-Up Lemma

Lemma 3. *If $T(n,c) \geq 1$, then $T(n, 2^c - 2) \leq T(n,c) - 1$.*

Proof. Let $t = T(n,c)$ and $A\colon [n]^{t+1} \to [c]$ be a one-sided c-colouring algorithm. We will construct a faster one-sided algorithm B as follows. Consider a node u and its successor v. In $t - 1$ rounds, node u can find out the colours of its $t - 1$ predecessors and its own colour, that is, some vector $(x_0, \ldots, x_{t-1}) \in [n]^t$. In particular, node u now knows what information node v can gather in t rounds *except* the colour of v since A is one-sided. However, u can enumerate all the possible outputs of v which give the set

$$B(x_0, \ldots, x_{t-1}) = \{A(x_0, \ldots, x_{t-1}, y)\colon y \neq x_{t-1}, y \in [n]\} \subseteq [c].$$

Clearly $B(x_0, \ldots, x_{t-1}) \neq \emptyset$. We also have $B(x_0, \ldots, x_{t-1}) \neq [c]$: For the sake of contradiction, suppose otherwise. This would imply that v could output any value in $[c]$. In particular, if u outputs $A(z, x_0, \ldots, x_{t-1}) = a$ for some $z \in [n]$, we could pick $y \in [n]$ such that $A(x_0, \ldots, x_{t-1}, y) = a$ as well. However, this would contradict the fact that A was a colouring algorithm. Hence there exists an injection f that maps any possible set $B(\cdot)$ to a value in $[2^c - 2]$.

It remains to argue that no two adjacent nodes construct the same set. Suppose a node u outputs set X and its successor v also outputs X. Now we can pick $k \in X$ such that

$$A(x_0, \ldots, x_{t-1}, y) = k = A(x_1, \ldots, x_{t-1}, y, y')$$

for some $x_{t-1} \neq y \neq y'$ contradicting that A outputs a proper colouring. Therefore, $f \circ B$ is a one-sided $(2^c - 2)$-colouring algorithm that runs in time $t - 1 = T(n,c) - 1$.

Lemma 4. *For any $r > 0$, we have $T(^{r+3}2, 16) \geq r + 1$.*

Proof. Fix $r > 0$. We repeatedly apply Lemma 3. Now suppose we have an algorithm that reduces the number of colours from n to $16 = {}^3 2$ in r rounds. That is, $T(n, {}^3 2) \leq r$ holds for some $n \geq 3$. From Lemma 3 it follows that

$$T(n, {}^3 2) \leq r \implies T(n, {}^4 2 - 2) \leq r - 1 \implies \cdots$$
$$\implies T(n, {}^{3+r} 2 - 2) \leq 0,$$

but as $T(k, k-1) \geq 1$ for any k it follows that $n < {}^{3+r} 2$. Thus, $T({}^{r+3} 2, 16) \geq r+1$. $\blacksquare$

4.2   Proof of Theorem 3

In addition to the speed-up lemma, we need a few more lemmas that bound $T(n, 3)$ below for small values of n.

Lemma 5. $T(4, 3) \geq 2$.

Proof. Let B' be a one-sided 3-colouring algorithm that runs in one round. Now B' yields a partitioning of the possible input pairs (u, v) where $u \neq v$. It is simple to check that there always exists a pair (u, v) with $u \neq v$ such that there also exists some $w \neq v$ satisfying $B'(u, v) = B'(v, w)$. $\blacksquare$

Lemma 6. $T(16, 3) \geq 3$.

Proof. As observed by Linial [13], we can show $C(n, c) = t$ if the so-called *neighbourhood graph* $\mathcal{N}_{n,t}$ has a chromatic number of c. While Linial analytically bounded the chromatic number of such graphs, we can also compute their chromatic numbers exactly for small values of n, c, and t; see [18] for a detailed discussion. We use the latter technique to show the claimed bound. That is, the neighbourhood graph $\mathcal{N}_{7,1}$ is not 3-colourable.

The neighbourhood graph $\mathcal{N}_{7,1} = (V, E)$ is defined as follows. The set of vertices is
$$V = \{(x_0, x_1, x_2) \in [n]^3 : x_0 \neq x_1 \neq x_2, x_0 \neq x_2\},$$
where $n = 7$ and the set of edges is
$$E = \{\{u, v\} : u, v \in V, u = (x_0, x_1, x_2), v = (x_1, x_2, x_3)\}.$$

It is easy to check with a computer (e.g. using any off-the-shelf SAT or an IP solver) that the graph $\mathcal{N}_{7,1}$ is not 3-colourable. Therefore, $C(7, 3) > 1$ and in particular $T(16, 3) \geq T(7, 3) > 2$. $\blacksquare$

To get a lower bound for 3-colouring, we show in the following sections that the existence of a t-time one-sided 3-colouring algorithm implies a $(t - 2)$-time one-sided 16-colouring algorithm.

Lemma 7. *For any $n \geq 16$, it holds that $T(n, 16) \leq T(n, 3) - 2$.*

Now we have all the results for showing the lower bound.

Theorem 3. *For any $h > 1$, we have $T(^h2, 3) \geq h$.*

Proof. The cases $r = 2$ and $r = 3$ follow from Lemmas 5 and 6. For the remaining cases, let $h = r + 3$ for some $r > 0$. Suppose $T(^h2, 3) = T(^{r+3}2, 3) < h$. Then by Lemma 7 we would get that $T(^{r+3}2, 16) < h - 2 = r + 1$ which contradicts Lemma 4.

4.3   Proof of Lemma 7 via Successor Graphs

To prove Lemma 7, we analyse the chromatic number of so-called *successor graphs*—a notion similar to Linial's neighbourhood graphs [13]. In the following, given a binary relation R, we will write $x \in R(y)$ to mean $(y, x) \in R$.

Colouring Relations. Suppose $A = A_0$ is a one-sided 3-colouring algorithm that runs in t rounds. Let $A_1, \ldots, A_t$ denote the one-sided algorithms given by iterating Lemma 3 and $C_{k+1} \subseteq 2^{C_k}$ be the set of colours output by algorithm A_{k+1}. As before, we can interpret a set of colours $X \in C_{k+1}$ as a colour in $2^{|C_k|}$ using an appropriate injection.

In the following, let $t' = t - k$. Define the *potential successor relation* $S_k \subseteq C_k \times C_k$ to be a binary relation such that $(x, y) \in S_k$ if there exist $x_0, \ldots, x_{t'+1} \in C_{k-1}$ where $x_i \neq x_{i+1}$ such that

$$A_k(x_0, \ldots, x_{t'}) = x \text{ and } A_k(x_1, \ldots, x_{t'+1}) = y.$$

That is, in the output of algorithm A_k there can be an x-coloured node with a successor of colour y. Moreover, define the *output relation* $R_k \subseteq C_k \times C_{k+1}$ such that $(x, X) \in R_k$ if

$$A_{k+1}(x_0, \ldots, x_{t'}, x) = X$$

for some $x_0, \ldots, x_{t'}$ where $x_i \neq x_{i+1}$. That is, a node with colour x can output colour X when executing A_{k+1}. From the construction of A_{k+1} given in Lemma 3, we get that $R_k = \{(x, X) : X \subseteq S_k(x), X \neq \emptyset\}$.

Lemma 8. *Suppose $X \in R_k(x)$, $Y \in R_k(y)$, and $y \in X$ for some $x, y \in C_k$, then $(X, Y) \in S_{k+1}$ holds. Moreover, the converse holds.*

Proof. As we have $y \in X \subseteq S_k(x)$, this means that a node with colour x may have a successor of colour y after executing algorithm A_k. Moreover, as $X \in R_k(x)$ and $Y \in R_k(y)$ hold, then a node with colour x may output X and node with colour y may output Y when executing A_{k+1}. Thus, after executing A_{k+1} we may have a node with colour X that has a successor with colour Y. Therefore, $(X, Y) \in S_{k+1}$.

To show the converse, suppose that $(X, Y) \in S_{k+1}$, that is, in some output of A_{k+1} a node u with colour X having a successor v with colour Y. Now there must exist some colour x that $X \in R_k(x)$ and some colour y such that $Y \in R_k(y)$. As v is a successor of u, the algorithm A_{k+1} outputs a set X consisting of all possible colours for any successor of u, and thus, we have $y \in X$.

Successor Graphs. For any choice of $A = A_0$, we can construct the successor relation S_k and using this relation, we can define the *successor graph* of A to be the graph $\mathcal{S}_k(A) = (C_k, E_k)$, where $E_k = \{\{x, y\} : (x, y) \in S_k\}$. These graphs have the following property:

Lemma 9. *Let $\mathcal{S}_k = (C_k, S_k)$ be the successor graph of A, and let t be the running time of A. If $f : C_k \to [\chi]$ is a proper colouring of $\mathcal{S}_k$, then $f \circ A_k$ is a one-sided χ-colouring algorithm that runs in $t - k$ rounds. That is, $T(n, \chi) \leq t - k$.*

Proof. Let u be the predecessor of v on a directed path. Now by definition,

$$A_k(x_0, \ldots, x_{t-k-1}, u) = x \neq y = A_k(x_1, \ldots, x_{t-k-1}, u, v)$$
$$\implies (x, y) \in S_k \implies f(x) \neq f(y).$$

Therefore, $f \circ A_k$ is a one-sided χ-colouring algorithm.

In the next section, we show the following lemma from which Lemma 7 follows.

Lemma 10. *For any t-time 3-colouring algorithm A, the successor graph $\mathcal{S}_2(A)$ can be coloured with 16 colours.*

In particular, this holds for an optimal algorithm A with a running time of $t = T(n, 3)$. Together with Lemma 9, this implies Lemma 7. We next show how to prove Lemma 10 in two ways: with computers, and without them.

4.4   A Human-Readable Proof of Lemma 10

We start by sketching a traditional human-readable proof for Lemma 10. The main argument is that for any one-sided 3-colouring algorithm $A = A_0$ the successor graph $\mathcal{S}_2(A)$ can be coloured with 16 colours. Later in Section 4.5, we give a computational proof of the same result. In the following, we fix A and denote $\mathcal{S}_2 = \mathcal{S}_2(A)$ for brevity.

Structural Properties. We start with the following observations.

Remark 1. Sets C_0 and C_1 satisfy

$$C_0 \subseteq \{1, 2, 3\},$$
$$C_1 \subseteq \{\{1\}, \{2\}, \{3\}, \{1, 2\}, \{1, 3\}, \{2, 3\}\}.$$

Remark 2. Relation S_1 satisfies

$$S_1(i) \subseteq \{X \in C_1 : i \notin X\},$$
$$S_1(\{i, j\}) \subseteq \{X \in C_1 : \{i, j\} \not\subseteq X\}.$$

Remark 3. Consider any $X \subseteq C_1$ with $\{\{1, 2\}, \{1, 3\}, \{2, 3\}\} \subseteq X$. Then there is no $x \in C_1$ with $X \subseteq S_1(x)$. Therefore A_2 cannot output colour X, and hence $X \notin C_2$.

Hence graph $\mathcal{S}_2$ has $|C_2| \leq 55$ nodes: out of the $2^6 = 64$ candidate colours, we can exclude the empty set and 8 other sets identified in Remark 3. We will now partition the remaining nodes in 16 colour classes (independent sets).

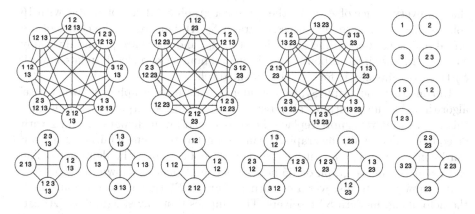

Fig. 3. This illustrations shows the *complement* of a graph we call S_2^*. For any algorithm A, the successor graph $S_2(A)$ is a subgraph of S_2^*, and hence, a proper colouring of S_2^* is a proper colouring of $S_2(A)$. Each clique in the figure corresponds to a colour class in S_2^*. We use a shorthand notation: for example, the circle labelled with "1 2 12" is the node $\{\{1\}, \{2\}, \{1, 2\}\}$.

Colour Classes. There are four types of colour classes. First, for each $\emptyset \neq X \subseteq [3]$ we define a singleton colour class

$$\mathcal{X}_0(X) = \Big\{\{\{x\} : x \in X\}\Big\},$$

that is, an independent set of size 1. Then for each triple

$$(i, j, k) \in \big\{(1, 2, 3), (1, 3, 2), (2, 3, 1)\big\}$$

we have three colour classes:

$$\mathcal{X}_1(i, j, k) = \Big\{X \in C_2 : \{\{i, j\}, \{i, k\}\} \subseteq X \subseteq \{\{i, j\}, \{i, k\}, \{i\}, \{j\}, \{k\}\}\Big\}$$

$$\mathcal{X}_2(i, j, k) = \Big\{X \in C_2 : \{\{i, j\}, \{k\}\} \subseteq X \subseteq \{\{i, j\}, \{i\}, \{j\}, \{k\}\}\Big\},$$

$$\mathcal{X}_3(i, j, k) = \Big\{X \in C_2 : \{\{i, j\}\} \subseteq X \subseteq \{\{i, j\}, \{i\}, \{j\}\}\Big\}.$$

In total, there are 7 singleton colour classes, and 3×3 other colour classes, giving in total 16 colour classes. Figure 3 shows the complement of a supergraph of S_2; each of the above colour classes correspond to a clique in the complement graph.

It can be verified that each of the 55 possible nodes of S_2 is included in exactly one of the colour classes. Now it suffices to show that each colour class is indeed an independent set of S_2. This is a relatively straightforward, albeit slightly tedious task to do by hand.

4.5 Computational Proof of Lemma 10

We now give a *computational proof* of Lemma 10, that is, we show how to easily verify with a computer that the claim holds. Essentially this amounts to checking that for every choice of $A = A_0$, the successor graph $\mathcal{S}_2(A)$ is colourable with 16 colours. However, since any successor graph $\mathcal{S}_2(A)$ depends on the choice of the initial one-sided 3-colouring algorithm $A = A_0$, and there are potentially many choices for A, we instead bound the chromatic number of a closely-related graph $\mathcal{S}_2^*$ that contains $\mathcal{S}_2(A)$ for any A as a subgraph.

To construct the graph $\mathcal{S}_2^*$, we consider the successor graph of a "worst-case" algorithm that may output "all possible" colours in its output set. Specifically, this means that we simply replace the subset relation in Remarks 1 and 2 with an equality. Therefore, the graph $\mathcal{S}_2^*$ can be constructed using a fairly straightforward computer program, with a mechanical application of the definitions. The end result is a dense graph on 55 nodes; its complement is shown in Figure 3.

It is now easy to discover a colouring of graph $\mathcal{S}_2^*$ that uses 16 colours with the help of e.g. modern SAT solvers. This implies that any subgraph $\mathcal{S}_2(A)$ can also be coloured with 16 colours and Lemma 10 follows.

5   Main Theorems

We now have all the pieces for proving Theorem 1:

Theorem 1. *For infinitely many values of n, it takes exactly $\frac{1}{2}\log^* n$ rounds to compute a 3-colouring of a directed n-coloured path.*

Proof. Let $n = {}^{2k+1}2 + 1$ for any $k \geq 2$. By Lemma 1 we have the identity

$$C(n,3) = \lceil T(n,3)/2 \rceil \tag{1}$$

and from Theorems 2 and 3 we get that

$$2k + 1 \leq T(n,3) \leq 2k + 2,$$

which together with (1) yields $C(n,3) = k + 1$. Since $\log^* n = 2k + 2$ it follows that $C(n,3) = k + 1 = \log^* n/2$.

For the remaining values of n we get almost-tight bounds. There remains a slack of *one* communication round in the upper and lower bounds for $C(n,3)$.

Theorem 4. *For any $n \geq 4$,*

$$\left\lceil \frac{1}{2}\left(\log^* n - 1\right) \right\rceil \leq C(n,3) \leq \left\lceil \frac{1}{2}\left(\log^* n + 1\right) \right\rceil.$$

Proof. For $n = 4$, we have shown that $T(4,3) = 2$ so the bounds follow. Fix $n > 4$. Now there exists some $h > 1$ such that $n \in \{{}^h2 + 1, \ldots, {}^{h+1}2\}$ and $h = \log^* n - 1$. Theorems 2 and 3 give us the bounds

$$\log^* n - 1 = h \leq T(n,3) \leq h + 2 = \log^* n + 1$$

and since $C(n,3) = \lceil T(n,3)/2 \rceil$, the claimed bounds follow.

6   Conclusions and Discussion

In this work we gave exact and near-exact bounds on the complexity of distributed graph colouring. The key result is that the complexity of colour reduction from n to 3 on directed paths and cycles is exactly $\frac{1}{2} \log^* n$ rounds for infinitely many values of n, and very close to it for all values of n.

In essence, we have shown that the colour reduction algorithm by Naor and Stockmeyer (and Szegedy and Vishwanathan) is near-optimal, while the algorithm by Cole and Vishkin is suboptimal. We have also seen that Linial's lower bound had still some room for improvements.

One of the novel techniques of this work was the use of **computers in lower-bound proofs**. Two key elements are results of a computer search:

- Lemma 6: The proof of $T(16, 3) \geq 3$ is based on the analysis of the chromatic number of the neighbourhood graph $\mathcal{N}_{7,1}$.
- Lemma 7: The proof of $T(n, 16) \leq T(n, 3) - 2$ is based on the analysis of the chromatic number of the successor graph $\mathcal{S}_2$.

In both cases we used computers to analyse the chromatic numbers of various successor graphs and neighbourhood graphs, in order to find the right parameters for our needs.

The idea of analysing **neighbourhood graphs** and their chromatic numbers is commonly used in the context of human-designed lower-bound proofs [7,10,13, 14]. It is also fairly straightforward to construct neighbourhood graphs so that we can use computers and graph-colouring algorithms to discover new upper bounds [18], and the same technique can be used to prove lower bounds on $T(n, c)$ for small, fixed values of n and c; in our case we used it to bound $T(16, 3)$. However, this does not yield bounds on, e.g., $T(n, 3)$ for large values of n.

The key novelty of our work is that we can use the chromatic number of **successor graphs** to give improved bounds on $T(n, 3)$ for all values of n. To do that, it is sufficient to find a successor graph $\mathcal{S}_k$ with a small chromatic number, and apply Lemma 9. The same technique can be also used to study $T(n, c)$ for any fixed $c \geq 3$.

Acknowledgements. We thank Juho Hirvonen and anonymous reviewers for helpful comments. Parts of this work are based on the first author's MSc thesis [18]. Computer resources were provided by the Aalto University School of Science "Science-IT" project, and by the Department of Computer Science at the University of Helsinki.

References

1. Åstrand, M., Suomela, J.: Fast distributed approximation algorithms for vertex cover and set cover in anonymous networks. In: Proc. 22nd Annual ACM Symposium on Parallelism in Algorithms and Architectures (SPAA 2010), pp. 294–302. ACM Press (2010)

2. Barenboim, L., Elkin, M.: Distributed graph coloring: Fundamentals and recent Developments. Morgan & Claypool (2013)
3. Cole, R., Vishkin, U.: Deterministic coin tossing with applications to optimal parallel list ranking. Information and Control 70(1), 32–53 (1986)
4. Cormen, T.H., Leiserson, C.E., Rivest, R.L.: Introduction to Algorithms. The MIT Press, Cambridge (1990)
5. Czygrinow, A., Hańćkowiak, M., Wawrzyniak, W.: Fast distributed approximations in planar graphs. In: Taubenfeld, G. (ed.) DISC 2008. LNCS, vol. 5218, pp. 78–92. Springer, Heidelberg (2008)
6. Fich, F.E., Ramachandran, V.: Lower bounds for parallel computation on linked structures. In: Proc. 2nd Annual ACM Symposium on Parallel Algorithms and Architectures (SPAA 1990), pp. 109–116. ACM Press (1990)
7. Fraigniaud, P., Gavoille, C., Ilcinkas, D., Pelc, A.: Distributed computing with advice: Information sensitivity of graph coloring. In: Arge, L., Cachin, C., Jurdziński, T., Tarlecki, A. (eds.) ICALP 2007. LNCS, vol. 4596, pp. 231–242. Springer, Heidelberg (2007)
8. Garay, J.A., Kutten, S., Peleg, D.: A sublinear time distributed algorithm for minimum-weight spanning trees. SIAM Journal on Computing 27(1), 302–316 (1998)
9. Goldberg, A.V., Plotkin, S.A., Shannon, G.E.: Parallel symmetry-breaking in sparse graphs. SIAM Journal on Discrete Mathematics 1(4), 434–446 (1988)
10. Kuhn, F., Wattenhofer, R.: On the complexity of distributed graph coloring. In: Proc. 25th Annual ACM Symposium on Principles of Distributed Computing (PODC 2006), pp. 7–15. ACM Press (2006)
11. Laurinharju, J., Suomela, J.: Brief announcement: Linial's lower bound made easy. In: Proc. 33rd ACM SIGACT-SIGOPS Symposium on Principles of Distributed Computing (PODC 2014), pp. 377–378. ACM Press (2014)
12. Lenzen, C., Patt-Shamir, B.: Improved distributed Steiner forest construction. In: Proc. 33rd ACM SIGACT-SIGOPS Symposium on Principles of Distributed Computing (PODC 2014), pp. 262–271. ACM Press (2014)
13. Linial, N.: Locality in distributed graph algorithms. SIAM Journal on Computing 21(1), 193–201 (1992)
14. Naor, M.: A lower bound on probabilistic algorithms for distributive ring coloring. SIAM Journal on Discrete Mathematics 4(3), 409–412 (1991)
15. Naor, M., Stockmeyer, L.: What can be computed locally? SIAM Journal on Computing 24(6), 1259–1277 (1995)
16. Panconesi, A., Rizzi, R.: Some simple distributed algorithms for sparse networks. Distributed Computing 14(2), 97–100 (2001)
17. Peleg, D.: Distributed Computing: A Locality-Sensitive Approach. SIAM Monographs on Discrete Mathematics and Applications. Society for Industrial and Applied Mathematics, Philadelphia (2000)
18. Rybicki, J.: Exact bounds for distributed graph colouring. Master's thesis, University of Helsinki, May 2011. http://urn.fi/URN:NBN:fi-fe201106091715.
19. Suomela, J.: Distributed Algorithms (2014). http://users.ics.aalto.fi/suomela/da/
20. Szegedy, M., Vishwanathan, S.: Locality based graph coloring. In: Proc. 25th Annual ACM Symposium on Theory of Computing (STOC 1993), pp. 201–207. ACM Press (1993)
21. Wattenhofer, R.: Lecture notes on principles of distributed computing (2013). http://dcg.ethz.ch/lectures/podc_allstars/

Essential Traffic Parameters for Shared Memory Switch Performance

Patrick Eugster[1,2,*], Alex Kesselman[3],
Kirill Kogan[4], Sergey Nikolenko[5,6,**], and Alexander Sirotkin[7,8]

[1] Purdue University, Lafayette, IN, USA
p@cs.purdue.edu
[2] Technical University of Darmstadt, Darmstadt, Germany
[3] Google Inc., California, USA
alx@google.com
[4] IMDEA Networks Institute, Madrid, Spain
kirill.kogan@imdea.org
[5] National Research University Higher School of Economics, St. Petersburg, Russia
[6] Steklov Institute of Mathematics at St.Petersburg, Russia
sergey@logic.pdmi.ras.ru
[7] International Laboratory for Applied Network Research
National Research University Higher School of Economics, Moscow, Russia
[8] St. Petersburg Institute for Informatics and Automation of the RAS, St. Petersburg, Russia
alexander.sirotkin@gmail.com

Abstract. Cloud applications bring new challenges to the design of network elements, in particular accommodating for the burstiness of traffic workloads. Shared memory switches represent the best candidate architecture to exploit buffer capacity; we analyze the performance of this architecture. Our goal is to explore the impact of additional traffic characteristics such as varying processing requirements and packet values on objective functions. The outcome of this work is a better understanding of the relevant parameters for buffer management to achieve better performance in dynamic environments of data centers. We consider a model that captures more of the properties of the target architecture than previous work and consider several scheduling and buffer management algorithms that are specifically designed to optimize its performance. In particular, we provide analytic guarantees for the throughput performance of our algorithms that are independent from specific distributions of packet arrivals. We furthermore report on a comprehensive simulation study which validates our analytic results.

1 Introduction

Cloud data centers are faced with workloads which evolve rapidly, driven by high volumes of end users, application types, cluster nodes, and overall data movement (e.g., big

---

[*] P. Eugster was partially supported by the German Research Foundation (DFG) under project MAKI ("Multi-mechanism Adaptation for the Future Internet").
[**] The work of Sergey Nikolenko was partially supported by the Government of the Russian Federation grant 14.Z50.31.0030 and the Presidential Grant for Leading Scientific Schools, NSh-3856.2014.1.

C. Scheideler (Ed.): SIROCCO 2015, LNCS 9439, pp. 61–75, 2015.
DOI: 10.1007/978-3-319-25258-2_5

data processing [4,10]). A primary design challenge in this context consists in selecting and deploying network switches that scale application performance, in a way which is robust and cost-effective. A network switch receives packets on ingress ports, applies specific policies to them, identifies destination ports, and sends them out through egress ports. When application-induced traffic bursts create an imbalance between incoming and outgoing packet rates for a given port, packets must be queued in the switch packet buffer. The available queue size on a port determines the port's ability to hold packets until the egress port can emit it. When buffer queue is full, packets are dropped. The allocation and availability of buffer resources to the ports, determined by the buffering architecture, affects burst absorption capabilities and performance characteristics of the network switch. Overprovisioning in terms of buffer capacity at each network node to absorb bursty behavior is not viable, as networks do not have unlimited resources; conversely, cloud data centers can only scale out as fast as the effective per-port cost and power consumption. These factors, in turn, are driven by the chosen buffering architecture. The *shared memory switch* allows to absorb traffic bursts in the best way since the whole buffer can be utilized by a same output port if needed. Since this is an actual choice in practice [13], here we focus our efforts on this type of buffer architecture.

The *buffer management policy* is a key element in meeting network design challenges. It directly impacts a switch's ability to transfer data at line rate and optimize desired objectives during congestion under various traffic conditions. Most existing buffer management policies are based on a simple characteristic such as buffer occupancy [9, 16], whereas traffic workloads have additional important characteristics such as processing requirements or value that are not explicitly taken into account. Efficient methods for buffer management incorporating new characteristics in admission decisions beyond *fairness* objective functions lead to new challenges in performance and implementation for traditional switch architectures. Inherited from the Internet, fairness is in fact a design choice which can conflict with other objectives in various economic models (e.g., utilization of network infrastructure or profit [7, 19, 20]).

We thus consider a shared memory switch where a buffer of size B is shared among all types of traffic. Each arriving packet is labeled with an output queue. Arrivals can be adversarial. During arrival, packets concurrently "access" the shared memory. Each input port decides if its arriving packets should be admitted based on information computed by output ports in a distributed manner. In this work we consider (possibly weighted) throughput optimization since relevant objectives such as better reuse of underlying infrastructure or profit maximization can be reduced to throughput optimization [19, 20]. However, in stark contrast to the seminal work of Aiello et al. [1] where packets have uniform values and processing requirements, in our model each arriving packet has an intrinsic value ("worth") or processing requirement. Moreover, we remove a strong constraint from the recent work of Eugster et al. [15] by allowing packets for the same output port to have distinct processing requirements. We consider the paradigm of *competitive analysis* [8, 36]: an algorithm ALG is α-competitive for some $\alpha \geq 1$ if for any arrival sequence the number of packets transmitted by ALG is at least $1/\alpha$ times the number of packets transmitted by an optimal offline clairvoyant algorithm OPT. Worst case analysis shows whether additional workload characteristics should be taken into account in buffer management. Since all policies we consider are greedy (they accept and transmit all traffic if there is no congestion), we need to consider extreme cases during congestion periods that can actually happen in scenarios like big data processing [12,37,38]. Worst case results help define buffer management "rules of thumb" that are independent of specific arrival distributions.

The goal of this paper is to offer designs with proven performance guarantees for the shared memory switch; we analyze the performance of buffer management policies and provide guarantees for their worst-case throughput. We consider two different traffic characteristics: (a) required processing and (b) values for packet transmission. Intuitively, they should have similar impacts on the desired objective. In the case of a single queue, both of them reach optima when all packets are ordered by required processing [21] or values with push-out[1]. However, generalizing them to a shared memory buffering architecture is challenging: the case of heterogeneous values was presented as an open question in SIGACT News [18, p. 22].

In the first part of this paper every incoming packet has unit value (and an output port label) but has processing requirement varying from 1 to k. In this case the objective comes down to maximizing the number of transmitted packets. We show that LQD is at least $(n/2 - o(n))$-competitive for sufficiently large buffer size B and maximal required processing k; besides that we show that Biggest-Packet-Drop (BPD, a policy that pushes out packets with maximal processing requirement in case of congestion) degrades to at least $(n + 1)/2$-competitiveness. In addition we introduce a natural Biggest-Average-Drop policy (that pushes out a packet with maximal required processing from a queue with maximal average processing requirements in the case of congestion) that achieves the same lower bound as BPD. All lower bounds hold even for PQ (priority queueing) processing order, where packets are ordered according to processing requirements. The main result of this work is the 2-competitiveness of a semi-greedy variant of the Longest-Work-Drop (LWD) policy of [15] that holds in our general model when packets with heterogeneous processing requirements are processed in PQ order in each queue (Section 5). In addition we show that in the FIFO case of our general model, LWD is at least $(\log_{B/n} k)(1 - 1/B) + 1$-competitive. In the second part of this paper (b) we consider a model where each incoming packet has, in addition to an output port label, a heterogeneous value from 1 to V (and uniform processing). In this case the objective is to maximize transmitted value. Intuitively, the model with values should be similar to the model with required processing. However, we show that the Maximal-Total-Value-Drop (MTVD) policy, which is similar to LWD, is at least V-competitive. We also turn to policies that combine several characteristics and consider the Minimal-Ratio-Drop (MRD) policy introduced in [15] that considers both queue occupancy and the average value in the same queue. MRD was conjectured in [15] to have constant competitiveness. We show that the model with values has a different nature as a generalization from the single-queue case (where both models have optimal online algorithms) to the shared memory switch, and it is not enough to simply consider the total value. In particular, we prove that MRD is at least V-competitive.

The paper is organized as follows. Section 2 discusses related prior art. Section 3 details the model underlying our work. Section 4 considers lower bounds of several

---

[1] In case of packets with values, the optimality of the greedy algorithm with pushout is trivial: order the queue by value.

algorithms for packets with heterogeneous processing requirements to understand properties of an "ideal" policy. The main result of this paper — 2-competitiveness of LWD policy for packets with heterogeneous requirements — is presented in Section 5. Section 6 considers a model with heterogeneous packet values. Section 7 concludes the paper.

2 Related Work

Aiello et al. [27] propose a non-push-out buffer management policy called Harmonic that is at most $O(\log n)$-competitive and establish a lower bound of $\Omega(\frac{\log n}{\log \log n})$ on the performance of any online non-push-out deterministic policy, where n is the number of output ports. Kesselman and Mansour [1] demonstrate that the LQD policy is at most 2- and at least $\sqrt{2}$-competitive. Both works consider homogeneous packet processing, i.e., each packet requires a single processing cycle. Eugster et al. [15] consider a limited variant of our model where all incoming packets for the same output port have identical processing requirements. But even in this case it was shown there that LQD is at least $\left(\sqrt{k} - o(\sqrt{k})\right)$-competitive. Fortunately, in [15] a generalization of LQD, namely Longest-Work-Drop (LWD), was proposed for this limited model; LWD is at most 2-competitive in case when packets are processed with minimal current required processing first (PQ order). Besides, it was shown in [15] that Biggest-Packet-Drop (BPD, a policy that pushes out packets with maximal processing requirement in case of congestion) is at least $\log k$ competitive for $B > \frac{k(k+1)}{2}$. Unlike the model in [15] the model we consider in this paper however allows for packets with heterogeneous processing requirements to be admitted to the same queue. This generalization over the model in [15] has a significant impact on the efficiency of considered policies and applicability to real-world scenarios. In particular, we can open a separate queue per processing requirement per output port but in this case the scalability of maximal number of supported queues can become a strong constraint once k and n are growing. Our current work can be viewed as part of a larger research effort concentrated on studying competitive algorithms for management of bounded buffers. Surveys by Goldwasser [18] and later by Nikolenko and Kogan [35] provide an excellent overview of this field. Initiated in [26, 34], this line of research has received tremendous attention over the past decade. Various models have been proposed and studied, including QoS-oriented models where packets have individual weights [2, 14, 26, 34]. A related field that has recently attracted much attention focuses on various switch architectures and aims to design competitive algorithms for various scenarios therein (cf. [3, 5, 6, 22–25, 33]). However, none of these models cover the case of packets with heterogeneous processing requirements, and our work extends and generalizes previous models to heterogeneous processing. The single queue case with heterogeneous processing requirements is considered in [21, 31, 32]. Kogan et al. considered the multiple separated queues case with heterogeneous processing requirements in [29]. The single queue case with packets containing a combination of heterogeneous processing with packet lengths or values has considered in [11, 30].

3   Model Description

We consider an $n \times n$ shared memory switch with n input and n output ports and a buffer of size B, that is, the total length of all queues is bounded by B. We assume that $B \geq n$. Each output port manages a single output queue, denoted Q_i for port i, $1 \leq i \leq n$; the number of packets in Q_i is denoted by $|Q_i|$. Each packet $p(d, w)$ arriving at an input port is labeled with the output port number d and its required work w in processing cycles ($1 \leq d \leq n$ and $1 \leq w \leq k$), where k denotes the global upper bound on required work per packet. Each Q_i implements either (i) priority queueing (PQ) processing order, where packets are ordered in non-decreasing order of required processing, or (ii) first-in-first-out (FIFO) processing order, where packets are ordered in the order of arrival. In what follows, we denote by $\boxed{w \mid i}$ a packet with required work w intended for output port i; by $h \times \boxed{w \mid i}$, a burst of h $\boxed{w \mid i}$ packets arriving at the same time. We also denote by $r_t(p)$ the remaining required processing of a packet p at time unit t. Time is slotted; we divide each time slot into two phases (see Fig. 1). During the (1) *arrival phase* a burst of new packets arrives at each input port that decides which ones should be admitted based on the state computed by each output port in the distributed manner. The arrivals are adversarial and do not assume any specific traffic distribution (more than n arrivals are allowed at the same time slot). An accepted packet can be later dropped from the buffer when another packet is accepted instead; in this case we say that a packet p is *pushed out* by another packet q, and a policy that allows this is called a *push-out* policy. During the (2) *transmission phase*, required work of the head-of-line packet according to the supported processing order (PQ or FIFO) at each non-empty queue is reduced by one, and every packet with zero residual work is transmitted.

To facilitate our proofs, we use some properties of ordered (multi-)sets. These notions, as well as the properties we recall in this section, will enable us to compare the performance of our proposed algorithms. In the following, we consider multi-sets of real numbers, where we assume each multi-set is ordered in non-decreasing order. We will refer to such multi-sets as *ordered sets*. For every $1 \leq i \leq |A|$, we will further refer to element $a_i \in A$ or to $A[i]$ as the i-th element in the set A, as induced by the order. Given two ordered sets A and B, we say $A \geq B$, if for every i for which both a_i and b_i exist, $a_i \geq b_i$. The following lemma, and its corollary, will be used in our analysis; their proofs can be found in [28] (Lemma 1 and Corollary 2).

Lemma 1. *For any two ordered sets A and B satisfying $A \geq B$, and any two real numbers a, b such that $a \geq b$, if (i) $b \leq b_{|B|}$ or (ii) $|A| \leq |B|$ then the ordered sets $A' = A \cup \{a\}$, $B' = B \cup \{b\}$ satisfy $A' \geq B'$.*

Corollary 1. *For any two ordered sets A, B satisfying $A \geq B$, and any real number b, if (i) $b \leq b_{|B|}$ or (ii) $|A| \leq |B|$ then the ordered set $B' = B \cup \{b\}$ satisfies $A \geq B'$.*

4   The Quest for an Ideal Policy with Heterogeneous Processing

In this section, we consider several possible candidates for the "ideal" policy that might provide constant competitiveness in the model presented above. These algorithms either look like natural candidates or have been proven to be efficient for uniform process-ing [1, 27]. Note that in our model, each algorithm has two versions, with PQ and FIFO processing order in each output queue. By default, we assume that every queue imple-ments PQ order. Lower bounds on the competitive ratio represent specific sequences of packets on which the optimal algorithm is much better than the one in question. they are easier to prove than upper bounds since it suffices to present a hard instance of an input sequence, but lower bounds can still provide important information regarding the comparative quality of online algorithms.

Longest-Queue-Drop (LQD): during the arrival of a packet p with output port i, denote by $j^* = \arg \max_j \{|Q_j| + [i = j]\}$ where $[i = j] = 1$ if $i = j$ and 0 otherwise (i.e., Q_{j^*} is the longest queue once we virtually add p to Q_i; we choose one with largest required processing if there are several); then do the following: (1) if the buffer is not full, accept p into Q_i; (2) if the buffer is full and $i \neq j^*$, push out last packet from Q_{j^*} and accept p into Q_i; else drop p.

Note that the proposed here version of LQD is not fully oblivious to processing requirements since it will drop a packet with a maximal processing from the longest queue in the case of congestion. In case of homogeneous processing, LQD is at least $\sqrt{2}$- and at most 2-competitive [1]. For heterogeneous required processing, the situation is worse. Proofs of all theorems in this section are given in the Appendix.

Theorem 1. *For sufficiently large B and $k \geq n(n-1)$, LQD is at least $(n/2 - o(n))$-competitive.*

Proof. Over the first burst, there arrive B packets of each of the following kinds: $\boxed{1 \mid 1}$, $\boxed{k \mid 2}$, $\boxed{k \mid 3}$, ..., $\boxed{k \mid n}$. LQD evenly distributes the packets among queues and has B/n packets in each of its nonempty queues (throughout the proof we assume that B is large and is divisible by everything we need it to be). OPT accepts $(B - n + 1) \times \boxed{1 \mid 1}$ and one each in the remaining queues. Every k processing cycles there arrive $1 \times \boxed{k \mid 2}$, $1 \times \boxed{k \mid 3}$, ..., $1 \times \boxed{k \mid n}$, so OPT always has packets in these queues to work on, but there are no more $\boxed{1 \mid 1}$s. OPT spends $(B - n + 1)$ time to process all $\boxed{1 \mid 1}$s (after that the arrival iteration is restarted); let us estimate the number of processed packets by this time. OPT will have processed $(B - n + 1)$ in queue 1 and $(B - n + 1)/k$ in each one of $n - 1$ other queues. LQD will have the same $(B - n + 1)/k$ processed packets in each queue but the first, and in the first queue LQD will have processed B/n since there are no more packets in first queue. Thus, the overall competitive ratio is $\frac{(B-n+1)+(n-1)\times\frac{B-n+1}{k}}{\frac{B}{n}+(n-1)\times\frac{B-n+1}{k}}$, and for $k = n(n-1)$ we get $\frac{(B-n+1)+(n-1)\times\frac{B-n+1}{n(n-1)}}{\frac{B}{n}+(n-1)\times\frac{B-n+1}{n(n-1)}} = \frac{(B-n+1)+\frac{B-n+1}{n}}{\frac{B}{n}+\frac{B-n+1}{n}} \geq \frac{(B-n+1)}{2\frac{B}{n}} \approx n/2$.

The next two algorithms drop packets with the largest processing requirement in case of congestion.

Biggest-Packet-Drop (BPD): during the arrival of a packet p with required work w and output port i, denote by Q_j the nonempty queue that contains a packet $p_{\max}$ with the

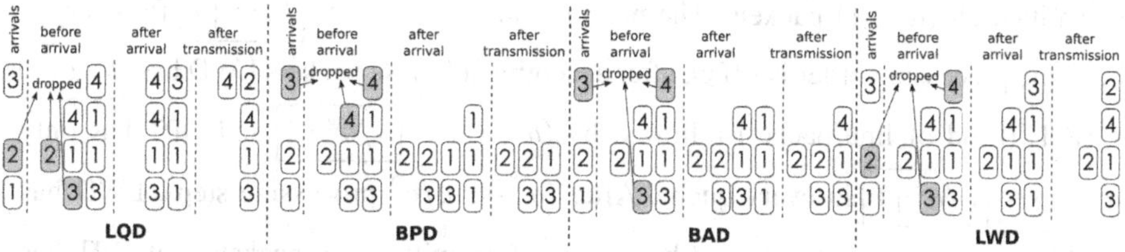

Fig. 1. A sample time slot of Longest Queue Drop (LQD), Biggest Packet Drop (BPD), Biggest Average Drop (BAD), and Largest Work Drop (LWD) policies with maximal processing $k = 4$, $n = 4$ output ports, and a shared buffer of size $B = 8$. Queues for each output port are shown horizontally. Shaded packets are dropped during arrival.

largest processing requirement w_{max}; then do the following: (1) if the buffer is not full, accept p into Q_i; (2) if the buffer is full and $w < w_{max}$, push out p_{max} from Q_j and accept p into Q_i; (3) if the buffer is full and $w > w_{max}$, drop p.

Biggest-Average-Drop (BAD): during the arrival of a packet p with required work w and output port i, denote by Q_j the nonempty queue with largest average processing requirement $\bar{w}_{max}$; then do the following: (1) if the buffer is not full, accept p into Q_i; (2) if the buffer is full and $w < \bar{w}_{max}$, push out packet with maximal work from Q_j and accept p into Q_i; (3) if the buffer is full and $w > \bar{w}_{max}$, drop p.

Theorem 2. *BPD and BAD are both at least $(n + 1)/2$-competitive.*

Proof. The counterexample is as follows: every time slot, there arrive $B \times \boxed{1 \mid 1}$ followed by $B \times \boxed{2 \mid 2}, \ldots, B \times \boxed{2 \mid n}$ (a full set of packets); BPD and BAD both accept only $B \times \boxed{1 \mid 1}$ and keep processing one packet per time slot, i.e., 2 packets per 2 time slots, while OPT is free to accept the packets evenly and get $2 + 2/2 + \ldots + 2/2$ packets per 2 time slots, getting the bound as $\frac{n+1}{2}$.

Largest-Work-Drop (LWD): during the arrival of a packet p with output port i and required processing w, denote by $j^* = \arg \max_j \{W_j + \mathbb{1}_{i=j} w\}$ where $\mathbb{1}_{i=j} = 1$ if $i = j$ and 0 otherwise, and W_j is the total required processing of all packets in queue Q_j (i.e., Q_{j^*} is the queue with the largest total required processing once we virtually add p to Q_i; we choose the one with the largest single packet if there are several queues with largest work); then do the following: (1) if the buffer is not full, accept p into Q_i; (2) if the buffer is full and w is smaller than the required processing of at least one packet in Q_i, push out the largest packet from Q_{j^*} and accept p into Q_i; else drop p.

Theorem 3. *LWD with FIFO processing order is at least $(\log_{B/n} k)(1 - n/B) + 1$-competitive.*

Proof. Consider LWD with n output ports and suppose that n divides B. Let $a = B/n$. For every output port i, there arrive $1 \times \boxed{k \mid i}$ followed by $(a - 1) \times \boxed{k/a \mid i}$. OPT discards $\boxed{k \mid i}$ and accepts all $\boxed{k/a \mid i}$. After $(a - 1)k/a$ processing steps, LWD has $a \times \boxed{k/a \mid i}$ in every queue and has not yet transmitted any packets, while OPT has transmitted all $(a - 1)$ packets. The next arrival is $(a - 1) \times \boxed{k/a^2 \mid i}$ for every i. Since the processing order is FIFO, after accepting all these packets LWD has $\boxed{k/a \mid i}$ as HOL (head of line packet) followed by $(a - 1) \times \boxed{k/a^2 \mid i}$ and OPT has only $(a - 1) \times \boxed{k/a^2 \mid i}$ in every queue. After $(a - 1)k/a^2$ processing steps, LWD has $a \times \boxed{k/a^2 \mid i}$ in each queue and has not yet transmitted any packets, but OPT has transmitted all $(a - 1)$ packets. Next we repeat the above arrival sequence for packets of size $k/a^3, \ldots, k/a^m$, until $k/a^m = 1$, i.e., for $log_a(k)$ steps. On every step, OPT transmits $(a-1) \times n$ packets and LWD transmits nothing. After all these steps, $a \times \boxed{1 \mid i}$ has arrived in every queue, so after a processing cycles both OPT and LWD transmit B packets and finish with empty buffers. Thus, the total number of packets that LWD transmits is B, and the total number of packets transmitted by OPT is $n(a - 1) \log_a k + B$, getting the ratio $\frac{n \cdot ((a-1) \log_a k)}{B} + 1$. Recall that we had $a = B/n$, so the final ratio is $\frac{n \cdot (B/n-1) \log_{B/n} k}{B} + 1 = (\log_{B/n} k)(1 - n/B) + 1$.

5   Scheduling with Heterogeneous Processing

To avoid ambiguity during the arrival phase, a reference time t should be interpreted as the arrival of a single packet. If several packets arrive at the same time slot, we consider them independently, in the sequence in which they arrive. A time slot is divided into time units; arrival of each packet is a separate time unit (so the arrival phase takes up several time units), while processing and transmission phases both use only a single time unit (we do not separate them). We introduce the class of *semi-greedy* algorithms $\mathcal{SG}$. A semi-greedy algorithm $G \in \mathcal{SG}$ accepts a packet if G's buffer is not full; G is defined by an *iteration*. An iteration begins during the first time unit t_s when G's buffer is congested and ends on the first time unit t_e when G has transmitted at least B packets since t_s. To simplify analysis, G drops the content of its buffer at the end of an iteration at time t_d, $t_e \leq t_d < t_e + 1$, without gain to its throughput; in this section we show an upper bound, so weakening the algorithm only makes things worse for us.

In what follows we consider an artificially enhanced version of OPT: (1) OPT never pushes out admitted packets (since OPT is offline, it is clear that this property can be satisfied); (2) at the end of an iteration, OPT flushes out all packets residing in its buffer with extra gain to its throughput (in this case, the throughput of OPT is no worse than any other optimal algorithm); (3) if at time t G transmits out of port i, the first packet q (in PQ order) is transmitted out of the i-th port of OPT (if q exists) regardless of its remaining work value $r_t(q)$ with extra gain to OPT's throughput (again, clearly we only make OPT better). Note that by definition, for a given sequence of inputs all algorithms in $\mathcal{SG}$ with the same processing order accept and transmit the same number of packets between starting with an empty buffer and the first moment of congestion. With PQ processing order, moreover, no algorithm can transmit more packets from this sequence over this time. And, by definition, at the end of an iteration an $\mathcal{SG}$ algorithm has an empty buffer. The difference in the number of packets remaining at the end of an iteration (just before t_d) is irrelevant since all these packets are dropped at time t_d. Since during $[t_s, t_d)$ any semi-greedy algorithm G transmits B plus at most $n - 1$ packets, dropping all buffered packets at time t_d adds at most 1 to the competitiveness of G. The general idea of our analysis here is similar to [27] but the definition of an iteration and the analysis of what happens during an iteration and between two consecutive iterations are completely new. We denote by t_b the first time unit after the end of a previous iteration or the time unit of the first arrival in the system. Since a semi-greedy G and OPT both clean their buffers at time t_d, it suffices to compare performance of G versus OPT only during $[t_b, t_e]$. The class of semi-greedy algorithms is based on a well-structured accounting infrastructure that significantly simplifies analysis of online buffer management policies with various characteristics. The major question that we will soon answer is: is there a policy with a constant competitiveness in the model where each packet has both required processing and output port (admission of heterogeneous packets to the same queue is allowed)? Note that the processing order implemented in each queue has significant impact on the performance of a scheduling policy. We assume that every queue implements priority queueing (PQ), where all packets in the same queue are ordered in non-decreasing order of required processing. For simplicity, we denote queue Q_i of an algorithm ALG by A_i, where A is the the first letter of the name of the considered algorithm (e.g, O_i and G_i are the i-th queue of OPT and a semi-greedy G). We treat queues as ordered sets in the sense of Lemma 1 and correspondingly write $A_i \leq B_i$ for two queues if for every slot in the queue where both A_i and B_i have packets p_A and p_B respectively, $w(p_A) \leq w(p_B)$.

The *latency* $\mathrm{lat}_t^A(p)$ of a packet $p \in A_i$ at time t is the number of time slots currently needed to transmit p out of A_i. We define the latency of an already transmitted packet as -1 and the latency of a packet that has not yet arrived as ∞. An i-th port or queue of ALG's buffer is called *active* at time unit t if it transmits during t; otherwise, it is called *idle*. To show that OPT does not transmit more packets than a semi-greedy algorithm G during $[t_b, t_s)$, we formulate the following lemma (proven in the Appendix). Actually, we prove an even stronger result that will be used in the proof of the key Lemma 3. Consider an interval of time I, $I \subseteq [t_b, t_d)$. We denote by S_I^A the set of packets transmitted by an algorithm A during I.

Lemma 2. *For a semi-greedy algorithm G with PQ processing and time unit $t \in [t_b, t_s)$ between two consecutive iterations, (1) $S_{[t_b, t]}^{OPT} \leq S_{[t_b, t]}^{G}$; (2) for any $i \in [1, n]$, at time t $G_i \leq O_i$ and $|G_i| \geq |O_i|$.*

Proof. The proof proceeds by induction on the number of time units. *Base:* During the first arrival of a packet p to Q_i at time t_b, since G is greedy, G accepts p, so the induction base follows. *Hypothesis:* Assume that the lemma holds during $[t_b, t)$, $t \in [t_b, t_s - 1]$. *Step:* We are to show that the lemma holds during t.

Processing and Transmission: the induction step holds by induction hypothesis for all empty queues or queues with head-of-line packet whose remaining processing is at least one. Consider any active queue Q_j in OPT or G, $1 \leq j \leq n$. If both O_j and G_j are nonempty just before t, by the induction hypothesis we have $O_j[1] \geq G_j[1]$. If G_j is active at time unit $t + 1$, then by definition of OPT (property (3)) O_j is also active (even if there are additional processing cycles in the HOL packet), and the induction step follows.

A Packet p Arrives to Q_i: during the arrival phase, the number of transmitted packets is unchanged, so condition (1) follows. Since there is no congestion during $[t_b, t_s)$, G accepts all arrivals. By the induction hypothesis, at the end of time unit $t - 1$ we had $G_i \leq O_i$ and $|G_i| \geq |O_i|$. Thus, if OPT accepts (G accepts since G is greedy and there is no congestion between two consecutive iterations), by Lemma 1(ii) condition (2) follows at time unit t. If OPT does not accept p to O_i, condition (2) follows by Corollary 1(ii).

Note that due to property (3) in the definition of OPT, at this point it is unclear if our version of OPT can transmit more packets than a semi-greedy G during $[t_b, t_s)$, and theoretically it can happen, so we have to prove (1) in Lemma 2. Part (2) of Lemma 2 will be used in the proof of Lemma 3.

The Largest Work Drop (LWD) policy belongs to the $\mathcal{SG}$ class. The rationale behind LWD is to minimize the duration of an iteration. It can be done by optimizing a "local" state of LWD buffer, and that is why we suggest to drop packets from a queue with the largest total required processing. Our plan is as follows. By Lemma 2, between two consecutive iterations OPT does not transmit more than LWD. We denote by T the number of packets transmitted by LWD between two consecutive iterations. Later we are to show that during $[t_s, t_d)$ LWD transmits B packets no later than OPT transmits B packets. Since at t_d OPT contains at most B packets, during $[t_b, t_e]$ OPT transmits at most $T + 2B$, whereas LWD transmits $T + B$ packets. For any time interval $I' = [t_s - 1, t], t \in [t_s - 1, t_d)$, during an iteration we say that $B - S_{I'}^{LWD}$ packets with minimal latency in LWD buffer are colored in red; any other packet in LWD buffer is colored in white. Note that packets that ceased to be red are immediately recolored in white again. We denote by R_i the set of all red packets in L_i. Lemma 3 contains the main ideas of this upper bound; due to space constraints, its proof is given in the Appendix.

Observation 4. *If a packet $p_j \in L_i$ is red then every $p_l \in L_i$ is red for $l \in [1, j - 1]$.*

Lemma 3. *For every OPT packet $p_j \in O_i$, $j \in [1, |O_i|]$ and $i \in [1, n]$, at time unit $t \in (t_s, t_e)$ either (1) there is a red packet $q_j \in L_i$, $r_t(p_j) \geq r_t(q_j)$ ($R_i \leq O_i$), or (2) for any red packet q at LWD buffer, $r_t(p_j) + W(R_i) \geq \text{lat}^t(q)$.*

Proof. The proof is by induction on time units. *Base:* Consider time unit $t_s - 1$. By definition of iteration, at the end of $t_s - 1$ LWD's buffer is full. Since LWD is semi-greedy, by Lemma 2 at time $t_s - 1$ $L_i \leq O_i$ and, therefore, $R_i \leq O_i$ for every $i \in [1, n]$. Thus, the induction base follows. *Hypothesis:* Assume that the lemma holds for every time unit $t' \in [t_{s-1}, t), t < t_d$. We are to show that it holds at the t-th time unit.

Induction step. **Processing and Transmission:** suppose that the t-th time unit is devoted to processing all HOL packets and transmitting fully processed packets. In this case, either every nonempty queue L_j is active (in this case O_j is active too regardless of how many processing cycles remains in HOL packet of O_j by definition of OPT (property (3))) or the processing cycles of HOL packets of L_i and O_i are decreased by one. Assume that during $t \in (t_s, t_e)$, O_j is active and transmits a packet p; while L_i is idle. In this case by condition (2) LWD's buffer does not contain any red packet that means the iteration is already over, hence, $t \geq t_e$, which is a contradiction.

Arrival of a Packet p **to** Q_i**:** Note that if OPT accepts p, its buffer has free space since by definition OPT never pushes out already accepted packets.

OPT and LWD Reject p**:** The induction hypothesis holds at time t.

OPT Accepts p**, but LWD Rejects:** LWD's buffer is congested. Furthermore, since p is rejected by LWD, its required processing exceeds that of any packet in L_i. Suppose that p is at the l's position in O_i after acceptance, $l \leq |O_i|$. If $q_l \in L_i$ is red, condition (1) holds (the required processing of p is at least the required processing of any packet in L_i, including all red packets in L_i). If $q_l \in L_i$ is white or $l > |L_i|$, assume that there is a red packet whose latency is more than $r_t(p) + W(R_i)$. If $l > |L_i|$, $r_t(p) + W(R_i) = r_t(p) + W(L_i)$ that is (by definition of LWD) at least $W(L_j)$ since p is rejected. Thus, condition (2) holds. If $q_l \in L_i$ is white then $r_t(q_l) + W(R_i) \geq W(R_j)$, for any $j \in [1, n]$ (by definition of red packet); $r_t(q_l) \leq r_t(p)$ (otherwise, LWD will not drop p). Therefore, condition (2) holds, and the induction hypothesis holds too.

OPT and LWD Accept p**:** 1. If $r_t(p)$ is less than at least one red packet in L_i then p is recolored in red and the last red packet in L_i is recolored in white. Since no new red packets are added to the queues other than L_i, condition (1) holds in these queues. By Theorem 1(i), condition (1) holds for any red packet in R_i. Next we show that condition (2) continues to hold for any OPT packet that is not covered by condition (1). Since the maximal latency among red packets does not increase for any queue except j, condition (2) holds. Consider a packet $u_l \in O_j$ corresponding to q_l recolored from red to white; by condition (1) of the induction hypothesis, $r_t(u_l) \geq r_t(q_l)$. Therefore, $r_t(u_l) + W(R_j) \geq r_t(q_l) + W(R_j)$, and (2) holds.

2. If the value of $r_t(p)$ is at least the required processing of any red packet in L_i then if $r_t(p) + W(R_i)$ is less than the latency of some red packet in LWD's buffer, recolor p in red, but the red packet q_l with a maximal latency in LWD's buffer recolor in white. Otherwise, p remains white.

If p is white then condition (1) follows by induction hypothesis. Since p is white, $r_t(p) + W(R_i)$ is at least the latency of any red packet in LWD's buffer (otherwise, p is recolored in red). If p is recolored in red, condition (2) follows similar to case 1. Since only Q_i is affected, condition (1) is satisfied for any Q_m, $m \neq i$ and holds for Q_i by Lemma 1(ii).

OPT Rejects, LWD Accepts: 1. Consider the case when LWD's buffer is not congested. (i) If $r_t(p)$ is at least the remaining processing of some white packet in Q_i, the set of the red packets is not changed. Also since OPT rejects p the set of OPT's packets is not changed also. Hence, conditions (1) and (2) hold. (ii) Otherwise, if $r_t(p) + W(R_i)$ is less than the latency of the red packet q with a maximal latency in LWD's buffer then recolor p in red and q in white. Denote by p an OPT packet in the position $|R_i| + 1$ of O_i. If $|O_i| > |R_i|$ just before p is arrived, $r_t(p) + W(R_i)$ is more than the latency of q. Hence, $r_t(p) + W(R_i) > r_t(p) + W(R_i)$ and therefore, $r_t(p) > r_t(p)$. Thus, condition (1) holds. Condition (2) holds similar to case 1.2. LWD's buffer is congested. If a white packet is pushed out, we can drop it and run the case when the congestion did not occur as in case 1. If the pushed out packet is red then recolor a new packet p in red and apply case (ii).

The main result of this section is the following theorem (see proof in Appendix).

Theorem 5. *For a shared memory* $n \times n$ *switch with a buffer B, LWD is at most* $1 + \frac{B}{T+B}$*-competitive, where T is the minimal number of packets transmitted between any two consecutive iterations.*

Proof. By Lemma 3, during (t_s, t_e) OPT cannot transmit more packets than LWD. Note that during t_e it is possible that OPT transmits L more packets than LWD, $0 \leq L < N$. By definition of OPT, at the end of an iteration OPT gets all remaining $B - L$ packets for free, and its buffer is empty. By Lemma 2, between two consecutive iterations OPT cannot transmit more than LWD. So if OPT transmits $T \geq 0$ packets between two consecutive iterations, P packets during the iteration, the OPT's throughput is at most $T + P + B - L = T + 2B$, whereas LWD transmits $T + P - L = T + B$. Thus, LWD is at most $1 + \frac{B}{T+B}$-competitive.

6   Scheduling with Heterogeneous Values

In this section, we consider a model with values: each incoming packet has an output port from 1 to n and an intrinsic value from 1 to V; in this model all packets have uniform processing requirements. The objective is to maximize the total transmitted value. Similar to the model with heterogeneous processing requirements, the work [15] showed that in the model with values LQD is at least $\left(\sqrt[3]{k} - o\left(\sqrt[3]{k} \right) \right)$-competitive. In Section 5, we have shown that LWD with PQ processing is 2-competitive in the model with heterogeneous processing requirements. Therefore, we begin with LWD's counterpart for this model: the Minimal-Total-Value-Drop policy (MTVD) that has packets in each queue sorted in non-increasing order of values; MTVD tries to process and transmit packets with maximal value first but in case of congestion MTVD drops a packet with minimal value. Proofs of all results in this section can be found in the Appendix.

Minimal-Total-Value-Drop (MTVD): during the arrival of a packet p with output port i and value v, (1) if the buffer is not full, accept p into Q_i; (2) if the buffer is full and v exceeds the minimal value of some packet, push out a packet with the smallest value from the buffer and accept p into Q_i; else drop p.

For a single queue, MTVD is optimal by reasoning similar to LWD. Unfortunately, this does not generalize to the shared memory switch, as the following theorem shows.

Theorem 6. *The Minimal-Total-Value-Drop (MTVD) algorithm is at least* $\frac{Vn-(n-1)}{V}$*-competitive in the model with values (this is* $n - o(n)$ *unless* $V = o(n)$*).*

Proof. In the first burst, there arrive B packets with value V for output port 1 and B packets with value $V - 1$ for every other output port $2..n$. MTVD accepts B packets to the first queue, while OPT accepts B/n packets to each queue. In B/n steps, MTVD will have transmitted total value BV/n, while OPT will have transmitted total value $(V + (V - 1)(n - 1))B/n$, and the first burst repeats, getting the bound.

Theorem 6 shows that in the model with values the total value characteristic is insufficient and additional parameters should be included if an "ideal" online policy that achieves a constant competitiveness exists. This is why the work [15] introduced the

Maximal-Ratio-Drop policy that considers both buffer occupancy and values as a potential policy that achieves constant competitiveness.

Maximal-Ratio-Drop (MRD): during the arrival of a packet p with output port i and value v, denote $j^* = \arg \max_j\{|Q_j|/V_j\}$, where V_j is the total value of packets in queue j and $|Q_j|$ is the queue length; then: (1) if buffer is not full, accept p into Q_i; (2) if buffer is full and v exceeds the minimal value of a packet from queue Q_{j^*}, push out a packet from Q_{j^*} with minimal value and accept p into Q_i; else drop p.

Theorem 7. *The Maximal-Ratio-Drop (MRD) algorithm is at least V-competitive if $n \geq B - V^2 + 1$.*

Proof. In the first burst, there arrive $2(m-1)$ packets of value 1 destined to output ports $[1, m-1]$, 2 packets per port, followed by B packets of value V destined to output port m, where $B > V$ is the buffer size and $m = B - V^2 + 1$. OPT accepts only packets of value V accruing the total value of BV. On the other hand, MRD accepts just V packets of value V at which point the ratio of the length to the average value becomes 1 and it retains $m - 1$ packets of value 1 gaining the total value of $V^2 + m - 1$. Thus, the competitive ratio of MRD is $\frac{BV}{V^2+m-1} = V$.

Unfortunately, the MRD example shows that even both values and buffer occupancy together are not enough to achieve constant competitiveness. As a result, we are more pessimistic regarding the existence of a policy in this model with constant competitiveness (the open problem posed in SIGACT News [18, p. 22]).

7 Conclusion

Over the recent years, there has been a growing interest in understanding the impact of buffer architecture on network performance. The needs and (bursty) behavior of many modern data center applications further add incentive to fill this knowledge gap. In this work, we study the tradeoffs inevitable on the path to a "perfect" policy in a shared memory switch, both analytically and with simulations. Recent research advocates smaller buffers in routers, aiming to reduce queueing delay in the presence of (mostly) TCP traffic; however, it sidesteps the issue that as buffers get smaller, the effect of processing delay becomes much more pronounced. The majority of currently deployed admission control policies do not take into account (at least explicitly) the importance of heterogeneous packet processing. In this work, we study the impact of heterogeneous processing on throughput in the shared memory switch architecture. We demonstrate that policies attractive under uniform processing requirements perform poorly in the worst case, which provides new insights to the practice of admission control policies. Our main result is a constant upper bound on the competitiveness of the LWD policy that drops packets from the queues with largest total processing in case of congestion; this is a significant improvement over [15], as our generalized model requires different proof methods. In addition, we consider a model with heterogeneous packet values and provide preliminary results on whether a policy with constant competitiveness can exist. Simulations confirm our analytical findings and in particular demonstrate the relevance of worst-case analysis results for understanding overall (average) performance.

References

1. Aiello, W., Kesselman, A., Mansour, Y.: Competitive buffer management for shared-memory switches. ACM Transactions on Algorithms 5(1) (2008)
2. Aiello, W., Mansour, Y., Rajagopolan, S., Rosén, A.: Competitive queue policies for differentiated services. J. Algorithms 55(2), 113–141 (2005)
3. Albers, S., Schmidt, M.: On the performance of greedy algorithms in packet buffering. SIAM Journal on Computing 35(2), 278–304 (2005)
4. Alizadeh, M., Edsall, T., Dharmapurikar, S., Vaidyanathan, R., Chu, K., Fingerhut, A., Lam, V.T., Matus, F., Pan, R., Yadav, N., Varghese, G.: CONGA: distributed congestion-aware load balancing for datacenters. In: ACM SIGCOMM 2014 Conference, pp. 503–514 (2014)
5. Azar, Y., Litichevskey, A.: Maximizing throughput in multi-queue switches. Algorithmica 45(1), 69–90 (2006)
6. Azar, Y., Richter, Y.: An improved algorithm for CIOQ switches. ACM Transactions on Algorithms 2(2), 282–295 (2006)
7. BBC News. US Watchdog to Propose New Net Neutrality Rules (2014). http://www.bbc.com/news/technology-27141121.
8. Borodin, A., El-Yaniv, R.: Online Computation and Competitive Analysis. Cambridge University Press (1998)
9. Feng, W.C., Kandlur, D.D., Saha, D., Shin, K.G.: Stochastic fair blue: A queue management algorithm for enforcing fairness. In: INFOCOM, pp. 1520–1529 (2001)
10. Chowdhury, M., Zhong, Y., Stoica, I.: Efficient coflow scheduling with varys. In: SIGCOMM, pp. 443–454 (2014)
11. Chuprikov, P., Nikolenko, S.I., Kogan, K.: Priority queueing with multiple packet characteristics. In: INFOCOM, pp. 1–9 (2015)
12. Costa, P., Donnelly, A., Rowstron, A.I.T., O'Shea, G.: Camdoop: Exploiting in-network aggregation for big data applications. In: Proc. 9th USENIX Symposium on Networked Systems Design and Implementation (NSDI 2012), pp. 29–42 (2012)
13. Das, S., Sankar, R.: Broadcom smart-buffer technology in data center switches for cost-effective performance scaling of cloud applications (2012). https://www.broadcom.com/collateral/etp/SBT-ETP100.pdf
14. Englert, M., Westermann, M.: Lower and upper bounds on FIFO buffer management in QoS switches. Algorithmica 53(4), 523–548 (2009)
15. Eugster, P., Kogan, K., Nikolenko, S., Sirotkin, A.: Shared memory buffer management for heterogeneous packet processing. In: ICDCS (2014)
16. Floyd, S., Jacobson, V.: Random early detection gateways for congestion avoidance, pp. 397–413 (1993)
17. CAIDA The Cooperative Association for Internet Data Analysis. http://www.caida.org/
18. Goldwasser, M.: A survey of buffer management policies for packet switches. SIGACT News 41(1), 100–128 (2010)
19. Hong, C.-Y., Kandula, S., Mahajan, R., Zhang, M., Gill, V., Nanduri, M., Wattenhofer, R.: Achieving high utilization with software-driven WAN. In: ACM SIGCOMM 2013 Conference, pp. 15–26 (2013)
20. Jain, S., Kumar, A., Mandal, S., Ong, J., Poutievski, L., Singh, A., Venkata, S., Wanderer, J., Zhou, J., Zhu, M., Zolla, J., Hölzle, U., Stuart, S., Vahdat, A.: B4: experience with a globally-deployed software defined wan. In: ACM SIGCOMM 2013 Conference, pp. 3–14 (2013)
21. Keslassy, I., Kogan, K., Scalosub, G., Segal, M.: Providing performance guarantees in multipass network processors. IEEE/ACM Trans. Netw. 20(6), 1895–1909 (2012)

22. Kesselman, A., Kogan, K., Segal, M.: Packet mode and QoS algorithms for buffered crossbar switches with FIFO queuing. Distributed Computing 23(3), 163–175 (2010)
23. Kesselman, A., Kogan, K., Segal, M.: Improved competitive performance bounds for CIOQ switches. Algorithmica 63(1-2), 411–424 (2012)
24. Kesselman, A., Kogan, K., Segal, M.: Best Effort and Priority Queuing Policies for Buffered Crossbar Switches. Chicago Journal of Theoretical Computer Science (2012)
25. Kesselman, A., Kogan, K.: Nonpreemptive Scheduling of Optical Switches. IEEE Transactions on Communications 55(6), 1212–1219 (2007)
26. Kesselman, A., Lotker, Z., Mansour, Y., Patt-Shamir, B., Schieber, B., Sviridenko, M.: Buffer overflow management in QoS switches. SIAM Journal on Computing 33(3), 563–583 (2004)
27. Kesselman, A., Mansour, Y.: Harmonic buffer management policy for shared memory switches. Theor. Comput. Sci. 324(2-3), 161–182 (2004)
28. Kogan, K., López-Ortiz, A., Nikolenko, S., Scalosub, G., Segal, M.: Large profits or fast gains: A dilemma in maximizing throughput with applications to network processors. CoRR, abs/1202.5755 (2013)
29. Kogan, K., López-Ortiz, A., Nikolenko, S., Sirotkin, A.: Multi-queued network processors for packets with heterogeneous processing requirements. In: COMSNETS, pp. 1–10 (2013)
30. Kogan, K., López-Ortiz, A., Nikolenko, S., Scalosub, G., Segal, M.: Balancing work and size with bounded buffers. In: COMSNETS, pp. 1–8 (2014)
31. Kogan, K., López-Ortiz, A., Nikolenko, S.I., Sirotkin, A.V., Tugaryov, D.: FIFO queueing policies for packets with heterogeneous processing. In: Even, G., Rawitz, D. (eds.) MedAlg 2012. LNCS, vol. 7659, pp. 248–260. Springer, Heidelberg (2012)
32. Kogan, K., López-Ortiz, A., Nikolenko, S., Sirotkin, A.: A taxonomy of semi-FIFO policies. In: IPCCC, pp. 295–304 (2012)
33. Kogan, K., Nikolenko, S., Keshav, S., López-Ortiz, A.: Efficient demand assignment in multi-connected microgrids with a shared central grid. In: SustainIT, pp. 1–5 (2013)
34. Mansour, Y., Patt-Shamir, B., Lapid, O.: Optimal smoothing schedules for real-time streams. Distributed Computing 17(1), 77–89 (2004)
35. Nikolenko, S.I., Kogan, K.: Single and multiple buffer processing. In: Encyclopedia of Algorithms. Springer (2015)
36. Sleator, D.D., Tarjan, R.E.: Amortized efficiency of list update and paging rules. Communications of the ACM 28(2), 202–208 (1985)
37. Yang, H.-C., Dasdan, A., Hsiao, R.-L., Parker Jr., D.S.: Map-reduce-merge: simplified relational data processing on large clusters. In: Proc. ACM SIGMOD International Conference on Management of Data, pp. 1029–1040 (2007)
38. Yu, Y., Gunda, P.K., Isard, M.: Distributed aggregation for data-parallel computing: interfaces and implementations. In: SOSP, pp. 247–260 (2009)

Scheduling Multipacket Frames
with Frame Deadlines*

Łukasz Jeż[1,2], Yishay Mansour[3,4], and Boaz Patt-Shamir[5]

[1] Eindhoven University of Technology, Eindhoven, The Netherlands
[2] Institute of Computer Science, University of Wrocław, Wrocław, Poland
[3] Blavatnik School of Computer Science, Tel Aviv University, Tel Aviv, Israel
[4] Microsoft Research, Hertezelia, Israel
[5] School of Electrical Engineering, Tel Aviv University, Tel Aviv, Israel

Abstract. We consider scheduling information units called frames, each with a delivery deadline. Frames consist of packets, which arrive on-line in a roughly-periodic fashion, and compete on allocation of transmission slots. A frame is deemed useful only if all its packets are delivered before its deadline. Using standard techniques, one can derive polylog-competitive algorithms for this model; in this paper we study special cases which allow for better results. Specifically, we present constant-competitive algorithms for two important cases: in one, the value of a frame is proportional to its size and all frames have (roughly) the same period, and in the other, each frame may have its own period but all frames have the same value and size. The former result also implies better polylog-competitive algorithm for the general case.

1 Introduction

In many networking settings the ingress flows to the network has a nice periodic, or almost periodic, structure. The network would like to guarantee the flows a pre-specified Quality of Service (QoS), where one of the most basic QoS guarantees is a deadline by which the transfer would be completed. The uncertainty regarding the arrival of future flows motivates the online setting. We study this setting from the competitive analysis viewpoint. Let us start by giving a few motivating examples.

Consider a switch with multiple incoming video streaming flows competing for the same output link. Each flow consists of *frames*, and each frame consists of a variable number of *packets*. The video source is completely periodic, but due

---

* The 1st author is partially supported by the NWO Vidi grant 639.022.211, the Israeli Centers of Research Excellence (I-CORE) program, Center No.4/11, and the Polish National Science Center (NCN) Grant DEC-2013/09/B/ST6/01538. The 2nd author is partially supported by the Israeli Centers of Research Excellence (I-CORE) program, Center No. 4/11, a grant from the Israel Science Foundation (ISF), and a grant from United States-Israel Binational Science Foundation (BSF). The 3rd author is partially supported by the Israel Science Foundation (grant No. 1444/14) and by a grant from Israel Ministry of Science and Technology. This work was carried out while the first author was visiting Tel Aviv University.

© Springer International Publishing Switzerland 2015
C. Scheideler (Ed.): SIROCCO 2015, LNCS 9439, pp. 76–90, 2015.
DOI: 10.1007/978-3-319-25258-2_6

to compression, different frames may consist of a different number of packets. On top of that, asynchronous network transfer typically adds some jitter, so the input at the switch is only approximately periodic. In order for a frame to be useful, all its packets must be delivered before the frame's deadline. A frame is considered completed if all its packets are delivered before the frame's deadline, and the goal of a scheduling algorithm is to maximize the number of completed frames. Partially completed frames are considered worthless.

As another example, consider a Voice over IP (VoIP) setting. Voice calls generate samples at a relatively fast rate. Samples are wrapped in packets which are aggregated in logical frames with lower-granularity deadlines. Frames deadlines are more lax due to the tolerance of the human ear. Completed frames are reconstructed and replayed at the receiver's side; incomplete frames are discarded, resulting in an audible interruption (click) of the call. Our focus is on an oversubscribed link on the path of many such calls.

As a last example, consider a database (or data center) engaged in transferring truly huge files (e.g., petabytes of data) for replication purposes. It is common in such a scenario that the transfer must be completed by a certain given deadline. Typically, the transmission of such files is done piecemeal by breaking the file into smaller units, which are transmitted periodically so as to avoid overwhelming the network resources. We are interested in scenarios where multiple such transfers cross a common congested link.

Motivated by the above examples, we define the following abstract model. There are data units called *frames*, each with a *deadline* and a *value*. Each frame consists of several *packets*. Time is slotted. Packets arrive in an approximately periodic rate at a link, and can be transmitted (served) one packet at a step. A *scheduling algorithm* needs to decide which packet to transmit at each time slot. The goal of the algorithm is to maximize the total value of *delivered* frames, where a frame is considered delivered only if all its packets are transmitted before the frame's deadline.

The scheduling algorithm may be *preemptive* or *non-preemptive*. An algorithm is called non-preemptive if any packet it transmits belongs to a frame which is eventually delivered, whereas a preemptive algorithm may transmit a packet from some frame but later decide not to complete that frame.

Our performance measure is the competitive ratio, i.e., the worst case ratio between the value delivered by the online algorithm and the best possible value that can be delivered by an optimal (offline) schedule for a given arrival sequence.

Our Approach and Results. Our model assumes that the arrival sequence is *not arbitrary*. Studying restricted instance classes and/or adversaries is common, and related work typically assumes specific order of frames and packets or restricted bursts. Instead, we assume that once the first packet of a frame arrives, the arrival times of the remaining packets are predictable within a given bounded jitter. Under this assumption, using the *classify and select* technique [1], it is relatively straightforward to guarantee a poly-logarithmic competitive ratio, cf. Section 2.2. The conceptual contribution of this work is to identify interesting and important special cases where a *constant* competitive ratio can be achieved. Moreover, one of them results in improved polylog guarantees for the general case, cf. Section 3. Technically, the main results in this paper are constant-competitive, deterministic algorithms for the following cases.

- All frames have (roughly) the same period but arbitrary sizes, where the size of the frame is the number of its packets. The frame value is its size.
- All frames have the same size but possibly different periods, and they are perfectly periodic (no jitter, frame deadline determined by its period; cf. Section 2). The value of all frames is identical (say, 1).

In fact, the first result is more general: the periods can be arbitrary but the competitive ratio is proportional to the min-to-max period ratio. (And clearly, the same holds in general for "densities" of frames, i.e., their value-to-size ratios.)

We also consider similar case (common period, different size) assuming unit value per frame. By same token, there is a simple randomized algorithm whose competitive ratio (reciprocal) is logarithmic in the maximum number of packets in a frame. We show that in this case a few natural algorithms, such as Earliest Deadline First (EDF) or Shortest Remaining Processing Time (SRPT), cannot guarantee significantly better competitive ratio.

Related Work. The first multipacket-frame on-line model was introduced in [7], and further studied in [10]. Emek at al. [3] consider the basic model where the main difficulty is not deadlines but rather limited buffer space. Their results express the competitive ratio as a function of the maximum burst size and the number of packets in a frame. Subsequent work considered extension to the basic model, including redundancy [8], and hierarchically structured frames [8,10].

Possibly the work closest to ours is [9], which essentially uses the same model, except that in [9], each *packet* has its deadline, and the packet arrivals may be arbitrary (whereas we assume that packets arrive approximately periodically). It is shown in [9] that the competitive ratio of the problem (both a lower and an upper bound) is exponential in the number of packets in a frame. One can view our results as showing that adding the extra assumptions that (1) packet arrival is approximately periodic, and that (2) the deadlines are per frame rather than per packet, allows for significantly better competitive ratio, namely *constant*.

We note that the classic preemptive job scheduling problem of maximizing (weighted) throughput on a single machine [5,6,2] corresponds to a special case of the problem we study in which all frames have period 1 and have no jitter. Thus strong upper bounds (almost tight in the job scheduling problem) follow for the general setting of our problem if frame values are either unit or arbitrary [2]. However, none of the known results, neither upper nor lower bounds, apply or easily extend to special cases of our problem motivated by network applications.

Paper Organization: Section 2 introduces the model and a few basic properties. In Section 3, we study the model where the value of a frame is proportional to its size. In Section 4, we consider frames with common size and value but different periods. Section 5 discusses the case of different number of packets for each frame, assuming unit value and identical period. Some proofs are omitted due to lack of space.

2   Model and Preliminary Observations

We consider a standard scheduling model at the ingress of a link. Time is slotted, packets arrive on-line, and in each time slot at most one packet can be transmitted (meaning implicitly that we assume that all packets have the same length). The idiosyncrasies of our model are our assumptions about the arrival pattern and about the way the algorithm is rewarded for delivering packets.

Input: Packets and Frames. The basic entities in our model are *frames* and *packets*. Each frame f consists of $k_f \in \mathbb{N}$ packets, and has a *value* $v_f \in \mathbb{N}$. We assume that packets of frame f arrive with *periodicity* d_f and *jitter* Δ_f, namely if packet 1 of f arrives at time t, then packet $i \in \{2, \ldots, k_f\}$ arrives in the time interval $t + (i-1)d_f \pm \Delta_f$. Each frame f has a *slack* $s_f \geq 1$, which determines the *deadline* of f (see "output" paragraph below). A frame f is called *perfectly periodic* if $\Delta_f = 0$ and $s_f = d_f$. The parameters of a frame f (i.e., size k_f, value v_f, period d_f, jitter Δ_f and slack s_f) are made known to the algorithm when the first packet of f arrives; it is also convenient to introduce a frame's *density*, $\rho_f := v_f/k_f$. We denote the actual arrival time of the i-th packet of frame f, for $i \in \{1, \ldots, k_f\}$, by $t_i(f) \in \mathbb{N}$. The arrival time of the first packet of frame f, $t_1(f)$, is also called the arrival time of f.

We assume that the algorithm knows nothing about a frame f before its arrival, and even then, it does not know the exact arrival times of the remaining packets: let $\tau_i(f) \stackrel{\text{def}}{=} t_1(f) + (i-1)d_f$. Then the guarantee is that the actual arrival time satisfies that $t_i(f) \in [\tau_i(f) - \Delta_f, \tau_i(f) + \Delta_f]$ for $i > 1$.

For a given instance, and a parameter $\pi \in \{\Delta, s, k, d, v, \rho\}$, we let $\pi_{\max} = \max_f(\pi_f)$ and $\pi_{\min} = \min_f(\pi_f)$, both taken over all frames in the instance, and extend these to instance classes. We assume that there is a constant $c \geq 0$ such that $\Delta_f \leq c \cdot s_f$ holds for all frames f (cf. Section 2.1 for its necessity).

Output: Delivered Frames. A *schedule* says which packet is transmitted in each time step. The *deadline* of frame f is $D_f \stackrel{\text{def}}{=} \tau_{k_f}(f) + \Delta_f + s_f$, and a frame f is said to be *delivered* in a given schedule if all its packets are transmitted before the frame deadline (we use s_f instead of $s_f - 1$ to reduce clutter later.) Given a schedule, the value delivered by that schedule is the sum of values of frames delivered by that schedule. A schedule is called *work conserving* if it always transmits a packet if some packet is pending .

Algorithms. The duty of an algorithm is to produce a schedule for any given arrival sequence, and the goal is to maximize the sum of values of delivered frames. An algorithm is called *on-line* if its decision at any time t depends only on the arrivals and transmissions before time t. We assume that the buffer space is unbounded, which means that the only contention is for the transmission slots.

The *competitive ratio* of an algorithm A is the worst-case ratio, over all arrival sequences σ, between the value delivered on σ by A and by the optimal off-line schedule . Formally, the competitive ratio of A is

$$\rho(A) \stackrel{\text{def}}{=} \inf_{\sigma \in M(s_{\max}, \Delta_{\max})} \frac{A(\sigma)}{\text{OPT}(\sigma)}$$

where $A(\sigma)$ and $\text{OPT}(\sigma)$ denote the gain of A on σ and the optimum gain on σ respectively, $M(s_{\max}, \Delta_{\max})$ is the set of arrival sequences with jitter at most $\Delta_{\max}$ and slack of at most $s_{\max}$. Note that $\rho(A) \in [0,1]$ by definition.

2.1   On the Relation between s and Δ

Some settings of the parameters are uninteresting. In particular, we observe that if $s \ll \Delta$ (in words: the slack is much smaller than the input jitter), then one cannot expect good worst-case performance from any on-line algorithm, even if all frames have identical period, jitter, slack, value, and size. Specifically, we show that in such a case, denoting the common frame size by k, every on-line algorithm has competitive ratio at most $O(1/k)$, and that $\Omega(1/k)$-competitiveness is easily achievable if $k < s$ (see appendix for proofs).

Theorem 1. *No randomized algorithm on instances with all frames of size k, jitter Δ, slack s, and period $d \geq 2\Delta + s$ has competitive ratio larger than $\frac{s + 2\Delta/k}{2\Delta + s}$.*

Note that Theorem 1 is meaningless for instances with 0 jitter.

Theorem 2. *If all frames have size k and each frame f has slack $s_f > k$, then there exists a $1/(2k)$-competitive deterministic on-line algorithm.*

Theorems 1 and 2 motivate our assumption that Δ_f/s_f is bounded from above by a constant: otherwise there is no way to attain a non-trivial competitive ratio.

2.2   Uniform Instances and Polylog Competitiveness

For a tuple $\Pi_n = (\pi_1, \pi_2, \ldots, \pi_n)$ of frame parameters, such as size, value, period, or density, and a set $\Gamma_n = (\gamma_1, \gamma_2, \ldots, \gamma_n)$ of real numbers no smaller than 1, we call an instance (Π, Γ)-*uniform* if for every $1 \leq i \leq n$, the ratio of the max-to-min value of parameter π_i over all frames in the instance is at most γ_i. In case of uniform instances, we generally assume that the extreme values of frame parameters in Π are known to the algorithm. In such case, using the *classify and randomly select* paradigm [1] extends any algorithm for nearly uniform instances to general instances in the following sense.

Lemma 1. *Let $\gamma > 1$ and let A be a ρ-competitive algorithm for $((\pi), (\gamma))$-uniform instances. Then, given a $((\pi), (\gamma'))$-uniform class of instances $\mathcal{I}$ with $\pi_{\min}$ and $\pi_{\max}$ the minimum and maximum values of π in $\mathcal{I}$, $B(A, \pi_{\min}, \pi_{\max})$ (defined below) is a $(\rho/(\lfloor \log_\gamma \gamma' \rfloor + 1))$-competitive randomized algorithm for $\mathcal{I}$.*

Algorithm $B(A, \pi_{\min}, \pi_{\max})$:
1. *Classify* each frame f in *class* $\lfloor \log_\gamma \pi_f - \log_\gamma \pi_{\min} \rfloor$.
2. *Randomly select* a class i with uniform distribution, and run A on frames of this class only, discarding packets of frames from all other classes.

Proof. The expected contribution of the chosen class of frames to the optimum throughput is clearly $1/\xi$, where $\xi = \lfloor \log_\gamma \gamma' \rfloor + 1$ is the number of classes. $\square$

Lemma 1 can be applied iteratively over successive parameters, yielding ratio $\Omega(1/\prod_{i=1}^{n}\log_{\gamma_i}\gamma_i')$ for (Π_n, Γ_n')-uniform instances if only we have a constant-competitive algorithm for (Π_n, Γ_n)-uniform instances. Fortunately, there is a simple $\Omega(1)$-competitive deterministic algorithm for instances that are nearly uniform in terms of frame size, value, and period. As a warm-up, to illustrate our approach, we state such algorithm instances with no jitter and slack $s \geq d_{\min}$.

The state of the algorithm consists of a set of up to $d_{\min}$ *active frames*, initially empty. (Recall that $d_{\min}$, the minimum period of frames in the instance, is known to the algorithm.) When a new frame arrives, it enters the set of active frames iff there are strictly less than $d_{\min}$ active frames at the time. A frame remains active until its deadline. The algorithm transmits available packets of active frames in FIFO order, and discards all packets of all inactive frames.

Theorem 3. *The algorithm above is* $\left(\left(\frac{2 \cdot k_{\max} \cdot d_{\max} \cdot v_{\max}}{k_{\min} \cdot d_{\min} \cdot v_{\min}} + 1\right)^{-1}\right)$-*competitive. Moreover, each packet of an active frame is transmitted within $d_{\min}$ steps of its arrival.*

Proof. We begin with proving that each packet of an active frame is transmitted within $d_{\min}$ steps of its arrival. Suppose it does not hold, and let p be the first packet for which it fails. Then p is delayed by at least $d_{\min}$ active packets that were already in the buffer when it arrived. This implies that there are more than $d_{\min}$ active packets (counting p as well), so two of them must belong to the same frame f. This is a contradiction to the choice of p, since the earlier of those packets could not have been transmitted within $d_f \geq d_{\min}$ steps of its arrival.

We prove the competitive ratio by a charging scheme. For simplicity, we ignore frame values: as the worst case is that each frame of OPT has value $v_{\max}$ whereas each frame of the algorithm $v_{\min}$, this contributes the $\frac{v_{\min}}{v_{\max}}$ factor to the competitive ratio. Firstly, each frame completed by both OPT and the algorithm is charged to itself. Moreover, each active frame f, accepted at its arrival time t, provides a credit of $(k_{\min} \cdot d_{\min})^{-1}$ to each time slot in $[t, D_f + k_{\max} \cdot d_{\max})$. Each f thus provides a credit of $\frac{k_f \cdot d_f + k_{\max} \cdot d_{\max}}{k_{\min} \cdot d_{\min}} \leq 2 \cdot \frac{k_{\max} \cdot d_{\max}}{k_{\min} \cdot d_{\min}}$. Taking the self-charges into account, this establishes the ratio. It remains to show how frames completed by OPT but rejected by the algorithm are charged to the credit. Let f' be such frame and t' be its arrival time. Then each packet of f' charges $1/k_{f'}$ to the credit of the time slot in which OPT sends it out. As f' is rejected, there were $d_{\min}$ active frames at time t', each of them contributing credit to each slot in $[t', t' + k_{\max} \cdot d_{\max}) \supseteq [t', D_{f'})$. Hence, each slot that f' may charge to receives a credit of at least $1/k_{\min} \geq 1/k_{f'}$. The theorem follows. $\square$

In the next section, we give an improved algorithm for instances that may have (larger) jitter and smaller slack, and that are nearly uniform in frame period and density. I.e., not only is the class of instances less restrictive in terms of jitter and slack, but also the extension to general instances via iterative application of Lemma 1 results in improved competitive ratio. Specifically, rather than losing a $\log(v_{\max}/v_{\min}) \cdot \log(k_{\max}/k_{\min})$ term for value and size parameters, we only lose a $\log(\rho_{max}/\rho_{min}) = \log(v_{max}/v_{min}) + \log(k_{max}/k_{min})$ term in the competitive ratio, for density, which combines size and value.

3 Similar Periods, Uniform Density

In this section we consider $((d, \rho), (\delta, 1))$-uniform instances, i.e., with periods between d_{min} and $d_{max} = \delta d_{min}$ and uniform density, assumed to be 1. We give an algorithm with competitive ratio depending on c, δ, and $\alpha := \max\{0, \frac{\Delta_{max} + s_{max}}{d_{max}} - 1)\}$, i.e., $\Omega(1)$-competitive when all these are bounded by constants.

3.1 The Algorithm

Our approach is as follows. A packet is said to be of *type 1* if it must be transmitted in less than d_{min} steps since its latest possible arrival time; other packets are *type 2*. Type 1 packets are exactly all last packets of frames whose slack is smaller than d_{min}. Packets of the two types will be scheduled differently. We extend these types to frames and let them inherit the types of their last packets. At every point in time, the algorithm maintains up to $d_{min}/2$ *active frames*. The algorithm guarantees that each type 2 packet of an active frame is delivered within the d_{min} steps following its latest possible arrival time. Limiting the number of active frames makes this invariant easy to maintain using greedy scheduling, but this cannot be applied to type 1 packets, because these must be transmitted in fewer than d_{min} steps after their latest possible arrival. To schedule type 1 packets, the algorithm maintains explicit slot reservations. To make sure that these do not interfere with type 2 packets, type 1 frames remain (quasi-)active for a short time after their completion and prevent accepting new type 1 size 1 frames, which could result in delaying type 2 packets too much.

Algorithm Specification. The algorithm maintains a set Act of up to $d_{min}/2$ *active frames*. Each active frame f with $s_f < d_{min}$ has a reserved slot for its last packet in the interval $[\tau_k(f) + \Delta_f, \tau_k(f) + \Delta_f + s_f)$. The algorithm consists of two subroutines. Subroutine A decides, for each new frame f, whether to add it to Act or not. In the former case we say that f is accepted, and in the latter that f is rejected. When a frame f is accepted, the algorithm may remove a previously active frame f' from Act, in which case we say that f' is preempted. For conciseness, Subroutine A always preempts some f' when a new frame f is accepted, but f' may be "virtual", in which case so is the preemption. We also maintain a set Act_1 where active frames of type 1 remain for $d_{min} - 1$ steps after they have been completed. This set, rather than Act, determines whether an arriving type 1 frame of size 1 is accepted. All packets of non-active frames (those rejected or preempted) are dropped. Subroutine S schedules packets, deciding which one to transmit next. The following notions are used in the subroutines:

$$S_f \stackrel{\text{def}}{=} \left[\tau_{k_f}(f) + \Delta_f, \ \tau_{k_f}(f) + \Delta_f + s_f\right) \qquad \textit{slack interval} \text{ of type 1 frame } f$$

$$D_f(i) \stackrel{\text{def}}{=} \tau_i(f) + \Delta_f + d_{\min} \qquad\qquad \textit{deadline} \text{ of packet } i \text{ of type 2 frame } f$$

$$I_f(i) \stackrel{\text{def}}{=} [\tau_i(f) + \Delta_f, D_f(i)) \qquad \textit{designated interval} \text{ of packet } i \text{ of type 2 frame } f$$

Subroutine A. Upon arrival of a new frame f:

- If f is type 1 and $k_f = 1$ and $|\mathsf{Act}_1| \geq d_{\min}/2$: reject f and **return**.
- (Otherwise) If f is type 1 (and $k_f > 1$) and all slots in S_f are reserved:
  - let f' be the smallest frame with a reserved slot in S_f
- Else:
  - let f' be a virtual type 2 frame of size 0
- If f' has size 0 and $|\mathsf{Act}| \geq d_{\min}/2$, let f' be the smallest frame in Act.
- If $k_f < 2k_{f'}$: reject f and **return**
- (Otherwise):
  - If f' is type 1, remove f' from Act_1 and cancel its reservation for last packet
  - If f is type 1, add it to Act_1 and make a reservation for its last packet in S_f
  - remove f' (if real) from Act, add f to Act, and **return**

Subroutine S. In each step t:

- If slot t is reserved for the last packet p of a frame f:
  - remove f from Act now and mark it for deletion from Act_1 at time $t + d_{\min}$
  - transmit p and **return**
- Else:
  - let p be the earliest deadline packet in $\{$packet i of $f \mid f \in \mathsf{Act} \wedge t \in I_f(i)\}$
  - if p is the last packet of a frame f, remove f from Act
  - transmit p and **return**

3.2 Analysis

Intuitively, the analysis is an extension of Theorem 3, whose two claims correspond to Theorem 4 and Lemma 2 respectively. Proving these is somewhat more involved: the latter due to the extra constraints and special treatment of type 1 packets, and the former due to varying sizes (and values) of frames.

Lemma 2. *Every packet p of an active frame is sent out during its reserved slot if it is type 1 or during its designated interval I if it is type 2.*

To analyze the competitive ratio, we define *chains* of frames inductively as follows. Each completed frame f is in a distinct chain C_f, and if a frame f' was preempted by a frame f, and f is in a chain C, then f' belongs to C as well, preceding f in it. All chains start with a frame that did not preempt any other frame, and end with a frame that was not preempted. We note that our chains are virtually the same as in the analyses of online interval scheduling [11,4], and part of our analysis is reminiscent of those.

The high level overview of the charging scheme is as follows. There are three kinds of charges: a self-charge of f to itself if both OPT and the algorithm completed it and two further kinds of charges for the frames completed only by OPT. Here, we distinguish the cause of rejection. If f is a type 2 frame or a type 1 frames of size 1, it has been rejected due to too many active frames in Act and Act_1 respectively. Then each active frame from the respective set had at least half the size of f, so f can be charged to any of such frames. If f is type 1 of size greater than 1, then f has been rejected due to lack of slots for its last packet in S_f. Namely, each slot in S_f was reserved for a last packet of another frame of at least half f's size, since otherwise f would preempt the smallest of those. Thus f can be charged to one of those frames. Note that in both cases the frame we charge to may not be completed by the algorithm in the end. But as it is a part of some chain, and frame sizes in a chain increase geometrically, all charges can be relayed to the last frames of chains, which the algorithm completes. For both kinds of charges, we show that globally there are sufficiently many active frames to be charged, rather than identify a particular active frame to be charged. To this end, both charges are towards a "credit" that the chain(s) provide, and in the end, this credit is charged to the last frame of a chain. We note that the jitter of last packets of frames effectively contribute to the frame sizes; as the jitter does not scale with frame size, the maximum effective sizes of frames preceding the last one in a chain do not form an exact geometric progression.

Theorem 4. *The algorithm is* $(2\,(5 + 2c + 4\delta + 2\alpha\delta))^{-1}$-*competitive.*

Proof. We define *chains* of frames. Each completed frame f defines a chain C_f that ends with f. Moreover, if a frame f that belongs to a chain C preempted a frame f', then f' belongs to C as well, preceding f in it; if f did not preempt any frame, then the chain C starts with f.

Let us now define the credits associated with chains. For a given chain C, let f'_C and f_C denote its first and last frame respectively, and let $T(f_C)$ denote the time f_C was removed from both Act and Act_1. In other words, $T(f_C)$ is the completion time of f_C if it is type 2, or its completion time plus $d_{\min} - 1$ if it is type 1. We give a credit of $2/d_{\min}$ to all time slots since the arrival of f'_C until $2(k_{f_C} - 1)d_{\max} + \Delta_{\max} + s_{\max}$ time slots past $T(f_C)$, i.e., to $[t_1(f'_C),\ T(f_C) + 2(k_{f_C} - 1)d_{\max} + (\Delta_{\max} + s_{\max}))$. We stress that the credits granted to a time slot from different chains add up.

We are now ready to describe the preliminary charging scheme, i.e., the charges that are later relayed to last frames of chains. Let f be a frame delivered by OPT. The charging is as follows:

1. If f was accepted by the algorithm, f is charged to itself.
2. If f was rejected by the algorithm due to lack of slot for its last packet, f is charged to the frames that prevented its acceptance; details are given later.
3. If f was rejected by the algorithm due to too many active frames, f is charged as follows: for each packet p of f, charge p to the credit associated with the time slot in which OPT sends p. Each such slot has a credit of at least 1: When f arrived at time $t_1(f)$, the algorithm had $d_{\min}/2$ active frames, each of size at least $k_f/2$. (If f is type 1 of size 1, these are the frames from $\mathtt{Act}_1$.) Thus our credit rule guarantees that each slot in $[t_1(f), t_1(f) + (k_f - 1)d_{\max} + \Delta_f + s_f)$, i.e., from $t_1(f)$ until the deadline of f, receives a credit of $2/d_{\min}$ from each of the $d_{\min}/2$ chains corresponding to the active frames.

We now describe the charging for a frame f that was rejected due to lack of reservation space for the last packet. Then at f's arrival time, $t_1(f)$, all the slots that f's last packet could have used were already reserved for other frames, all of size at least $k_f/2$. We charge f to those frames as follows.

Let A_i denote the set of frames of size at least i that OPT delivers and the algorithm rejects due to lack of slot for their last packets. Consider all maximal intervals $L_1^i, L_2^i, \ldots, L_{m_i}^i$ of time such that $\bigcup_j L_j^i$ is the (maximal) set of slots that the algorithm had ever reserved (i.e., these reservations may have been canceled later) for last packets of frames of size at least i. For each interval L_j^i, let $L_j^i = [a_j^i, b_j^i)$ and $|L_j^i| = b_j^i - a_j^i$.

Let t_0 be the time when OPT delivered f's last packet. Then f is charged to the $L_j^{k_f}$ where j is minimum such that $t_0 < b_j^{k_f}$, i.e., to the $L_j^{k_f}$ whose right end is the first one after t_0. (Note that we are not guaranteed that $t_0 \in L_j^{k_f}$ since OPT might deliver the last packet before $\tau_{k_f}(f) + \Delta_f$.) Next, for each L_j^i, we distribute the charge it receives evenly between all the frames of size at least i that ever made reservation for their last packets within L_j^i. Denote the set of frames charged to L_j^i by F_j^i, and let $f_0 = \mathrm{argmax}_{g \in F_j^i} \Delta_g$. Then for any $g \in F_j^i$, the following hold: $D_g \leq b_j^i$, $t_{k_g}(g) \geq a_j^i - 2\Delta_{f_0}$, and $|L_j^i| \geq s_{f_0}$. Thus $|F_j^i|/|L_j^i| \leq (s_{f_0} + 2\Delta_{f_0})/s_{f_0} \leq 1 + 2c$.

To summarize, for each A_i, there is a corresponding set B_i of frames of size at least $i/2$ that made reservations for last packets in the union of intervals allowed for the last packets of frames in A_i such that $|A_i| \leq (1+2c)|B_i|$. We charge $\bigcup A_i$ to $\bigcup B_i$. Despite different frame sizes, the charging ratio is at most $2(1+2c)$, as

$$\sum_{i=1}^{k_{\max}} i|A_i \setminus A_{i+1}| = \sum_{i=1}^{k_{\max}} i(|A_i| - |A_{i+1}|) = \sum_{i=1}^{k_{\max}} |A_i| \leq \sum_{i=1}^{k_{\max}} (1+2c)|B_i|$$

$$= (1+2c)\sum_{i=1}^{k_{\max}} i(|B_i| - |B_{i+1}|)(1+2c)\sum_{i=1}^{k_{\max}} i(|B_i \setminus B_{i+1}|) \ .$$

We now bound the total charge that the last frame f_C of a chain C can receive. Each frame f belonging to the chain may receive a charge of the first type (a self-charge) of value k_f and a charge of the second type (from frames rejected due to lack of slots for their last packet) of value at most $2(1 + 2c)k_f$. For each f in C, these are relayed to f_C. As each frame in C is at least twice as large as its predecessor (the one it preempted), the total charge of the first two types relayed to f_C is at most $2(5 + 2c)k_f$.

It remains to do similar calculations for the charges of the last type, namely frames that are rejected due to too many active frames. These are slightly different, because now instead of summing the sizes of all frames in a chain, we need to determine to how many slots a chain might grant credit. I.e., we need to account for gaps between successive frames of the chain, which could be as large as $\Delta_{\max} + s_{\max}$, and the extra credit that is granted past the end of a chain.

Each frame f that belongs to a chain C may provide credit of $2/d_{\min}$ per slot for up to $(k_f - 1) \cdot d_f + \Delta_f + s_f \le (k_f - 1) \cdot d_{\max} + \Delta_{\max} + s_{\max}$ time slots, plus additional $2(k_{f_C} - 1)d_{\max} + \Delta_{\max} + s_{\max}$ slots in case of f_C, and $d_{\min} - 1$ more slots if f_C is type 1, due to f_C's remaining longer in Act_1 — we call this last term *spare type 1 credit* and ignore it for the time being. As each frame in C is at least twice as large as the one it preempted, the total credit provided by the chain C of length i_C is at most

$$\frac{2}{d_{\min}}\Big(4k_{f_C}d_{\max} + (i_C + 1)(\Delta_{\max} + s_{\max} - d_{\max})\Big)$$

$$= 8\delta k_{f_C} + \frac{2}{d_{\min}}(i_C + 1)(\Delta_{\max} + s_{\max} - d_{\max})$$

$$\le 2\delta\big(4k_{f_C} + \alpha(i_C + 1)\big) \ ,$$

since $\Delta_{\max} + s_{\max} - d_{\max} \le \alpha d_{\max} = \alpha \delta d_{\min}$. We can now justify why the spare type 1 credit can be ignored: the term $4k_{f_C}d_{\max}$ in the above bound is an (over-)estimation of $k_{f_C}(2 + 1 + \frac{1}{2}\ldots)$, which corresponds to sum of sizes of frames in C. However, all frames have integer sizes, and thus their total size if at most $4k_{f_C} - 1$. Thus, we are overestimating the credit by at least $\frac{2}{d_{\min}}d_{\max}$, which is larger than the unaccounted for spare type 1 credit.

Overall, the total charge to f_C is thus at most

$$2\left(k_{f_C}(5 + 2c + 4\delta) + \alpha\delta(i_C + 1)\right) \le 2\left(k_{f_C}(5 + 2c + 4\delta) + 2k_{f_C}\alpha\delta\right)$$

$$= 2k_{f_C}\left(5 + 2c + 4\delta + 2\alpha\delta\right) \ ,$$

since $i_C \le 1 + \lfloor \log_2 k_{f_C} \rfloor$ due to the sizes of successive frames in a chain, and finally since $\lfloor \log_2 k_{f_C} \rfloor + 2 \le 2k_{f_C}$ for every positive integer k_{f_C}.   □

4 Common Size, Different Periods

In this section, we consider instances in which all frames have the same size k and same value v (w.l.o.g., $v = 1$), but each frame f has a possibly different period d_f, focusing on the perfectly periodic instances. Surprisingly, we were unable to provide any impossibility result for this setting. Instead, we propose a $\Theta(1)$-competitive non-preemptive algorithm. We assume that each and every packet of a frame has a deadline that coincides with the deadline of the frame.

4.1   A Non-preemptive Algorithm

As in Section 3, our algorithm consists of two subroutines. The first decides, for each newly arriving frame, whether to accept or reject it, and the second schedules for transmission packets of accepted frames. Unlike the algorithm in Section 3, however, accepted frames are never preempted. The algorithm classifies every frame as either *completed*, *accepted*, or *rejected*.

- *Frame Arrival:* When a new frame f arrives, the algorithm accepts it if and only if the set of all accepted frames together with f has a feasible schedule.
- *Packet Transmission:* The algorithm always transmits the packet with the earliest deadline from the set of all pending packets of accepted frames. Once all packets of a frame have been sent, the frame is marked "completed."

Let us comment briefly on the feasibility test and the algorithm's correctness (i.e., why the deadlines are met). The feasibility test considers packets rather than frames: the set of packets in question is that of all pending packets and those yet to arrive that belong either to an accepted frame or the frame f whose status is being decided. Note that by our assumption of perfectly periodic instances, the exact arrival time of all packets considered is known. Thus testing the feasibility of a set of packets (which are just unit-length jobs) can be done by running EDF on that set, since EDF produces a (single machine) feasible schedule if there is one. Similarly, our algorithm observes all deadlines because it produces an EDF schedule for a feasible set of packets.

Alternatively, the schedule for packets can be viewed as a *bipartite matching* of packets to time slots. Hence, one can test for feasibility with a new arriving frame f by using any dynamic matching algorithm that checks whether the current matching (schedule) can be augmented to match all packets of f as well. If so, the resulting schedule can then be reordered to become an EDF schedule.

The algorithm is non-preemptive. As only packets belonging to accepted frames are ever transmitted, the algorithm never "wastes" a slot. This, and the fact that all frames have the same size, allows for counting the number of transmitted packets instead of frames in the analysis.

We further note that the algorithm is "eager" in the sense that acceptance of an arriving frame is decided immediately. One can also consider a similar "lazy" algorithm that decides to either accept or reject a frame only when its first packet would be scheduled by EDF. At such point, if the set of accepted frames together with f is feasible, then f is accepted and the packet is transmitted. Otherwise, f is rejected and another EDF packet is chosen for inspection. Intuitively, the lazy algorithm should perform no worse than the eager one. However, we analyze the eager variant due to its immediate decisions. Moreover, in the next section we show that neither variant is 1-competitive.

4.2   Upper Bound for the Algorithm

We do not know of any impossibility result for perfectly periodic instances. However, we can show that neither variant of our algorithm is 1-competitive.

Theorem 5. *On perfectly periodic instances with periods d and $d/2$ such that $k > 2(d+1)$, both variants of the algorithm have competitive ratios at most $1 - 1/d$. Moreover, no non-preemptive work-conserving algorithm is 1-competitive.*

4.3   Analysis of the Algorithm

For convenience, we extend the arrival time and deadline notation to packets: for a packet p, these are denoted $t(p)$ and D_p respectively; recall that a packet is assigned the deadline of its frame. To reason about intervals, we denote the left and the right endpoint of an interval I by $l(I)$ and $r(I)$ respectively. Moreover, for any family of intervals $\mathcal{F}$, we let $u(\mathcal{F}) = |\bigcup_{I \in \mathcal{F}} I|$ and $s(\mathcal{F}) = \sum_{I \in \mathcal{F}} |I|$.

Analysis Outline. To analyze the algorithm, we establish a charging scheme. As before, we charge a frame f delivered both by OPT and the algorithm to itself. Thus we can restrict our attention to frames delivered by OPT that the algorithm rejected. We observe in Lemma 3 that for every rejected frame f, there is an interval I_f that *covers* f, i.e., spans both its arrival time and deadline, such that the algorithm delivers a packet in roughly a constant fraction of I_f's slots. We call such an I_f a *busy interval*.

Intuitively, this should yield a constant competitive ratio since we can count packets rather than frames as noted in Section 4.1. Specifically, every frame f delivered by OPT that is not covered by a busy interval is delivered by the algorithm as well. And in each busy interval I, OPT can deliver at most $|I|$ packets, which is proportional to the number of packets that the algorithm delivers in I.

However, there are two issues. First, Lemma 3 states that the algorithm sends packets in $|I|/2 - k$ slots of a busy interval I, which means that we have a constant ratio on a packet basis only if I is sufficiently large. Fortunately, it follows from Lemma 3 that short busy intervals correspond to rejected frames of small periods, and we can deal with such frames separately.

Second, busy intervals may overlap, leading to overcounting the packets delivered by the algorithm (and OPT). Thus, we need a claim similar to Lemma 3 for the union of all busy intervals. We remedy this by showing that there is a subset of the busy intervals that covers every rejected frame, with an additional property that, when ordered by either endpoint, no three successive intervals in the subset intersect. Clearly, the number of packets that OPT sends in any busy interval is no larger than the total length of the intervals in the subset. Thus, if we charge these packets of OPT to those sent by the algorithm in either all odd-numbered or all or even-numbered intervals from the subset, whichever maximizes the total length, we do not charge a single slot twice, as these intervals are disjoint, and we lose only a factor of 2 in the total length of the intervals.

We note that each rejected frame is covered a "busy" interval.

Lemma 3. *If the algorithm rejects a frame f_0 upon its arrival at time $t_1(f_0)$, then there exists $T \geq D_{f_0}$ such that in $[t_1(f_0), T)$, i.e., the interval of $T - t_1(f_0)$ slots starting at $t_1(f_0)$, the algorithm delivers strictly more than $(T - t_1(f_0))/2 - k$ packets, each with a deadline no larger than T, within $[t_1(f_0), T)$. Moreover, if*

$d_{f_0} = 1$, then the algorithm delivers strictly more than $T - t_1(f_0) - k$ packets, each with a deadline no larger than T, within $[t_1(f_0), T)$.

It is an intriguing question whether the theorem can be strengthened: is it true that there exists a $T \geq D_{f_0}$ such that the algorithm delivers strictly more than $T - t_1(f_0) - k$ packets in the interval $[t_1(f_0), T)$?

Next, we construct a *good family of busy intervals* that underpins our analysis. Again, one of the properties we guarantee is covering all rejected frames. Note that when we say that a family $\mathcal{F}$ of intervals covers a frame, we mean that the frame is covered by $\bigcup_{I \in \mathcal{F}} I$, rather than a particular $I \in \mathcal{F}$.

Lemma 4. *There exists a family $\mathcal{I}_0$ of busy intervals of length at least $3k$ and a subset $\mathcal{I}_0' \subseteq \mathcal{I}_0$ with the following properties.*

1. *Every rejected frame of period at least 3 is covered by $\bigcup \mathcal{I}_0$.*
2. *$u(\mathcal{I}_0) \leq s(\mathcal{I}_0) \leq 2 \cdot u(\mathcal{I}_0)$.*
3. *$u(\mathcal{I}_0') = s(\mathcal{I}_0') \geq \frac{1}{2} \cdot s(\mathcal{I}_0)$.*

In particular, the last property implies that $\mathcal{I}_0'$ is a family of disjoint intervals.

Together, Lemmas 3 and 4 imply the following.

Theorem 6. *The algorithm is $\frac{1}{17}$-competitive on perfectly periodic instances.*

5 Common Period, Unit Value

In this section we consider instances in which all frames have the same period d and unit value, but arbitrary sizes. Combining Lemma 1 with either of the algorithms from Sections 2.2 or 3 yields the following result.

Corollary 1. *There is a $\Omega(1/\log k_{\max})$-competitive randomized algorithm for instances with common period and unit value.*

We could not find a better algorithm. In fact, two natural algorithms, EDF and SRPT, cannot perform much better: we prove an $O(\log \log k_{\max}/\log k_{\max})$ upper bound on their competitive ratios. We do not provide any guarantees for either of them. One could expect SRPT to be $\Omega(1/\log k_{\max})$-competitive as it attains this ratio for single machine preemptive throughput maximization [6,2], which corresponds exactly to our setting with $d = 1$ and arbitrary s_f values. However, we do not know if its analysis can be extended to our problem.

EDF and SRPT are defined as follows. At any given time t, we say that a frame f with deadline D_f is *feasible* if the number of remaining packets of f (ones that were not yet transmitted, including those that did not arrive yet) is no more than $D_f - t$. Clearly, an infeasible frame cannot be delivered. At step t, both algorithms examine the set of all available packets of feasible frames, and transmits one chosen as follows. EDF chooses a packet of the frame with the earliest deadline. SRPT chooses a packet of the frame with the smallest number of remaining packets. Ties can be broken arbitrarily in both algorithms.

Since a frame's deadline is roughly its arrival time plus d times its size, these algorithms behave similarly. In particular, they share the following property: If the algorithm starts transmitting packets of a frame whose deadline is t_f, then by time t_f at least one frame is completed. However, ignoring long frames may not be the right choice, as the following theorem, which also stated a rather weak impossibility result for any algorithm, shows.

Theorem 7. *The competitive ratio of any randomized algorithm on perfectly uniform instances is at most 0.75. Moreover, the competitive ratios of both EDF and SRPT on such instances are* $O(\log \log k_{\max} / \log k_{\max})$.

References

1. Awerbuch, B., Bartal, Y., Fiat, A., Rosén, A.: Competitive non-preemptive call control. In: Proc. of the 5th Annual ACM-SIAM Symp. on Discrete Algorithms (SODA), pp. 312–320 (1994)
2. Dürr, C., Jeż, Ł., Thang, N.K.: Online scheduling of bounded length jobs to maximize throughput. J. Scheduling 15(5), 653–664 (2012). Also appeared in Proc. of the 7th Workshop on Approx. and Online Algorithms (WAOA), pp. 116–127 (2009)
3. Emek, Y., Halldórsson, M.M., Mansour, Y., Patt-Shamir, B., Radhakrishnan, J., Rawitz, D.: Online set packing. SIAM J. Comput 41(4), 728–746 (2010). Also appeared in Proc. of the 29th ACM Symp. on Principles of Distributed Comput. (PODC), pp. 440–449 (2010)
4. Epstein, L., Jeż, Ł., Sgall, J., van Stee, R.: Online Scheduling of Jobs with fixed start times on related machines. In: Gupta, A., Jansen, K., Rolim, J., Servedio, R. (eds.) APPROX/RANDOM 2012. LNCS, vol. 7408, pp. 134–145. Springer, Heidelberg (2012), To appear in Algorithmica:
   http://dx.doi.org/10.1007/s00453-014-9940-2
5. Kalyanasundaram, B., Pruhs, K.: Speed is as powerful as clairvoyance. J. ACM 47(4), 617–643 (2000). Also appeared in Proc. of the 36th Symp. on Foundations of Comp. Sci (FOCS), pp. 214–221 (1995)
6. Kalyanasundaram, B., Pruhs, K.: Maximizing job completions online. J. Algorithms 49(1), 63–85 (1998). Also appeared in Proc. of the 6th European Symp. on Algorithms (ESA), pp. 235–246 (1998)
7. Kesselman, A., Patt-Shamir, B., Scalosub, G.: Competitive buffer management with packet dependencies. Theor. Comput. Sci. 489-489, 75–87 (2013). Also appeared in 23rd IEEE Int. Parallel and Distributed Processing Symp. (IPDPS), pp. 1–12 (2009)
8. Mansour, Y., Patt-Shamir, B., Rawitz, D.: Overflow management with multipart packets. Computer Networks 56(15), 3456–3467 (2011). Also appeared in Proc. of the 30th IEEE Int. Conf. on Computer Communications (INFOCOM), pp. 2606–2614 (2011)
9. Markovitch, M., Scalosub, G.: Bounded delay scheduling with packet dependencies. In: Proc. of the IEEE INFOCOM Workshops, pp. 257–262 (2014)
10. Scalosub, G., Marbach, P., Liebeherr, J.: Buffer management for aggregated streaming data with packet dependencies. IEEE Trans. Parallel Distrib. Syst. 24(3), 439–449 (2010). Also appeared in Proc. of the 29th IEEE Int. Conf. on Computer Communications (INFOCOM), pp. 241–245 (2010)
11. Woeginger, G.J.: On-line scheduling of jobs with fixed start and end times. Theor. Comput. Sci. 130(1), 5–16 (1994)

A Randomized Algorithm for Online Scheduling with Interval Conflicts*

Marcin Bienkowski, Artur Kraska, and Paweł Schmidt

Institute of Computer Science, University of Wrocław, Poland

Abstract. In the contiguous variant of the Scheduling with Interval Conflicts problem, there is a universe $\mathcal{U}$ consisting of elements being consecutive positive integers. An input is a sequence of conflicts in the form of intervals of length at most σ. For each conflict, an algorithm has to choose at most one surviving element, with the ultimate goal of maximizing the number of elements that survived all conflicts. We present an $O(\log \sigma / \log \log \sigma)$-competitive randomized algorithm for this problem, beating known lower bound of $\Omega(\log \sigma)$ that holds for deterministic algorithms.

Keywords: online algorithms, competitive analysis, interval conflicts, online scheduling.

1 Introduction

In the contiguous variant of the Scheduling with Interval Conflicts problem (SIC), an algorithm is given a universe $\mathcal{U}$ consisting of n consecutive positive integers. All elements are initially active. The input consists of *conflicts* in the form of intervals $[a, b]$ where $a, b \in \mathcal{U}$. The conflict means that at most one element from the set $[a, b]$ may remain active; it is up to an algorithm to choose this element. Once an element becomes inactive, it remains in this state till the end. The goal of the algorithm is to maintain, in online manner, a set of active elements, with the ultimate goal of maximizing their number.

This problem, introduced by Halldórsson, Patt-Shamir and Rawitz [9], has an interesting set of applications: choosing transmissions in wireless stations, maximizing the number of processed tasks by bounded-capacity servers, or maximizing the goodput (number of data or video frames) that are successfully forwarded by a router.

The problem is analyzed in the framework of competitive-analysis [3], where the gain of an online algorithm (the final number of surviving elements) is compared to the gain of the optimal offline algorithm OPT. The ratio of these two is called *competitive ratio* and is subject to minimization.

The authors of [9] presented a deterministic $O(\log \sigma)$-competitive algorithm PRIORITY for this problem, where σ is an upper bound on the number of elements in any conflict. Their algorithm is *oblivious*, i.e., an active element is chosen

---

* Supported by Polish National Science Centre grant DEC-2013/09/B/ST6/01538.

© Springer International Publishing Switzerland 2015
C. Scheideler (Ed.): SIROCCO 2015, LNCS 9439, pp. 91–103, 2015.
DOI: 10.1007/978-3-319-25258-2_7

for each conflict independently of other conflicts. They also showed that the competitive ratio of any deterministic strategy is at least $\Omega(\log \sigma)$.

1.1   Our Result

One of the open questions of [9] was whether the competitive ratio can be improved using randomization. We answer this question affirmatively, presenting a $O(\log \sigma / \log \log \sigma)$-competitive algorithm RANDOM PRIORITY (RAND) for the contiguous variant of SIC. We also show that our analysis is asymptotically tight.

1.2   Related Work

Another variant of SIC is a so-called non-contiguous model, where $\mathcal{U}$ consists of integers that are not necessarily consecutive. Halldórsson et al. [9] presented an $O(\log \sigma)$-competitive deterministic algorithm also for this variant. It is worth noting that in contrast to the contiguous model, here the competitive ratio of any oblivious algorithm is at least $\Omega(n)$. A further generalization, called Online Set Packing, where conflicts are not required to be intervals, was considered by Emek et al. [5].

A natural minimization problem, dual to SIC, is to find minimum number of points intersecting all input intervals. For this problem, an optimal 2-competitive algorithm is known [11]. A variant in which intervals arrive sorted from left to right is equivalent to the deadline variant of the TCP acknowledgement problem and is solvable optimally online [4].

Another related maximization problem, known as the call admission problem on a line graph [1, 2, 6, 7, 12], is to choose maximum subset of non-intersecting intervals. Algorithms achieving logarithmic competitive ratios were given for various flavors of this problem.

The randomized algorithm presented in this paper is *barely random*, i.e., it uses random bits only at the beginning and their number is independent of the length of an input sequence (number of conflicts). The algorithm is basically a random shift of the deterministic algorithm PRIORITY [9]. It is worth noting that a natural randomized approach for this problem, where we make random independent decisions for each conflict separately, leads to an algorithm that is not competitive at all. Similar phenomenon was observed also in other areas of competitive analysis. For example, for the list accessing problem [14, 10], there are two natural randomizations of the optimal deterministic algorithm MOVE-TO-FRONT [14]: a barely random algorithm BIT [13] and an algorithm RANDOM-MOVE-TO-FRONT [8] that makes a random decision for each input element. The former approach substantially outperforms the latter.

1.3   Preliminaries

Throughout this text, for two integers $a < b$, $[a, b]$ denotes the set $\{a, a+1, \ldots, b\}$ and is called *interval* $[a, b]$. Let $\mathcal{U}$ be our universe, consisting of n initially active elements being consecutive positive integers.

An input is a sequence of conflicts, each represented by an interval $[a, b]$. The choice of an algorithm is to *pick* an (active or inactive) element $x \in [a, b]$, and the remaining elements from $[a, b]$ become inactive. If x is active, it is called *surviving*. (It can be assumed without loss of generality that if a conflict contains at least one active element, an algorithm picks it.) The gain of an algorithm is the number of elements that are active at the end of the input sequence.

A conflict $c = [a, b]$ is usually treated as a set of respective elements from $[a, b]$. In particular, $|c| = b - a + 1$ is the number of elements that are in conflict c, and is called *size* of conflict c.

Let σ denote the maximum size of any conflict occurring in the input. We emphasize that σ is not known a priori to an online algorithm. Throughout this paper, we assume that $\sigma \geq 4$.

2 Randomized Algorithm

Our randomized algorithm RANDOM PRIORITY is closely related to the $O(\log \sigma)$-competitive deterministic algorithm by Halldórsson et al. [9] and can be viewed as a random shift variant of their algorithm.

As already stated, without loss of generality, one may assume that an algorithm picks an active element whenever a conflict contains at least one active element. However, to ensure that the order of the intervals in the input does not matter (cf. Observation 1), RAND will not have this property.

We assume that the algorithm knows M that is a power of two and is an upper bound on n. (In the following subsection, we show how to get rid of this assumption.) At the beginning, RAND chooses r uniformly at random from the set $\{0, 1, \ldots, M - 1\}$. To each element $u \in \mathcal{U}$, it assigns a *priority*

$$p(u) = \max\{\ell \in \mathbb{Z} : u + r \text{ is divisible by } 2^\ell\} . \tag{1}$$

Note that $u + r$ is always positive, so $p(u)$ is well defined. When processing a conflict $[a, b]$, RAND *picks* (possibly inactive) x to be the element from $[a, b]$ with the maximum priority. Note that the maximum priority element is unique for each conflict (cf. Observation 2 from [9]), and thus RAND is well defined.

2.1 Unknown Value of n

When n is not known a priori to the algorithm, the algorithm may still set M to be the smallest power of two that is larger than any conflict element seen so far; the initial value of M is set to be 1.

What remains is to show how to choose appropriate values of r. When $M = 2^a$, r is equivalent to an a-bit string. When RAND increases M from 2^a to 2^b, r has to become a b-bit string. The algorithm leaves a least significant bits of r intact and chooses randomly remaining $b - a$ most significant ones. It is easy to observe that the choices made by RAND so far using the a-bit value of r would have been the same if it had worked with the b-bit value of r from the very beginning.

3   Analysis

In this section, we bound the competitive ratio of RAND. We start with a few basic observations.

As elements that are not in any conflict remain active in a solution of any algorithm, for the analysis, we assume that each element of $\mathcal{U}$ belongs to some conflict. Then, the gain of an algorithm can be also defined as the number of elements that survived in all conflicts they belonged to.

Once RAND picks the random shift r, it becomes the deterministic algorithm PRIORITY by Halldórsson et al. [9]. In particular, this implies the following observation that holds also for their algorithm.

Observation 1. RAND *is order-oblivious, i.e., the final set of active elements does not depend on the order in which conflicts are presented to the algorithm.*

Observation 1 has some immediate consequences. First, we may assume that the input sequence does not contain two intervals that are equal or properly contained in each other. Indeed, if $[a', b'] \subseteq [a, b]$, then RAND may process $[a, b]$ first and then conflict $[a', b']$ does not change the set of active elements.

In the following, we assume that the input sequence $\mathcal{I}$ contains the intervals sorted by their left ends (by the observation above, we may assume that their left ends are all different and their right ends are then sorted as well). After sorting, we may assume that any two consecutive conflicts overlap. Otherwise, we may treat the corresponding multiple disjoint sets of intervals separately as acting on disjoint parts of $\mathcal{U}$ and derive competitiveness of RAND on each part separately.

3.1   Core Subsequence

From any sorted input sequence $\mathcal{I}$, we pick a sparse subset $\mathsf{core}(\mathcal{I})$ with the following properties, and analyze both OPT and RAND using this subset.

Definition 2. *For any sorted input sequence* $\mathcal{I} = \{d_1, d_2, \ldots, d_{|\mathcal{I}|}\}$*, a subsequence* $\mathsf{core}(\mathcal{I}) = \{c_1, c_2, \ldots, c_m\} \subseteq \mathcal{I}$ *satisfies the following three properties:*

1. $\bigcup_{i=1}^{m} c_i = \bigcup_{d \in \mathcal{I}} d$.
2. *Any* $u \in \mathcal{U}$ *belongs to at most two conflicts from* $\mathsf{core}(\mathcal{I})$.
3. *For* $c_i \in \mathsf{core}(\mathcal{I})$*, a conflict from* $\mathcal{I}$ *intersecting* c_i *is contained in* $\bigcup_{j=i-2}^{i+1} c_j$.

We assumed that $c_{-1} = c_0 = c_{m+1} = \emptyset$.

Lemma 3. *Set* $\mathsf{core}(\mathcal{I})$ *exists for any sorted input sequence* $\mathcal{I}$.

Proof. We include elements in $\mathsf{core}(\mathcal{I})$ starting from d_1 and iterating over conflicts $d_2, d_3, \ldots, d_{|\mathcal{I}|}$. Assume that we already added elements $c_1, \ldots, c_i = d_j$ to $\mathsf{core}(\mathcal{I})$ and $j < |\mathcal{I}|$. Then, c_{i+1} is chosen as the rightmost interval from the set $\{d_{j+1}, \ldots, d_{|\mathcal{I}|}\}$ that has a nonempty intersection with c_i or, equivalently, as the interval from the set $\{d_{j+1}, \ldots, d_{|\mathcal{I}|}\}$ that has minimal nonempty intersection with c_i.

The first property of Definition 2 holds trivially for the chosen subset $(c_i)_i$; it remains to show that $(c_i)_i$ satisfies also the remaining two properties.

Assume that the second property does not hold, i.e., that $c_i \cap c_{i+2} \neq \emptyset$ for some i. Conflict c_{i+2} has smaller intersection with c_i than c_{i+1}, which contradicts the choice of c_{i+1}.

For showing the third property, we choose any conflict $[p, q] \in \mathcal{I}$ intersecting c_i. For any conflict c_j let a_j and b_j denote its beginning and end, respectively, i.e., $c_j = [a_j, b_j]$. We consider two cases.

1. Interval $[p, q]$ is before c_i in the sorted sequence $\mathcal{I}$, i.e. $p < a_i \leq q < b_i$. Assume that the third property does not hold, i.e. $p < a_{i-2}$. As $c_{i-2} \cap c_i = \emptyset$, $b_{i-2} < a_i$, and thus $p < a_{i-2} < b_{i-2} < a_i < q$, which means that c_{i-2} is completely contained in $[p, q]$, a contradiction. (We showed that $[p, q] \subseteq c_{j-2} \cup c_{j-1} \cup c_j$; note that $[p, q] \subseteq c_{j-1} \cup c_j$ need not hold for our construction of $(c_i)_i$.)
2. Interval $[p, q]$ is after c_i in the sorted sequence $\mathcal{I}$, i.e. $a_i < p \leq b_i < q$. As c_{i+1} has the minimum possible overlap with c_i among all intervals from $\mathcal{I}$ that are after c_i, $p \leq a_{i+1}$. This means that $q \leq b_{i+1}$, and thus $[p, q] \subseteq c_i \cup c_{i+1}$. □

Lemma 4. *For any input sequence $\mathcal{I}$ with subsequence $\mathsf{core}(\mathcal{I})$, it holds that* $\mathrm{OPT}(\mathcal{I}) \leq |\mathsf{core}(\mathcal{I})|$.

Proof. From the first property of Definition 2, it follows that we can cover $\mathcal{U} = \bigcup_{d \in \mathcal{I}} d$ with $|\mathsf{core}(\mathcal{I})|$ intervals. OPT may choose at most one element from each of these conflicts, and hence $\mathrm{OPT}(\mathcal{I}) \leq |\mathsf{core}(\mathcal{I})|$. □

3.2 Crucial Lemma

In this section, we lower-bound the gain of RAND (the number of surviving elements) from an interval that corresponds to a conflict in $\mathsf{core}(\mathcal{I})$. We start with the following technical observation on the priorities of elements picked from any conflict by RAND.

Lemma 5. *Fix a conflict c and an integer $\ell \leq \log M$, such that $|c| \leq 2^\ell$. For an element u picked by RAND from c, it holds that $\Pr[p(u) \geq \ell] = |c|/2^\ell$.*

Proof. Fix any element $x \in c$ and let

$$S_x = \{r \in [0, M-1] : x + r \text{ is divisible by } 2^\ell\}.$$

That is, S_x is the set of all random shifts for which $p(x) \geq \ell$. It means that the element picked from c by RAND has priority at least ℓ if and only if r belongs to $\bigcup_{x \in c} S_x$. Hence,

$$\Pr[p(u) \geq \ell] = \frac{|\bigcup_{x \in c} S_x|}{M}.$$

For $\ell \leq \log M$ each S_x contains exactly $M/2^\ell$ elements and for $|c| \leq 2^\ell$ all these sets are disjoint. Therefore, $\Pr[p(u) \geq \ell] = (|c| \cdot M/2^\ell)/M = |c|/2^\ell$. □

Lemma 6. *Fix an input $\mathcal{I}$, its subsequence* $\mathsf{core}(\mathcal{I}) = (c_i)_{i=1}^m$, *and any conflict* $c_i \in \mathsf{core}(\mathcal{I})$. *The expected number of surviving elements from the interval* c_i *is at least* $\frac{1}{2} \cdot |c_i| / \sum_{j=i-2}^{i+1} |c_j|$.

Proof. Let A be the set of all intervals from $\mathcal{I}$ (including c_i itself) that intersect with c_i and let $\mathsf{span}(A) = \bigcup_{d \in A} d$ be the smallest interval containing all these conflicts. Let ℓ be an integer satisfying $2^{\ell-1} < \mathsf{span}(A) \leq 2^\ell$.

The expected number of surviving elements from c_i is equal to the probability that the element picked from c_i by RAND, say u, survives, i.e., it is picked also in other conflicts it belongs to. A sufficient condition for u's survival is that its priority is at least ℓ. (In such case, u is the only element in $\mathsf{span}(A)$ with this property, any conflict containing u is contained in $\mathsf{span}(A)$, and hence u is picked by RAND from any conflict u belongs to.) Using Lemma 5, the probability that u survives is then at least

$$\Pr\left[p(u) \geq \ell\right] = \frac{|c_i|}{2^\ell} \geq \frac{|c_i|}{2 \cdot |\mathsf{span}(A)|} \geq \frac{1}{2} \cdot \frac{|c_i|}{\left|\bigcup_{j=i-2}^{i+1} c_j\right|} \geq \frac{1}{2} \cdot \frac{|c_i|}{\sum_{j=i-2}^{i+1} |c_j|} \ .$$

Above, we used the third property of Definition 2, i.e., $\mathsf{span}(A) \subseteq \bigcup_{j=i-2}^{i+1} c_j$. □

3.3   Bounding the Gain of RAND

For any sequence $(x_j)_{j=1}^m$ of positive integers and any index $i \in \{1, \ldots, m\}$, we define

$$\gamma_x(i) = \frac{x_i}{\sum_{j=i-2}^{i+1} x_j} \ , \tag{2}$$

where we assume that $x_{-1} = x_0 = x_{m+1} = 0$. In the next section, we prove the following lemma.

Lemma 7. *For any sequence of positive integers* $(x_i)_{i=1}^m$ *whose all elements are at most* σ, *it holds that* $\sum_{i=1}^m \gamma_x(i) \geq 2^{-8} \cdot m \cdot \log\log\sigma / \log\sigma$.

Theorem 8. *On input sequences where the size of any conflict is at most* σ, RANDOM PRIORITY *is* $O(\log\sigma / \log\log\sigma)$-*competitive.*

Proof. Fix any sorted input $\mathcal{I}$ with set $\mathsf{core}(\mathcal{I}) = (c_i)_{i=1}^m$ consisting of m conflicts. By Lemma 6, in the schedule of RAND, the expected number of surviving elements from any conflict $c_i \in \mathsf{core}(\mathcal{I})$ is at least $\gamma_x(i)/2$. If we sum it over all conflicts, we calculate the gain from each element of the universe at most twice (because of the second property of Definition 2), and hence the number of surviving elements is at least

$$\mathbf{E}[\mathrm{RAND}(\mathcal{I})] \geq \frac{1}{2} \cdot \sum_{i=1}^m \gamma_x(i)/2 \geq \frac{\log\log\sigma}{2^{10} \cdot \log\sigma} \cdot m \geq \frac{\log\log\sigma}{2^{10} \cdot \log\sigma} \cdot \mathrm{OPT}(\mathcal{I}) \ ,$$

where the second inequality above follows by Lemma 7 and the third one follows by Lemma 4. □

3.4   Integer Sequences (Proof of Lemma 7)

Proof Plan. To sketch our proof plan, we assume for a while that $\gamma_x(i)$ is defined as $x_i / \sum_{j=i-1}^{i+1} x_j$. In our description, we say that $\gamma_x(i)$ is γ-value of the i-th element.

For the analysis, we cover the sequence $(x_i)_{i=1}^{m}$ with monotonic subsequences; two consecutive subsequences share a single element. We now focus on a single increasing subsequence $(x_i)_{i=g}^{g+t-1}$ of length t and neglect constant factors. Lemma 7 would follow by summing over all monotonic subsequences if we could show the following relation: $\sum_{i=g}^{g+t-1} \gamma_x(i) = \Omega(t \cdot \log\log \sigma / \log \sigma)$.

As x_{g+t-1} is a local maximum of the sequence, $\gamma_x(g + t - 1) = \Omega(1)$. Hence, if $t = O(\log \sigma / \log\log \sigma)$, then the relation follows trivially. Otherwise, $t = \Omega(\log \sigma / \log\log \sigma)$ and we analyze the sum $\sum_{i=g}^{g+t-2} \gamma_x(i) = \Omega(\sum_{i=g}^{g+t-2} x_i / x_{i+1})$. As there are many elements in the increasing subsequence, many pairs of consecutive elements will be quite close to each other. For example, if the subsequence $(x_i)_{i=g}^{g+t-1}$ increased geometrically, then $x_i / x_{i+1} = \Omega(\log\log \sigma / \log \sigma)$ for any $i \in \{g, \ldots, g + t - 2\}$ and the relation would follow. In fact, we are able to show that $\sum_{i=g}^{g+t-2} x_i / x_{i+1} = \Omega(t \cdot \log\log \sigma / \log \sigma)$ for any increasing sequence, not necessarily geometrically growing.

Up to this point, in our informal description, we assumed that $\gamma_x(i) = x_i / \sum_{j=i-1}^{i+1} x_j$. For the actual definition of $\gamma_x(i)$, i.e., $\gamma_x(i) = x_i / \sum_{j=i-2}^{i+1} x_j$, it turns out that the bound $\gamma_x(g + t - 1) = \Omega(1)$ is no longer true. For example, in an increasing subsequence (x_g, x_{g+1}) (of length two), it may happen that $\gamma_x(g + 1) = o(1)$, because $x_{g-1} \gg x_{g+1}$.

To alleviate this issue, we preprocess $(x_i)_{i=1}^{m}$, dropping some elements from this sequence. We ensure that the sum of γ-values does not change much and the resulting sequence $(y_i)_{i=1}^{m'}$ has the desired property: $\gamma_x(k) = \Omega(1)$ for the maximal element x_k from each monotonic subsequence of $(y_i)_{i=1}^{m'}$. This will allow us to show that the sums of γ-values for any monotonic subsequence of length t can be lower bounded by $\Omega(t \cdot \log\log \sigma / \log \sigma)$.

Zigzags. In the reasoning below, we assume that if we refer to a sequence element with index that does not exist in the given sequence, then the corresponding element is equal to zero. In particular, for a sequence $(x_i)_{i=1}^{m}$, we assume that $x_{-1} = x_0 = x_{m+1} = 0$.

Definition 9. *Let $(x_i)_{i=1}^{m}$ be a sequence of positive integers. A (contiguous) subsequence $x_g, x_{g+1}, \ldots, x_h$ is called **proper** if*

- *it is either non-decreasing or non-increasing and*
- *$x_k \geq \max\{x_{k-2}, x_{k-1}, x_{k+1}\}$, where x_k is the first element for non-increasing subsequence and the last element for non-decreasing subsequence (if there are no ties, then x_k is simply the maximum element).*

Definition 10. *Let $(x_i)_{i=1}^{m}$ be a sequence of positive integers. A zigzag at position i is a subsequence of four integers $x_i, \ldots, x_{i+3}$, such that $x_i > x_{i+2} > x_{i+1}$ and $x_{i+2} \geq x_{i+3}$. We call element x_{i+1} a zigzag **dent**.*

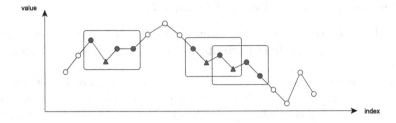

Fig. 1. An example sequence of integers with three zigzags (in rectangles). Zigzag dents are marked with triangles.

Removing Zigzags. An example sequence with zigzags marked is given in Fig. 1. Note that two zigzags can share at most two points. We now verify what happens if we remove a zigzag dent (marked with a triangle in Fig. 1) from the sequence.

Lemma 11. *Let $(x_j)_{j=1}^m$ be a sequence of positive integers with a zigzag at position $i - 1$. Assume we remove zigzag dent (x_i) from $(x_j)_{j=1}^m$ and we denote the resulting sequence $(y_j)_{j=1}^{m-1}$. Then, the following relations hold:*

- $\gamma_x(j) \geq \gamma_y(j)$ *for all $j < i$;*
- $\gamma_x(i) \geq 0$;
- $\gamma_x(i+1) \geq \gamma_y(i)/2$;
- $\gamma_x(j) \geq \gamma_y(j-1)$ *for all $j > i + 1$.*

Proof. By the definition of the sequence $(y_j)_{j=1}^{m-1}$, it holds that $x_j = y_j$ for any $j < i$ and $x_j = y_{j-1}$ for any $j > i$. Therefore, $\gamma_x(j) = \gamma_y(j)$ for any $j < i - 1$ and $\gamma_x(j) = \gamma_y(j-1)$ for any $j > i + 2$.

Now we consider the remaining four indices. As $(x_j)_{j=1}^m$ contains a zigzag at position $i - 1$, $x_{i-1} > x_{i+1} > x_i$, and thus

- $\gamma_x(i-1) \geq x_{i-1}/(x_{i-3} + x_{i-2} + x_{i-1} + x_{i+1}) = \gamma_y(i-1)$,
- $\gamma_x(i) \geq 0$,
- $\gamma_x(i+1) \geq x_{i+1}/(2x_{i-1} + x_{i+1} + x_{i+2}) \geq \gamma_y(i)/2$,
- $\gamma_x(i+2) \geq x_{i+2}/(x_{i-1} + x_{i+1} + x_{i+2} + x_{i+3}) = \gamma_y(i+1)$.   □

Lemma 12. *From any sequence $(x_j)_{j=1}^m$ of positive integers, it is possible to remove at most half of the elements, so that the resulting sequence $(y_j)_{j=1}^{m'}$ does not contain zigzags and $\sum_{i=1}^{m'} \gamma_y(i) \leq 2 \cdot \sum_{i=1}^m \gamma_x(i)$.*

Proof. We first identify all the places in $(x_i)_{i=1}^m$ where some zigzag exist. Note that the position of each zigzag is uniquely defined by the position of its dent. Furthermore, each dent is followed by an element that is not a dent of another zigzag. Therefore, the number of dents is at most $m/2$. We show that the sequence $(y_i)_{i=1}^{m'}$ created by removing all dents from $(x_i)_{i=1}^m$ satisfies the conditions of the lemma.

We proceed iteratively from right to left and we remove dents, one at a time. We show that such removal destroys a current zigzag and does not create a new one. Assume now that there was a zigzag starting at position i and that we removed its dent at position $i + 1$. If a new zigzag appears, then certainly it contains a pair of (now adjacent) elements x_i and x_{i+2}. We consider three cases:

- There is no zigzag $x_i, x_{i+2}, x_{i+3}, x_{i+4}$, because $x_{i+2} \geq x_{i+3}$.
- There is no zigzag $x_{i-1}, x_i, x_{i+2}, x_{i+3}$, because $x_i > x_{i+2}$.
- If $x_{i-2}, x_{i-1}, x_i, x_{i+2}$ is a zigzag, then $x_{i-2}, x_{i-1}, x_i, x_{i+1}$ was already a zigzag before removal of element x_{i+1}.

By Lemma 11, when we remove a single dent at position i, the γ-values do not increase at any element but the removed one and at element at $i+1$, where it may increase twofold. After such removal, to find the next dent, we have to move at least two positions to the left in the sequence, and thus the next possible increase of γ-value occurs at a different element. Hence, $\sum_{i=1}^{m'} \gamma_y(i) \leq 2 \cdot \sum_{i=1}^{m} \gamma_x(i)$. $\square$

Zigzag-free Sequences. We now show that a sequence without zigzags (obtained for example using Lemma 12) can be covered with proper sequences. Later, we analyze the sum of γ-values on a single proper sequence and finally we combine the pieces to obtain the proof of Lemma 7.

Lemma 13. *Any sequence $(y_j)_{j=1}^{m'}$ not containing zigzags can be covered by proper subsequences, so that each element is in at least one and at most two such subsequences.*

Proof. We assume that not all elements of $(y_j)_{j=1}^{m'}$ are equal as otherwise the whole $(y_j)_{j=1}^{m'}$ would be a single proper sequence. We first cover the sequence with monotonic subsequences and later show that they are proper.

A straightforward routine for choosing a monotonic contiguous subsequence of a given type (non-increasing or non-decreasing) starts from a chosen element y_g and greedily adds as many consecutive elements as possible. To cover $(y_j)_{j=1}^{m'}$ with monotonic subsequences, we start with finding the first two consecutive non-equal pair of elements: the first subsequence will be of non-decreasing type if this pair is increasing and non-increasing type otherwise. We use the greedy routine above to choose a monotonic sequence starting from y_1. If the subsequence ends at y_g, we choose the next sequence (of the opposite type) starting from y_g (that is, two consecutive subsequences share exactly one element). We proceed this way, till all elements of $(y_j)_{j=1}^{m'}$ are covered.

It now remains to show that the chosen subsequences are proper. We only have to prove the second property of Definition 9, i.e., that for the maximal element y_k of any sequence, it holds that $y_k \geq \max\{y_{k-2}, y_{k-1}, y_{k+1}\}$. The condition follows trivially for the two special cases: when the first subsequence is a non-increasing one (and $k = 1$) and when the last subsequence is a non-decreasing one (and $k = m'$).

Otherwise, we have to verify this condition for an element y_k that is ending a non-decreasing subsequence and starting the following non-increasing subsequence. By the choice of k, $y_k \geq y_{k-1}$ and $y_k \geq y_{k+1}$. For showing $y_k \geq y_{k-2}$, we consider two cases. If y_{k-2}, y_{k-1}, y_k belong to a single non-decreasing subsequence, we are done. Otherwise, y_{k-2}, y_{k-1} is a part of a non-increasing sequence and y_{k-1}, y_k is a two element non-decreasing sequence. In such case, $y_{k-1} < y_k$ as otherwise the non-decreasing sequence containing y_{k-2} and y_{k-1} would also contain y_k. But as the sequence does not contain a zigzag starting at y_{k-2}, it holds that $y_{k-2} \leq y_k$. Thus, the corresponding subsequences are proper. $\square$

Lemma 14. *Fix a sequence* $(y_i)_{i=1}^{m'}$ *of positive integers not greater than σ. For any proper subsequence* $y_g, y_{g+1}, \ldots, y_h$, *it holds that* $\sum_{i=g}^{h} \gamma_y(i) \geq (1/32) \cdot (h - g + 1) \cdot \log \log \sigma / \log \sigma$.

Proof. First, we observe that

$$\gamma_y(i) = \frac{y_i}{\sum_{j=i-2}^{i+1} y_j} \geq \frac{y_i}{4 \cdot \max\{y_{i-2}, y_{i-1}, y_i, y_{i+1}\}} \ .$$

Let $k = h$ be the rightmost element if the subsequence is non-decreasing and $k = g$ if the subsequence is non-increasing. By the definition of a proper sequence, $\gamma_y(k) \geq 1/4$. If $h - g + 1 \leq 8 \log \sigma / \log \log \sigma$, then the lemma follows trivially.

Hence, from now on, we assume that $h - g + 1 \geq 8 \log \sigma / \log \log \sigma$. We use the relation between geometric and harmonic means: for any t positive real numbers $a_1, \ldots, a_t$ it holds that $\prod_{i=1}^{t} a_i^{1/t} \geq t / \sum_{i=1}^{t} (1/a_i)$, or equivalently,

$$\sum_{i=1}^{t} (1/a_i) \geq t \cdot \prod_{i=1}^{t} a_i^{-1/t} \ . \tag{3}$$

Let $t = h - g - 2 \geq (h - g + 1)/2 \geq 4 \log \sigma / \log \log \sigma$. Our goal is now to lower-bound the sum of t terms, $\sum_{i=g+2}^{h-1} \gamma_y(i)$. If the subsequence is non-decreasing, then $\gamma_y(i) \geq (1/4) \cdot (y_i/y_{i+1})$ for any $i \in \{g+2, \ldots, h-1\}$, and thus

$$\sum_{i=g}^{h} \gamma_y(i) \geq \sum_{i=g+2}^{h-1} \gamma_y(i) \geq \frac{1}{4} \cdot \sum_{i=g+2}^{h-1} \frac{y_i}{y_{i+1}}$$

$$\geq \frac{t}{4} \cdot \prod_{i=g+2}^{h-1} \left(\frac{y_{i+1}}{y_i}\right)^{-1/t} \geq \frac{t}{4} \cdot \left(\frac{y_h}{y_{g+2}}\right)^{-1/t} \ . \tag{4}$$

Similarly, if the subsequence is non-increasing, then $\gamma_y(i) \geq (1/4) \cdot (y_i/y_{i-2})$ for any $i \in \{g+2, \ldots, h-1\}$, and thus

$$\sum_{i=g}^{h} \gamma_y(i) \geq \sum_{i=g+2}^{h-1} \gamma_y(i) \geq \frac{1}{4} \cdot \sum_{i=g+2}^{h-1} \frac{y_i}{y_{i-2}}$$

$$\geq \frac{t}{4} \cdot \prod_{i=g+2}^{h-1} \left(\frac{y_{i-2}}{y_i}\right)^{-1/t} \geq \frac{t}{4} \cdot \left(\frac{y_g \cdot y_{g+1}}{y_{h-2} \cdot y_{h-1}}\right)^{-1/t} \ . \tag{5}$$

We combine both cases ((4) and (5)), using that $1 \le y_j \le \sigma$ for all j, obtaining

$$\sum_{i=g}^{h} \gamma_y(i) \ge \frac{t}{4} \cdot \sigma^{-2/t} \ge \frac{t}{4} \cdot \sigma^{-\frac{\log \log \sigma}{2 \log \sigma}} = \frac{t}{4 \cdot \sqrt{\log \sigma}} \ge \frac{h-g+1}{16 \cdot \frac{\log \sigma}{\log \log \sigma}} .$$

The last inequality follows as $2 \cdot \sqrt{\log \sigma} \ge \log \log \sigma$ for $\sigma \ge 4$. $\square$

Putting Pieces Together. We may now combine the results above to show Lemma 7.

Proof (of Lemma 7). Let $(x_j)_{j=1}^{m}$ be a sequence of positive integers smaller than σ. We want to lower-bound $\sum_{i=1}^{m} \gamma_x(i)$.

By Lemma 12, it is possible to construct a sequence $(y_j)_{j=1}^{m'}$ of length $m' \ge m/2$ not containing zigzags, such that $\sum_{i=1}^{m} \gamma_x(i) \ge (1/2) \cdot \sum_{i=1}^{m'} \gamma_y(i)$.

By Lemma 13, we may cover sequence $(y_j)_{j=1}^{m'}$ with proper subsequences. For any such proper subsequence $y_g, y_{g+1}, \ldots, y_h$, Lemma 14 yields the relation $\sum_{i=g}^{h} \gamma_y(i) \ge (1/32) \cdot (h-g+1) \cdot \log \log \sigma / \log \sigma$. If we sum this inequality over all proper subsequences covering $(y_j)_{j=1}^{m'}$, the left hand side is at most $2 \cdot \sum_{i=1}^{m'} \gamma_y(i)$ because each element is in at most two subsequences, and the right hand side is at least $(1/32) \cdot m' \cdot \log \log \sigma / \log \sigma$. Hence,

$$\sum_{i=1}^{m} \gamma_x(i) \ge \frac{1}{2} \cdot \sum_{i=1}^{m'} \gamma_y(i) \ge \frac{1}{4 \cdot 32} \cdot m' \cdot \frac{\log \log \sigma}{\log \sigma} \ge 2^{-8} \cdot m \cdot \frac{\log \log \sigma}{\log \sigma} ,$$

which concludes the proof. $\square$

4   Lower Bound

In this section, we show that our analysis of the algorithm RANDOM PRIORITY presented in the previous section is asymptotically tight, i.e., the competitive ratio of RAND is $\Omega(\log \sigma / \log \log \sigma)$.

To this end, we need the following technical observation that bounds the probability of picking an element from a subset of a given conflict.

Lemma 15. *Fix a conflict c and let b be a subset of elements of c. For an element u picked by RAND from c, it holds that $\Pr[u \in b] \le 2 \cdot |b|/|c|$.*

Proof. Fix an element $x \in c$. Let k be the maximal integer, such that $2^k \le |c|$. Let $S_x = \{r \in [0, M-1] : x + r$ is divisible by $2^k\}$. Note that there is always an element of priority k inside c. A condition necessary for an element x to be picked by RAND is that $p(x) \ge k$, i.e. $r \in S_x$. Therefore, $\Pr[u = x] \le |S_x|/M = 1/2^k$. Summing over all elements from b, we obtain that

$$\Pr[u \in b] = \sum_{x \in b} \Pr[u = x] \le |b|/2^k \le 2 \cdot |b|/|c| ,$$

which concludes the proof. $\square$

Lemma 16. *On input sequences where the size of any conflict is at most σ, the competitive ratio of* RANDOM PRIORITY *is* $\Omega(\log \sigma / \log \log \sigma)$.

Proof. Fix any $\sigma \geq 4$ and let $R = \lfloor \log \sigma / \log \log \sigma \rfloor$. We show that there exists an input sequence $\mathcal{I}$, such that the size of any conflict in $\mathcal{I}$ is at most $R^R \leq \sigma$ and $\text{OPT}(\mathcal{I}) = \Omega(R) \cdot \text{RAND}(\mathcal{I})$.

Let $\mathcal{U} = \{2, \ldots, R^R\}$ be the universe. The sequence $\mathcal{I}$ contains $R - 1$ conflicts $c_1, \ldots, c_{R-1}$ where $c_i = [R^{i-1}+1, R^{i+1}]$ for each $i \in \{1, \ldots, R-1\}$. Note that any conflict c_i intersects only c_{i-1} and c_{i+1}. Elements that are right ends of conflicts $c_1, c_3, c_5, \ldots$ form a feasible solution set: there is no conflict that contains two such elements. Hence, $\text{OPT}(\mathcal{I}) \geq \lceil R/2 \rceil \geq R/2$.

Now we analyze the gain of RAND. Each element of the universe is contained in some conflict, i.e. $\mathcal{U} = \bigcup_{i=1}^{R-1} c_i$. By the definition of conflicts, it is possible to partition $\mathcal{U}$ into disjoint chunks:

$$c_1 \setminus c_2, \quad c_1 \cap c_2, \quad c_2 \cap c_3, \quad \ldots \quad c_{R-3} \cap c_{R-2}, \quad c_{R-2} \cap c_{R-1}, \quad c_{R-1} \setminus c_{R-2} \ .$$

At most one element survives from the first and the last chunk. Now we bound the expected gain on any other chunk $c_i \cap c_{i+1}$. An element from $c_i \cap c_{i+1}$ survives only if it is picked by RAND both in c_i and in c_{i+1}. The probability of the latter event can be upper-bounded using Lemma 15 by $2 \cdot |c_i \cap c_{i+1}| / |c_{i+1}| = 2 \cdot (R^{i+1} - R^i)/(R^{i+2} - R^i) \leq 2/R$, and thus the expected gain of RAND on chunk $c_i \cap c_{i+1}$ is at most $2/R$.

Summing up, the total gain of RAND on $\mathcal{I}$ is $\mathbf{E}[\text{RAND}(\mathcal{I})] \leq 1 + (R-2) \cdot 2/R + 1 \leq 4$, and hence the competitive ratio of RAND is at least $\text{OPT}(\mathcal{I}) \geq (R/8) \cdot \text{RAND}(\mathcal{I})$.   □

References

[1] Awerbuch, B., Bartal, Y., Fiat, A., Rosén, A.: Competitive non-preemptive call control. In: Proc. of the 5th ACM-SIAM Symp. on Discrete Algorithms (SODA), pp. 312–320 (1994)

[2] Bachmann, U.T., Halldórsson, M.M., Shachnai, H.: Online selection of intervals and t-intervals. Information and Computation 233, 1–11 (2013); Also appeared in Proc. of the 12th SWAT, pp. 383–394 (2010)

[3] Borodin, A., El-Yaniv, R.: Online Computation and Competitive Analysis. Cambridge University Press (1998)

[4] Chrobak, M.: Online aggregation problems. SIGACT News 45(1), 91–102 (2014)

[5] Emek, Y., Halldórsson, M.M., Mansour, Y., Patt-Shamir, B., Radhakrishnan, J., Rawitz, D.: Online set packing. SIAM Journal on Computing 41(4), 728–746 (2012); Also appeared as Online set packing and competitive scheduling of multi-part tasks. In: Proc. of the 29th PODC, pp. 440–449 (2010)

[6] Garay, J.A., Gopal, I.S.: Call preemption in communication networks. In: Proc. of the 11th IEEE Int. Conference on Computer Communications (INFOCOM), pp. 1043–1050 (1992)

[7] Garay, J.A., Gopal, I.S., Kutten, S., Mansour, Y., Yung, M.: Efficient on-line call control algorithms. Journal of Algorithms 23(1), 180–194 (1997)

[8] Garefalakis, T.: A new family of randomized algorithms for list accessing. In: Burkard, R.E., Woeginger, G.J. (eds.) ESA 1997. LNCS, vol. 1284, pp. 200–216. Springer, Heidelberg (1997)

[9] Halldórsson, M.M., Patt-Shamir, B., Rawitz, D.: Online scheduling with interval conflicts. Theory of Computing Systems 53(2), 300–317 (2013); Also appeared in Proc. of the 28th STACS, pp. 472–483 (2011)

[10] Irani, S.: Two results on the list update problem. Information Processing Letters 38(6), 301–306 (1991)

[11] Jaromczyk, J.W., Pezarski, A., Ślusarek, M.: An optimal competitive on-line algorithm for the minimal clique cover problem in interval and circular-arc graphs. In: Proc. of the 19th European Workshop on Computational Geometry, EWCG (2003)

[12] Lipton, R.J., Tomkins, A.: Online interval scheduling. In: Proc. of the 5th ACM-SIAM Symp. on Discrete Algorithms (SODA), pp. 302–311 (1994)

[13] Reingold, N., Westbrook, J., Sleator, D.D.: Randomized competitive algorithms for the list update problem. Algorithmica 11(1), 15–32 (1994)

[14] Sleator, D.D., Tarjan, R.E.: Amortized efficiency of list update and paging rules. Communications of the ACM 28(2), 202–208 (1985)

Online Admission Control
and Embedding of Service Chains[*]

Tamás Lukovszki[1] and Stefan Schmid[2]

[1] Faculty of Informatics, Eötvös Loránd University, Budapest, Hungary
lukovszki@inf.elte.hu
[2] TU Berlin & Telekom Innovation Laboratories, Berlin, Germany
stefan.schmid@tu-berlin.de

Abstract. The virtualization and softwarization of modern computer networks enables the definition and fast deployment of novel network services called *service chains*: sequences of virtualized network functions (e.g., firewalls, caches, traffic optimizers) through which traffic is routed between source and destination. This paper attends to the problem of admitting and embedding a maximum number of service chains, i.e., a maximum number of source-destination pairs which are routed via a sequence of ℓ to-be-allocated, capacitated network functions. We consider an Online variant of this maximum Service Chain Embedding Problem, short *OSCEP*, where requests arrive over time, in a worst-case manner. Our main contribution is a deterministic $O(\log \ell)$-competitive online algorithm, under the assumption that capacities are at least logarithmic in ℓ. We show that this is asymptotically optimal within the class of deterministic and randomized online algorithms. We also explore lower bounds for offline approximation algorithms, and prove that the offline problem is APX-hard for unit capacities and small $\ell \geq 3$, and even Poly-APX-hard in general, when there is no bound on ℓ. These approximation lower bounds may be of independent interest, as they also extend to other problems such as Virtual Circuit Routing. Finally, we present an exact algorithm based on 0-1 programming, implying that the general offline SCEP is in NP and, by the above hardness results, it is NP-complete for constant ℓ.

Keywords: Computer Networks, Network Virtualization, Virtual Circuit Routing, Online Call Admission, Competitive Analysis.

1 Introduction

Today's computer networks provide a rich set of in-network functions, including access control, firewall, intrusion detection, network address translation, traffic shaping and optimization, caching, among many more. While such functionality

---

[*] Supported by the FP7 EU project UNIFY and the DFG SFB 901 project: Part of this research was done when the first author visited the Heinz Nixdorf Institute, Germany.

C. Scheideler (Ed.): SIROCCO 2015, LNCS 9439, pp. 104–118, 2015.
DOI: 10.1007/978-3-319-25258-2_8

is traditionally implemented in hardware middleboxes, computer networks become more and more virtualized [12,24]: *Network Function Virtualization (NFV)* enables a flexible instantiation of network functions on network nodes, e.g., running in a virtual machine on a commodity x86 server.

Modern computer networks also offer new flexibilities in terms of how traffic can be routed through such network functions. In particular, using *Software-Defined Networking (SDN)* [19] technology, traffic can be steered along arbitrary routes, i.e., along routes which depend on the application [13], and which are not necessarily shortest paths or destination-based, or not even loop-free [11].

These trends enable the realization of interesting new in-network communication services called *service chains* [8,14,25,26]: sequences of network functions which are allocated and stitched together in a flexible manner. For example, a service chain c_i could define that traffic originating at source s_i is first steered through an intrusion detection system for security (1st network function), next through a traffic optimizer (2nd network function), and only then is routed towards the destination t_i. Such advanced network services open an interesting new market for Internet Service Providers, which can become "miniature cloud providers" [27], specialized for in-network processing.

1.1 Paper Scope

In this paper, we study the problem of how to optimally admit and embed service chain requests. Given a redundant distribution of network functions and a sequence $\sigma = (\sigma_1, \sigma_2, \ldots, \sigma_k)$, where each $\sigma_i = (s_i, t_i)$ for $i \in [1, k]$ defines a source-destination pair (s_i, t_i) which needs to be routed via a sequence of network function instances, we ask: Which requests σ_i to admit and where to allocate their service chains c_i? The service chain embedding should respect capacity constraints as well as constraints on the length (or stretch) of the route from s_i to t_i via its service chain c_i.

Our objective is to maximize the number of admitted requests. We are particularly interested in the *Online Service Chain Embedding Problem (OSCEP)*, where σ is only revealed over time. We assume that a request cannot be delayed and once admitted, cannot be preempted again. Sometimes, we are also interested in the general (offline) problem, henceforth denoted by SCEP.

1.2 Our Contribution

We formulate the online and offline problems OSCEP and SCEP, and make the following contributions:

1. We present a deterministic online algorithm ACE[1] which, given that node capacities are at least logarithmic, achieves a competitive ratio $O(\log \ell)$ for OSCEP. This result is practically interesting, as the number of to be traversed network functions ℓ is likely to be small in practice. In our analysis, we adapt a proof strategy known from virtual circuit routing [22]. Note however that in contrast to virtual circuit routing, where the end nodes have to

---

[1] **A**dmission control and **C**hain **E**mbedding.

be connected by a path in the network, in the SCEP, the path must traverse a sequence of ℓ nodes, such that the ith node of this sequence hosts network function f_i. Furhermore, in the SCEP, the path length must be bounded by r hops. So far, only heuristic and offline approaches to solve the service chain embedding problem have been considered [6,4,20,26].

2. We prove that ACE is asymptotically optimal in the class of both deterministic and randomized online algorithms, by adapting a proof strategy from virtual circuit routing in [2]. Moreover, we initiate the study of lower bounds for the offline version of our problem, and show that no good approximation algorithms exist, unless $P = NP$: for unit capacities and already small ℓ, the offline problem SCEP is APX-hard. For arbitrary ℓ, the problem can even become Poly-APX-hard. These results also apply to the offline version of classic online call control problems, which to the best of our knowledge have not been studied before.

3. We present a 0-1 program for SCEP, which also shows that SCEP is in NP for constant ℓ and, taking into account our hardness result, that SCEP is NP-complete for constant ℓ. More precisely, if the number of all possible chains that can be constructed over the network function instances is polynomial in the network size n, then the number of variables in the 0-1 program is also polynomial, and thus the problem is in NP. If m_i is the number of instances of network function f_i in the network, $i = 1, ..., \ell$, and $m = \max_i \{m_i\}$, then the size of the 0-1 program is polynomial for $m^\ell = \text{poly}(n)$. For example, this always holds for constant ℓ. When m is constant, then it holds for $\ell = O(\log n)$.

1.3   Outline

This paper is organized as follows. Section 2 introduces our model and puts the model into perspective with respect to classic online optimization problems. Section 3 presents and analyzes the $O(\log \ell)$-competitive algorithm, Section 4 presents our lower bound, and in Section 5 we present the 0-1 linear program. We summarize our results and conclude our work in Section 6.

2   Model

We are given an undirected network $G = (V, E)$ with $n = |V|$ nodes and $m = |E|$ edges. On this graph, we need to route a sequence of requests $\sigma = (\sigma_1, \sigma_2, \ldots, \sigma_k)$: σ_i for any i represents a node pair $\sigma_i = (s_i, t_i) \in V \times V$. Each pair σ_i needs to be routed (from s_i to t_i) via a sequence of ℓ network functions $(F_1, \ldots, F_\ell)$. For each network function type F_i, there exist multiple instantiations $f_i^{(1)}, f_i^{(2)}, \ldots$ in the network. (We will omit the superscript if it is irrelevant or clear in the context.) Each of these instances can be applied to σ_i along the route from s_i to t_i. However, in order to minimize the detour via these functions and in order to keep the route from s_i to t_i short, a "nearby instance" $f_i^{(j)}$ should be chosen, for each i. A service chain instance for (s_i, t_i) is denoted by $c_i = (f_1^{(x_1)}, f_2^{(x_2)}, \ldots, f_\ell^{(x_\ell)})$, for some function instances $f_j^{(x_y)}$, $j \in [1, \ell]$.

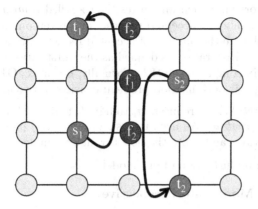

Fig. 1. Illustration of the model: The communication from s_1 to t_1 and from s_2 to t_2 needs to be routed via a service chain (F_1, F_2). In this example, function F_1 is instantiated once, and function F_2 is instantiated twice. Resources for (s_1, t_1) are allocated only at the second instance of F_2 (the upper one).

For ease of presentation, we will initially assume that requests σ_i are of infinite duration. We will later show how to generalize our results to scenarios where requests can have arbitrary and unknown durations.

Concretely, in order to satisfy a request $\sigma_i = (s_i, t_i)$, a route of the following form must be computed:

1. The route must start at s_i, traverse a sequence of network functions $(f_1^{(x_1)}, f_2^{(x_2)}, \ldots, f_\ell^{(x_\ell)})$, and end at t_i. Here, $f_j^{(x_y)}$, $j \in [1, \ell]$ is an instance of the network function of type F_j.
2. The route must not violate capacity constraints on any node $v \in V$. Nodes $v \in V$ are capacitated and resources need to be allocated for each network function which is used, for any (s_i, t_i) pair. Multiple network functions may be available on the same physical machine, and only consume resources once they are used in certain service chains. The capacity $\kappa(v)$ of each node $v \in V$ hence defines the maximum number of requests σ_i for which v can apply its network functions. However, node v can always simply serve as a regular forwarding node for other requests, without applying the function.
3. The route should be of (hop) length at most r (or have a bounded stretch).

Otherwise, a request σ_i must be rejected. For ease of notation, in the following, we will sometimes assume that for a rejected request σ_i, $c_i = \emptyset$. Also note that the resulting route may not form a simple path, but more generally describes a *walk*: it may contain forwarding loops (e.g., visit a network function and come back).

Our objective is to maximize the number of satisfied requests σ_i, resp. to embed a maximum number of service chains. We are mainly interested in the online variant of the problem, where σ is revealed over time. More precisely, and as usual in the realm of online algorithms and competitive analysis, we seek to devise an online algorithm which minimizes the so-called *competitive ratio*: Let $\mathbf{ON}(\sigma)$ denote the number of accepted requests of a given online algorithm for σ and let $\mathbf{OFF}(\sigma)$ denote the number of accepted requests of an optimal offline algorithm. The competitive ratio ρ is defined as the worst ratio (over all possible σ) of the value of $\mathbf{ON}$ compared to $\mathbf{OFF}$. Formally, $\rho = \max_\sigma \mathbf{OFF}(\sigma)/\mathbf{ON}(\sigma)$.

Note that solving this optimization problem consists of two subtasks:

1. *Admission control:* Which requests σ_i to admit, and which to reject?
2. *Assignment and routing:* We need to assign $\sigma_i = (s_i, t_i)$ pairs to a sequence of network functions and route the flow through them accordingly.

See Figure 2 for an illustration of our model.

2.1   Putting the Model into Perspective

From an algorithmic perspective, the models closest to ours occur in the context of online call admission respectively virtual circuit routing. There, the fundamental problem is to decide, in an online manner, which "calls" resp. "virtual circuits" or entire networks, to admit and how to route them, in a link-capacitated graph. [2,3,9,10,22]

Instead of routes, in our model, service functions have to be allocated and connected to form service chains. In particular, in our model, nodes have a limited capacity and can only serve as network functions for a bounded number of source-destination pairs. The actual routes taken in the network play a secondary role, and may even contain loops. In particular, our model supports the specification of explicit constraints on the length of a route, but also on the stretch: the factor by which the length of a route from a source to a destination can be increased due to the need to visit certain network functions.

Nevertheless, as this paper shows, several techniques from classic literature on online call control can be applied to our model. At the same time, to the best of our knowledge, some of our results also provide new insights into the classic variants of call admission control. For example, our lower bounds on the approximation ratio also translate to classic problems, which so far have mainly been studied from an online perspective.

3   Competitive Online Algorithm

We present an online algorithm ACE for OSCEP. ACE admits and embeds at least a $\Omega(\log \ell)$-fraction of the number of requests embedded by an optimal offline algorithm $\mathbf{OFF}$.

Let us first introduce some notation. Let A_j be the set of indices of the requests admitted by ACE just *before* considering the jth request σ_j. The index set of all admitted requests after processing all k requests in σ, will be denoted by A_{k+1} resp. A.

The relative load $\lambda_v(j)$ at node v before processing the jth request, is defined by the number of service chains c_i in which v participates, divided by v's capacity:

$$\lambda_v(j) = \frac{|\{c_i \; : \; i \in A_j, v \in c_i\}|}{\kappa(v)}.$$

We seek to ensure the invariant that capacity constraints are enforced at each node, i.e., $\forall\, v \in V, j \leq k+1 : \lambda_v(j) \leq 1$.

We define $\mu = 2\ell + 2$, and in the following, will assume that

$$\min_{v}\{\kappa(v)\} \geq \log \mu \tag{1}$$

3.1 Algorithm

In a preprocessing step we compute the length $d(u, v)$ of the shortest path between all pairs of nodes $u, v \in V$ in the network G. Then we compute the set of all possible chains $\mathcal{C}$ that can be constructed from the network function instances $\mathcal{C} = \{c = (f_1, ..., f_\ell) \ : \ \sum_{i=2}^{\ell} d(f_{i-1}, f_i) \leq r, \ f_1 \in F_1, ..., f_\ell \in F_\ell\}$. For a request $\sigma_j = (s_j, t_j)$, let C_j be the set of chains, such that σ_j can be routed through the chains $c \in C_j$ on a path of length at most r, i.e. $C_j = \{c = (f_1, ..., f_\ell) \in \mathcal{C} \ : \ d(s_j, f_1) + d(f_\ell, t_j) + \sum_{i=2}^{\ell} d(f_{i-1}, f_i) \leq r\}$.

The key idea of ACE is to assign to each node, a cost which is exponential in the relative node load. More precisely, with each node we associate a cost $w_v(j)$ just before processing the jth request σ_j:

$$w_v(j) = \kappa(v)(\mu^{\lambda_v(j)} - 1).$$

Our online algorithm ACE simply proceeds as follows:

- When request σ_j arrives, ACE checks if there exists a chain $c_j \in C_j$ satisfying the following condition:

$$\sum_{v \in c_j} \frac{w_v(j)}{\kappa(v)} \leq \ell \tag{2}$$

- If such a chain c_j exists, then *admit* σ_j and assign it to c_j. Otherwise, reject σ_j.

In order to ensure that chains selected for Condition 2 also fulfill the constraint on the maximal route length, ACE simply uses preprocessing. We maintain at each node its relative load. When a new request arrives, ACE has to test the costs of at most $O(n^\ell)$ chains, and the cost can be computed in $O(\ell)$ time per chain. The overall runtime of ACE per step is hence bounded by $O(\ell \cdot n^\ell)$, which is polynomial for constant ℓ.

3.2 Analysis

For the analysis of ACE, we adapt the proof strategy used in [22] in the context of virtual circuit routing. First, in Lemma 1 we prove that the set A of requests admitted by ACE are feasible and respect capacity constraints. Second, in Lemma 2, we show that at any moment in time, the sum of node costs is within a factor $O(\ell \cdot \log \mu)$ of the number of requests already admitted by ACE.

Third, in Lemma 3, we prove that the number of requests admitted by the optimal offline algorithm **OFF** but rejected by the online algorithm, is bounded by the sum of node costs after processing all requests.

Let W be the sum of the node costs after ACE processed all k request, let $A_{\mathbf{OFF}}$ be the indices of the requests admitted by **OFF**, and let $A^* = A_{\mathbf{OFF}} \setminus A$. Then, from Lemma 2 we will obtain a bound $|A| \geq W/(2\ell \cdot \log \mu)$, and from Lemma 3 that $|A^*| \leq W/\ell$.

Thus, even by conservatively ignoring all the requests which ACE might have admitted which **OFF** did not, we obtain that the competitive ratio of ACE is at most $O(\log \ell)$.

Let us now have a closer look at the first helper lemma.

Lemma 1. *For all nodes $v \in V$:*

$$\sum_{j \in A : v \in c_j} 1 \leq \kappa(v).$$

Proof. Let σ_j be the first request admitted by ACE, such that the relative load $\lambda_v(j+1)$ at some node $v \in c_j$ exceeds 1. By definition of the relative load we have $\lambda_v(j) > 1 - 1/\kappa(v)$.

By the assumption that $\log \mu \leq \kappa(v)$, we get

$$\frac{w_v(j)}{\kappa(v)} = \mu^{\lambda_v(j)} - 1 > \mu^{1 - 1/\log \mu} - 1 = \mu/2 - 1 = \ell.$$

Therefore, by Condition (2), the request σ_j could not be assigned to c_j. We established a contradiction.   □

Next we show that the sum of node costs is within an $O(\ell \cdot \log \mu)$ factor of the number of already admitted requests.

Lemma 2. *Let A be the set of indices of requests admitted by the online algorithm. Let k be the index of the last request. Then*

$$(2\ell \log \mu)|A| \geq \sum_v w_v(k+1).$$

Proof. We show the claim by induction on k. For $k = 0$, both sides of the inequality are zero, thus the claim is trivially true. Rejected requests do not change either side of the inequality. Thus, it is enough to show that, for each $j \leq k$, if we admit σ_j, we get:

$$\sum_v (w_v(j+1) - w_v(j)) \leq 2\ell \log \mu.$$

Consider a node $v \in c_j$. Then by definition of the costs:

$$\begin{aligned}
w_v(j+1) - w_v(j) &= \kappa(v)(\mu^{\lambda_v(j)+1/\kappa(v)} - \mu^{\lambda_v(j)}) \\
&= \kappa(v)(\mu^{\lambda_v(j)}(\mu^{1/\kappa(v)} - 1)) \\
&= \kappa(v)(\mu^{\lambda_v(j)}(2^{(\log \mu) \cdot 1/\kappa(v)} - 1))
\end{aligned}$$

By Assumption (1), $1 \leq \kappa(v)/\log \mu$. Since $2^x - 1 \leq x$, for $0 \leq x \leq 1$, it follows:

$$w_v(j+1) - w_v(j) \leq \mu^{\lambda_v(j)} \log \mu = \log \mu (w_v(j)/\kappa(v) + 1).$$

Summing up over all the nodes and using the fact that the request σ_j was admitted and chain c_j was assigned, and that the number of nodes $|c_j|$ in c_j is ℓ, we get:

$$\sum_v (w_v(j+1) - w_v(j)) \leq \log \mu (\ell + |c_j|) = 2\ell \log \mu.$$

This proves the claim. □

We finally prove that ℓ times the number of requests rejected by ACE but admitted by the optimal offline algorithm **OFF** is bounded by the sum of node costs after processing all requests.

Lemma 3. *Let $A_{\mathbf{OFF}}$ be the set of indices of the requests that were admitted by the optimal offline algorithm, and let $A^* = A_{\mathbf{OFF}} \setminus A$ be the set of indices of requests admitted by $A_{\mathbf{OFF}}$ but rejected by the online algorithm. Then:*

$$|A^*| \cdot \ell \leq \sum_v w_v(k+1).$$

Proof. For $j \in A^*$, let c_j^* be the chain assigned to request σ_j by the optimal offline algorithm. By the fact that σ_j was rejected by the online algorithm, we have:

$$\ell < \sum_{v \in c_j^*} \frac{w_v(j)}{\kappa(v)}.$$

Since the costs $w_v(j)$ are monotonically increasing in j, we have

$$\ell < \sum_{v \in c_j^*} \frac{w_v(j)}{\kappa(v)} \leq \sum_{v \in c_j^*} \frac{w_v(k+1)}{\kappa(v)}.$$

Summing over all $j \in A^*$, we get

$$|A^*|\ell \leq \sum_{j \in A^*} \sum_{v \in c_j^*} \frac{w_v(k+1)}{\kappa(v)} \leq \sum_v w_v(k+1) \cdot \sum_{j \in A^*: v \in c_j^*} \frac{1}{\kappa(v)} \leq \sum_v w_v(k+1).$$

The last inequality follows from the fact that capacity constraints need to be met at any time. □

Theorem 1. ACE *is $O(\log \ell)$-competitive.*

Proof. By Lemma 1, capacity constraints are never violated. It remains to show that the number of requests admitted by the online algorithm is at least $1/(2\log 2\mu)$ times the number of requests admitted by the optimal offline algorithm. The number of requests admitted by the optimal offline algorithm $|A_{\mathbf{OFF}}|$

can be bounded by the number of requests admitted by the online algorithm $|A|$ plus the number of requests in $A^* = A_{\mathbf{OFF}} \setminus A$. Therefore,

$$|A_{\mathbf{OFF}}| \leq |A| + |A^*|.$$

By Lemma 3 this is bounded by

$$|A_{\mathbf{OFF}}| \leq |A| + \frac{1}{\ell} \sum_v w_v(k+1).$$

By Lemma 2 this is bounded by

$$|A_{\mathbf{OFF}}| \leq |A| + 2 \cdot (\log \mu) \cdot |A| = (1 + 2 \log \mu)|A|$$

Therefore, the number of requests admitted by the optimal offline algorithm is at most $(1 + 2 \log \mu)$ times the number of requests admitted by ACE. $\square$

Remarks. We conclude with some remarks. First, we note that our approach leaves us with many flexibilities in terms of constraining the routes through the network functions. For instance, we can support maximal path length requirements: the maximal length of the route from s to t *via the network functions*. A natural alternative model is to define a limit on the *stretch*: the factor by which the "detour" via the network functions can be longer than the shortest path from s to t. Moreover, so far, we focused on a model where requests, once admitted, stay forever. Our approach can also be used to support service chain requests of bounded or even unknown duration. In particular, by redefining μ to take into account the duration of a request, we can for example apply the technique from [22] to obtain competitive ratios for more general models.

4   Optimality and Approximation

It turns out that ACE is asymptotically optimal within the class of online algorithms (Theorem 2). This section also initiates the study of lower bounds for (offline) approximation algorithms, and shows that for low capacities, the problem is APX-hard even for short chains (Theorem 3), and even Poly-APX-hard in general, that is, it is as hard as any problem that can be approximated to a polynomial factor in polynomial time (Theorem 4).

Theorem 2. *Any deterministic or randomized online algorithm for OSCEP must have a competitive ratio of at least $\Omega(\log \ell)$.*

Proof. We can adapt the proof strategy of Lemma 4.1 in [2] for our model. We consider a capacity of $\kappa \geq \log \ell$, and we divide the requests in σ into $\log \ell + 1$ phases. We assume that $n \geq 2\ell^2$, and only focus on a subset L of $\ell = |L|$ nodes which are connected as a chain $(v_1, \ldots, v_\ell)$ and at which the different service chains will overlap. In phase 0, a group of κ service chains are requested, all of which need to be embedded across the nodes $L = \{v_1, \ldots, v_\ell\}$. In phases $i \geq 1$, 2^i groups of κ identical requests will need to share subsets of L of size $\ell/2^i$, that

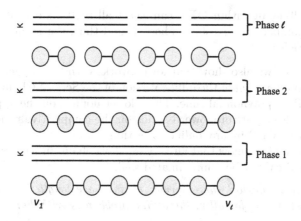

Fig. 2. Illustration of lower bound construction: The adversary issues service chain requests in $1+\log\ell$ phases, where each phase i consists of 2^i groups of $\kappa \geq \log\ell$ requests. In phase 0 the adversary issues requests that can be assigned to $L = (v_1, ..., v_\ell)$. As intersections of chains in phase i with L are becoming shorter over time, the online algorithm needs to decide whether to admit service service chain requests in phases, where each phase consists of groups with κ chains. As chains are becoming shorter over time, the online algorithm faces the problem whether to admit service chains early (and hence block precious resources), or late (in which case the adversary stops issuing new requests).

is, the jth group, $0 \leq j < 2^i$, consists of κ requests to be embedded across nodes $[v_{j\ell/2^i+1}, v_{(j+1)\ell/2^i}]$. See Figure 2 for an illustration.

Let x_i denote the number of requests an online algorithm **ON** admits in phase i. Each request accepted in phase i will occupy $\ell/2^i$ units of capacities of nodes in L. Overall, the nodes in L have a capacity of $\ell \cdot \kappa$, so it must hold that

$$\sum_{i=0}^{\log\ell} \frac{\ell}{2^i} \cdot x_i \leq \ell \cdot \kappa.$$

Now, for $0 \leq j \leq \log\ell$, define $S_j = \frac{\ell}{2^j} \cdot \sum_{i=0}^{j} x_i$. S_j is a lower bound on the occupied capacity on the nodes of L after phase j. Then:

$$\sum_{j=0}^{\log\ell} S_j = \sum_{j=0}^{\log\ell} \frac{\ell}{2^j} \sum_{i=0}^{j} x_i = \sum_{i=0}^{\log\ell} x_i \sum_{j=i}^{\log\ell} \frac{\ell}{2^j} \leq \sum_{i=0}^{\log\ell} x_i 2\frac{\ell}{2^i} = 2\ell\kappa.$$

Hence there must exist a j such that $S_j \leq 2\ell\kappa/\log\ell$. Then after phase j, the number of requests admitted by the online algorithm **ON** is

$$\sum_{i=0}^{j} x_i = \frac{2^j}{\ell} S_j \leq \frac{2^j}{\ell} 2\ell\kappa/\log\ell = 2 \cdot 2^j \kappa/\log\ell.$$

The optimal offline algorithm **OFF** can reject all requests except for those of phase j. The number of requests in phase j, and thus, the number of requests admitted by **OFF** is $2^j \kappa$.   □

In the following, we also show that for networks with low capacities, it is not even possible to *approximate* the offline version of the Service Chain Embedding Problem, SCEP, in polynomial time. These lower bounds on the approximation ratio naturally also constitute lower bounds on the competitive ratio which can be achieved for OSCEP by any online algorithm.

In particular, we first show that already for short chains in scenarios with unit capacities, SCEP cannot be approximated well.

Theorem 3. *In scenarios where service chains have length $\ell \geq 3$ and where capacities are $\kappa(v) = 1$, for all v, the offline problem is APX-hard.*

Proof. The proof follows from an approximation-preserving reduction from *Maximum k-Set Packing Problem (KSP)*. The *Maximum Set Packing (SP)* is one of Karp's 21 NP-complete problems, where for a given collection C of finite sets a collection of disjoint sets $C' \subseteq C$ of maximum cardinality has to be found. The KSP is the variation of the SP in which the cardinality of all sets in C are bounded from above by any constant $k \geq 3$, is APX-complete [15]. We refer to such sets as k-sets.

KSP can be reduced to our problem as follows. Let U be the universe and C be a collection of k-sets of U in the KSP. W.l.o.g., we assume that each k-set contains exactly k elements, otherwise we can add disjoint auxiliary elements to the sets in order to obtain exactly k elements in each set in C. For each $u \in U$ in the KSP instance we construct a node v_u in the SCEP instance. Furthermore, for each k-set S in C, we construct a service chain c_S, such that c_S contains exactly the nodes $\{v_u : u \in S\}$. Let C be the set of obtained service chains. For the set of requests σ we require that $|\sigma| \geq |C|$ and that each request can be assigned to each service chain. Due to the unit capacity assumption, the set of admitted request must be assigned to mutually disjoint service chains. Thus, the maximum number of admitted requests is at most the maximum number of disjoint service chains. Since each request can be assigned to each service chain and $|\sigma| \geq |C|$, an optimal solution for the SCEP determines a maximum set of mutually disjoint service chains. This maximum set of disjoint service chains determines a maximum number of disjoint k-sets, and thus, an optimal solution for the KSP.   □

It turns out that in general, with unit capacities, SCEP cannot even be approximated within polylogarithmic factors.

Theorem 4. *In general scenarios where capacities are $\kappa(v) = 1$, for all nodes v, and chain lengths $\ell \geq 3$, the SCEP is APX-hard, and not approximable within ℓ^ε for some $\varepsilon > 0$. Without a bound on the chain length the SCEP with $\kappa(v) = 1$, for all nodes v, is Poly-APX-hard.*

Proof. We reduce the *Maximum Independent Set (MIS)* problem with maximum degree ℓ to the SCEP with capacity $\kappa(v) = 1$, for all $v \in V$ and chain length ℓ.

For graphs with bounded degree $\ell \geq 3$, the MIS is APX-complete [21] and cannot be approximated within ℓ^ε for some $\varepsilon > 0$ [1]. By our reduction we obtain the APX-hardness and non-approximability within ℓ^ε for some $\varepsilon > 0$ for the SCEP. In general, for graphs without degree bound, the MIS is Poly-APX-complete [5], i.e., it is as hard as any problem that can be approximated to a polynomial factor. By our reduction we obtain that the SCEP without chain length bound is Poly-APX-hard.

For an instance $G = (V, E)$ of the MIS problem with maximum degree ℓ, we construct an instance of the SCEP with capacity $\kappa = 1$ and chain length ℓ as follows. For each node $v \in G$, let c_v be the chain whose nodes correspond to the edges in G incident to v. If $\deg_G(v) < \ell$ then we complete the chain with $\ell - \deg_G(v)$ unique auxiliary nodes, in order to have ℓ nodes in the chain. The chain set is $C = \{c_v \ : \ v \in G\}$. For the set of requests σ, we require that $|\sigma| \geq |C|$ and each request $\sigma_i \in \sigma$ can be assigned to each $c \in C$. Assigning a σ_i to a chain $c \in C$ fills the capacity of all nodes in c and the capacity of all chains $c' \in C$ that contain a common node with c. Therefore, no further request σ_j, $j \neq i$, can be assigned to those chains. The chains having a common node with c_v correspond exactly the neighbors of v in G. Therefore, nodes u and v are independent in the MIS instance iff chains c_u and c_v do not have a common node in the SCEP instance. Since each request σ_i can be assigned to each $c \in C$ and $|\sigma| \geq |C|$, a maximum number of admitted requests is determined by a maximum chain set C', such that for all $c_u, c_v \in C'$, c_u and c_v do not contain a common node. Therefore, C' determines a maximum independent set in G. Consequently, an α-approximation for the SCEP would imply an α-approximation for the MIS problem. $\qquad\square$

5   Optimal 0-1 Program and NP-Completeness

SCEP can be formulated as a 0-1 integer linear program. If the number of all possible chains that can be constructed over the network function instances is polynomial in the network size, then the number of variables in the 0-1 program is also polynomial, and thus the problem is in NP. 0-1 integer linear programming is one of Karp's NP-complete problems [17]. This together with our hardness results also proves NP-completeness for constant ℓ.

Let $\sigma = \{\sigma_i = (s_i, t_i) : s_i, t_i \in V\}$ be the set of requests, and let $\mathcal{C}$ be the set of possible chains over the network function instances, respecting route length constraints. We refer by $c \in \mathcal{C}$ to a potential chain. For all potential chains $c \in \mathcal{C}$, let S_c be the set of connection requests in σ that can be routed through c on a path of length at most r, i.e., for $c = (v_1, ..., v_\ell)$, let $S_c = \{\sigma_i = (s_i, t_i) \in \sigma : d(s_i, v_1) + \sum_{i=2}^{k} d(v_{i-1}, v_i) + d(v_k, t_i) \leq r\}$, where $d(u, v)$ denotes the length of the shortest path between nodes $u, v \in V$ in the network G. The shortest paths between nodes can be computed in a preprocessing step.

For all connection requests $\sigma_i \in \sigma$, we introduce the binary variable $x_i \in \{0, 1\}$. The variable $x_i = 1$ indicates that the request i is admitted in the solution. For all potential network function chains $c \in \mathcal{C}$, we introduce the binary variable

$x_c \in \{0, 1\}$. The variable $x_c = 1$ indicates that c is selected in the solution. For all $c \in \mathcal{C}$ and $\sigma_i \in \sigma$, we introduce the binary variable $x_{c,i} \in \{0, 1\}$. The variable $x_{c,i}$ indicates that the request $\sigma_i = (s_i, t_i) \in \sigma$ is routed through the nodes of c, such that the length of the walk from s_i to t_i through c has length at most r.

$$\text{maximize} \quad \sum_{\sigma_i \in \sigma} x_i \tag{3}$$

$$\text{s.t.} \quad x_i - \sum_{c \in \mathcal{C}} x_{c,i} = 0 \quad \forall\, \sigma_i \in \sigma \tag{4}$$

$$\sum_{c \in \mathcal{C}:\sigma_i \notin S_c} x_{c,i} = 0 \quad \forall\, \sigma_i \in \sigma \tag{5}$$

$$x_c \leq x_v \quad \forall\, v \in V, \forall\, c \in \mathcal{C} : v \in c \tag{6}$$

$$\sum_{c \in \mathcal{C}:v \in c} x_c \geq x_v \quad \forall\, v \in V \tag{7}$$

$$\sum_{\sigma_i \in \sigma} \sum_{c \in \mathcal{C}:v \in c} x_{c,i} \leq \kappa(v) \cdot x_v \quad \forall\, v \in V \tag{8}$$

$$x_i, x_v, x_c, x_{c,i} \in \{0, 1\} \quad \forall\, v \in V, \forall\, c \in \mathcal{C}, \forall\, \sigma_i \in \sigma \tag{9}$$

The objective function (3) asks for admitting a request set of maximum cardinality. The Constraints (4) enforce that each admitted request $\sigma_i \in \sigma$ is assigned to exactly one chain $c \in \mathcal{C}$, and rejected requests are not assigned to any chain, i.e., for each σ_i with $x_i = 1$, there is exactly one chain c with $x_{c,i} = 1$, and for each i with $x_i = 0$, we have $x_{c,i} = 0$ for all c. Constraints (5) state that each $\sigma_i \in \sigma$ can only be assigned to a chain $c \in \mathcal{C}$ with $\sigma_i \in S_c$. By definition of S_c, the nodes s_i and t_i can be routed through c by a path of length at most r. Constraints (6) ensure that if a node $v \in V$ is contained in a selected chain c (i.e., $x_c = 1$), then $x_v = 1$. Constraints (7) enforce that if a node $v \in V$ is not contained in any selected chain, i.e., $x_c = 0$ for all chains c with $v \in c$, then $x_v = 0$. Therefore, Constraints (6) and (7) together imply that $x_v = 1$ iff v is contained in a selected chain c. Constraints (8) describe that the number of requests routed through a node v of a selected chain is limited by the capacity $\kappa(v)$ of v. Furthermore, (8) ensures that if v is not contained in any selected chain (i.e., $x_v = 0$) then no request q is assigned to any chain c with $v \in c$.

The solution of this 0-1 program defines a maximum cardinality set of admitted requests $\sigma_{admit} = \{\sigma_i : x_i = 1\}$, and an assignment of each request $\sigma_i \in \sigma_{admit}$ to a chain $c \in \mathcal{C}$. Each request $\sigma_i \in \sigma_{admit}$ is assigned to a chain $c \in \mathcal{C}$ iff $x_{c,i} = 1$. This assignment guarantees that (i) the request $\sigma_i = (s_i, t_i)$ can be routed through c on a path of length at most r, (ii) the number of pairs routed through any node $v \in V$ of a selected chain is limited by the capacity $\kappa(v)$ of v, and (iii) none of the requests $\sigma_i \in \sigma_{admit}$ are assigned to a non selected chain. Furthermore, it is guaranteed that rejected requests $\sigma_i \in \sigma \setminus \sigma_{admit}$ are not assigned to any chain.

6 Summary and Conclusion

Over the last decades, a large number of middleboxes have been deployed in computer networks, to increase security and application performance, as well as to offer new services in the form of static and dynamic in-network processing (see the services by Akamai, Google Global Cache, Netflix Open Connect). However, the increasing cost and inflexibility of hardware middleboxes (slow deployment, complex upgrades, lack of scalability), motivated the advent of Network Function Virtualization (NFV) [7,12,16,18,23], which aims to run the functionality provided by middleboxes as software on commodity hardware. The transition to NFV is discussed within standardization groups such as ETSI, and we currently also witness first deployments, e.g., TeraStream [28]. Especially the possibility to chain individual network functions to form more complex services has recently attracted much interest, both in academia [20,26], as well as in industry [25].

Our paper made a first step towards a better understanding of the algorithmic problem underlying the embedding of service chains. Our main contribution is a deterministic and asymptotically optimal online algorithm ACE which achieves a competitive ratio of $O(\log \ell)$ for OSCEP. This is an encouraging result, as the number ℓ of to-be-chained network functions is likely to be a small constant in practice.

References

1. Alon, N., Feige, U., Wigderson, A., Zuckerman, D.: Derandomized graph products. Computational Complexity 5, 60–75 (1995)
2. Awerbuch, B., Azar, Y., Plotkin, S.A.: Throughput-competitive on-line routing. In: Proc. 34th Annual Symposium on Foundations of Computer Science (FOCS), pp. 32–40 (1993)
3. Awerbuch, B., Azar, Y., Plotkin, S.A., Waarts, O.: Competitive routing of virtual circuits with unknown duration. In: Proc. 5th Annual ACM-SIAM Symposium on Discrete Algorithms (SODA), pp. 321–327 (1994)
4. Bari, F., Chowdhury, S.R., Ahmed, R., Boutaba, R.: On orchestrating virtual network functions in NFV. CoRR (2015)
5. Bazgan, C., Escoffier, B., Paschos, V.T.: Completeness in standard and differential approximation classes: Poly-(d)apx- and (d)ptas-completeness. Theoretical Computer Science 339(2-3), 272–292 (2005)
6. Dietrich, D., Abujoda, A., Papadimitriou, P.: Network Service Embedding Across Multiple Providers with Nestor. In: Proc. IFIP Networking (2015)
7. Dobrescu, M., Egi, N., Argyraki, K., Chun, B.G., Fall, K., Iannaccone, G., Knies, A., Manesh, M., Ratnasamy, S.: Routebricks: Exploiting parallelism to scale software routers. In: Proc. ACM SOSP, pp. 15–28 (2009)
8. ETSI: Network functions virtualisation (nfv); use cases (2014), http://www.etsi.org/deliver/etsi_gs/NFV/001_099/001/01.01.01_60/ gs_NFV001v010101p.pdf
9. Even, G., Medina, M.: A nonmonotone analysis with the primal-dual approach: Online routing of virtual circuits with unknown durations. In: Moscibroda, T., Rescigno, A.A. (eds.) SIROCCO 2013. LNCS, vol. 8179, pp. 104–115. Springer, Heidelberg (2013)

10. Even, G., Medina, M., Schaffrath, G., Schmid, S.: Competitive and deterministic embeddings of virtual networks. Elsevier Theoretical Computer Science (TCS) (2013)
11. Fayazbakhsh, S., et al.: Flowtags: Enforcing network-wide policies in the presence of dynamic middlebox actions. In: Proc. ACM HotSDN (2013)
12. Gember-Jacobson, A., et al.: OpenNF: Enabling innovation in network function control. In: Proc. ACM SIGCOMM (2014)
13. Gupta, A., Vanbever, L., Shahbaz, M., Donovan, S.P., Schlinker, B., Feamster, N., Rexford, J., Shenker, S., Clark, R., Katz-Bassett, E.: Sdx: A software defined internet exchange. In: Proc. ACM SIGCOMM, pp. 551–562 (2014)
14. Hartert, R., et al.: Declarative and expressive approach to control forwarding paths in carrier-grade networks. In: Proc. ACM SIGCOMM (2015)
15. Hazan, E., Safra, S., Schwartz, O.: On the complexity of approximating k-set packing. Comput. Complex. 15(1), 20–39 (2006)
16. Joseph, D., Stoica, I.: Modeling middleboxes. IEEE Network: The Magazine of Global Internetworking 22(5), 20–25 (2008)
17. Karp, R.M.: Reducibility among combinatorial problems. In: Complexity of Computer Computations (1972)
18. Martins, J., Ahmed, M., Raiciu, C., Huici, F.: Enabling fast, dynamic network processing with clickos. In: Proc. HotSDN, pp. 67–72 (2013)
19. McKeown, N., Anderson, T., Balakrishnan, H., Parulkar, G., Peterson, L., Rexford, J., Shenker, S., Turner, J.: Openflow: Enabling innovation in campus networks. SIGCOMM Comput. Commun. Rev. 38(2), 69–74 (2008)
20. Mehraghdam, S., Keller, M., Karl, H.: Specifying and placing chains of virtual network functions. In: Proc. 3rd IEEE International Conference on Cloud Networking (CloudNet), pp. 7–13 (2014)
21. Papadimitriou, C.H., Yannakakis, M.: Optimization, approximation, and complexity classes. J. Comput. System Sci. 43, 425–440 (1991)
22. Plotkin, S.A.: Competitive routing of virtual circuits in ATM networks. IEEE Journal on Selected Areas in Communications 13(6), 1128–1136 (1995)
23. Schulz-Zander, J., et al.: OpenSDWN: Programmatic control over home and enterprise WiFi. In: ACM Sigcomm Symposium on SDN Research, SOSR (2015)
24. Sekar, V., Ratnasamy, S., Reiter, M.K., Egi, N., Shi, G.: The middlebox manifesto: Enabling innovation in middlebox deployment. In: Proc. HotNets, pp. 21:1–21:6 (2011)
25. Skoldstrom, P., et al.: Towards unified programmability of cloud and carrier infrastructure. In: Proc. European Workshop on Software Defined Networking, EWSDN (2014)
26. Soulé, R., Basu, S., Marandi, P.J., Pedone, F., Kleinberg, R., Sirer, E.G., Foster, N.: Merlin: A language for provisioning network resources. In: Proc. 10th ACM International on Conference on Emerging Networking Experiments and Technologies (CoNEXT), pp. 213–226 (2014)
27. Stoenescu, R., Popovici, M., Olteanu, V., Martins, J., Bifulco, R., Huici, F., Ahmed, M., Smaragdakis, G., Handley, M., Raiciu, C.: In-net: Enabling in-network processing for the masses. In: Proc. ACM EuroSys (2015)
28. Telekom, D.: Terastream (2013), http://www.a10networks.com/resources/files/A10-CS-80103-EN.pdf#search=

Optimizing Spread of Influence
in Social Networks via Partial Incentives

Gennaro Cordasco[2], Luisa Gargano[1], Adele A. Rescigno[1], and Ugo Vaccaro[1]

[1] Department of Informatics, University of Salerno, Italy,
[2] Department of Psychology, Second University of Naples, Italy

Abstract. A widely studied process of influence diffusion in social networks posits that the dynamics of influence diffusion evolves as follows: Given a graph $G = (V, E)$, representing the network, initially *only* the members of a given $S \subseteq V$ are influenced; subsequently, at each round, the set of influenced nodes is augmented by all the nodes in the network that have a sufficiently large number of already influenced neighbors. The general problem is to find a small initial set of nodes that influences the whole network. In this paper we extend the previously described basic model in the following ways: firstly, we assume that there are non negative values $c(v)$ associated to each node $v \in V$, measuring how much it costs to initially influence node v, and the algorithmic problem is to find a set of nodes of *minimum total cost* that influences the whole network; successively, we study the consequences of giving *incentives* to member of the networks, and we quantify how this affects (i.e., reduces) the total costs of starting an influence diffusion process that influence the whole network. For the two above problems we provide both hardness results and algorithms. We also experimentally validate our algorithms via extensive simulations on real life networks.

1   Introduction

Social influence is the process by which individuals adjust their opinions, revise their beliefs, or change their behaviors as a result of interactions with other people. It has not escaped the attention of advertisers that the natural human tendency to conform can be exploited in *viral marketing* [24]. Viral marketing refers to the spread of information about products and behaviors, and their adoption by people. For what strictly concerns us, the intent of maximizing the spread of viral information across a network naturally suggests many interesting optimization problems. Some of them were first articulated in the seminal papers [22, 23], under various adoption paradigms. The recent monograph [8] contains an excellent description of the area. In the next section, we will explain and motivate our model of information diffusion, state the problems that we plan to investigate, describe our results, and discuss how they relate to the existing literature.

© Springer International Publishing Switzerland 2015
C. Scheideler (Ed.): SIROCCO 2015, LNCS 9439, pp. 119–134, 2015.
DOI: 10.1007/978-3-319-25258-2_9

1.1   The Model

Let $G = (V, E)$ be a graph modeling a social network. We denote by $\Gamma_G(v)$ and by $d_G(v) = |\Gamma_G(v)|$, respectively, the neighborhood and the degree of vertex v in G. Let $S \subseteq V$, and let $t : V \to \mathbb{N} = \{1, 2, \ldots\}$ be a function assigning integer thresholds to the vertices of G; we assume w.l.o.g. that $1 \leq t(u) \leq d(u)$ holds for all $v \in V$. For each node $v \in V$, the value $t(v)$ quantifies how hard it is to influence node v, in the sense that easy-to-influence elements of the network have "low" $t(\cdot)$ values, and hard-to-influence elements have "high" $t(\cdot)$ values [21]. An *activation process in G starting at $S \subseteq V$* is a sequence

$\mathsf{Active}_G[S, 0] \subseteq \mathsf{Active}_G[S, 1] \subseteq \ldots \subseteq \mathsf{Active}_G[S, \ell] \subseteq \ldots \subseteq V$

of vertex subsets[1], with $\mathsf{Active}_G[S, 0] = S$, and such that for all $\ell > 0$,

$\mathsf{Active}_G[S, \ell] = \mathsf{Active}_G[S, \ell - 1] \cup \Big\{ u \, : \, \big|\Gamma_G(u) \cap \mathsf{Active}_G[S, \ell - 1]\big| \geq t(u) \Big\}.$

In words, at each round ℓ the set of active (i.e, influenced) nodes is augmented by the set of nodes u that have a number of *already* activated neighbors greater or equal to u's threshold $t(u)$. We say that v *is activated* at round $\ell > 0$ if $v \in \mathsf{Active}_G[S, \ell] \setminus \mathsf{Active}_G[S, \ell - 1]$. A target set for G is a set S such that it will activate the whole network, that is, for which it holds that $\mathsf{Active}_G[S, \ell] = V$, for some $\ell \geq 0$. The classical Target Set Selection (TSS) problem (see e.g. [1, 13]) is defined as follows:

TARGET SET SELECTION.
Instance: A network $G = (V, E)$ with thresholds $t : V \longrightarrow \mathbb{N}$.
Problem: Find a target set $S \subseteq V$ of *minimum* size for G.

The TSS Problem has roots in the general study of the *spread of influence* in Social Networks (see [8, 18]). For instance, in the area of viral marketing [17], companies wanting to promote products or behaviors might initially try to target and convince a set of individuals (by offering free copies of the products or some equivalent monetary rewards) who, by word-of-mouth, can successively trigger a cascade of influence in the network leading to an adoption of the products by a much larger number of individuals. In order to make the model more realistic, we extend the previously described basic model in two ways: First, we assume that there are non negative values $c(v)$ associated to each vertex $v \in V$, measuring how much it costs to initially convince the member v of the network to endorse a given product/behavior. Indeed, that different members of the network have different activation costs (see [2], for example) is justified by the observation that celebrities or public figures can charge more for their endorsements of products. Therefore, we are lead to our first extension of the TSS problem:

WEIGHTED TARGET SET SELECTION (WTSS).
Instance: A network $G = (V, E)$, thresholds $t : V \to \mathbb{N}$, costs $c : V \to \mathbb{N}$.
Problem: Find a target set $S \subseteq V$ of *minimum* cost $C(S) = \sum_{v \in S} c(v)$

Our second, and more technically challenging, extension of the classical TSS problem is inspired by the recent interesting paper [16]. In it, the authors observed that the basic model misses a crucial feature of practical applications.

---

[1] We will omit the subscript G whenever the graph G is clear from the context.

Indeed, it forces the optimizer to make a binary choice of either zero or complete influence on each individual (for example, either not offering or offering a free copy of the product to individuals in order to initially convince them to adopt the product and influence their friends about it). In realistic scenarios, there could be more reasonable and effective options. For example, a company promoting a new product may find that offering for free ten copies of a product is far less effective than offering a discount of ten percent to a hundred of people. Therefore, we formulate our second extension of the basic model as follows.

Targeting with Partial Incentives. An assignment of partial incentives to the vertices of a network $G = (V, E)$, with $V = \{v_1, \ldots, v_n\}$, is a vector $\mathbf{s} = (s(v_1), \ldots, s(v_n))$, where $s(v) \in \mathbb{N}_0 = \{0, 1, 2, \ldots\}$ represents the amount of influence we initially apply on $v \in V$. The effect of applying incentive $s(v)$ on node v is to decrease its threshold, i.e., to make individual v more susceptible to future influence. It is clear that to start the process, there should be an initial number of nodes v's to which the amount of exercised influence $s(v)$ is at least equal to their thresholds $t(v)$. Therefore, an *activation process in G starting with incentives* $\mathbf{s}$ is a sequence of vertex subsets

$$\mathsf{Active}[\mathbf{s}, 0] \subseteq \mathsf{Active}[\mathbf{s}, 1] \subseteq \ldots \subseteq \mathsf{Active}[\mathbf{s}, \ell] \subseteq \ldots \subseteq V,$$

with $\mathsf{Active}[\mathbf{s}, 0] = \{v \mid s(v) \geq t(v)\}$, and such that for all $\ell > 0$,

$$\mathsf{Active}[\mathbf{s}, \ell] = \mathsf{Active}[\mathbf{s}, \ell - 1] \cup \Big\{ u : \big| \Gamma_G(u) \cap \mathsf{Active}[\mathbf{s}, \ell - 1] \big| \geq t(u) - s(u) \Big\}.$$

A *target vector* $\mathbf{s}$ is an assignment of partial incentives that triggers an activation process influencing the whole network, that is, such that $\mathsf{Active}[\mathbf{s}, \ell] = V$ for some $\ell \geq 0$. The Targeting with Partial Incentive problem can be defined as follows:

TARGETING WITH PARTIAL INCENTIVES (TPI).
Instance: A network $G = (V, E)$, thresholds $t : V \longrightarrow \mathbb{N}$.
Problem: Find target vector $\mathbf{s}$ which minimizes $C(\mathbf{s}) = \sum_{v \in V} s(v)$.

Notice that the Weighted Target Set Selection problem, when the costs $c(v)$ are always equal to the thresholds $t(v)$, for each $v \in V$, can be seen as a particular case of Targeting with Partial Incentives in which the incentives $s(v)$ are set either to 0 or to $t(v)$. Therefore, in a certain sense, the Targeting with Partial Incentives can be seen as a kind of "fractional" counterpart of the Weighted Target Set Selection problem (notice, however, that the incentives $s(v)$ are integer as well). In general, the two optimization problems are quite different since arbitrarily large gaps are possible between the costs of the solutions of the WTSS and TPI problems, as the following example shows.

Example 1. Consider the complete graph on n vertices $v_1, \ldots, v_n$, with thresholds $t(v_1) = \ldots = t(v_{n-2}) = 1$, $t(v_{n-1}) = t(v_n) = n - 1$. An optimal solution to the WTSS problem consists of either vertex v_{n-1} or vertex v_n, hence of total cost equal to $n - 1$. On the other hand, if partial incentives are possible one can assign incentives $s(v_1) = s(v_n) = 1$ and $s(v_i) = 0$ for $i = 2, \ldots, n - 1$, and have an optimal solution of value equal to 2. Indeed, we have

$\mathsf{Active}[\mathbf{s}, 0] = \{v_1\}$, since $t(v_1) = s(v_1)$, $\mathsf{Active}[\mathbf{s}, 1] = \{v_1, v_2, \ldots, v_{n-2}\}$, since $t(v_i) = 1$ for $i = 2, \ldots, n - 2$, $\mathsf{Active}[\mathbf{s}, 2] = \{v_1, v_2, \ldots, v_{n-2}, v_n\}$, since $t(v_n) - s(v_n) = n - 2$, and $\mathsf{Active}[\mathbf{s}, 3] = \{v_1, v_2, \ldots, v_{n-1}, v_n\}$, since $t(v_{n-1}) = n - 1$.

1.2   Related Works

The algorithmic problems we have articulated have roots in the general study of the *spread of influence* in Social Networks (see [8, 18] and references quoted therein). The first authors to study problems of spread of influence in networks from an algorithmic point of view were Kempe *et al.* [22, 23]. They introduced the Influence Maximization problem, where the goal is to identify a set $S \subseteq V$ such that its cardinality is bounded by a certain budget β and the activation process activates as much vertices as possible. However, they were mostly interested in networks with randomly chosen thresholds. Chen [10] studied the following minimization problem: Given a graph G and fixed arbitrary thresholds $t(v)$, $\forall v \in V$, find a target set of minimum size that eventually activates all (or a fixed fraction of) nodes of G. He proved a strong inapproximability result that makes unlikely the existence of an algorithm with approximation factor better than $O(2^{\log^{1-\epsilon} |V|})$. Chen's result stimulated a series of papers [1, 3–6, 12, 11, 13, 14, 7, 20, 27, 28, 30] that isolated many interesting scenarios in which the problem (and variants thereof) become tractable. The Influence Maximization problem with partial incentives was introduced in [16]. In this model the authors assume that the thresholds are randomly chosen values in $[0, 1]$ and they aim to understand how a fractional version of the Influence Maximization problem differs from the original version. To that purpose, they introduced the concept of partial influence and show that, from a theoretical point of view, the fractional version retains essentially the same computational hardness as the integral version but, on the practical side, the solutions computed, using heuristics, are more efficient in the fractional setting.

1.3   Our Results

Our main contributions are the following. We first show, in Section 2, that there exists a (gap-preserving) reduction from the classical TSS problem to our TPI and WTSS problems (for the WTSS problem, the gap preserving reduction holds also in the case particular case in which $c(v) = t(v)$, for each $v \in V$). Using the important results by [10], this implies the TPI and WTSS problems cannot be approximated to within a ratio of $O(2^{\log^{1-\epsilon} n})$, for any fixed $\epsilon > 0$, unless $NP \subseteq DTIME(n^{polylog(n)})$ (again, for the latter problem this inapproximability result holds also in the case $c(v) = t(v)$, for each $v \in V$). Moreover, since the WTSS problem is equivalent to the TSS problem when all thresholds are equal, the reduction also show that the particular case in which $c(v) = t(v)$, for each $v \in V$, of the WTSS problem is NP-hard. Again, this is due to the corresponding hardness result of TSS given in [10]. In Section 3 we present a polynomial time algorithm that, given a weighted network and vertices thresholds, computes a cost efficient target set. Our algorithm exhibits the following features: **1)** for general graphs, it always return a solution of cost at most equal to $\sum_{v \in V} \frac{c(v)t(v)}{d_G(v)+1}$. It is interesting to note that, when $c(v) = 1$ for each $v \in V$, we recover the same upper bound on the cardinality of an optimal target set given in [1], and proved therein by means of the probabilistic method. **2)** For complete graphs our algorithm always returns a solution of *minimum* cost. In Section 4 we turn our attention to the problem with incentives and we propose a polynomial time algorithm that, given a network and vertices thresholds, computes a cost efficient target vector. Our algorithm has the following properties: **1)** for general graphs, it always return a solution **s** (i.e., a target vector) for G such that $C(\mathbf{s}) = \sum_{v \in V} s(v) \leq \sum_{v \in V} \frac{t(v)(t(v)+1)}{2(d_G(v)+1)}$. **2)** For trees and complete graphs our algorithm always returns an *optimal* target vector. Finally, in Section 5 we experimentally validate our algorithms by running them on real life networks, and we compare the obtained results with that of well known heuristics in the area (especially tuned to our scenarios). The experiments shows that our algorithms consistently outperform those heuristics.

Due to the space limit, some proofs are omitted and given in the Appendix.

2   Hardness of WTSS and TPI

We prove the following result

Theorem 1. *WTSS and TPI cannot be approximated within a ratio of $O(2^{\log^{1-\epsilon} n})$ for any fixed $\epsilon > 0$, unless $NP \subseteq DTIME(n^{polylog(n)})$.*

Proof. We first construct a gap-preserving reduction from the TSS problem. The claim of the theorem follows from the inapproximability of TSS proved in [10]. In the following, we give details for only for the TPI problem. Starting from an arbitrary graph $G = (V, E)$ and threshold function t, input instances of the TSS problem, we build a graph $G' = (V', E')$ as follows:

- $V' = \bigcup_{v \in V} V'_v$ where $V'_v = \{v', v'', v_1, \ldots v_{d_G(v)}\}$. In particular,
  - we replace each $v \in V$ by the gadget Λ_v (cfr. Fig. 1) in which the vertex set is V'_v and v' and v'' are connected by the disjoint paths (v', v_i, v'') for $i = 1, \ldots, d_G(v)$;
  - the threshold of v' in G' is equal to the threshold $t(v)$ of v in G, while each other vertex in V'_v has threshold set to 1.
- $E' = \{(v', u') \mid (v, u) \in E\} \bigcup_{v \in V} \{(v', v_i), (v_i, v''), \text{ for } i = 1, \ldots, d_G(v)\}$.

Summarizing, G' is constructed in such a way that for each gadget Λ_v, the vertex v' plays the role of v and is connected to all the gadgets representing neighbors of v in G. Hence, G corresponds to the subgraph of G' induced by the set $\{v' \in V'_v \mid v \in V\}$. It is worth mentioning that during an activation process if any vertex that belongs to a gadget Λ_v is active, then all the vertices in Λ_v will be activate within the next 3 rounds.
We claim that there is a target set $S \subseteq V$ for G of cardinality $|S| = k$ if and only if there is a target vector **s** for G' and $C(\mathbf{s}) = \sum_{u \in V'} s(u) = k$.
Assume that $S \subseteq V$ is a target set for G, we can easily build an assignation of partial incentives **s** as follows:

$$s(u) = \begin{cases} 1 & \text{if } u \text{ is the extremal vertex } v'' \text{ in the gadget } \Lambda_v \text{ and } v \in S; \\ 0 & \text{otherwise.} \end{cases}$$

Clearly, $C(\mathbf{s}) = \sum_{v \in S} 1 = |S|$. To see that $\mathbf{s}$ is a target vector we notice that
$\mathsf{Active}_{G'}[\mathbf{s}, 2] = \{u \mid u \in V_v', v \in S\}$, consequently since S is a target set and G is
isomorphic to the subgraph of G' induced by $\{v' \in V_v' \mid v \in V\}$, all the vertices
$v \in V'$ will be activated.

On the other hand, assume that $\mathbf{s}$ is a target vector for G' and $C(\mathbf{s}) = k$, we can
easily build a target set S

$$S = \{v \in V \mid \exists u \in V_v' \text{ such that } s(u) > 0\}.$$

By construction $|S| \leq \sum_{u \in V'} s(u) = C(\mathbf{s})$. To see that S is a target set for G,
for each $v \in V$ we consider two cases on the values $s(\cdot)$:

If there exists $u \in V_v'$ such that $s(u) > 0$ then, by construction $v \in S$.

Suppose otherwise $s(u) = 0$ for each $u \in V_v'$. We have that in order to activate
v' (and then any other vertex in Λ_v) there must exist a round i such that
$\mathsf{Active}_{G'}[\mathbf{s}, i-1] \cap (V' - V_v')$ contains $t(v)$ neighbors of v'. Recall that G is the
subgraph of G' induced by the set $\{v' \in V_v' \mid v \in V\}$. Then each round $i \geq 0$
and for each $v' \in \mathsf{Active}_{G'}[\mathbf{s}, i]$, we get that the set $\mathsf{Active}_G[S, i]$ contains the
corresponding vertex v. Consequently v will be activated in G. One can see
that the same graph G' can be used to derive a similar reduction from TSS to
WTSS.   □

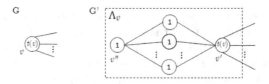

Fig. 1. The gadget Λ_v: (left) a generic vertex $v \in V$ having degree $d_G(v)$ and threshold
$t(v)$; (right) the gadget Λ_v, having $d_G(v) + 2$ vertices, associated to v.

3   An Algorithm for Weighted Target Set Selection

The algorithm works by iteratively deleting vertices from the input graph G. At
each iteration, the vertex to be deleted is chosen as to maximize a certain function
(Case 3). During the deletion process, some vertex v in the surviving graph may
remain with less neighbors than its threshold; in such a case (Case 2) v is added
to the target set and deleted from the graph while its neighbors thresholds are
decreased by 1 (since they receive v's influence). It can also happen that the
surviving graph contains a vertex v whose threshold has been decreased down
to 0 (e.g., the deleted vertices are able to activate v); in such a case (Case 1) v
is deleted from the graph and its neighbors thresholds are decreased by 1 (since
once v activates, they will receive vs influence). The proofs of the following
Theorems are given in Appendix A.

Theorem 2. *For any graph G and threshold function t, the algorithm $WTSS(G)$
outputs a target set for G. The algorithm can be implemented so to run in
$O(|E| \log |V|)$ time.*

Theorem 3. *For any $G = (V, E)$, the algorithm WTSS(G) returns a target set S with*

$$C(S) \leq \sum_{v \in V} \frac{c(v)t(v)}{d_G(v) + 1}. \tag{1}$$

Theorem 4. *The algorithm WTSS(G) outputs an optimal target set if G is a complete graph such that $c(v) \leq c(u)$ whenever $t(v) \leq t(u)$.*

Algorithm WTSS(G)
Input: A graph $G = (V, E)$ with thresholds $t(v)$ and costs $c(v)$, for $v \in V$.
Output: A target set S for G.
1. $S = \emptyset$; $U = V$
2. **for** each $v \in V$ **do** $\{ \delta(v) = d_G(v);$ $k(v) = t(v);$ $N(v) = \Gamma_G(v) \}$
3. **while** $U \neq \emptyset$ **do**
4.   *[Select one vertex and eliminate it from the graph]*
5.   **if** there exists $v \in U$ s.t. $k(v) = 0$ **then**
6.   *[Case 1: The selected vertex v is activated by the influence of its*
7.   *neighbors in $V - U$ only;*
8.   *it can then influence its neighbors in U]*
9.   **for** each $u \in N(v)$ **do** $k(u) = \max\{0, k(u) - 1\}$
10.   **else**
11.   **if** there exists $v \in U$ s.t. $\delta(v) < k(v)$ **then**
12.   *[Case 2: The vertex v is added to S, since no sufficient neighbors*
13.   *remain in U to activate it;*
14.   *v can then influence its neighbors in U]*
15. $S = S \cup \{v\}$
16.   **for** each $u \in N(v)$ **do** $k(u) = k(u) - 1$
17.   **else** *[Case 3: The selected vertex v will be activated by*
18.   *its neighbors in U]*
19. $v = \text{argmax}_{u \in U} \left\{ \frac{c(u) \, k(u)}{\delta(u)(\delta(u) + 1)} \right\}$
20.   *[Remove the selected vertex v from the graph]*
21.   **for** each $u \in N(v)$ **do** $\{ \delta(u) = \delta(u) - 1;$ $N(u) = N(u) - \{v\} \}$
22. $U = U - \{v\}$

---

4   Targeting with Partial Incentives

In this section, we design an algorithm to efficiently allocate incentives to nodes of a network, in such a way that it triggers an influence diffusion process that influences the whole network. The algorithm is close in spirit to Algorithm WTSS(G), with some crucial differences. Again the algorithm proceeds by iteratively deleting nodes from the graph and at each iteration the vertex to be deleted it is chosen as to maximize a certain parameter (Case 2). If, during the deletion process, a vertex v in the surviving graph remains with less neighbors than its remaining threshold (Case 1), then v's partial incentive is increased so that the v's remaining threshold is at least as large as the number of v's neighbors.

Example 2. Consider a complete graph on 7 vertices with thresholds $t(v_1) = \ldots = t(v_5) = 1$, $t(v_6) = t(v_7) = 6$. A possible execution of the algorithm is summarized below. At each iteration of the while loop, the algorithm considers the vertices in the order shown in the table below, where we also indicate for each vertex whether Cases 1 or 2 applies and the updated value of the partial incentive for the selected vertex:

Iteration	1	2	3	4	5	6	7	8
vertex	v_7	v_6	v_6	v_1	v_2	v_3	v_4	v_5
Case	2	1	2	2	2	2	2	1
Incentive	$s(v_7)=0$	$s(v_6)=1$	$s(v_6)=1$	$s(v_1)=0$	$s(v_2)=0$	$s(v_3)=0$	$s(v_4)=0$	$s(v_5)=1$

The algorithm $TPI(G)$ outputs the vector of partial incentives having non zero elements $s(v_5) = s(v_6) = 1$, for which we have
Active$[s, 0] = \{v_5\}$ *(since $s(v_5) = 1 = t(v_5)$)*,
Active$[s, 1] =$ Active$[s, 0] \cup \{v_1, v_2, v_3, v_4\} = \{v_1, v_2, v_3, v_4, v_5\}$,
Active$[s, 2] = \{v_1, v_2, v_3, v_4, v_5, v_6\}$ *(since $s(v_6) = 1$)*,   Active$[s, 3] = V$.

Algorithm TPI(G)
Input: A graph $G = (V, E)$ with thresholds $t(v)$, for each $v \in V$.
Output: s a target vector for G.
  1. $U = V$
  2. **for** each $v \in V$ **do**
  3. $s(v) = 0$ *[Partial incentive initially assigned to v]*
  4. $\delta(v) = d_G(v)$
  5. $k(v) = t(v)$
  6. $N(v) = \Gamma_G(v)$
  7. **while** $U \neq \emptyset$ **do**
  8. *[Select one vertex and either update its incentive or remove it from the graph]*
  9.   **if** there exists $v \in U$ s.t. $k(v) > \delta(v)$
 10.   **then**   *[Case 1: Increase $s(v)$ and update $k(v)$]*
 11. $s(v) = s(v) + k(v) - \delta(v)$
 12. $k(v) = \delta(v)$
 13.   **if** $k(v) = 0$ **then** $U = U - \{v\}$ *[here $\delta(v) = 0$]*
 14.   **else** *[Case 2: Choose a vertex v to eliminate from the graph]*
 15. $v = \text{argmax}_{u \in U} \left\{ \frac{k(u)(k(u)+1)}{\delta(u)(\delta(u)+1)} \right\}$
 16.   **for** each $u \in N(v)$ **do** $\{\delta(u) = \delta(u) - 1; N(u) = N(u) - \{v\}\}$
 17. $U = U - \{v\}$

We first prove the algorithm correctness, next we give a general upper bound on the size $\sum_{v \in V} s(v)$ of its output and prove its optimality for trees and cliques.

To this aim we will use the following notation.

Let ℓ be the number of iterations of the while loop in TPI(G). For each iteration j, with $1 \leq j \leq \ell$, of the while loop we denote

 – by U_j the set U at the beginning of the j-th iteration (cfr. line 8 of $TPI(G)$), in particular $U_1 = V(G)$ and $U_{\ell+1} = \emptyset$;
 – by $\mathcal{G}(j)$ the subgraph of G induced by the vertices in U_j,

- by v_j the vertex selected during the j-th iteration[2],
- by $\delta_j(v)$ the degree of vertex v in $\mathcal{G}(j)$,
- by $k_j(v)$ the value of the remaining threshold of vertex v in $\mathcal{G}(j)$, that is, as it is updated at the beginning of the j-th iteration, in particular $k_1(v) = t(v)$ for each $v \in V$,
- by $s_j(v)$ the partial incentive collected by vertex v in $\mathcal{G}(j)$ starting from the j-th iteration, in particular we set $s_0(v) = 0$ for each $v \in V$; and
- by σ_j the increment of the partial incentives during the j-th iteration, that

$$\text{is,} \quad \sigma_j = s_j(v_j) - s_{j-1}(v_j) = \begin{cases} 0 & \text{if } k_j(v_j) \leq \delta_j(v_j), \\ k_j(v_j) - \delta_j(v_j) & \text{otherwise.} \end{cases}$$

According to the above notation, we have that if vertex v is selected during the iterations $j_1 < j_2 < \ldots < j_{a-1} < j_a$ of the while loop in TPI(G), where the last value j_a is the iteration when v has been eliminated from the graph, then

$$s_j(v) = \begin{cases} \sigma_{j_1} + \sigma_{j_2} + \ldots + \sigma_{j_a} & \text{if } j \leq j_1, \\ \sigma_{j_b} + \sigma_{j_{b+1}} + \ldots + \sigma_{j_a} & \text{if } j_{b-1} < j \leq j_b, \\ 0 & \text{if } j > j_a. \end{cases}$$

The following results are immediate.

Proposition 1. *Consider the vertex v_j, selected during the iteration j, for $1 \leq j \leq \ell$, of the while loop in the algorithm TPI(G),*
1.1) *If Case 1 of TPI(G) holds and $\delta_j(v_j) = 0$, then $k_j(v_j) > \delta_j(v_j) = 0$ and the isolated vertex v_j is eliminated from $\mathcal{G}(j)$. Moreover,*
$$U_{j+1} = U_j \setminus \{v_j\}, \quad s_{j+1}(v_j) = s_j(v_j) - \sigma_j, \quad \sigma_j = k_j(v_j) - \delta_j(v_j) > 0, \text{ and}$$
$$s_{j+1}(v) = s_j(v), \quad \delta_{j+1}(v) = \delta_j(v), \quad k_{j+1}(v) = k_j(v), \text{ for each } v \in U_{j+1}.$$
1.2) *If Case 1 of TPI(G) holds with $\delta_j(v_j) > 0$, then $k_j(v_j) > \delta_j(v_j) > 0$ and no vertex is deleted from $\mathcal{G}(j)$, that is, $U_{j+1} = U_j$. Moreover, $\sigma_j = k_j(v_j) - \delta_j(v_j) > 0$ and for each $v \in U_{j+1}$ it holds*
$$s_{j+1}(v) = \begin{cases} s_j(v_j) - \sigma_j & \text{if } v = v_j \\ s_j(v) & \text{if } v \neq v_j \end{cases}, \quad \delta_{j+1}(v) = \delta_j(v),$$
$$k_{j+1}(v) = \begin{cases} \delta_j(v) & \text{if } v = v_j \\ k_j(v) & \text{if } v \neq v_j \end{cases}$$
2) *If Case 2 of TPI(G) holds then $k_j(v_j) \leq \delta_j(v_j)$ and v_j is pruned from $\mathcal{G}(j)$. Hence, $U_{j+1} = U_j \setminus \{v_j\}$, $\sigma_j = 0$, and for each $v \in U_{j+1}$ it holds*
$$s_{j+1}(v) = s_j(v), \quad \delta_{j+1}(v) = \begin{cases} \delta_j(v) - 1 & \text{if } v \in \Gamma_{\mathcal{G}(j)}(v_j) \\ \delta_j(v) & \text{otherwise.} \end{cases}, \quad k_{j+1}(v) = k_j(v)$$

Lemma 1. *For each iteration $j = 1, 2, \ldots, \ell$, of the while loop in the algorithm TPI(G),*
1) if $k_j(v_j) > \delta_j(v_j)$ then $\sigma_j = k_j(v_j) - \delta_j(v_j) = 1$;
2) if $\delta_j(v_j) = 0$ then $s_j(v_j) = k_j(v_j)$.

---

[2] A vertex can be selected several times before being eliminated; indeed in Case 1 we can have $U_{j+1} = U_j$.

Theorem 5. *For any graph G the algorithm TPI(G) outputs a target vector for G.*

Proof. We show that for each iteration j, with $1 \leq j \leq \ell$, the assignation of partial incentives $s_j(v)$ for each $v \in U_j$ activates all the vertices of the graph $\mathcal{G}(j)$ when the distribution of thresholds to its vertices is $k_j(\cdot)$. The proof is by induction on j.

If $j = \ell$ then the unique vertex v_ℓ in $\mathcal{G}(\ell)$ has degree $\delta_\ell(v_\ell) = 0$ and $s_\ell(v_\ell) = k_\ell(v_\ell) = 1$ (see Lemma 1).

Consider now $j < \ell$ and suppose the algorithm be correct on $\mathcal{G}(j+1)$ that is, the assignation of partial incentives $s_{j+1}(v)$, for each $v \in U_{j+1}$, activates all the vertices of the graph $\mathcal{G}(j+1)$ when the distribution of thresholds to its vertices is $k_{j+1}(\cdot)$.

Recall that v_j denotes the vertex the algorithm selected from U_j (to obtain U_{j+1} that is the vertex set of $\mathcal{G}(j+1)$). To prove the theorem we analyzes the three cases according to the current degree and threshold of the selected vertex v_j.

- Let $k_j(v_j) > \delta_j(v_j) = 0$. By Lemma 1 we have $k_j(v_j) = s_j(v_j)$. Furthermore, recalling that 1.1) in Proposition 1 holds and by using the inductive hypothesis on $\mathcal{G}(j+1)$, we get the correctness on $\mathcal{G}(j)$.
- Let $k_j(v_j) > \delta_j(v_j) \geq 1$. By recalling that 1.2) in Proposition 1 holds we get $k_j(v) - s_j(v) = k_{j+1}(v) - s_{j+1}(v)$, for each vertex $v \in U_j$. Hence the vertices that can be activated in $\mathcal{G}(j+1)$ can be activated in $\mathcal{G}(j)$ with thresholds $k_j(\cdot)$ and partial incentives $s_j(\cdot)$. So, by using the inductive hypothesis on $\mathcal{G}(j+1)$, we get the correctness on $\mathcal{G}(j)$.
- Let $k_j(v_j) \leq \delta_j(v_j)$. By recalling that 2) in Proposition 1 holds and the inductive hypothesis on $\mathcal{G}(j+1)$ we have that all the neighbors of v_j in $\mathcal{G}(j)$ that are vertices in U_{j+1} gets active; since $k_j(v_j) \leq \delta_j(v_j)$ also v_j activates in $\mathcal{G}(j)$.   □

Theorem 6. *For any graph G the algorithm TPI(G) returns a target vector* **s** *for G such that* $C(\mathbf{s}) = \sum_{v \in V} s(v) \leq \sum_{v \in V} \frac{t(v)(t(v)+1)}{2(d_G(v)+1)}$

Theorem 7. *TPI(K) returns an optimal target vector for any complete graph K.*

Theorem 8. *TPI(T) outputs an optimal target vector for any tree T.*

We can also explicitly evaluate the cost of an optimal solution for any tree.

Theorem 9. *The cost of the optimal target vector* $\mathbf{s}^*$ *on a tree T, having n vertices with thresholds* $t : V \longrightarrow \mathbb{N}$ *is* $C(\mathbf{s}^*) = n - 1 + \sum_{v \in V} t(v) - d_T(v)$.

5   Experiments

We have experimentally evaluated both our algorithms WTSS(G) and TPI(G) on real-world data sets and found that they perform quite satisfactorily. We conducted experiments on several real networks of various sizes from the Stanford

Large Network Data set Collection (SNAP) [25], the Social Computing Data Repository at Arizona State University [29] and Newman's Network data [26]. The data sets we considered include both networks for which "low cost" target sets exist and networks needing an expensive target sets (due to a community structure that appears to block the diffusion process).

The Competing Algorithms. We compare the performance of our algorithms toward that of the best, to our knowledge, computationally feasible algorithms in the literature [16]. It is worth to mention that the following competing algorithms were initially designed for the Maximally Influencing Set problem, where the goal is to identify a set $S \subseteq V$ such that its cost is bounded by a certain budget β and the activation process activates as much vertices as possible. In order to compare such algorithms toward our strategies, for each algorithm we performed a binary search in order to find the smallest value of β which allow to activate all the vertices of the considered graph. We compare the WTSS algorithm toward the following two algorithms:

- *DegreeInt*, a simple greedy algorithm, which selects vertices in descending order of degree [22, 9];
- *DiscountInt*, a variant of DegreeInt, which selects a node v with the highest degree at each step. Then the degree of nodes in $\Gamma(v)$ is decreased by 1 [9].

Moreover, we compare the TPI algorithm toward the following two algorithms:

- *DegreeFrac*, which selects each node fractionally proportional to its degree. Specifically, given a graph $G = (V, E)$ and budget β this algorithm spend on each node $v \in V$, $s(v) = \left\lfloor \frac{d(v) \times \beta}{2|E|} \right\rfloor$ [16]. Remaining budget, if any, is assigned increasing by 1 the budget assigned to some nodes (in descending order of degree).
- *DiscountFrac*, which in each step, selects the node v with the highest degree and assigns to it a budged $s(v) = max(0, t(v) - |\Gamma(v) \cap S|))$, which represent the minimum amount that allows to activate v (S denotes the set of already selected nodes). As for the DiscountInt algorithm, after selecting a node v, the degree of nodes in $\Gamma(v)$ is decreased by 1 [16].

Thresholds Values. We tested with three categories of threshold function: *Random thresholds* where $t(v)$ is chosen uniformly at random in the interval $[1, d(v)]$; *Constant thresholds* where the thresholds are constant among all vertices (precisely the constant value is an integer in the interval $[2, 10]$ and for each vertex v the threshold $t(v)$ is set as $min(t, d(v))$ where $t = 2, 3, \ldots, 10$ (nine tests overall); *Proportional thresholds* where for each v the threshold $t(v)$ is set as $\alpha d(v)$ with $\alpha = 0.1, 0.2, \ldots, 0.9$ (nine tests overall). Notice that for $\alpha = 0.5$ we are considering a particular version of the activation process named "majority" [19].

Node Costs. We report experiments results for the WTSS problem in case the costs are equal to the thresholds, that is $c(v) = t(v)$ for each vertex $v \in V$. Similar results hold for different cost choices.

Table 1. Random Threshold Results.

	Targeting with Partial Incentives			Weighted Target Set Selection with $c(\cdot) = t(\cdot)$		
Name	PTI	DiscountFrac	DegreeFrac	WTSS	DiscountInt	DegreeInt
Amazon0302	52703	328519 (623%)	879624 (1669%)	85410	596299 (698%)	890347 (1042%)
BlogCatalog	21761	824063 (3787%)	980670 (4507%)	82502	1799719 (2181%)	2066014 (2504%)
BlogCatalog2	16979	703383 (4143%)	178447 (1051%)	67066	1095580 (1634%)	1214818 (1811%)
BlogCatalog3	161	3890 (2416%)	3113 (1934%)	3925	3890 (99%)	3890 (99%)
BuzzNet	50913	1154952 (2268%)	371355 (729%)	166085	1838430 (1107%)	2580176 (1554%)
ca-AstroPh	4520	67189 (1486%)	198195 (4385%)	13242	183121 (1383%)	198195 (1497%)
ca-CondMath	5694	31968 (561%)	94288 (1656%)	10596	76501 (722%)	94126 (888%)
ca-GrQc	1422	5076 (357%)	15019 (1056%)	2141	12538 (586%)	15019 (701%)
ca-HepPh	4166	42029 (1009%)	120324 (2888%)	11338	118767 (1048%)	120324 (1061%)
ca-HepTh	2156	9214 (427%)	26781 (1242%)	3473	25417 (732%)	26781 (771%)
Douban	51167	140676 (275%)	345036 (674%)	91342	194186 (213%)	252739 (277%)
Facebook	1658	29605 (1786%)	54508 (3288%)	5531	77312 (1398%)	86925 (1572%)
Flikr	31392	2057877 (6555%)	134017 (427%)	110227	5359377 (4862%)	5879532 (5334%)
Hep	4122	11770 (286%)	33373 (810%)	5526	33211 (601%)	33373 (604%)
LastFM	296083	1965839 (664%)	4267035 (1441%)	631681	2681610 (425%)	4050280 (641%)
Livemocha	26610	861053 (3236%)	459777 (1728%)	57293	1799468 (3141%)	2189760 (3822%)
Power grid	767	2591 (338%)	4969 (648%)	974	3433 (352%)	4350 (447%)
Youtube2	313786	1210830 (386%)	3298376 (1051%)	576482	2159948 (375%)	3285525 (570%)

Results. In our experiments we compare the cost of the target set (or target vector) generated by six algorithms (PTI, DiscountFrac, DegreeFrac, WTSS, DiscountInt, DegreeInt) on 18 networks, fixing the thresholds in 19 different ways (Random, Constant with $t = 2, 3, \ldots, 10$ and Proportional with $\alpha = 0.1, 0.2, \ldots, 0.9$). Overall we performed $6 \times 18 \times 19 = 2052$ tests.

Random Thresholds. Table 1 depicts the results of the Random threshold test setting. Each number represents the cost of the target vector (left side of the table) or the target set (right side of the table) generated by each algorithm on each network using random thresholds (the same thresholds values have been used for all the algorithms). The value in bracket represents the overhead percentage compared to our algorithms (TPI for DiscountFrac and DegreeFrac and WTSS for DiscountInt and DegreeInt). Analyzing the results Table 1, we notice that in all the considered cases, with the exception of the network BlogCatolog3, our algorithms always outperform their competitors. In the network BlogCatalog3, the WTSS algorithm is slightly worse than its competitors but PTI performs much better than the other algorithms.

Constant and Proportional Thresholds. The following figure depicts the results of Constant and Proportional thresholds settings. For each network the results are reported in two separated figures: Proportional thresholds (left-side), the value of the α parameter appears along the X-axis, while the cost of the solution appears along the Y-axis; Constant thresholds (right-side), in this case the X-axis indicates the value of the thresholds. We present the results only for four networks because of space limitations; the experiments performed on the other networks exhibit similar behaviors. Analyzing the results from Figs. 2-3, we can make the following observations: In all the considered case our

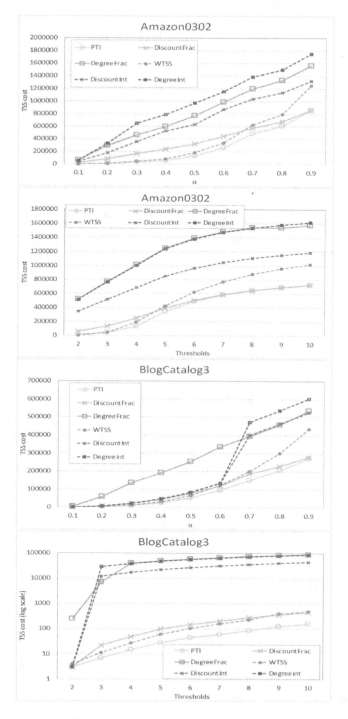

Fig. 2. Amazon0302 and BlogCatalog3.

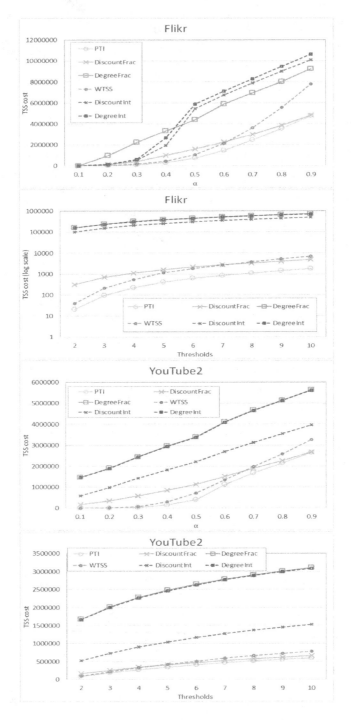

Fig. 3. Flikr and YouTube2.

algorithms always outperform their competitors; the only algorithm that provides performance close to our algorithms is the DiscountFrac algorithm. However, for intermediate values of the α parameter, the gap to our advantage is quite significant. In general, in case of partial incentives we have even better results, the gap to our advantage increases with the increase of the parameter α.

References

1. Ackerman, E., Ben-Zwi, O., Wolfovitz, G.: Combinatorial model and bounds for target set selection. Theoretical Computer Science 411, 4017–4022 (2010)
2. Bakshy, E., Hofman, J.M., Mason, W.A., Watts, D.J.: Everyone's an influencer: quantifying influence on twitter. In: Proceedings of the Fourth ACM International Conference on Web Search and Data Mining (WSDM 2011), pp. 65–74 (2011)
3. Bazgan, C., Chopin, M., Nichterlein, A., Sikora, F.: Parametrized Approximability of Maximizing the Spread of Influence in Networks. In: Du, D.-Z., Zhang, G. (eds.) COCOON 2013. LNCS, vol. 7936, pp. 543–554. Springer, Heidelberg (2013)
4. Ben-Zwi, O., Hermelin, D., Lokshtanov, D., Newman, I.: Treewidth governs the complexity of target set selection. Discrete Optimization 8, 87–96 (2011)
5. Centeno, C.C., et al.: Irreversible conversion of graphs. Theoretical Computer Science 412(29), 3693–3700 (2011)
6. Chopin, M., Nichterlein, A., Niedermeier, R., Weller, M.: Constant Thresholds Can Make Target Set Selection Tractable. In: Even, G., Rawitz, D. (eds.) MedAlg 2012. LNCS, vol. 7659, pp. 120–133. Springer, Heidelberg (2012)
7. Coja-Oghlan, A., Feige, U., Krivelevich, M., Reichman, D.: Contagious Sets in Expanders. In: Proceedings of SODA 2015 (1987)
8. Chen, W., Lakshmanan, V.S., Castillo, C.: Information and Influence Propagation in Social Networks. Morgan & Claypool (2013)
9. Chen, W., Wang, Y., Yang, S.: Efficient Influence Maximization in Social Networks. In: Proc. 15th ACM SIGKDD Intl. Conf. on Know. Dis. and Data Min. (2009)
10. Chen, N.: On the approximability of influence in social networks. SIAM J. Discrete Math. 23, 1400–1415 (2009)
11. Chiang, C.-Y., Huang, L.-H., Yeh, H.-G.: Target Set Selection Problem for Honeycomb Networks. SIAM J. Discrete Math. 27(1), 310–328 (2013)
12. Chiang, C.-Y., Huang, L.-H., Li, B.-J., Wu, J., Yeh, H.-G.: Some results on the target set selection problem. Journal of Comb. Opt. 25(4), 702–715 (2013)
13. Cicalese, F., Cordasco, G., Gargano, L., Milanič, M., Peters, J.G., Vaccaro, U.: Spread of Influence in Weighted Networks under Time and Budget Constraints. Theoretical Computer Science 586, 40–58 (2015)
14. Cicalese, F., Cordasco, G., Gargano, L., Milanič, M., Vaccaro, U.: Latency-Bounded Target Set Selection in Social Networks. Theoretical Computer Science 535, 1–15 (2014)
15. Christakis, N.A., Fowler, J.H.: The collective dynamics of smoking in a large social network. N. Engl. J. Med., 2249–2258 (2008)
16. Demaine, E.D., Hajiaghayi, M.T., Mahini, H., Malec, D.L., Raghavan, S., Sawant, A., Zadimoghadam, M.: How to influence people with partial incentives. In: Proc. of WWW 2014, pp. 937–948 (2014)
17. Domingos, P., Richardson, M.: Mining the network value of customers. In: Proc. of 7th ACM SIGKDD Int. Conf. on Know. Disc. and Data Min., pp. 57–66 (2001)

18. Easley, D., Kleinberg, J.: Networks, Crowds, and Markets: Reasoning About a Highly Connected World. Cambridge University Press (2010)
19. Flocchini, P., Královic, R., Ruzicka, P., Roncato, A., Santoro, N.: On time versus size for monotone dynamic monopolies in regular topologies. J. Discrete Algorithms 1, 129–150 (2003)
20. Gargano, L., Hell, P., Peters, J., Vaccaro, U.: Influence Diffusion in Social Networks under Time Window Constraints. In: Moscibroda, T., Rescigno, A.A. (eds.) SIROCCO 2013. LNCS, vol. 8179, pp. 141–152. Springer, Heidelberg (2013)
21. Granovetter, M.: Thresholds Models of Collective Behaviors. American Journal of Sociology 83(6), 1420–1443 (1978)
22. Kempe, D., Kleinberg, J.M., Tardos, E.: Maximizing the spread of influence through a social network. In: Proc. of 9th ACM SIGKDD Int. Conf. on Know., Disc. and Data Min., pp. 137–146 (2003)
23. Kempe, D., Kleinberg, J.M., Tardos, É.: Influential vertices in a Diffusion Model for Social Networks. In: Caires, L., Italiano, G.F., Monteiro, L., Palamidessi, C., Yung, M. (eds.) ICALP 2005. LNCS, vol. 3580, pp. 1127–1138. Springer, Heidelberg (2005)
24. Leskovic, H., Adamic, L.A., Huberman, B.A.: The dynamic of viral marketing. ACM Transactions on the WEB 1 (2007)
25. Leskovec, J., Krevl, A.: SNAP Datasets: Stanford Large Network Dataset Collection (2015), http://snap.stanford.edu/data
26. Newman, M.: Network data (2015), http://www-personal.umich.edu/~mejn/netdata/
27. Nichterlein, A., Niedermeier, R., Uhlmann, J., Weller, M.: On Tractable Cases of Target Set Selection. Social Network Analysis and Mining, 1–24 (2012)
28. Reddy, T.V.T., Rangan, C.P.: Variants of spreading messages. J. Graph Algorithms Appl. 15(5), 683–699 (2011)
29. Zafarani, R., Liu, H.: Social Computing Data Repository at ASU, http://socialcomputing.asu.edu
30. Zaker, M.: On dynamic monopolies of graphs with general thresholds. Discrete Mathematics 312(6), 1136–1143 (2012)

Approximation Algorithms for Multi-budgeted Network Design Problems

Georgios Stamoulis

LAMSADE
PSL* Research University, Université Paris-Dauphine
CNRS UMR 7243, France
Universitá della svizzera Italiana, Lugano, Switzerland
stamoulis.georgios@gmail.com

Abstract. We study the multi-budgeted version of the *Survivable Network Design Problem* [3] where, besides the usual connectivity requirements between pairs of points, we also need to satisfy a set of linear constraints (the budgets). For this case, we provide a polynomial time $(3, 3)$ bi-criteria approximation algorithm for the problem which is based on combinatorial properties of the extreme point solutions of the natural linear relaxation of the problem.

1 Introduction

In the Survivable Network Design (SND) problem, we are given a simple, non-directed graph $G = (V, E)$ with a weight function $w : E \to \mathbb{Q}^+$ and a set of *connectivity requirements* $\mathcal{R} = \{\{s_i, t_i\}\}$, $v_i, t_i \in V(G)$ such that $v_i \neq t_i$. Let $|\mathcal{R}| = q \in \mathbb{Z}^+$. With each pair of source-sink vertices $\{s_i, t_i\} \in \mathcal{R}$ is associated a positive integer r_i. The goal is to find a minimum weight (cost) *Steiner* network $S \subseteq G$ such that in S there are at least r_i disjoint paths connecting each pair $\{s_i, t_i\} \in \mathcal{R}$. Note that, since the graph is non-directed, we do not required the source-sink pairs to be ordered pairs.

This is a very important problem, both from a theoretical and from a practical perspective: from a practical perspective, this problem is motivated by telecommunication applications where we want to find a *sparse* network of a initially given dense one that can tolerate failures on its links. In other words, even if any $r_j - 1$ edges fail in the new network, pair $i \in \mathcal{R}$ will still be connected in the new network. Highly important links (pairs $\{s_i, t_i\}$) have high connectivity requirement.

The problem probably is even more interesting from a *theoretical* point of view. Prior to the seminal work of Kamal Jain [3] who derived a 2-approximation for the above problem, the best approximation guarantee was $2 \cdot \max_{i \in q}\{r_i\}$ [14] which was improved to an approximation guarantee of $2H_k$ [2], where H_k is the k-th harmonic $1 + \frac{1}{2} + \frac{1}{3} + \cdots + \frac{1}{k} \approx \log k$. As we said, in the seminal work of Jain [3] a 2-approximation algorithm as given. Again, this is very interesting for two reasons: the obvious reason is that it significantly improved the approximation

© Springer International Publishing Switzerland 2015
C. Scheideler (Ed.): SIROCCO 2015, LNCS 9439, pp. 135–148, 2015.
DOI: 10.1007/978-3-319-25258-2_10

guarantee (providing the first constant approximation ratio for this problem) for a very well known and well studied combinatorial problem. The second reason, is that this result was based on the concept of *iterated* rounding of linear programs. The novel ideas introduced in this paper triggered a host of other very related results and the approach was so successful that even a whole book is dedicated the the study of iterative methods in combinatorial optimization [7]. The approach of Jain is the following:

First, model the problem as an integer linear program where we want to minimize $w^T x \equiv \sum_{e \in E} w_e x_e$

$$\text{subject to} \quad \sum_{e \in \delta(S)} x_e \geq \max_{s_i \in S, t_i \notin S} \{r_i\}, \quad \forall S \subseteq V \tag{1}$$

$$x_e \in \{0, 1\}, \quad \forall e \in E(G) \tag{2}$$

The intuition of the first set of constraints is that in every feasible solution we require that for every subset of vertices S, at least $\max\{r_i\}$ (for $s_i \in S, t_i \notin S$) edges cross this set (i.e., have one endpoint in S and the other not in S).

As usual, the integrality constraints $x_e \in \{0, 1\}$ are replaced by the $x_e \in [0, 1]$ and the above linear program is solved to optimality in polynomial time using a separation oracle and the Ellipsoid algorithm (since it has an exponential number of constraints), to produce an optimal fractional vector $x^* \in [0, 1]^E$. The cost of this vector, i.e. $opt = \sum_{e \in E} x_e^* w_e$ is a lower bound on the cost of the true integral optimal solution. Then, a polyhedral characterization of *any* extreme point solution[1] is given i.e., that every extreme point solution of the bounded polyhedron defined by the above linear program (after we drop the integrality constraints) satisfies the following property: $\exists e \in E$ such that $x_e \geq \frac{1}{2}$. In other words, any basic (extreme point) feasible solution x must always have at least one variable with fractional value greater than $\frac{1}{2}$.

If we define $\max_{s_i \in S, t_i \notin S} \{r_i\} = f(S)$ for any $S \subseteq V$, then the constraint of the above LP becomes $\sum_{e \in \delta(S)} x_e \geq f(S)$. The above characterization of the extreme point solutions holds also whenever the function $f(S)$ is *weakly supermodular*: A function $f : 2^V \to \mathbb{Z}$ is weakly supermodular if $f(\emptyset) = f(V) = 0$ and, $\forall A, B \in 2^V$ one of the following holds:

1. $f(A) + f(B) \leq f(A \cap B) + f(A \cup B)$,
2. $f(A) + f(B) \leq f(A - B) + f(B - A)$

It can be shown using standard arguments that $f(S) = \max_{s_i \in S, t_i \notin S} \{r_i\}$ is a weakly supermodular function [13]. So, in fact, the result of Jain concerns a more general class of problems, of which one special case happens to be the SND problem when $f(S)$ has the previous form.

---

[1] Extreme point, or basic feasible solution is any point y belonging in the polyhedron defined by the corresponding inequalities such that y *cannot* be written as a convex combination of two other feasible points

Given the characterization of all extreme point solutions x of the polyhedron defined by the relaxed LP introduced above, a 2-approximation algorithm is-almost-immediate: solve the LP to get a fractional optimal vector x^*. Let I be the set of indexes (edges) whose fractional value is "high": $I = \{e \in E : x_e^* \geq \frac{1}{2}\}$. Include all these edges into the solution (i.e., round their fractional value to 1). Re-define the connectivity requirements of all source-sink pairs $\mathcal{R}$ and resolve the linear program to obtain a new optimal basic feasible solution for the reduced instance. Since each time we round up variables by a factor of at most 2, the cost of the final solution would be at most twice the cost of the optimal fractional LP solution.

Generalizations of the SND Problem

Several generalizations of the SND problem have been studied in the literature. Most common among these is the *Minimum Bounded degree Steiner Network* which is identical with the SND plus degree constraints for a subset of vertices $W \subseteq V$. This means that for every feasible solution $x \in \{0,1\}^E$ and every vertex $v \in W$, we must have that $\sum_{e \in \delta(v)} x_e \leq b_v$, where $b_v \in \mathbb{Z}^+$ is the degree bound of vertex v.

A $(2, 2b_v + 3)$ bi-criteria algorithm (i.e., a polynomial time algorithm that returns a solution at most twice as costly as the optimal while the degree of each vertex $v \in W$ is at most $2b_v+3$) is known for the problem [6] which was improved to $(2, 2b_v+2)$ in [9]. In [8] an *additive* $(2, b_v+6r_{\max}+3)$ ($r_{\max} = \max_{j \in \mathcal{R}}\{r_j\}$) bi-ctiteria algorithm was given which, although additive in nature, does not directly compares with the previous bi-criteria algorithms. In [4], [10], [1] the problem was studied for *directed* input graphs G.

In [5] the author defined and studied the *budgeted* version of the SND problem which is the standard SND problem complicated with an extra linear constraint $\text{lin} = \ell^T x = \sum_{e \in E} x_e \ell_e \leq L \in \mathbb{Q}^+$. The author notices that a $(4,4)$ bi-criteria algorithm (4-approximation on the objective function value plus a violation of the packing constraint by a factor of at most 4) can be easily derived by a direct application of the famous Caratheodory's theorem on convex polytopes. Solve the LP defined by (1) and (2) + lin and get a basic fractionaly feasible solution x^* that does not violate the linear budget. By Caratheodory's theorem, x^* can be written as a convex combination of at most two points of the SND polyhedron: $x^* = \lambda y + (1 - \lambda)z$ such that $\lambda \in [0,1]$ and $y, z \in [0,1]^E$ are extreme point solutions of the fractional SND polyhedron defines by (1) plus (2). By Jain's result [3] each of y, z has a variable with fractional value greater than half. This means that at least one variable in one of the λy or $(1 - \lambda)z$ has value $\geq \frac{1}{4}$. Apply Jain's algorithm and a $(4,4)$ bi-criteria algorithm is immediate.

Moreover, the author proved that for every basic feasible solution x of the generalized LP (that spells out the connectivity requirements plus the linear budget), there is always a coordinate with fractional value greater than $\frac{1}{3}$. This is proved by the following way: a basic feasible solution is obtained for the relaxed LP. Then, the author notices that among all *tight* sets (sets that are satisfied with equality by the corresponding bfs) of vertices that define the basic feasible solution, there must be at least one such set S with degree ≤ 3. Since the sum of these edges must sum at least to $f(S) = \max_{s_i \in S, t_i \notin S}\{r_i\}$ which is at least 1, we have that there must be an edge crossing S with fractional value $\geq \frac{1}{3}$. This is done by considering only the tight constraints of subsets of vertices defining the obtained basic feasible solution x (by dropping the linear budget). Since there is only one linear budget, this drops the dimension of the vector space defined by these sets by 1 and then a counting argument on the remaining tight sets is done to produce the desired result.

Our Contribution

In this paper we study the SND problem with an *unbounded* number of extra linear budgets. In particular, we assume that each edge has a unique color. Let $\mathcal{C} = \{C_1, \ldots, C_k\}$ be the set of all color classes and let $E_j = \{e \in C_j\}$ be the set of edge of color C_j. In other words $\mathcal{C}$ forms a partition on the edges and let $c(e) = C_j$ to be the color (constraint) of e. Let the function $c : E \to \mathcal{C}$ be the function that assigns colors to the edges. For each color class $C_j \in \mathcal{C}$ we have a linear budget

$$\ell_j^T x = \sum_{e \in E_j} \ell_j(e)x(e) \leq L_j$$

and our task is to find a minimum cost Steiner network that satisfies (2) plus all the linear budgets. We note that a usual token redistribution argument is unlikely to work for this case, due to the nature of the constraints. Nevertheless, we provide an adaptation of the *fractional* token argument (first introduced in [11]) which seems to better facilitate the structure of the problem and we prove that even in this generalization of the problem considered in [5], a similar claim can be made: for any basic feasible solution to the relaxed LP defined by (2) plus all the linear budgets, there always exists a fractional entry with value $\geq \frac{1}{3}$.

As it is the case with all generalization of SND problem [6,9,8,5], our algorithm relies on existing tools developed originally by Jain, although the technical details provide extra challenge. On the other hand, since the usual approach (employed in [5]) seems that it cannot give us any meaningful result (because with k budgets the dimension of the vector space defined by the remaining sets after dropping these budgets could drop by as much as k), our contribution is to show how we can employ a particular *fractional* charging scheme that let us prove our result for a generalization of the problem defines in [5] and which the author leaves as open problem. Also, we note that an application of Caratheodory's theorem in our case does not yield any meaningful insight on the fractional values of the extreme point solutions.

2   Characterization of Extreme Point Solutions

We consider the polyhedron that includes all the feasible points for the multi-budgeted version of SDN problem:

$$\mathcal{P}^I = \left\{ y \in \{0,1\}^E : \sum_{e \in \delta(S)} y_e \geq f(S), \forall S \subseteq V \bigwedge \sum_{e \in E_j} \ell_j(e) y(e) \leq L_j, \forall C_j \in \mathcal{C} \right\}$$

where as usual for all $S \subseteq V$, $f(S) = \max_{s_i \in S, t_i \notin S}\{r_i\}$ (and we use the notation y_e and $y(e)$ interchangably). We drop the integral constraints on the variables and this defines the fractional polyhedron $\mathcal{P}^f_{SND-b}$ which from now on we will call it simply $\mathcal{P}_f$ to emphasize we are talking about the fractional polyhedron of the budgeted Survivable Network Design problem. We start with some essential definitions:

Definition 1. *Let $E' \subseteq E$ be a subset of the edges of the graph. Then, we define the **characteristic vector** of E' to be the binary vector $\chi_{E'} \in \{0,1\}^E$ such that $\chi_{E'}(e) = 1 \Leftrightarrow e \in E'$ i.e. the i-th component of $\chi_{E'}$ is 1, if the i-th edge belongs to E' and zero otherwise.*

*Let y be a real-valued vector in a n-dimensional space. Define the **support** of y to be the indices of all the non-zero components of y i.e. $supp(y) = \{i \in [n] : y_i \neq 0\}$.*

*A family $\mathcal{L}$ of subsets of some universe U is called **laminar** if it is not intersecting i.e. for any two subsets $L_1, L_2 \in \mathcal{L}$ it is not the case that all of the $L_1 \setminus L_2$, $L_2 \setminus L_1$ and $L_1 \cap L_2$ are non empty.*

Also we say that a constraint is *tight* if it is satisfied by a solution vector with equality. Our study begins with a characterization of all extreme point (or basic) solutions of $\mathcal{P}_f$ (which carry over to the integral polytope as well):

Lemma 1. *Let the connectivity requirement function $f(S) = \max_{s_i \in S, t_i \notin S}\{r_i\}$ be weakly super-modular and let x be an extreme point solution of $\mathcal{P}_f$ such that $0 < x_e < 1$ for all $e \in E$. Then, there exists a laminar family $\mathcal{L} \subseteq 2^V$, and a set $\mathcal{Q} \subseteq \mathcal{C}$ such that:*

1. *$\sum_{e \in \delta(S)} x_e = f(S)$, $\forall S \in \mathcal{L}$,*
2. *$\sum_{e \in E_j} \ell_j(e) x(e) = L_j$, $\forall C_j \in \mathcal{Q}$,*
3. *the characteristic vectors of $\chi_{\delta(S)}$ for $S \in \mathcal{L}$ and χ_{C_j} for $C_j \in \mathcal{Q}$ are all linearly independent, and*
4. *$|\mathcal{L}| + |\mathcal{Q}| = |E| = |supp(x)|$.*

This Lemma is proved by standard properties of basic feasible solutions [12]: indeed we can form a basic feasible solution by selecting $|E|$ linearly independent constraints, set them to equality and solve the linear system. The last item of the lemma simply says that the number of non-zero variables is the number of linearly independent constraints set to equality. The fact that we can take the family of tight constraints of subsets of vertices $\mathcal{L}$ to be laminar, follows from standard un-crossing arguments [3], [7].

Armed with Lemma 1, we now prove the main theorem of this paper that asserts that for any given bfs of $\mathcal{P}_f$ there exists variable with "high" fractional value:

Theorem 1. *Take any basic feasible solution (a vertex) of the polyhedron $\mathcal{P}_f$ such that the connectivity requirement function is weakly super-modular. Then there always exists a variable e such that $x_e \geq \frac{1}{3}$.*

Proof. The proof will be given by contradiction. In particular, assume we have a basic feasible solution $x \in (0,1)^E$ of $\mathcal{P}_f$ and for all $e \in \mathrm{supp}(x)$ we have that $x_e < \frac{1}{3}$. We will employ a fractional charging scheme which works as follows: Initially each non-zero edge e is given a positive charge of value at most 1. Then, this edge will redistribute fractionally this charge to sets $S \in \mathcal{L}$ and $C_j \in \mathcal{Q}$, where $C_j = c(e)$, in such a way that at the end each tight set $S \in \mathcal{L}$ and $C_j \in \mathcal{Q}$ will receive charge at least one for a total receiving charge of at least $|\mathcal{L}| + |\mathcal{Q}| = |E|$. On the other hand, we will show that the total charge distributed must be *strictly* less than $|E|$, which will give the desired contradiction.

For each edge $e \in \mathrm{supp}(x)$, with endpoints u, v, the distribution scheme we carry on is the following:

1. Edge e gives charge of x_e to its color class $C_j = c(e)$ only if $C_j \in \mathcal{Q}$.
2. For each endpoint u of e, edge e assigns charge of x_e to the minimal set (inclusion-wise) $S \in \mathcal{L}$ that contain u, if such sets exists and
3. Assigns charge $1 - 3x_e$ to the minimal set $S \in \mathcal{L}$ that contains *both* of its endpoints, if such a set exists.

Note that by the hypothesis, $1 - 3x_e > 0$ so, for every edge, each charge distributed is positive. The total charge per edge is at most $1 3x_e + 2x_e + x_e = 1$. We first that not all edges can spend charge of exactly 1:

Claim. There exists at least one edge $e \in \mathrm{supp}(x)$ that assigns charge strictly less than 1.

Proof of Claim: There exists a natural way to represent the laminar family $\mathcal{L}$ as a *forest* which we will also use in the following: we say that $S_i \in \mathcal{L}$ is *child* of $S \in \mathcal{L}$ if S is the smallest (minimal) set in $\mathcal{L}$ containing S_i. In other words S_i is a child of S if there is no other set $S' \subseteq S$ such that $S_i \in S$. Such S is unique. Symmetrically, S is the *parent* of S_i. Any parentless set $S \in \mathcal{S}$ is called *root* of the subtree rooted at S. And childless sets S are called *leaves*. The subtree rooted at S contains all the sets that are descendants (the transitive, reflexive closure of the relation child) of S. Any root $R_i \in \mathcal{L}$ must contain, by definition, edges that cross R_i, i.e., $\delta(R_i) \neq \emptyset$ because $\sum_{e \in \delta(R_i)} = f(R_i)$ and if $f(R_i)$ was equal to zero, then the vector $\chi_{R_i} = 0^E$ is not linearly independent, a contradiction. So, there must be edges crossing any root R_i. Each such edge $e \in \delta_{R_i}$ wastes a charge of $1 - 3x_e > 0$ because it is not the case that both endpoints of such edge belong to the same set $S \in \mathcal{L}$.

So, there are edges that give charge strictly less than 1, for a total assigned charge of strictly less than $|E| = |\mathrm{supp}(x)|$. In the following we will show how each tight object $S \in \mathcal{L}$ and $C_j \in \mathcal{Q}$ receives charge of at least one for a total receiving charge of at least $|\mathcal{L}| + |\mathcal{Q}| = |E|$, which will be a contradiction of the hypothesis of the lemma that such a charging scheme exists which is implied by the hypothesis that $\exists e : x_e \geq 1/3$.

We begin with the case of the color tight classes $C_j \in \mathcal{Q}$, which is (relatively) easier: each such class C_j will receive a charge of x_e from every edge e such that $c(e) = C_j (\Leftrightarrow e \in E_j)$, for a total charge of $\sum_{e \in E_j} x_e$. We must show that this quantity is ≥ 1, which is not immediately obvious:

Claim. Let $C_j \in \mathcal{Q}$ be a color budget constraint that is met with equality by the basic feasible solution $x \in \mathcal{P}_f$ i.e., $\sum_{e \in E_j} x_e \ell_j(e) = L_j$. Then $\ell_j(e) \leq L_j$ $\forall e \in E_j$.

Proof of Claim: We can easily see that it is not the case that $L_j = 0$, since otherwise we would have that $\chi_{C_j} = 0^E$ which is not linearly independent. So, $L_j > 0$ and since the constraint is met with equality we know that $\exists e \in E_j$ such that $x_e > 0$. On the other hand, we see that $\ell_j(e) \leq L_j, \forall e \in E_j$. Indeed, assume not, i.e., $\exists e \in E$ such that $\ell_j(e) > L_j$: then any feasible *integral* solution must have $x_e = 0$, $\forall e \in E_j$, and in such a case we can drop the constraint after we delete all such edges of color C_j. In other words, if $\ell_j(e) > L_j$ for some edge of color C_j, delete the edge from the graph without any loss. All the remaining edges satisfy the desired property that $\ell_j(e) \leq L_j$ $\forall e \in E_j$, for all color classes $C_j \in \mathcal{Q}$. Take such a constraint:

$$\sum_{e \in E_j} x(e)\ell_j(e) = L_j \implies \sum_{e \in E_j} \underbrace{\left\lceil \frac{\ell_j(e)}{L_j} \right\rceil}_{\leq 1} x_e = 1 \implies \sum_{e \in E_j} x_e \geq 1$$

where the last inequality follows by an immediate averaging argument.

The previous claim is enough to prove that the charge received by any color class $C_j \in \mathcal{Q}$ is at least 1.

We now continue to prove that also any set $S \in \mathcal{L}$ receives charge at least 1. As a first case, consider the scenario where $S \in \mathcal{L}$ is a leaf of some tree in the tree-representation of $\mathcal{L}$. Since $S \in \mathcal{L}$, it is the case that $\sum_{e \in \delta(S)} x_e = f(S)$ and by definition $f(S)$ is an integer at least 1. This set will receive charge from all the crossing edges $e \in \delta(S)$ for a total charge $\sum_{e \in \delta(S)} x_e = f(S) \geq 1$ (remember that edge e with one endpoint on S will give S charge x_e if S is the smallest set containing one endpoint of e. Since S is a leaf, it is in fact the smallest set that contains one endpoint of all the crossing edges).

Now assume that $S \in \mathcal{L}$ is not a leaf and that it has q children $Z_1, Z_2, \ldots Z_q$ all of which are in $\mathcal{L}$ (by definition). This means that $\sum_{e \in \delta(Z_i)} x_e = f(Z_i)$, a positive integer and the same is of course true for S. The charge received by S comes from different set of edges:

1. Let A be the set of edges that cross between two children of S, i.e., $A = \{e = \{u, v\} \in E : u \in Z_i \land v \in Z_j\}$ for two distinct $i, j \in [q]$. Since S is the

smallest (minimal) set $S \in \mathcal{L}$ that contains *both* endpoints of all these edges in A, each such edge will assign charge of $1 - 3x_e$ to S, for a total charge of $|A| - 3\sum_{e \in A} x_e$.

2. Let $B = \{e = \{u, v\} \in E : u \in Z_i \wedge v \in S\}$, i.e., B contains all edges with one endpoint on one of S's children and the other endpoint in S (not in any other child of S). By the charging scheme, these edges will give to S charge of x_e (one endpoint of e is "owned" by S and thus charge of x_e is received) plus a charge of $1 - 3x_e$ because S is the minimal set containing both of e's endpoints for a total charge of $\sum_{e \in B} 1 - 2x_e = |B| - 2\sum_{e \in B} x_e$.

3. Let $\Gamma = \{e \in \delta(S) : e \cap \{S \setminus \cup_{i \in q} Z_i\} \neq \emptyset\}$. In other words, Γ contains all edges that cross S with one endpoint in $S \setminus \cup_{i \in q} Z_i$. Each of these edges $e \in \Gamma$ give S a charge of x_e (because S is the smallest set that contains exactly one endpoint of e) for a total charge of $\sum_{e \in \Gamma} x_e$.

4. Let Δ be the rest of edges, i.e., $\Delta = \{\cup_{i \in [q]} \{\delta(Z_i)\}\}$ such that $e = \{u, v\} \cap S = \emptyset\}$, i.e., Δ contains all edges that cross Z_j *and* S at same time. These edges do not assign any charge to S.

We note that if $|A| = |B| = |\Gamma| = 0$, then $\chi_S = \sum_{i \in [q]} \chi_{Z_i}$, i.e., the characteristic vectors of $S, Z_1, \ldots, Z_q$ are not linearly independent, a contradiction. So at least one of A, B, Γ must be nonempty and so the charge received from S is strictly greater than zero. By the previous four cases, S receives total charge of

$$\Omega = |A| - 3\sum_{e \in A} x_e + |B| - 2\sum_{e \in B} x_e + \sum_{e \in \Gamma} x_e =$$

$$|A| + |B| - \left(\sum_{e \in A} x_e + \sum_{e \in B} x_e\right) + \underbrace{\left(\sum_{e \in \Gamma} x_e - 2\sum_{e \in A} x_e - \sum_{e \in B} x_e\right)}_{f(S) - \sum_{i \in [q]} f(Z_i) \ = \ \Psi \in \mathbb{Z}} > 0$$

We want to show that the above expression (which we already know is positive) is in fact at least 1. Lets call it Ω and lets call the quantity $f(S) - \sum_{i \in [q]} f(Z_i)$ as Ψ. Note that Ψ, although obviously an integer as a sum and subtraction of integer quantities, might not always be positive. Finally, let $\Xi = \Omega - \Psi = |A| + |B| - (\sum_{e \in A} x_e + \sum_{e \in B} x_e)$. We divide our task into several subcases:

First Case: $|A| = |B| = 0$. Then of course we have also $\sum_{e \in A} x_e + \sum_{e \in B} x_e = 0$ and so $\Omega = \Psi = f(S) - \sum_{i \in [q]} f(Z_i) > 0$. But the last expression is a positive expression involving only integers and so is ≥ 1.

Second Case: $|A| + |B| \geq 1$. In this case notice we have that $|A| - \sum_{e \in A} x_e > \frac{2}{3}|A|$ and $|B| - \sum_{e \in B} x_e > \frac{2}{3}|B|$ since, by hypothesis, every edge has fractional value $< \frac{1}{3}$. There are several subcases to consider:

$\Psi > 0$: This actually means (since Ψ is an integer) that $\Psi \geq 1 \implies \Omega \geq 1$ (in fact $\Omega > 1$, but for our purposes the inequality is enough).

$\Psi = 0$: In this case, we first note that if $|A| + |B| \geq 2$ then we have that $\Omega > 2 \cdot \frac{2}{3} > 1$ and we are done. The interesting case is of course when

$|A| + |B| = 1$. Assume, w.l.o.g., that $|A| = 1$ and $|B| = 0$ end let e' be the lonely edge in A (between two of the children of S) such that $0 < x_{e'} < 1/3$. Since $f(S) = \sum_{j \in [q]} f(Z_j) \implies \sum_{e \in \Gamma} x_e = \sum_{e \in A} x_e = x_{e'}$. Let Z_1, Z_2 the two children of S that contain the different endpoints of e'. We can assume without any loss that S has only two children: otherwise, since all edges of the rest of the children are part of $\delta(S)$ and they sum to an integer, we can safely ignore them, reducing appropriately $f(S)$. Let us define

$$\alpha = \sum_{e \in \delta(Z_1) \setminus \{e'\}} x_e \quad \text{and} \quad \beta = \sum_{e \in \delta(Z_2) \setminus \{e'\}} x_e.$$

We have that: $\alpha + x_{e'} = f(Z_1)$, $\beta + x_{e'} = f(Z_2)$, $\alpha + \beta + \sum_{e \in \Gamma} x_e = \alpha + \beta + x_{e'}$ $= f(S) = f(Z_1) + f(Z_2) = \alpha + \beta + 2x_{e'}$. Since all the previous quantities are positive integers, we conclude that $2x_{e'}$ is an integer $\implies x_{e'} \in \{0, \frac{1}{2}, 1\}$, a contradiction by the hypothesis. So, $|A| + |B|$ must be ≥ 2, and we are done.

$\Psi < 0$: Let's assume that $f(S) - \sum_{j \in [q]} f(Z_j) = -k$ for some positive integer k. From the definition of Ω we have that $\Omega = |A| - \sum_{e \in A} x_e + |B| - \sum_{e \in B} x_e - k > 0$. Then we have that

$$f(S) - \sum_{j \in [q]} f(Z_j) = \sum_{e \in \Gamma} x_e + \sum_{e \in \Delta} x_e - 2 \cdot \sum_{e \in A} x_e - \sum_{e \in B} x_e - \sum_{e \in \Delta} x_e = k$$

$$\implies 2 \sum_{e \in A} x_e + \sum_{e \in B} x_e - \sum_{e \in \Gamma} x_e = k \tag{3}$$

By using the definition of Ω we can easily derive a bound on $|A| + |B|$ as function of k:

$$\Omega = |A| - \sum_{e \in A} x_e + |B| - \sum_{e \in B} x_e - k > 0$$

$$\implies |A| + |B| \geq \left\lceil \frac{3k}{2} \right\rceil \tag{4}$$

First, assume that $|\Gamma| = 0$ and so $2 \sum_{e \in A} x_e + \sum_{e \in B} x_e = k$. Then we have that:

$$\Omega = |A| - \sum_{e \in A} x_e + |B| - \sum_{e \in B} x_e - k$$

$$= \underbrace{|A| + |B|}_{> 3k/2} - \underbrace{\left(2 \sum_{e \in A} x_e + \sum_{e \in B} x_e \right)}_{= k} + \underbrace{\sum_{e \in A} x_e}_{\geq 0} - k$$

$$\geq \sum_{e \in A} x_e - \frac{k}{2} > 0 \tag{5}$$

from which we immediately conclude that $2 \sum_{e \in A} x_e > k$, which is a contradiction because $2 \sum_{e \in A} x_e + \sum_{e \in B} x_e = k$ and all the quantities involved are positive. So, $|\Gamma| \neq 0$.

Again, as before we have that $-\Psi = 2\sum_{e \in A} x_e + \sum_{e \in B} x_e - \sum_{e \in \Gamma} x_e = k$ and so we have that

$$\Omega = |A| - \sum_{e \in A} x_e + |B| - \sum_{e \in B} x_e - k$$

$$= |A| + |B| - \underbrace{\left(2\sum_{e \in A} x_e + \sum_{e \in B} x_e - \sum_{e \in \Gamma} x_e\right)}_{=k} + \sum_{e \in A} x_e - \sum_{e \in \Gamma} x_e - k$$

$$= \underbrace{|A| + |B|}_{>3k/2} + \sum_{e \in A} x_e - \sum_{e \in \Gamma} x_e - 2k$$

$$> \sum_{e \in A} x_e - \sum_{e \in \Gamma} x_e - \frac{k}{2} > 0$$

Since $2\sum_{e \in A} x_e + \sum_{e \in B} x_e - \sum_{e \in \Gamma} x_e = k$, with the help of the previous calculations, we conclude that $\sum_{e \in A} x_e + \sum_{e \in B} x_e < \frac{k}{2}$. Using the definition of Ω we have that

$$\Omega = |A| + |B| - \left(\sum_{e \in A} x_e + \sum_{e \in B} x_e\right) - k > 0$$

$$\Rightarrow |A| + |B| - \frac{3k}{2} > 0 \tag{6}$$

Now, if $k \mod 2 = 0$, the above expression is a positive expression involving only integers, and so it must be ≥ 1, in which case we are done. So assume that k is an odd positive integer. The worst case (actually the only worst case) is when $|A| + |B| - \frac{3k}{2} = \frac{1}{2} \Rightarrow |A| + |B| = \frac{3k+1}{2}$.

From the definition of the fractional charge received by S ($\Omega = |A| - 3\sum_{e \in A} x_e + |B| - 2\sum_{e \in B} x_e + \sum_{e \in \Gamma} x_e$) we have that if either of $\sum_{e \in \Gamma} x_e$ or $|B| - 2\sum_{e \in B} x_e$ is greater or equal to 1, then $\Omega > 1$ (since $|A| - 3\sum_{e \in A} x_e > 0$). So we assume that (1) $\sum_{e \in \Gamma} x_e < 1$ and (2) $|B| - 2\sum_{e \in B} x_e < 1 \Rightarrow |B| \leq 2$ (because, as above, $|B| - 2\sum_{e \in B} x_e > \frac{1}{3}|B|$).

We will consider the case where $|B| = 0$. The other two cases ($|B| \in \{1, 2\}$) are simpler subcases which we omit in the current reading.

Since $|B| = 0$, we have that $2\sum_{e \in A} x_e - \sum_{e \in \Gamma} x_e = k$. Let's denote the corresponding summations as α and γ respectively, i.e., $2\alpha + \gamma = k \Rightarrow \alpha = \frac{k+\gamma}{2}$ and $0 < \gamma < 1$ by hypothesis. From the definition of Ω we have that

$$\Omega = |A| - 3\alpha + \gamma > 0$$

$$\Longrightarrow |A| > \frac{3k + \gamma}{2} \geq \lceil \frac{3k+1}{2} \rceil \tag{7}$$

If $|A| > \frac{3k+1}{2} \geq \frac{3k+3}{2}$ (since $|A|$ is odd) then we are done. This can be easily seen because we have that $\Omega = |A| - 3\alpha + \gamma \geq \frac{3k+3}{2} - \frac{3k+3\gamma}{2} + \gamma = \frac{3-\gamma}{2} > 1$

because $\gamma < 1$. So, assume that $|A| = \frac{3k+1}{2}$, which means that we have $\frac{3k+1}{2}$ edges summing up to exactly $\frac{k+\gamma}{2}$. Now, if $\gamma \geq \frac{1}{3}$, we see that the average fractional value of the edges in A is $\frac{k+\gamma}{3k+1} > \frac{1}{3}$, which is a contradiction because of the assumption that all edges have fractional value $x_e < \frac{1}{3}$. So, γ must be $< \frac{1}{3}$.

We are left with the case where $|A| = \frac{3k+1}{2}$ and $\gamma < 1/3$. We will show that this case is contradicting the initial assumptions. We will do this by providing a counting argument: in fact, we will show that in the reduced instance including S and all of its children $\{Z_i\}_{i \in [q]}$ the number of edges (which are in fact the edges in A, Γ and Δ) is *greater* that the number on tight linearly independent constraints giving rise to this reduced instance. This will be a contradiction since, according to Lemma 1, the number of edges and tight linearly independent constraints characterizing this solution should be equal. We will first show that if we restrict the initial basic feasible solution x to this sub-instance, we are still left with a basic feasible solution, so that we can apply in the next step Lemma 1 and derive our contradiction.

Claim. Let I be an instance of the budgeted Survivable Steiner Network problem and x be a basic feasible solution to the LP for I. Let S be a set belonging to the laminar family $\mathcal{L}$ characterizing x (together with $\mathcal{Q}$). Assume that S has q children $Z_j, j \in [q]$ in $\mathcal{L}$. Then the restriction of x, x_S, to the sub-instance I_S generated by S and its children is a basic feasible solution for LP to I_S.

Proof of Claim: The fact that x_S is a feasible solution for I_S its trivial from the feasibility of x for I. In fact we can create x_S from x by simply zeroing out all the entries of x that correspond to entries (edges) different from the sets A, Γ, Δ. Assume that x_S is not basic. This means that there are two basic feasible solutions x_S^1 and x_S^2 such that $x_S = \lambda x_S^1 + (1-\lambda) x_S^2$, $\lambda \in (0, 1)$. But then if we expand x_S^j by including all the entries of x that correspond to edges not included in the sub-instance, we have a feasible solution for I. So, x can be written as a convex combination of the two "expanded" solutions created x_S^1 and x_S^2 with the same multiplier λ contradicting the fact that x is basic.

Assume that S has q children $Z_j, j \in [q]$ and that there are $|A| = \frac{3k+1}{2}$ edges running between the the Z_j's, where $k = \sum_{j \in [q]} f(Z_j) - f(S)$. Moreover assume that $\gamma = \sum_{e \in \Gamma} x_e < \frac{1}{3}$. We see that if there exist a child $Z_p, p \in [q]$, such that all edges in $\delta(Z_p) \in \Delta$ then we can assume that this child does not belong to $\mathcal{L}$ without any loss of generality, by just reducing the connectivity requirement of S by $f(Z_p)$. This does not change feasibility. So, we assume that there exist at least one edge in A in every set $\delta(Z_j)$ for every child j of S.

We will deliver a lower bound lb on the number of edges and an upper bound ub on the number of tight linearly independent constraints in the sub-instance I_S. By the above Claim, we do not need to count the edges that are fully inside children of S neither we would take into consideration and children of children of S. We will see that $lb > ub$, which constitutes a contradiction by Lemma 1. Since $\gamma < 1/3$, we see that $|\Gamma| \geq 1$. We already know that $|A| = \frac{3k+1}{2}$. Each edge of $|A|$ runs between two children of S, so in the maximum case, S would have at most $3k+1$ children and this is the worst case (for our counting argument). We will count now the edges in Δ. We remind that these are the edges going directly from any of the children Z_j of S to outside S. If S has $3k+1$ children, then since each child of S has one edge in A and its connectivity requirement is an integer ≥ 1, then it must have at least 3 edges in Δ (since all edges are $< 1/3$) i.e. $|\delta(Z_j) \cap \Delta| \geq 3, \forall j \in [q]$ and so $|\Delta| \geq 3(3k+1)$. Moreover, each tight budget constraint requires at least 4 edges. So, the maximum number of such budget constraints is $\leq \frac{|A|+|\Gamma|+|\Delta|}{4} \leq \frac{|A|+1+|\Delta|}{4}$ for a total maximum number of tight objects in $\mathcal{L}_S$ and $\mathcal{Q}_S$ to be $\leq 1 + (3k+1) + \frac{|A|+1+|\Delta|}{4} < \frac{3k+1}{2} + 3(3k+1) + 1 = lb$. A contradiction.

This proves that $|A|$ should be strictly more than $\frac{3k+1}{2}$, and by the previous cases we are done.

This proves that every tight object $S \in \mathcal{L}$ and $C_j \in \mathcal{Q}$ receives a fractional token value at least one, for a total fractional token distributed of $|\mathcal{L}|+|\mathcal{Q}| = |E|$. But from the previous claim, we know that the total fractional token distributed is strictly less than $|E|$ (each edge distributes at least one but there must exist at least one edge distributing strictly less than one fractional token). Contradiction. So the hypothesis must be false, and so there must exists an edge $e \in E : x_e > 1/3$. □

3   A $(3,3)$ Bi-criteria Algorithm

Given Theorem 1, the algorithm is relatively simple and is depicted in Algorithm 3. Although the proof that 3 is a $(3,3)$ bi-criteria algorithm differs only slightly with other similar proofs, we will include it here for completeness.

Algorithm 1. Iterative Algorithm for the multi-budgeted SND problem

Input: An instance of the multi-budgeted Survivable Network Design problem.

Output: A subgraph of G in which all the connectivity requirements are satisfied.

1. **Initialize:** $f' = f$ and $H = \emptyset$.
2. **while** $f' \neq f$:
    (α) Solve the relaxed LP as defined by (2) with f' as connectivity requirement function plus the remaining budgets. Let x be the bfs obtained.
    (β) Let $X = \{e \in E : x_e \geq 1/3\}$. Include all $e \in X$ in H.
    (γ) **Update Budgets:** If $e \in X \cap C_j \Rightarrow L_j := L_j - x_e$.
    (δ) **Update Connectivity Requirements:** $f'(S) := f(S) - |\delta(S) \cap X|, \forall S \in \mathcal{S}$.
3. **Return:** H.

Step 2.(α) can be solved in polynomial time using the fact that the new connectivity requirement function f' is again *weakly* super-modular and using the Ellipsoid method with the usual severation oracle used to solve the initial LP, see [3,13] for implementation details. Step 2.(δ) can be easily implemented to run in polynomial time: we do not need to update all $S \in 2^V$, but rather all $S \in \mathcal{L}$ such that $S \cap X \neq \emptyset$, where $\mathcal{L}$ is the laminar family which, together with $\mathcal{Q}$, defines the current extreme point solution x, see Lemma 1.

Theorem 2. *Algorithm 3 returns a solution $\bar{x} \in \{0,1\}^E$ for which the following holds: (1) $c^T \bar{x} \geq 3opt$ where opt is the optimal fractional solution cost of the initial LP and(2) $\sum_{e \in E_j} \bar{x}(e)\ell_j(e) \leq 3 \cdot L_j$.*

Proof. . This is a standard inductive argument, but for completeness we provide it here as well. The correctness of the algorithm is implied by Theorem 1.

First of all, to prove the violation of the budgets, it suffices to notice that in each iteration for every edge we round to 1, we decrease its budget by at least $1/3$. More formally, take a budget $C_j : \sum_{e \in E_j} x(e)\ell_j(e) \leq L_j$ and let T be the set of edges which are greater than $1/3$ and we rounded up to 1 at some point in the algorithm. At the end, the budget will be $\sum_{e \in T} \ell_j(e) \leq \sum_{e \in T} 3\bar{x}(e)\ell_j(e) \leq 3L_j$.

To prove that the algorithm approximates within a factor of 3 the (lower bound of the) optimal objective function value, we employ an inductive argument: for the base case assume that the algorithm terminates only after one iteration. This means that X defines a feasible set of edges. Since $X = \{e \in E : \bar{x}_e \geq 1/3\} \Rightarrow \sum_{e \in X} c_e \leq \sum_{e \in X} 3\bar{x}_e c_e \leq 3opt$, where as usual $opt = \sum_e \bar{x}_e c_e$. Now assume that the claim is true for up to k iterations. We will show that the claim remains true if the algorithm takes one more, i.e., for $k+1$ iterations.

Let X' be the current set of edges we round to 1 in the current iteration and let f' be the residual connectivity requirement after we round all edges in X' to 1. Let X'' be all the set of edges picked in subsequent iterations to satisfy the new connectivity requirement function f'. It is not hard to observe that the current solution $\bar{x}$ at the k-th iteration is also a feasible solution for f' if we restrict $\bar{x}$ to the edges in $E \setminus \{X'\}$. By inductive hypothesis the cost of X'' is at most $3 \sum_{e \in E \setminus X'} x_e c_e$ and set $X''' = X'' \cup X'$. The cost of X''' is

$$\sum_{e \in X'''} c_e x_e = \sum_{e \in X''} c_e x_e + \sum_{e \in X'} c_e x_e$$

$$\leq 3 \sum_{e \in E \setminus X'} x_e c_e + c(X') \leq 3 \sum_{e \in E} c_e x_e$$

where the last inequality follows since all edges in X' have fractional value $\geq 1/3$ and this concludes that the cost of the solution is at most thrice the cost of the optimal solution cost. $\square$

Acknowledgements. The work of the author was supported by the Swiss National Science Foundation Early Post-Doc mobility grant P1TIP2_152282.

References

1. Gabow, H.N.: On the L_8-norm of extreme points for crossing supermodular directed network IPs. In: Jünger, M., Kaibel, V. (eds.) IPCO 2005. LNCS, vol. 3509, pp. 392–406. Springer, Heidelberg (2005)
2. Goemans, M.X., Goldberg, A.V., Plotkin, S.A., Shmoys, D.B., Tardos, É., Williamson, D.P.: Improved approximation algorithms for network design problems. In: Sleator, D.D. (ed.) Proceedings of the Fifth Annual ACM-SIAM Symposium on Discrete Algorithms, Arlington, Virginia, January 23-25, pp. 223–232. ACM/SIAM (1994)
3. Jain, K.: A factor 2 approximation algorithm for the generalized steiner network problem. Combinatorica 21(1), 39–60 (2001)
4. Khanna, S., Naor, J., Shepherd, F.B.: Directed network design with orientation constraints. In: Shmoys, D.B. (ed.) Proceedings of the Eleventh Annual ACM-SIAM Symposium on Discrete Algorithms, San Francisco, CA, USA, January 9-11, pp. 663–671. ACM/SIAM (2000)
5. Krysta, P.: Bicriteria network design via iterative rounding. In: Wang, L. (ed.) COCOON 2005. LNCS, vol. 3595, pp. 179–187. Springer, Heidelberg (2005)
6. Lau, L.C., Naor, J., Salavatipour, M.R., Singh, M.: Survivable network design with degree or order constraints. In: Johnson, D.S., Feige, U. (eds.) Proceedings of the 39th Annual ACM Symposium on Theory of Computing, San Diego, California, USA, June 11-13, pp. 651–660. ACM (2007)
7. Lau, L.C., Ravi, R., Singh, M.: Iterative Methods in Combinatorial Optimization. Cambridge University Press (2011)
8. Lau, L.C., Singh, M.: Additive approximation for bounded degree survivable network design. In: Dwork, C. (ed.) Proceedings of the 40th Annual ACM Symposium on Theory of Computing, Victoria, British Columbia, Canada, May 17-20, pp. 759–768. ACM (2008)
9. Louis, A., Vishnoi, N.K.: Improved algorithm for degree bounded survivable network design problem. In: Kaplan, H. (ed.) SWAT 2010. LNCS, vol. 6139, pp. 408–419. Springer, Heidelberg (2010)
10. Melkonian, V., Tardos, É.: Algorithms for a network design problem with crossing supermodular demands. Networks 43(4), 256–265 (2004)
11. Nagarajan, V., Ravi, R., Singh, M.: Simpler analysis of LP extreme points for traveling salesman and survivable network design problems. Oper. Res. Lett. 38(3), 156–160 (2010)
12. Schrijver, A.: Theory of Linear and Integer Programming. John Wiley & Sons (1998)
13. Vazirani, V.V.: Approximation Algorithms. Springer (2004)
14. Williamson, D.P., Goemans, M.X., Mihail, M., Vazirani, V.V.: A primal-dual approximation algorithm for generalized steiner network problems. Combinatorica 15(3), 435–454 (1995)

Simple Distributed $\Delta + 1$ Coloring
in the SINR Model*

Fabian Fuchs and Roman Prutkin

Karlsruhe Institute for Technology (KIT)
Karlsruhe, Germany
{fabian.fuchs,roman.prutkin}@kit.edu

Abstract. In wireless ad hoc networks, distributed node coloring is a fundamental problem closely related to establishing efficient communication through TDMA schedules. For networks with maximum degree Δ, a $\Delta + 1$ coloring is the ultimate goal in the distributed setting as this is always possible. In this work we propose a very simple 4Δ coloring along with a color reduction technique to achieve $\Delta + 1$ colors. All algorithms have a runtime of $\mathcal{O}(\Delta \log n)$ time slots. This improves on previous algorithms for the SINR model either in terms of the number of required colors or the runtime, and matches the runtime of local broadcasting in the SINR model (which can be seen as an asymptotical lower bound).

1 Introduction

One of the most fundamental problems in wireless ad hoc or sensor networks is efficient communication. Indeed, most algorithms concerned with the physical or *Signal-to-Interference-and-Noise-Ratio* (SINR) model consider algorithms to establish initial communication right after the network begins to operate. However, those initial methods of communication are not very efficient, as there are either frequent collisions and reception failures due to interference, or time is wasted in order to provably avoid such collisions and failures. If local broadcasting [10, 13, 18] is used, a multiplicative $\mathcal{O}(\Delta \log n)$ factor is required to execute message-passing algorithms in the SINR model, where Δ is the maximum degree in the network (we use a broadcasting range to define neighborhood in the SINR model, cf. Section 2). Thus, wireless networks often use a more refined transmission schedule as part of the Medium Access Control (MAC) layer. One of the most popular solutions to the medium access problem are Time-Division-Multiple-Access (TDMA) schedules, which provide efficient communication by assigning nodes to time slots. The main problem in establishing a TDMA schedule can be reduced to a distributed node coloring. Given a node coloring, we can establish a transmission schedule by simply associating each color with one time slot. The node coloring considered in this work ensures that two nodes capable of communicating directly with each other do not select the same color. Note that a TDMA schedule based on such a coloring is not yet feasible in the SINR model.

---

* The full version of this work is available as [6].

© Springer International Publishing Switzerland 2015
C. Scheideler (Ed.): SIROCCO 2015, LNCS 9439, pp. 149–163, 2015.
DOI: 10.1007/978-3-319-25258-2_11

However, a feasible TDMA schedule can be computed based on our coloring, for example as shown in [3, 7].

The problem of distributed node coloring dates back to the early days of distributed computing in the mid-1980s. In contrast to the centralized setting, a $\Delta + 1$ coloring is considered to be the ultimate goal in distributed node coloring as it is already NP-complete to compute the chromatic number (i.e., the minimum number of colors required to color the graph) in the centralized setting [9]. There is a rich line of research in this area, however, most of the work has been done for message-passing models like the $\mathcal{LOCAL}$ model. Such models are designed for wired networks and do not fit the specifics of wireless networks.

In the SINR model, also denoted as the physical model due to its common use in electrical engineering, wireless communication is modelled based on the signal transmission and a geometric decay of the signal strength. It improves on other models for wireless communication, such as the protocol model, which considers interference as a local and binary property by declaring a transmission to be successful iff it is not in the interference range of another transmitting node. It has been shown that such models are quite limited, as protocols designed for the SINR model surpass the theoretically achievable performance of protocols designed for the protocol model [15]. In this work, we use two simple and well-known algorithms (covered for example in [2]) designed for message-passing models, and show that we can *efficiently* execute the algorithms in the SINR model. However, this cannot be achieved by a simple simulation of each round of the message passing algorithm by one execution of local broadcasting as this results in a runtime of $\mathcal{O}(\Delta \log^2 n)$ time slots. Instead, we modify both the communication rounds in the SINR model and the algorithms to perfectly fit together. The synergy effect of our careful adjustments is that the coloring algorithm runs in $\mathcal{O}(\Delta \log n)$ time slots, which is asymptotically exactly the runtime of one local broadcast [10]. This matches the runtime of current $\mathcal{O}(\Delta)$ coloring algorithms [3], and improves on current $\Delta + 1$ coloring algorithms for the SINR model which require $\mathcal{O}(\Delta \log n + \log^2 n)$ or $\mathcal{O}(\Delta \log^2 n)$ time slots [19].

The communication between nodes in our algorithm is based on the local broadcasting algorithm proposed by Goussevskaia *et al.* [10]. Thus, we require the nodes to know an upper bound on the maximum number of nodes in a node's surroundings (which we call proximity area, cf. Section 2), an upper bound on the number of nodes in the network, as well as some model-related hardware constants in order to enable initial communication. All our results hold *with high probability* (w.h.p.), i.e., with probability at least $1 - \frac{1}{n^c}$, where n is the number of nodes, and $c \geq 1$ a constant. As union bounding a w.h.p. event only decreases the constant c, resulting in a constant increase in the runtime, we refrain from stating exact w.h.p. bounds in our analysis to simplify notation. Note that such requirements and assumptions are common in the SINR model.

1.1 Related Work and Contributions

Due to the rich amount of work on distributed node coloring in the message-passing model, we refer to a recent monograph by Barenboim and Elkin [2] for a thorough overview on distributed graph coloring. In wireless networks, the SINR model received increasing attention first in the electrical engineering community, and was picked up by the algorithms community due to a seminal work by Gupta and Kumar [12]. An overview of works regarding transmission scheduling in the SINR model can be found in a survey by Goussevskaia, Pignolet and Wattenhofer [11]. A coloring algorithm due to Moscibroda and Wattenhofer [14] has been adapted to the SINR model by Derbel and Talbi [3], and extended to support directed communication by Fuchs and Wagner [8]. Derbel and Talbi provide an algorithm that computes an $\mathcal{O}(\Delta)$ coloring in $\mathcal{O}(\Delta \log n)$ time slots. Their algorithm first computes a set of leaders using a maximal independent set (MIS, cf. Section 2) algorithm, then leader nodes assign colors to non-leaders, which again compete for their final color with a restricted number of neighboring nodes that may have received the same assignment. Yu *et al.* [19] propose two $\Delta + 1$ coloring algorithms that do not require the knowledge of the maximum node degree Δ. Their first algorithm runs in $\mathcal{O}(\Delta \log n + \log^2 n)$ time slots and assumes that nodes are able to increase their transmission power for the computation. This prevents conflicts between non-leader nodes by allowing the set of leaders to directly communicate to other leaders outside the transmission region and thus coordinating the assignment process. Their second algorithm does not require this assumption, and runs in $\mathcal{O}(\Delta \log^2 n)$ time slots.

Our main contributions are 1. a simple and efficient 4Δ coloring algorithm, requiring $\mathcal{O}(\Delta \log n)$ time slots; 2. an abstract method that has the potential of improving the runtime of other randomized algorithms in the SINR model by a $\log n$ factor; and 3. an asynchronous color reduction scheme, which, combined with known coloring algorithms computes a $\Delta + 1$ coloring in overall $\mathcal{O}(\Delta \log n)$ time slots. Also, the color reduction simplifies to an almost trivial color reduction scheme yielding the same results restricted to the synchronous setting.

The coloring algorithms improve current algorithms in the same setting (cf. Derbel and Talbi [3]) regarding the number of colors, and achieve the declared goal of $\Delta+1$ colors, while the runtime is matched. Other $\Delta+1$ coloring algorithms in the SINR model require at least $\mathcal{O}(\Delta \log n + \log^2 n)$ time slots (under noncomparable assumptions). Our new method to improve the runtime by a $\log n$ factor carefully combines the uncertainty in randomized algorithms with the uncertainty in the SINR model to handle them simultaneously in the analysis. For more details, we refer to the Analysis of Algorithm 1 in Section 3.1.

Roadmap: In the next section we state the model along with required definitions. In Section 3 the simple 4Δ coloring algorithm is described and analyzed. We introduce and analyse the color reduction scheme in Section 4.

2   Model and Preliminaries

The *Signal-to-Interference-and-Noise-Ratio* (SINR) model is used to model if a transmission in a wireless network can be successfully decoded at the intended receivers or not. We say that a transmission from a sender to a receiver is *feasible* if it can be decoded by the receiver. In the SINR model it depends on the ratio between the desired signal and the sum of interference from other nodes plus the background noise whether a certain transmission is successful. Let each node v in the network use the same transmission power P. Then a transmission from u to v is feasible if and only if $\frac{\frac{P}{\text{dist}(u,v)^\alpha}}{\sum_{w \in I} \frac{P}{\text{dist}(w,v)^\alpha} + \text{N}} \geq \beta$, where α, β are constants depending on the hardware, N reflects the environmental noise, $\text{dist}(u,v)$ the Euclidean distance between two nodes u and v, and $I \subseteq V$ is the set of nodes transmitting simultaneously to u. The *broadcasting range* r_B of a node v defines the range around v up to which v's messages should be received. We denote the set of *neighbors* of v by $N_v := \{w \in V \backslash \{v\} \mid \text{dist}(v,w) \leq r_B\}$ and $N_v^+ := N_v \cup \{v\}$. Based on the SINR constraint, the *transmission range* of $r_T \leq (\frac{P}{\beta \text{N}})^{1/\alpha}$ is an upper bound for the broadcasting range (with $r_B < r_T$ to allow multiple simultaneous transmissions). Let the *broadcasting region* B_v be the disk with range r_B centered at v. To prove successful communication within the broadcasting range, we need the concept of a *proximity range* $r_A > 2r_B$ around a node v as introduced in [10]. Let Δ_A^v be the number of nodes with distance less than r_A to v, and $\Delta^A := \max_{v \in V} \Delta_A^v$. It holds that $\Delta^A \in \mathcal{O}(\Delta)$. As further technical details of the proximity range are not required in our analysis, we refer to [10] or [6] for the exact definitions.

The communication graph $G = (V, E)$ is defined as follows. The set of vertices V in the graph corresponds to the set of nodes in the network, while there is an edge $(u, v) \in E$ if and only if u and v are neighbors (i.e., they are within each other's broadcasting range). The *maximum degree* in the network is $\Delta := \max_{v \in V} |N_v|$. Note that since $r_B < r_T$, a node v may successfully receive transmissions from nodes that are not its neighbors in the communication graph, although successful transmission from those nodes cannot be guaranteed. As the signal strength decreases geometrically in the SINR model, we assume that messages from outside the broadcasting range are discarded by considering the signal strength of a received message (usually provided by wireless receivers as the Received-Signal-Strength-Indication (RSSI) value [1]). Thus, the maximum degree is defined as for the $\Delta + 1$ coloring in [19]. In a more practical setting, one could also define the communication graph based on the actual communication between two nodes.

We call two nodes $v, u \in V$ *independent* if they are not neighbors. A set $S \subseteq V$ such that the nodes in S are pairwise independent is called *independent set*. Obviously, $S \subseteq V$ is a *maximal independent set* (MIS) if S is independent and there is no $v \in V \backslash S$ with $S \cup \{v\}$ independent. We denote the set of integers $\{0, \ldots, i\}$ by $[i]$. Let us now define the coloring problem. Given a set of nodes V so that each node $v \in V$ has a color c_v, and let d be an integer. Then V has a *valid* $d+1$ *coloring*, if for each node v holds $\forall w \in N_v : c_v \neq c_w$ and $c_v \in [d]$. Observe that in a valid coloring each color in the network forms a independent set.

In the *synchronous setting*, we assume that nodes start the algorithm at the same time. In the more realistic *asynchronous setting*, arbitrary wake-up of nodes is allowed, and we do not require synchronized time slots; precise clocks, however, are assumed. With the so-called ALOHA trick [16], e.g. as used in [10], we use time slots in our analysis, although the nodes do not assume common time slots.

The nodes use two different transmission probabilities in order to adapt to the requirements of the corresponding algorithms. Probability $p_1 := \frac{1}{2\Delta^\lambda}$ is used in Algorithm 1, while Algorithm 2 uses p_1 and $p_2 := \frac{1}{180}$. If $p \geq c$ for probability p and a constant c, we say that p is at least constant, or simply constant. Let c be an arbitrary constant with $c > 1$. Throughout the paper, we use the following definitions: $\kappa_\ell := c\lambda \ln n / p_\ell$ for $\ell = 1, 2$, $\kappa_0 := \lambda \ln 12 / p_1$, and λ a constant (for more details, we refer to [6]). Note that $\kappa_0 \in \mathcal{O}(\Delta)$, $\kappa_1 \in \mathcal{O}(\Delta \log n)$, and $\kappa_2 \in \mathcal{O}(\log n)$.

Extending Local Broadcasting: We show that local broadcasting with constant success probability in time reversely proportional to the transmission probability can be achieved. This extends known results regarding local broadcasting, which guarantee local broadcasting with *high* probability for a fixed number of time slots. Although we are the first to use local broadcasting with constant success probability, the proof of the following lemma is mainly based on standard techniques. Thus, we defer it to [6] due to space constraints.

Lemma 1. *Let v be a node transmitting with probability p_1, then it successfully transmits to its neighbors with probability $\geq 11/12$ within κ_0 time slots. Transmissions with probability p_ℓ for κ_ℓ time slots are successful w.h.p. for $\ell \in \{1, 2\}$.*

3 Simple 4Δ Coloring

The algorithm we propose is at its heart a very simple and well-known randomized coloring algorithm. The underlying approach is well-known, and for example covered in [2, Chapter 10]. Essentially, this kind of algorithms draw a random color whenever two neighboring nodes have the same color (i.e., there is a conflict between them). Our first algorithm, RAND4DELTACOLORING (Algorithm 1), is a simple, phase-based coloring algorithm. We say that two neighbors v, w have a *conflict* if $c_v = c_w$ and denote the temporary color of v in phase t by c_v^t. In each phase t the node v checks whether it knows of a conflict with one of its neighbors. The set of neighbors that are in a conflict with v in phase t is $X^t(v) := \{w \in N_v | c_v^t = c_w^t\}$. We call $X^t(v)$ the *conflict set* of v in phase t, and denote the event that v is in a conflict in phase t by $\mathcal{E}_{\text{confl}}^t(v) := \exists w \in X^t(v)$. Note that there may be nodes in $X^t(v)$, for which v is not aware of the conflict (due to the uncertainty in the nodes communication), however, this does not affect the event. If a conflict is detected by v, the node randomly draws a new

---

Algorithm 1. RAND4DELTACOLORING for node v

---
1 $F_v \leftarrow [4\Delta]$, $c_v^{-1} \leftarrow F_v.\text{rand}()$
2 **for** $t \leftarrow 0; t \leq 6(c+3)\ln n; t \leftarrow t+1$ **do**   // each one phase
3   **if** $c_v^{t-1} \notin F_v$ **then** $c_v^t \leftarrow F_v.\text{rand}()$ // if conflict, new color
4   **else** $c_v^t \leftarrow c_v^{t-1}$ // otherwise, keep it
5       $F_v \leftarrow [4\Delta]$
6   Transmit c_v^t with probability p_1 for κ_0 time slots
7   **foreach** *received color c_w^t from neighbor* $w \in N_v$ **do** $F_v \leftarrow F_v \backslash \{c_w^t\}$

---

color from the set F_v of colors not taken by a neighbor in the previous phase and transmits this color in the current phase. The event that a transmission from v to all neighbors N_v of v in phase t is *successful* is $\mathcal{E}^t_{\text{succ}}(v)$. A transmission from v to its neighbors in phase t is not successful or *fails* if at least one neighbor was unable to receive the message. The corresponding event is $\mathcal{E}^t_{\text{fail}}(v)$. We replace $\mathcal{E}$ by $\mathbb{P}$ to denote the probability of an event, e.g. $\mathbb{P}^t_{\text{succ}}(v)$ for $\mathcal{E}^t_{\text{succ}}(v)$. Note that although the events $\mathcal{E}^t_{\text{succ}}(v)$, and $\mathcal{E}^t_{\text{fail}}(v)$ may not be independent of events happening at other nodes, our bounds on the corresponding probabilities $\mathbb{P}^t_{\text{succ}}(v)$ and $\mathbb{P}^t_{\text{fail}}(v)$ are independent from the node v and possible events at other nodes. Also, our bounds $\mathbb{P}^t_{\text{succ}}(v)$ and $\mathbb{P}^t_{\text{fail}}(v)$ on these events include the event that v reaches some but not all of its neighbors, as $\mathbb{P}^t_{\text{fail}}(v) \leq 1 - \mathbb{P}^t_{\text{succ}}(v) \leq 1/12$ and $11/12 \leq \mathbb{P}^t_{\text{succ}}(v) \leq 1$ (see Lemma 1). Finally, the phase is concluded by transmitting the current color. This computes a valid coloring with 4Δ colors in $\mathcal{O}(\log n)$ phases, while each phase takes $\mathcal{O}(\Delta)$ time slots. In contrast to previous algorithms of this kind, we do not assume that successful communication is guaranteed by lower layers. Instead we allow the uncertainty in the randomized algorithm to be combined with the uncertainty in the communication in the SINR model, which is jointly handled in the analysis. Thereby we can reduce the number of time slots required for each phase by a $\log n$ factor (from $\mathcal{O}(\Delta \log n)$ for the trivial analysis to $\mathcal{O}(\Delta)$), making this simple approach viable in the SINR model. Thus, Algorithm 1 solves the node coloring problem using 4Δ colors in $\mathcal{O}(\Delta \log n)$ time slots, which matches the runtime of local broadcasting in the SINR model and improves the state-of-the-art $\mathcal{O}(\Delta)$ coloring in [3]. Let us now state the main results of this section.

Theorem 2. *Let all nodes start executing Algorithm 1 simultaneously. After the execution, all nodes have a valid color $c_v \leq 4\Delta$ w.h.p.*

For the asynchronous setting, the bound on the runtime holds for node v only after all nodes in v's $\log n$ neighborhood are awake

Corollary 3. *Let a node v execute Algorithm 1 in the asynchronous setting. Then v has a valid color $c_v \leq 4\Delta$ w.h.p., at most $\mathcal{O}(\Delta \log n)$ time slots after all nodes in its $\mathcal{O}(\log n)$-neighborhood started executing the algorithm.*

In the following section we prove the result for the synchronous setting. In Section 3.2 we briefly discuss extending it to the asynchronous setting. Our experiments in Section 3.3 show that the algorithm is very fast and robust even in the asynchronous setting.

3.1 Analysis of RAND4DELTACOLORING

Despite the fact that the underlying coloring algorithm is well-known, our analysis is new and quite involved. The main reason for this is the uncertainty in whether a message is successfully delivered in one phase of Algorithm 1. In contrast to guaranteed message delivery, based for example on local broadcasting, message delivery with constant probability can be achieved a logarithmic factor faster, see Lemma 1. However, this reduction in runtime comes at a cost: While in the guaranteed message delivery setting, a node v can finalize its color once a phase without a conflict at v happened, this is not possible in our setting. We cannot guarantee the validity of the colors even if a node did not receive a message implying a conflict in one phase, as message transmission is successful only with constant probability. Nevertheless, we can show that after $\mathcal{O}(\log n)$ phases of transmitting the selected color and resolving eventual conflicts, the coloring is valid in the entire network w.h.p.

In order to prove correctness of Algorithm 1 (RAND4DELTACOLORING) we shall first bound the probability of a conflict propagating from one phase of the algorithm to the next. This is the foundation for the result that our algorithm computes a valid 4Δ coloring in $\mathcal{O}(\Delta \log n)$ time slots w.h.p. for both the synchronous and the asynchronous setting. Assuming that a node v has a conflict in phase t, there are only two cases that may lead to a conflict at v in phase $t + 1$:

1. Node v had a conflict in phase t, and it did not get resolved (either due to being unaware of the conflict or since the new color implies a conflict as well).
2. A neighbor of v had a conflict in phase t and introduced the conflict by randomly selecting v's color.

We shall show that the probability for both cases is at most constant (see Lemma 4). Thus, after $\mathcal{O}(\log n)$ phases it holds with high probability that a valid color has been found. Note that the results in this section are restricted to the synchronous setting, however, they can be extended to the asynchronous case, cf. Section 3.2.

Lemma 4. *Let v be an arbitrary node and $\mathbb{P}^t_{\text{confl}}(v)$ the probability of a conflict at v in phase t. Then the probability of a conflict at v in phase $t + 1$ is at most*

$$\mathbb{P}^{t+1}_{\text{confl}}(v) \leq \frac{5}{6} \cdot \max_{w \in N_v} \mathbb{P}^t_{\text{confl}}(w).$$

Proof. We shall prove the lemma by considering the two cases that may lead to a conflict at node v in phase $t + 1$. The **first** case is that v has a conflict with at least one of its neighbors. Depending on which transmissions are successful there are 3 subcases. Note that $\rightarrow$ denotes $\mathcal{E}^t_{\text{succ}}(v)$, while $\leftarrow$ denotes $\exists w \in X^t(v) :$ $\mathcal{E}^t_{\text{succ}}(w)$—with negations accordingly[1].

(a) $\nrightarrow, \nleftarrow$: It is not guaranteed that any of the conflict partners know of the conflict, as the transmissions from v and the nodes in the conflict set $X^t(v) \neq \emptyset$ failed at least partially. There is at least one neighbor $u \in X^t(v)$ that failed to transmit its color successfully to v, which happens with probability $\mathbb{P}^t_{\text{fail}}(u)$. Combined with v's failure to transmit its color successfully, case 1(a) happens

---

[1] A partial success of transmission is often sufficient to trigger dealing with a conflict. We do not consider this in our notation, however, as we evaluate $\mathbb{P}^t_{\text{succ}}(v)$ to be at most 1 for all v and since $\Pr(\text{transmission from } v \text{ to } u \text{ fails}) \leq \mathbb{P}^t_{\text{fail}}(v) \leq 1/12$, our analysis covers this case.

with probability at most $\mathbb{P}^t_{\text{confl}}(v)(\mathbb{P}^t_{\text{fail}}(v)\Pr(\nrightarrow)) \leq \mathbb{P}^t_{\text{confl}}(v)\mathbb{P}^t_{\text{fail}}(v)\mathbb{P}^t_{\text{fail}}(u) \leq \mathbb{P}^t_{\text{confl}}(v)(1/12)^2$. If any conflict partner knows of the conflict, the conflict would be resolved with a certain probability (as in the following cases). However, as this is not guaranteed, we account for the worst case: the conflict is not resolved and propagates to the next phase. Note that since this case happens only with a small probability, it holds that the total probability of case (a) and conflict at v in phase $t+1$ is small.

(b) $\rightarrow, \nrightarrow$: All nodes in $X^t(v)$ failed to transmit successfully, but v transmitted successfully to all neighbors. Thus, all nodes in $X^t(v)$ know of the conflict, while v might be unaware of it. This case happens with probability at most $\mathbb{P}^t_{\text{confl}}(v) \cdot (\mathbb{P}^t_{\text{succ}}(v) \cdot \Pr(\nrightarrow))$. The probability that a node $w \in X^t(v)$ selects v's color in phase $t+1$ is at most $\sum_{w \in X^t(v)} 1/|F_w|$ (even if v knows of a conflict and itself selects a new color). This results in an overall probability of at most

$$\mathbb{P}^t_{\text{confl}}(v) \cdot (\mathbb{P}^t_{\text{succ}}(v) \cdot \Pr(\nrightarrow)) \cdot \sum_{w \in X^t(v)} \frac{1}{|F_w|}$$

$$\leq \mathbb{P}^t_{\text{confl}}(v) \left(\prod_{w \in X^t(v)} \mathbb{P}^t_{\text{fail}}(w) \right) \cdot \sum_{w \in X^t(v)} \frac{1}{|F_w|}$$

$$\overset{x:=|X^t(v)|}{\leq} \mathbb{P}^t_{\text{confl}}(v) \left(\mathbb{P}^t_{\text{fail}} \right)^x \cdot \frac{x}{3\Delta} \leq \frac{1}{3\Delta} \mathbb{P}^t_{\text{confl}} \cdot x \left(\frac{1}{12} \right)^x \leq \frac{1}{24} \mathbb{P}^t_{\text{confl}}$$

where the first inequality holds since the event $\nrightarrow$ is equivalent to $\forall w \in X^t(v) :$ $\mathcal{E}^t_{\text{fail}}(w)$ and $\mathbb{P}^t_{\text{succ}}(v) \leq 1$. The second inequality holds since $|F_w| \geq 3\Delta$ as w and v are uncolored and by setting $x = |X^t(v)|$. The last inequality holds since $x(1/12)^x \leq 1/12$ for all $x \in \{1, \dots, \Delta\}$, and $\Delta \geq 1$.

(c) $\leftarrow$: It holds that v knows of the conflict. Whether v's neighbors know of it or not is not guaranteed. This case happens with probability at most $\mathbb{P}^t_{\text{confl}}(v) \cdot$ $(\Pr(\leftarrow))$. The probability that at least one neighbor of v has or selects the same color as v is at most $\sum_{w \in N_v} \frac{1}{|F_v|} \leq |N_v| \frac{1}{3\Delta} \leq \frac{1}{3}$.

Using $\Pr(\leftarrow) \leq 1$, this results in a probability for a conflict at v in phase $t+1$ of at most $\mathbb{P}^t_{\text{confl}}(v) \cdot (1/144 + 1/24 + 1/3 \cdot \Pr(\leftarrow)) < \left(\frac{1}{2} \right) \cdot \mathbb{P}^t_{\text{confl}}$.

In the **second** case, there was no conflict at v in phase t, but a neighbor w of v selected v's color due to a conflict at w, which happens with probability at most

$$\sum_{w \in N_v} \underbrace{\Pr(c_v^{t+1} = c_w^{t+1})}_{v\text{'s neighbor } w \text{ selects } v\text{'s color}} \sum_{u \in N_w} \underbrace{\Pr(c_u^t = c_w^t)}_{\substack{u \in N(w) \text{ told } w \\ \text{about their conflict}}}$$

$$\leq \sum_{w \in N_v} \Pr(c_v^{t+1} = c_w^{t+1}) \mathbb{P}^t_{\text{confl}}(w)$$

$$\leq \sum_{w \in N_v} \frac{1}{|F_w|} \mathbb{P}^t_{\text{confl}}(w) \leq \left(\frac{1}{3} \right) \max_{w \in N_v} \mathbb{P}^t_{\text{confl}}(w)$$

The last inequality holds since $\sum_{w \in N_v} \frac{1}{|F_w|} \leq \sum_{w \in N_v} \frac{1}{3\Delta} \leq \frac{1}{3}$. Combining all events that could lead to a conflict at v in phase $t+1$ it holds that the probability of the union of the events is at most

$$\mathbb{P}^{t+1}_{\text{confl}}(v) \leq \left(\frac{1}{2}\right) \mathbb{P}^t_{\text{confl}}(v) + \left(\frac{1}{3}\right) \max_{w \in N_v} \mathbb{P}^t_{\text{confl}}(w) \leq \frac{5}{6} \cdot \max_{w \in N_v^+} \mathbb{P}^t_{\text{confl}}(w),$$

which concludes the proof. $\square$

Note that the second case could be avoided if message delivery in each phase would be guaranteed, as a node v that does not have a conflict in phase t, would simply finalize its current color and communicate this. Thus, v could not be forced into a conflict anymore. We shall now show that a set of nodes executing Algorithm 1 computes a valid coloring, and hence prove Theorem 2.

Proof (of Theorem 2). Let us consider the probability of a conflict at an arbitrary node $v \in V$ in phase $t = 6(c+3)\ln n$. It holds that

$$\mathbb{P}^t_{\text{confl}}(v) \leq \left(\frac{5}{6}\right) \max_{w \in N_v} \mathbb{P}^{t-1}_{\text{confl}}(w) \leq \left(\frac{5}{6}\right) \max_{w \in V} \mathbb{P}^{t-1}_{\text{confl}}(w)$$

$$\leq \left(\frac{5}{6}\right)^t \max_{w \in V} \mathbb{P}^0_{\text{confl}}(w) \leq \left(1 - \frac{1}{6}\right)^{6(c+3)\ln n} \leq \frac{1}{n^{c+3}},$$

where the first inequality is due to Lemma 4. The third inequality holds since all nodes are in the same phase due to the synchronous start of the algorithm. Note that the upper bound on the probability that a conflict propagates holds for all nodes. The fourth inequality holds as $\mathbb{P}^0_{\text{confl}}(v) \leq 1$ for all nodes v. The last inequality holds due to a well-known mathematical fact (cf. [6]). Thus, the probability for a conflict at an arbitrary node v is small. A union bound over all nodes in the network implies that the coloring is valid w.h.p. The runtime of Algorithm 1 is $\mathcal{O}(\Delta \log n)$, as it consists of $6(c+3)\ln n = \mathcal{O}(\log n)$ phases, and each phase takes $\kappa_0 = \mathcal{O}(\Delta)$ time slots according to Lemma 1. $\square$

3.2 Asynchronous Simple Coloring

Let us now briefly consider the asynchronous setting. For this section, we call all nodes that can reach v within $\mathcal{O}(\log n)$ rounds the neighborhood of v, and say that this neighborhood is stable if those nodes are all awake. If the neighborhood of a node v is stable, Lemma 4 holds as well, with only small changes to some constants in the proof [5]. Thus, once all nodes in v's neighborhood are awake, we can bound the probability using said lemma, and prove Corollary 3 analog to the proof of Theorem 2.

3.3 Experimental Evaluation

In our experiments, we evaluate RAND4DELTACOLORINGusing the well-known network simulator *sinalgo* [4]. We use between 500 and 2500 nodes, uniformly

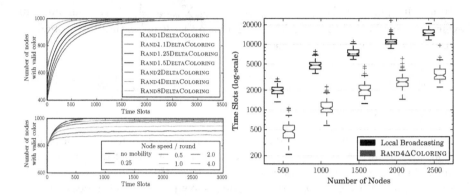

Fig. 1. Top left: Progress for varying number of available colors; Bottom left: Robustness under mobility constraints; Right: Runtime compared with local broadcasting

deployed on a square area of 1000×1000 meters. The SINR constants are set to $\alpha = 4$, $\beta = 10$, $N = 1^{-9}$, $P = 1$, resulting in a transmission range of 100 meters. We set the broadcasting range to 84 meters, with average degree values ranging from 10 to about 50. We generally use asynchronous simulation and the nodes start at a random within the first 10 time slots. However, as sinalgo requires synchronous simulation for the mobility models, this experiment uses synchronized time slots. The time required to transmit one message is set to 1 time slot. We measure the number of time slots, and the number of nodes that have a valid color. The nodes do not know the global Δ value, but use the number of neighbors (plus one) as an estimate. More experiments are shown in the full version [6]. We observe three main points in Fig. 1. First, RAND4DELTACOLORING is very fast, requiring less time than one round of local broadcasting. Second, the algorithm is relatively robust, even under moderate mobility values of 1 meter per time slot, more than 90% of the nodes have a valid color (note that using mobile nodes, some color conflicts cannot be avoided). Finally, we can see that although our theoretical guarantees hold only for 4Δ colors, the algorithm can compute a valid coloring with only $\Delta + 1$ colors in our setting.

4   Asynchronous Color Reduction

In the following section we assume a valid node coloring with $d \in \mathcal{O}(\Delta)$ colors to be given and reduce the number of colors to $\Delta + 1$ in $\mathcal{O}(d \log n)$ time slots. Let us first consider a very simple synchronous variant, which is also well known in the $\mathcal{LOCAL}$ model, cf. [2, Section 3.2]. In this variant, each node transmits its current color at the beginning of each round. Then, in the first round all nodes with color d select a color from the set $[\Delta]$, in the second round all those with color $d - 1$, etc. This translates to an almost trivial (but new) color reduction scheme for the synchronous case, which we defer to [6] due to space constraints. The algorithm we present in this section circumvents the synchronization problem, essentially, by using two levels of MIS executions. Our algorithm is illustrated in Fig. 2,

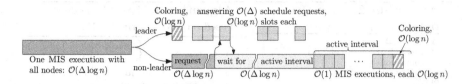

Fig. 2. Runtime. Overall $\mathcal{O}(\Delta \log n)$, given a $\mathcal{O}(\Delta)$ coloring.

the corresponding pseudocode can be found as Algorithms 2 to 5. We reference the MIS (Algorithm 3) executed with parameter $\ell = 1$ by *first level MIS*, and MIS($\ell = 2$) by *second level MIS*.

Let us now describe the algorithm in more detail. The algorithm starts by executing the first level MIS algorithm that determines a set of independent nodes, which we call leaders. Each leader node transitions to Algorithm 4, selects and transmits the color 0 it selected and initializes its periodic leader schedule. This schedule assigns each color an *active interval* of length $\mathcal{O}(\log n)$ time slots to allow the nodes of this color to select their final color from $[\Delta]$.

Each node v_i that is not in the first level MIS selects a leader from its broadcasting range and requests the relative time until it is v_i's turn to be *active*. Upon receipt of its active intervall, the node waits until the interval starts and then executes a second level MIS algorithm (which does not interfere with the first level MIS) for a constant number of times. In this second level MIS the algorithm benefits from fewer active nodes, and hence more efficient communication to allow each node to achieve successful transmission of a message to all neighbors in $\mathcal{O}(\log n)$ time slots. Moreover, we can speed up the MIS algorithm by the same factor of Δ to execute it in $\mathcal{O}(\log n)$ time slots, as only a constant number of nodes compete to be in each second level MIS. For each node that wins the second level MIS, there is no other node of the second level MIS in its broadcasting range. Thus, the winning node can select a valid color from $\{1, \ldots, \Delta\}$ and transmit its choice to its neighbors without a conflict. If a node does not succeed to be in the second level MIS, it simply executes MIS(2) again. As each node succeeds in such an MIS within its active interval, each node selects one of the $\Delta + 1$ colors.

4.1 MIS, and Notation for ASYNCCOLORREDUCTION

Let us now describe the notation used in the algorithm. We denote the set of available colors by F_v. Note that throughout the algorithm, each node deletes the final colors it received from F_v. The MIS algorithm (Algorithm 3) aims at allowing exactly one node in each neighborhood to succeed to Algorithm COLORED, select a color, and annouce its success in the MIS algorithm to its competitors. There are minor differences depending on the two levels $\ell = 1$ and $\ell = 2$, however, the algorithm remains the same. A description of the MIS algorithm can be found in [3], and the full version [6]. In Algorithm 4, v is a leader, col_v denotes the final color from $[\Delta]$, and Q is a queue used to store nodes w along with their initial color $\text{col}_w^{\text{tmp}}$ that request an active interval. The remaining time is based

Algorithm 2. ASYNCCOL-ORREDUCTION for node v

1 $F_v \leftarrow [\Delta] \backslash \{0\}$
2 **foreach** *received* col_w **do** continuously
3 $\quad \lfloor F_v \leftarrow F_v \backslash \{\text{col}_w\}$
4 MIS(1)

Algorithm 4. COLORED(ℓ) for node v

1 **if** $\ell = 1$ **then**   // Level 1 leader
2 $\quad$ $\text{col}_v \leftarrow 0$, $Q \leftarrow \emptyset$, $c'_v = 0$
3 $\quad$ announce $M_C^1(v, \text{col}_v)$ with prob. p_2 for κ_2 slots
4 $\quad$ Set $\tau(\text{col}, c_v) \equiv \text{col} \cdot \mu - c_v$ mod $\Delta\mu$ neg., max., with $|\tau(\text{col}, c_v)| > \kappa_2$
5 $\quad$ **while** *protocol is executed* **do**
6 $\quad\quad$ // serve requests
7 $\quad\quad$ $c'_v \leftarrow c'_v + 1$
8 $\quad\quad$ transmit $M_C^1(v, \text{col}_v)$ with probability p_1
9 $\quad\quad$ **foreach** *received request from neighbor* w: $M_R(w, v, \text{col}_w^{tmp})$ **do** continuously
10 $\quad\quad\quad$ $\lfloor Q.\text{push}((w, \text{col}_w^{tmp}))$
11 $\quad\quad$ **if** Q *not empty* **then**
12 $\quad\quad\quad$ $(w, \text{col}_w^{tmp}) \leftarrow Q.\text{pop}()$, $t \leftarrow \tau(\text{col}_w^{tmp}, c'_v)$
13 $\quad\quad\quad$ **for** $\mathcal{O}(\log n)$ *slots* **do**
14 $\quad\quad\quad\quad$ $\lfloor$ transmit $M_C^1(v, w, t)$ with probability p_2 // inc. c'_v, t

15 **else** // Level 2 / Non-leader node
16 $\quad$ $\text{col}_v \leftarrow F_v.\text{rand}()$ // valid
17 $\quad$ announce $M_C^2(c, \text{col}_v)$ with prob. p_2 for κ_2 slots
18 $\quad$ **while** *protocol is executed* **do**
19 $\quad\quad$ // keep color valid
20 $\quad\quad$ $\lfloor$ transmit col_v with prob. p_1

Algorithm 3. MIS(ℓ) for node v, based on MW-coloring [3,17]

1 $P_v = \emptyset$, NEXT $= \begin{cases} \text{LEVEL2} & \text{if } \ell = 1 \\ \text{MIS}(2) & \text{otherwise} \end{cases}$
2 **for** κ_ℓ *time slots* **do**   // Listen first
3 $\quad$ **foreach** $w \in P_v$ **do** $d_v(w) = d_v(w) + 1$
4 $\quad$ **if** $M_A^\ell(w, c_w)$ received **then** $P_v = P_v \cup \{w\}$; $d_v(w) = c_w$
5 $\quad$ $\lfloor$ **if** $M_C^\ell(w)$ received **then** NEXT(w)
6 $c_v = \Xi(P_v)$ // minimal, non-positive, not conflicting with competing counters in P_v
7 **while** *true* **do** // then compete for MIS
8 $\quad$ $c_v = c_v + 1$
9 $\quad$ **if** $c_v > \kappa_\ell$ **then** COLORED(ℓ) // success
10 $\quad$ **foreach** $w \in P_v$ **do** $d_v(w) = d_v(w) + 1$
11 $\quad$ **if** $M_C^\ell(w)$ received **then** NEXT(w)
11 $\quad$ transmit $M_A^\ell(v, c_v)$ with probability p_ℓ
12 $\quad$ **if** $M_A^\ell(w, c_w)$ *received* **then** // received competing counter
13 $\quad\quad$ $P_v = P_v \cup \{w\}$; $d_v(w) = c_w$
14 $\quad\quad$ $\lfloor$ **if** $|c_v - c_w| \le \kappa_\ell$ **then** $c_v = \Xi(P_v)$

Algorithm 5. LEVEL2(w) for node v with leader w

1 **while** *true* **do**
2 $\quad$ **if** $M_C^1(w, v, t)$ *received* **then**
3 $\quad\quad$ **while** $t < 0$ **do** // wait for interval
4 $\quad\quad\quad$ $\lfloor t \leftarrow t + 1$ // one time slot each
5 $\quad\quad$ **while** $t < 2k^2 \kappa_2$ **do**   // active interval
6 $\quad\quad\quad$ // increase t by one in each time slot during MIS(2)
7 $\quad\quad\quad$ $\lfloor$ MIS(2)
8 $\quad$ **else**   // transmit request
9 $\quad\quad$ $\lfloor$ transmit $M_R(v, w, \text{col}_v^{tmp})$ with probability p_1

---

on v's periodic schedule, which is defined by its counter value c'_v, and w's color. We set $k = 90$, which corresponds to the maximum number of active nodes in a broadcasting range, see Lemma 7. The function $\tau(\text{col}_w, c_v)$ intuitively sets t to the start of the next interval corresponding to w's color in v's schedule, so that the starting time of w can be communicated by v w.h.p. before w's active interval starts. During the transmission interval, t is decreased appropriately.

Adapting the MIS Algorithm. We assume in the analysis that the MIS algorithm indeed computes a maximal independent set. Algorithm 3 is a simplification of the MIS part of the coloring algorithm in [3,17], and therefore computes an MIS. Apart from constant changes, the lemma follows directly from Theorems 1 and 2 in [3] if $\ell = 1$, and from Lemma 1 along with setting Δ to a constant in the proofs of both theorems for $\ell = 2$.

Lemma 5. *Algorithm 3 computes an MIS among participating nodes in $\delta_\ell \kappa_2 \in$ $\mathcal{O}(\delta_\ell \log n)$ time slots, where $\delta_\ell = \begin{cases} \Delta & \text{if } \ell = 1 \\ k & \text{if } \ell = 2 \end{cases}$ w.h.p.*

4.2   Analysis

Let us first state the main result of this section.

Theorem 6. *Given a valid node coloring with $d \geq \Delta$ colors, Algorithm 2 computes a valid $\Delta + 1$ coloring in $\mathcal{O}(d \log n)$.*

As the algorithm is essentially a simple color reduction scheme, each node selects a valid color if the communication can be realized as claimed. To prove this we show that in the second level indeed only a constant number of nodes are active in each broadcasting range (cf. Fig. 3). We use this to achieve message transmission from active nodes to all their neighbors in $\mathcal{O}(\log n)$ time slots, and show that the second level MIS can be executed in $\mathcal{O}(\log n)$ time slots. Finally, we prove that each non-leader node v succeeds in a second level MIS, and thus colors itself with a color from $[\Delta]$, within the active interval v is assigned by its leader. The proofs of the following lemmas are only given in the full version [6].

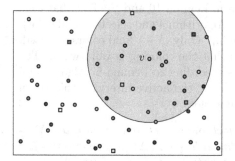 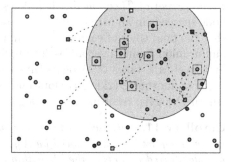

Fig. 3. Left: Node v with its broadcasting region in a network with valid coloring; Nodes in the first level MIS are squares. Right: Nodes in v's broadcasting range are connected to their selected leader by a dashed line. Nodes currently active in the second level are surrounded by a square.

Lemma 7. *In the second level, at most k nodes are active in each broadcasting range.*

The lemma follows from a geometric argument regarding the number of first level MIS nodes within a certain distance of each node. We use Lemma 7 to prove our bounds on the communication in Lemma 1. It allows us to increase the transmission probability in the second level MIS by a factor of Δ compared to classical local broadcasting, leading to a decrease in the time required for successful message transmission by the same factor of Δ to $\mathcal{O}(\log n)$. Based on this result we can bound the runtime of our algorithm, starting with Algorithm 5.

Lemma 8. *Let v execute Algorithm 5 with leader w. Then a) v transmits the request message successfully within κ_1 time slots w.h.p.; b) v receives its active interval after at most another κ_1 time slots w.h.p.; and c) the wait-time t until v's active interval starts is at most $\Delta 2k^2\kappa_2 \in \mathcal{O}(\Delta \log n)$.*

We shall now argue that each non-leader node succeeds to win a second level MIS in its active interval.

Lemma 9. *Given a node v executing Algorithm 5. Once $t = 0$, v wins a second level MIS set within $2k^2\kappa_2$ time slots.*

Essentially, this holds as there are multiple consecutive MIS executions, each allowing one node per broadcast range to win MIS, select a final color and withdraw. In the next MIS execution, another node wins, selects a color, etc. until all active nodes are colored. As a final step we show that the final color selected by each node is valid w.h.p.

Lemma 10. *Given a node v entering Algorithm 4. It holds that a) while v transmits its final color no neighbor of v succeeds in a second level MIS w.h.p.; and b) the color v selects is not selected by one of v's neighbors w.h.p.*

We are now able to prove the main theorem. Note that runtime bounds hold for each node once the node starts executing the algorithm.

Proof (Proof of Theorem 6). It follows from Lemma 10 and the fact that each node succeeds in an MIS (and hence enters Algorithm 4 and selects a final color), that the final color of each node is valid w.h.p. Only $\Delta + 1$ final colors are used, and a union bound over all nodes implies that the coloring is valid w.h.p. The first level MIS takes $\mathcal{O}(\Delta \log n)$ time slots according to Lemma 5. Algorithm 5 requires another $\mathcal{O}(\Delta \log n)$ slots until starting the active interval, which is of length $\mathcal{O}(\log n)$, resulting in $\mathcal{O}(\Delta \log n)$ time slots. □

Corollary 11. *Let each node in the asynchronous network execute the MW-coloring algorithm [3], followed by Algorithm 2. Then $\mathcal{O}(\Delta \log n)$ time slots after a node started executing the algorithms it selected a valid color from $[\Delta]$.*

5   Conclusion

We conclude that the proposed distributed 4Δ coloring algorithm is simple and very fast. RAND4DELTACOLORING performs well in our simulations, even in the asynchronous and mobile setting. Additionally, our color reduction scheme is the first $\Delta + 1$ coloring algorithm achieving a runtime of $\mathcal{O}(\Delta \log n)$, matching one round of local broadcasting.

Acknowledgements. We thank Magnús M. Halldórsson for helpful discussions on an early stage of this work, and the German Research Foundation (DFG), which supported this work within the Research Training Group GRK 1194 "Self-organizing Sensor-Actuator Networks".

References

1. Bardwell, J.: Converting signal strength percentage to dbm values. WildPackets' White Paper (2002)
2. Barenboim, L., Elkin, M.: Distributed Graph Coloring: Fundamentals and Recent Developments. Synthesis Lectures on Distributed Computing Theory. Morgan & Claypool Publishers (2013)
3. Derbel, B., Talbi, E.G.: Distributed Node Coloring in the SINR Model. In: Proc. 30th Internat. Conf. on Distributed Computing Systems (ICDCS 2010). pp. 708–717. IEEE (2010)
4. Distributed Computing Group, ETH Zurich: Sinalgo - simulator for network algorithms (2008), http://sourceforge.net/projects/sinalgo/, version 0.75.3
5. Fuchs, F.: On asynchronous node coloring in the SINR model (2015), http://i11www.iti.kit.edu/f-oancs-15.pdf (unpublished manuscript)
6. Fuchs, F., Prutkin, R.: Simple distributed delta + 1 coloring in the SINR model. CoRR abs/1502.02426 (2015), http://arxiv.org/abs/1502.02426
7. Fuchs, F., Wagner, D.: On Local Broadcasting Schedules and CONGEST Algorithms in the SINR Model. In: Proc. 9th Internat. Workshop on Algorithmic Aspects of WSN (ALGOSENSORS 2013). pp. 170–184. Springer (2013)
8. Fuchs, F., Wagner, D.: Local broadcasting with arbitrary transmission power in the SINR model. In: Proc. 21st Internat. Colloq. Structural Inform. and Comm. Complexity (SIROCCO 2014), pp. 180–193. Springer (2014)
9. Garey, M.R., Johnson, D.S.: Computers and Intractability: A Guide to the Theory of NP-Completeness. W. H. Freeman & Co. (1979)
10. Goussevskaia, O., Moscibroda, T., Wattenhofer, R.: Local Broadcasting in the Physical Interference Model. In: Proc. 5th ACM Internat. Workshop on Foundations of Mobile Computing (DialM-POMC 2008), pp. 35–44. ACM (2008)
11. Goussevskaia, O., Pignolet, Y.A., Wattenhofer, R.: Efficiency of wireless networks: Approximation algorithms for the physical interference model. Foundations and Trends in Networking 4(3) (November 2010)
12. Gupta, P., Kumar, P.R.: The capacity of wireless networks. IEEE Trans. on Inform. Theory 46(2), 388–404 (2000)
13. Halldórsson, M.M., Mitra, P.: Towards Tight Bounds for Local Broadcasting. In: Proc. 8th ACM Internat. Workshop on Foundations of Mobile Computing (FOMC 2012). ACM (2012)
14. Moscibroda, T., Wattenhofer, M.: Coloring Unstructured Radio Networks. J. Distr. Comp. 21(4), 271–284 (2008)
15. Moscibroda, T., Wattenhofer, R., Weber, Y.: Protocol design beyond graph-based models. In: Proc. of the ACM Workshop on Hot Topics in Networks (HotNets-V), pp. 25–30 (2006)
16. Roberts, L.G.: Aloha packet system with and without slots and capture. SIGCOMM Comput. Commun. Rev. 5(2), 28–42 (1975)
17. Schneider, J., Wattenhofer, R.: Coloring unstructured wireless multi-hop networks. In: Proc. 28th ACM Symp. on Principles of Distributed Computing (PODC 2009), pp. 210–219. ACM (2009)
18. Yu, D., Hua, Q.S., Wang, Y., Lau, F.C.M.: An $O(\log n)$ Distributed Approximation Algorithm for Local Broadcasting in Unstructured Wireless Networks. In: Proc. 8th Internat. Conf. on Distributed Computing in Sensor Systems (DCOSS 2012), pp. 132–139. IEEE (2012)
19. Yu, D., Wang, Y., Hua, Q.S., Lau, F.C.M.: Distributed $(\Delta + 1)$ Coloring in the Physical Model. Theoret. Comput. Sci. 553, 37–56 (2014)

Nearly Optimal Local Broadcasting in the SINR Model with Feedback

Leonid Barenboim[1,*] and David Peleg[2,**]

[1] Department of Mathematics and Computer Science,
The Open University of Israel, Raanana, Israel
leonidb@openu.ac.il
[2] Department of Computer Science and Applied Mathematics,
The Weizmann Institute of Science, Rehovot, Israel
david.peleg@weizmann.ac.il

Abstract. We consider the SINR wireless model with uniform power. In this model the success of a transmission is determined by the ratio between the strength of the transmission signal and the noise produced by other transmitting processors plus ambient noise. The *local broadcasting* problem is a fundamental problem in this setting. Its goal is producing a schedule in which each processor successfully transmits a message to all its neighbors. This problem has been studied in various variants of the setting, where the best currently-known algorithm has running time $O(\bar{\Delta} + \log^2 n)$ in n-node networks with feedback, where $\bar{\Delta}$ is the maximum neighborhood size [9]. In the latter setting processors receive free feedback on a successful transmission. We improve this result by devising a local broadcasting algorithm with time $O(\bar{\Delta} + \log n \log \log n)$ in networks with feedback. Our result is nearly tight in view of the lower bounds $\Omega(\bar{\Delta})$ and $\Omega(\log n)$ [13]. Our results also show that the conjecture that $\Omega(\bar{\Delta} + \log^2 n)$ time is required for local broadcasting [9] is not true in some settings.

We also consider a closely related problem of *distant-k coloring*. This problem requires each pair of vertices at geometrical distance of at most k transmission ranges to obtain distinct colors. Although this problem cannot be always solved in the SINR setting, we are able to compute a solution using an optimal number of *Steiner points* (up to constant factors). We employ this result to devise a local broadcasting algorithm that after a preprocessing stage of $O(\log^* n \cdot (\bar{\Delta} + \log n \log \log n))$ time obtains a local-broadcasting schedule of an optimal (up to constant factors) length $O(\bar{\Delta})$. This improves upon previous local-broadcasting algorithms in various settings whose preprocessing time was at least $O(\bar{\Delta} \log n)$ [3,10,5]. Finally, we prove a surprising phenomenon regarding the influence of the path-loss exponent α on performance of algorithms. Specifically, we show that in vacuum ($\alpha = 2$) any local broadcasting algorithm requires $\Omega(\bar{\Delta} \log n)$ time, while on earth ($\alpha > 2$) better results are possible as illustrated by our $O(\bar{\Delta} + \log n \log \log n)$-time algorithm.

* Part of this work has been performed while the author was a postdoctoral fellow at a joint program of the Simons Institute at UC Berkeley and I-CORE at Weizmann Institute.
** Supported in part by the Israel Science Foundation (grant 1549/13) and the I-CORE program of the Israel PBC and ISF (grant 4/11).

© Springer International Publishing Switzerland 2015
C. Scheideler (Ed.): SIROCCO 2015, LNCS 9439, pp. 164–178, 2015.
DOI: 10.1007/978-3-319-25258-2_12

1   Introduction

Setting and Problems.   We consider the SINR (Signal-to-Interference-plus-Noise-Ratio) wireless setting with uniform power. In this setting a set V of n processors (also called *vertices*) is placed on the plane in an arbitrary manner. The vertices perform local computations and send messages. A message sent from a vertex $x \in V$ to a vertex $y \in V$ successfully arrives if the transmitting signal is sufficiently strong with respect to the noise produced by other processors plus ambient noise. We assume all vertices transmit with the same fixed transmission power P, so the signal of x experienced at y depends only on the distance between x and y (denoted by d_{xy}), and on the path-loss exponent (denoted by α). The signal strength decreases as an inverse polynomial of the distance, where the polynomial degree is α. Specifically, the signal strength of x experienced at y is P/d_{xy}^α. Similarly, the noise level of another transmitting vertex $v \in V$ experienced at y is P/d_{vy}^α. Let $U \subseteq V$ be a set of processors that transmit in parallel, and $V \setminus U$ be the rest of the processors (the receivers). Whether a receiver $y \in V \setminus U$ succeeds in hearing the sender $x \in U$ is determined by the *SINR formula:*

$$\frac{P/d_{xy}^\alpha}{N + \sum_{v \in U \setminus \{x\}} P/d_{vy}^\alpha} \geq \beta.$$

Here N is the ambient noise, and $\beta \geq 1$ is the threshold for successful reception. The parameters α, β and N are constants whose values are defined by the environment. We assume that $\alpha \geq 2$, which is the case in practice, unless it is reduced artificially. Specifically, in vacuum it holds that $\alpha = 2$, and on earth it holds that $\alpha > 2$. The value of α usually ranges between 2 and 6.

The maximum transmission range R of a vertex is the maximum range to which a vertex can transmit if it is the only transmitter in the network (i.e., the only noise is N). Note that when the power level is the same for all vertices, R is the same for all vertices as well. Let Δ denote the maximum number of vertices in any disk of radius R centered at a vertex $v \in V$. Let $\rho < 1$ be a positive constant that is arbitrarily close to 1. We define $\bar{R} = \rho \cdot R$, and $\bar{\Delta}$ to be the maximum number of vertices in any disk of radius $\bar{R}$ centered at a vertex $v \in V$. We note that for a pair of vertices u, v that are exactly at distance R one from another, a successful transmission requires all other vertices in the network to be silent. Therefore, in order to allow parallel transmissions, we define a successful *local broadcasting of a vertex* as a transmission that is successfully received by all vertices within radius $\bar{R}$ rather than R.

The SINR setting has attracted considerable attention due to its more realistic assumptions comparing to other models, such as the radio network model or the unit disk graph model, which do not take into account the cumulative nature of interference. One of the most fundamental problems in the SINR setting is *local broadcasting*. The goal in this problem is to establish a schedule in which each vertex $u \in V$ successfully transmits to all vertices at distance at most $\bar{R}$ from u. We will henceforth refer to vertices at distance at most $\bar{R}$ from u as *neighbors* of u. Note that the problem requirement is that for each vertex there exists a transmission that is successfully received by all its neighbors. This is stronger than requiring that each vertex succeeds to deliver a message to all its neighbors, since such a delivery could be achieved by several transmissions, each covering a subset of neighbors. The latter requirement is sometimes referred to as *weak local broadcasting*. Although weak local broadcasting can be often used instead of (strong) local broadcasting, the disadvantage of weak local broadcasting is in the greater power consumption resulting from multiple transmissions. Therefore, the strong variant of local broadcasting is preferred, and this is the variant considered in the current paper.

The local broadcasting problem serves as a building block for many network tasks, and has numerous applications. One of the most notable applications is *Single Round Simulation*. Specifically, if we are given an algorithm that is designed for wired networks or networks with no interference, it can be simulated in the SINR setting using local broadcasting as follows. Each round of the original algorithm is simulated by performing local broadcasting in the SINR setting. Consequently, each vertex succeeds do communicate with all its neighbors, which make it possible to execute a single round of the original algorithm. If the original algorithm requires $T(n)$ time and local broadcasting requires $S(n)$ time, then the overall simulation time is $T(n) \cdot S(n)$.

Since the running time of local broadcasting affects significantly the time of tasks that employ it, designing efficient local broadcasting algorithms is crucial. There has been an intensive thread of research in this direction. The problem was introduced by Goussevskaia, Moscibroda and Wattenhofer [7] who studied several scenarios. In the harshest scenario the vertices are unaware of neighborhood sizes and do not have *feedback* on the success of a transmission (i.e., they cannot tell whether a transmission has successfully received by all their neighbors). For this scenario, an algorithm with time[1] $O(\bar{\Delta} \log^3 n)$ was devised in [7]. Later it was improved in a series of works due to Yu et al. [13,14], where currently the best known algorithm has time $O(\bar{\Delta} \log n + \log^2 n)$ [13]. On the other hand, several researchers have observed that by considering slightly less harsh settings, one can improve the performance of the algorithms significantly. Moreover, these slightly stronger settings are still feasible for practical use. Although they may require more advanced devices, such as a carrier-sense mechanism that measures signal strength, they still can be implemented in hardware at a reasonable cost [9]. Already in the work of [7] it was observed that if vertices have knowledge about their neighborhood size, then the running time of the $O(\bar{\Delta} \log^3 n)$-time algorithm can be improved to $O(\bar{\Delta} \log n)$. Another result of this nature was obtained by Halldórsson and Mitra [9] who showed that in networks with free feedback (but with other properties that are similar to the harshest setting) the running time becomes $O(\bar{\Delta} + \log^2 n)$. In the current work we continue this line of

---

[1] All running times mentioned in our paper refer to randomized algorithms and hold *with high probability*, unless stated otherwise. *High probability* is $1 - 1/n^c$, for an arbitrarily large constant c. Note that if we are given $O(n)$ independent events, each of which occurs with high probability, then the event that all of them occur holds with high probability as well.

research and devise significantly improved algorithms in settings that are slightly less harsh than the harshest setting.

Our Results. We devise a local broadcasting algorithm for networks with feedback requiring $O(\bar{\Delta} + \log n \log \log n)$ time. This improves the best previously-known result for networks with feedback that has running time $O(\bar{\Delta} + \log^2 n)$ [9]. Moreover, the running time of our algorithm is tight up to a $\log \log n$ factor, in view of the lower bounds $\Omega(\Delta)$ and $\Omega(\log n)$ [13]. In addition, it shows that the conjecture of [9] that the $\log^2 n$ term is necessary does not hold in some settings. (On the other hand, the conjecture may still be true in weaker settings, such as settings without feedback. This is an intriguing open problem.) We consider a slotted setting with simultaneous wake up. This is somewhat stronger than the settings of [7,9]. However, there are standard methods that allow to weaken these requirements [7]. Also, similarly to other works, we assume that vertices know (upper bounds on) n and $\bar{\Delta}$.

We also consider a closely related problem, called *distance-k coloring*. In this problem the goal is to color the vertices with $O(\bar{\Delta})$ colors, such that each pair of neighbors at *geometrical distance* at most $k \cdot \bar{R}$ from one another are assigned distinct colors. If k is a constant, then a distance-k coloring with $O(\bar{\Delta})$ colors always exists. If k is a sufficiently large constant, this coloring constitutes a feasible SINR schedule. Note, however, that this problem is more challenging than k-hop coloring in which the goal is to obtain a coloring such that any pair of vertices at *graph distance* at most k have distinct colors. Indeed, any distance-k coloring is a k-hop coloring, but not vice versa. Moreover, the vertices are not always able to compute a distance-k coloring. For example, if two vertices are at distance greater than R from one another, they may not be able to communicate. On the other hand, a distance-k coloring requires them to select distinct colors, which cannot be achieved without communication when the required probability is sufficiently large. To address this problem we propose to employ *helper vertices*, also known as Steiner points. We employ an optimal number of Steiner points (up to constant factors), and obtain a distance-k coloring with $O(\bar{\Delta})$ colors in $O(\log^* n \cdot (\bar{\Delta} + \log n \log \log n))$ time. This coloring gives rise to an optimal SINR schedule of length $O(\bar{\Delta})$ after a preprocessing stage of $O(\log^* n \cdot (\bar{\Delta} + \log n \log \log n))$ time.

An interesting question deals with the influence of the path-loss exponent α on the performance of algorithms. Intuitively, a lower path-loss exponent means less obstacles and better signal strength. Moreover, from the point of view of each vertex, its signal should be as strong as possible in order to allow a successful transmission. Therefore, it seems reasonable that a lower path-loss exponent implies a better SINR schedule. In other words, transmitting in vacuum where $\alpha = 2$ should be the option for best performance. Surprisingly, we prove that the opposite is true! Specifically, any feasible SINR schedule for an environment with $\alpha = 2$ has length $\Omega(\bar{\Delta} \log n)$. This is an unconditional lower bound, no matter how strong the setting is. We present a network in which any shorter schedule will certainly fail. Hence we illustrate a gap between settings with $\alpha = 2$, and settings with $\alpha > 2$, where an $O(\bar{\Delta} + \log n \log \log n)$-schedule can be achieved. This interesting phenomenon can be explained by noting that obstacles do not only weaken signals - they also weaken noise. Our findings demonstrate that modifying α may affect noise more significantly than transmission signals. Therefore, in some occasions it might be better to add obstacles in order to block noise, instead of removing them in order to strengthen the signal.

Our Techniques. The main idea of our local broadcasting algorithm is gradually reducing the sizes of neighborhoods. In other words, as the algorithm proceeds, more and more vertices succeed and terminate. Consequently, the remaining vertices have less competition, and they are able to perform transmission trials more intensively. Specifically, in each phase consisting of $O(\bar{\Delta})$ rounds a constant fraction of vertices in each neighborhood terminates, with high probability. (A vertex terminates once it has successfully transmitted to all its neighbors.) This reduces the bound $\bar{\Delta}$ on the maximum neighborhood size, which allows to execute each phase more efficiently than the previous one. These improvements, however, are only possible as long as $\bar{\Delta} > \log n$. Once $\bar{\Delta}$ reaches $\log n$ we cannot proceed in the same way, since the probability that all sizes of neighborhoods are reduced is no longer large enough. Hence we switch to another method that increases the number of trials after each phase. Although the number of trials becomes greater than $\bar{\Delta}$ it is still bounded by $O(\log n)$ per phase. The number of these phases is $O(\log \log n)$, contributing a factor $O(\log n \log \log n)$ to the running time. This is in addition to the $O(\bar{\Delta})$ term for the first part of the algorithm.

We employ our local broadcasting algorithm in order to compute distance-k colorings using Steiner points. Once appropriate Steiner points are deployed we can make sure that any pair of vertices in the network can communicate (not necessarily directly). We observe that the resulting communication graph G is a unit disk graph. Moreover, the graph G^k obtained by adding an edge between any pair of vertices at *geometrical distance* $O(\bar{R} \cdot k)$ from one another has bounded growth. We then employ an algorithm due to Schneider and Wattenhofer for $O(\bar{\Delta})$-coloring graphs with bounded growth in $O(\log^* n)$ time [11]. Invoking it on G^k results in the desired distance coloring. This algorithm, however, is designed for networks with no interference. Nevertheless, each round of the algorithm can be simulated using local broadcasting. More precisely, $O(k)$ executions of local broadcasting are required in order to propagate a message to distance k. The propagation is possible thanks to the Steiner points. Consequently, a single round of the algorithm of [11] is simulated within $O(k \cdot (\bar{\Delta} + \log n \log \log n)) = O(\bar{\Delta} + \log n \log \log n)$ rounds, since k is a constant. Thus we obtain an overall running time $O(\log^* n \cdot (\bar{\Delta} + \log n \log \log n))$. For a sufficiently large constant k, we show that all vertices of the same color can transmit in parallel in the SINR setting without interference. Thus we obtain an SINR schedule of length $O(\bar{\Delta})$.

For our lower bound in the scenario when $\alpha = 2$ we consider a grid of vertices of size roughly $\sqrt{n} \times \sqrt{n}$. Our goal is to show that in any partition of vertices into $o(\bar{\Delta} \log n)$ subsets, there must be a subset that causes too much noise that results in a failure of some transmission. By calculating the overall noise of all n vertices we conclude that whenever the length of a schedule is too short, there must be a subset generating noise that is too strong. We then show that this noise necessarily disturbs a certain transmission. Hence, in any $o(\bar{\Delta} \log n)$-schedule there must be a round in which the noise is too strong at a certain vertex that tries to receive a message. Consequently, at least one vertex will fail during the transmission.

Related Work. In their pioneering work Goussevskaia, Moscibroda and Wattenhofer [7] devised a local broadcasting algorithm with time $O(\bar{\Delta} \log n)$ when neighborhood sizes are known, and $O(\bar{\Delta} \log^3 n)$ time when the sizes are unknown. For the latter scenario, Yu et al. obtained improved local broadcasting algorithms that require $O(\bar{\Delta} \log^2 n)$ time [14] and $O(\bar{\Delta} \log n + \log^2 n)$ time [13]. By using carrier-sense (a mechanism that allows receiving feedback) Yu et al. [14] obtained an algorithm with time $O(\bar{\Delta} \log n)$. An improved algorithm for the latter scenario of networks with feedback was devised by Halldórsson and Mitra [9]. The running time of the algorithm of [9] is $O(\bar{\Delta} + \log^2 n)$. The feedback mechanism of [9] is similar to the one used in the current paper.

Several works obtained the optimal (up to constant factors) $O(\bar{\Delta})$-schedule at the expense of performing a preprocessing stage, and employing some additional mechanisms that are not available in the weaker settings mentioned above. Specifically, Derbel and Talbi [3] perform preprocessing of $O(\bar{\Delta} \log n)$ time and employ power-level adjustments. Jurdzinski and Kowalski [10] perform preprocessing of $O(\bar{\Delta} \log^3 n)$ time, do not require power-level adjustments, but require location information. In the latter setting, a better result was obtained recently by Fuchs and Wagner [5] whose algorithm has $O(\bar{\Delta} \log n)$ preprocessing time. Note that the result of [10] is deterministic, while the other results are randomized. It is natural to compare these results with our new randomized algorithm that obtains $O(\bar{\Delta})$-schedule with $O(\log^* n \cdot (\bar{\Delta} + \log n \log \log n))$ preprocessing time. Instead of employing power-level adjustments or location-information mechanisms, our algorithm employs Steiner points in networks with feedback. This allows us to break the $O(\bar{\Delta} \log n)$ barrier in the preprocessing time, and outperform the running time of the above-mentioned algorithms.

The problem of $O(\bar{\Delta})$-coloring is closely related to local broadcasting, and has been intensively studied in the SINR model as well. However, it is weaker than local broadcasting in the following sense. Given a feasible local-broadcasting schedule, no two vertices of the same neighborhood transmit in the same time. Therefore, all vertices that transmit in the same time form a proper color class. On the other hand, given a proper coloring, all vertices of the same color will not necessarily be able to transmit in parallel. In order to allow this, a geometrical distance-k coloring is required. Still, $O(\bar{\Delta})$-coloring has attracted much attention. Derbel and Talbi [3] devised an $O(\bar{\Delta})$-coloring algorithm with $O(\bar{\Delta} \log n)$ time, and a distance-coloring algorithm with the same time that requires power-level adjustments. Yu et al. [15] devised a $(\bar{\Delta}+1)$-coloring algorithm that requires power-level adjustments and runs in $O(\bar{\Delta} \log n + \log^2 n)$ time. They also devised an algorithm that does not require power-level adjustments and has running time $O(\bar{\Delta} \log^2 n)$. Fuchs and Prutkin [4] obtained a $(\bar{\Delta}+1)$-coloring in $O(\bar{\Delta} \log n)$ time. Coloring problems have been very intensively studied in additional settings, such as wireless radio networks and networks without interference. The best currently-known $(\Delta + 1)$-coloring algorithm for radio networks has time $O(\Delta + \log^2 n)$ [12]. The best currently-known $(\Delta + 1)$-coloring algorithm for networks without interference has running time $O(\log \Delta + 2^{O(\sqrt{\log \log n})})$ [2]. For an extensive overview of distributed coloring algorithms we refer the reader to [1].

2 Local Broadcasting in Networks with Feedback

In this section we devise a local broadcasting algorithm for networks with feedback that requires $O(\bar{\Delta} + \log n \log \log n)$ time. We start with the following claim. Suppose that all vertices $v \in V$ perform trials in which each vertex transmits with probability $1/(c \cdot \bar{\Delta})$, and listens with probability $1 - 1/(c \cdot \bar{\Delta})$, for a sufficiently large constant c. Then a transmitting node successfully performs its local broadcasting, with probability at least $1/2$. This is similar to a phenomenon observed in [7]. (We omit its prof from the current paper due to lack of space.) We refer to the set of vertices at distance at most $\bar{R}$ from the vertex $v \in V$ (excluding v) as the *neighborhood* of V, and denote it by $\Gamma_{\bar{R}}(v)$.

Lemma 1. *For a sufficiently large constant c, suppose that all vertices perform transmissions with probability $1/(c \cdot \bar{\Delta})$. Then a transmission of a sender $v \in V$ is successfully received in v's neighborhood $\Gamma_{\bar{R}}(v)$, namely, within radius $\bar{R}$ from v, with probability at least $1/2$.*

Next, we devise a procedure called *Feedback-Broadcasting* for performing local broadcasting in networks with feedback, namely, networks in which any vertex $v \in V$ can decide whether a transmission was successfully received by all vertices in its neighborhood $\Gamma_{\bar{R}}(v)$. The procedure consists of two phases. In the first phase, vertices repeatedly perform the following trials: each vertex transmits with probability $1/(c \cdot \bar{\Delta})$ for $\hat{c} \cdot \bar{\Delta}$ times, where $\hat{c} > c$ is a sufficiently large constant. If a vertex v discovers (using the feedback mechanism) that it has succeeded to transmit to its entire neighborhood $\Gamma_{\bar{R}}(v)$, then v terminates. If v has failed in all these $\hat{c} \cdot \bar{\Delta}$ trials, then it updates the bound on $\bar{\Delta}$ by setting $\bar{\Delta} := \frac{1}{2} \cdot \bar{\Delta}$, and performs another stage of $\hat{c} \cdot \bar{\Delta}$ trials. This continues as long as $\bar{\Delta} > \log n$, and then the first phase of the procedure terminates.

In the second phase, it holds that $\bar{\Delta} \leq \log n$, with high probability. This phase consists of $O(\log \log n)$ stages, each of which consists of $O(\log n)$ trials in which each vertex transmits with probability $1/(c \cdot \bar{\Delta})$. In the end of each stage, all unsuccessful vertices update $\bar{\Delta}$ by setting $\bar{\Delta} = \frac{1}{2} \cdot \bar{\Delta}$. We later prove that once the second stage has been completed, all vertices have succeeded with high probability. Next, we provide the pseudocode of the procedure. (Note that $\Gamma_{\bar{R}}(v)$ denotes all neighbors of v including those that have terminated. In other words, the feedback in lines 8 and 24 of the algorithm has to be received for all neighbors, namely, the active and the terminated ones.)

Algorithm 1. Procedure Feedback-Broadcasting$(V, \bar{\Delta})$ (code for vertex $v \in V$)

Let $c, \hat{c}$ be sufficiently large constants, and $\hat{c} > c$.

1: $success := F$
2: (* Phase 1 *)
3: **while** $\bar{\Delta} > \log n$ **do**
4:    (* Stage k *)
5:    **for** $i = 1, 2, ..., \hat{c} \cdot \bar{\Delta}$ **do**
6:       (* trial i of stage k *)
7:       transmit with probability $1/(c \cdot \bar{\Delta})$
8:       **if** all neighbors of v in $\Gamma_{\bar{R}}(v)$ receive the transmission successfully **then**
9:          $success := T$
10:       **end if**
11:    **end for**
12:    **if** $success = T$ **then**
13:       terminate
14:    **else**
15:       $\bar{\Delta} := \lfloor \frac{1}{2} \cdot \bar{\Delta} \rfloor$
16:    **end if**
17: **end while**
18: (* Phase 2 *)
19: **for** $k = 1, 2, ..., \lfloor \log \log n \rfloor$ **do**
20:    (* Stage k *)
21:    **for** $i = 1, 2, ..., \lfloor \hat{c} \cdot \log n \rfloor$ **do**
22:       (* trial i of stage k *)
23:       transmit with probability $1/(c \cdot \bar{\Delta})$
24:       **if** all neighbors of v in $\Gamma_{\bar{R}}(v)$ receive the transmission successfully **then**
25:          $success := T$
26:       **end if**
27:    **end for**
28:    **if** $success = T$ **then**
29:       terminate
30:    **else**
31:       $\bar{\Delta} := \max\{\lfloor \frac{1}{2} \cdot \bar{\Delta} \rfloor, 1\}$
32:    **end if**
33: **end for**

We say that a vertex is *active* if it has not terminated yet. The invariant that the algorithm attempts to preserve is Bound$(\bar{\Delta}) \equiv$ "the parameter $\bar{\Delta}$ is an upper bound on the maximum neighborhood size (counting only active vertices)". The correctness of the algorithm follows from the observation that this invariant holds at all stages of the algorithm, with high probability. This observation, in turn, follows from the fact that in each stage the number of active neighbors of each vertex is reduced by a factor of $1/2$. We prove this in the next lemma.

Lemma 2. *Suppose that Procedure Feedback-Broadcasting is invoked by all vertices with a parameter $\bar{\Delta}$ that satisfies* Bound$(\bar{\Delta})$. *Then the invariant* Bound$(\bar{\Delta})$ *holds throughout the entire execution, with high probability.*

Proof. The assertion holds trivially in the beginning of the execution of the procedure. We have to prove that each time the value of $\bar{\Delta}$ is updated, its new value indeed satisfies Bound($\bar{\Delta}$). Note that $\bar{\Delta}$ is updated only in the end of a stage. (Lines 4 - 16 constitute a stage of Phase 1; lines 20 - 32 are a stage of Phase 2.) We start by analyzing the first phase (lines 2 - 17). Assuming that Bound($\bar{\Delta}$) holds in the beginning of stage k of Phase 1, we show that in the end of stage k, Bound($\bar{\Delta}$) still holds, namely, for each vertex $v \in V$, the number of neighbors of v that are still active is at most $\frac{1}{2} \cdot \bar{\Delta}$, with high probability.

Suppose that in the beginning of stage k the number d of active neighbors of v is at least $\frac{1}{2} \cdot \bar{\Delta}$. (Otherwise, the assertion holds already in the beginning of the stage, and will hold in the end of the stage since the number of active neighbors can only decrease.) Let X_i, $i = 1, 2, ..., \hat{c} \cdot \bar{\Delta}$, be a random indicator variable that equals 1 if a neighbor of v succeeds in trial i of stage k, and 0 otherwise, and let $X = \sum_{i=1}^{\hat{c} \cdot \bar{\Delta}} X_i$. The probability that exactly one neighbor of v tries in trial i is $d \cdot (c \cdot \bar{\Delta} - 1)^{d-1}/(c \cdot \bar{\Delta})^d > d/(4 \cdot c \cdot \bar{\Delta})$. Thus, by lemma 1, the probability that it succeeds is at least $d/(8 \cdot c \cdot \bar{\Delta})$. Therefore, $\mathbb{E}(X) \geq (\hat{c}/c) \cdot d/8$. By Chernoff bound, as the X_i's are independent,

$$Pr(X < \mathbb{E}(X)/2) \leq e^{-\mathbb{E}(X)/8} \leq e^{-(\hat{c}/c) \cdot d/64}.$$

In other words, for a sufficiently large constant $\hat{c}$, we can obtain at least $2d$ successful trials in a stage, with high probability. (Recall that $d > \frac{1}{2}\bar{\Delta} > \frac{1}{2}\log n$.) However, the trials were performed with repetitions, and thus, the number of successful neighbors may be smaller than d. Next, we analyze the probability that it is smaller than $d/2$. Since in each iteration the vertices have equal chances of performing a trial, this problem is equivalent to balls-into-bins, where vertices are bins and successful trials are balls. The value $2d$ denotes the number of balls, d denotes the number of bins, and we would like to analyze the probability that more than $d/2$ bins contain balls. We calculate the probability of the complementary event, i.e., that at most $d/2$ bins contain balls. This probability is at most $\binom{d}{d/2} \cdot (1/2^{2d}) < (1/2^d)$. Note that by increasing the constant $\hat{c}$ we can have an arbitrarily large constant multiplicative factor, instead of the factor 2 in the term $2d$. Since $d = \Omega(\log n)$, at least $d/2$ neighbors succeed, with probability $1 - 1/poly(n)$. By the union bound, for all vertices, all neighborhoods are (at least) halved, with high probability. Thus the size of the maximum neighborhood is reduced by a factor of at least 2 in each stage of the first phase, with high probability.

Consequently, within $O(\log \bar{\Delta})$ stages of Phase 1, the maximum neighborhood size becomes at most $\log n$, with high probability. Therefore, once Phase 2 (lines 18 - 33) starts, it holds that $\bar{\Delta} \leq \log n$ is an upper bound on the maximum neighborhood size, as required. Denote again by d the number of active neighbors of a vertex $v \in V$ that is still active. In each trial of a stage of Phase 2 (lines 22 - 26), the probability that exactly one active neighbor of v succeeds is at least $d/(4 \cdot c \cdot \bar{\Delta}) = \Omega(1)$, if $d \geq \bar{\Delta}/2$. Consequently, the expected number of successful trials is $\mathbb{E}(X) = \Omega(\log n)$, where the constant hidden in the Ω-notation can be made as large as desired by choosing a sufficiently large constant

$\hat{c}$. Thus, by Chernoff bound, the number of successful trials is $\Omega(\log n)$, with high probability. Next, we analyze the probability that at least $d/2$ different neighbors have succeeded. Again we reduce the problem to balls-to-bins, where here we have $\Omega(\log n)$ balls and d bins. Therefore, this probability is $1 - \binom{d}{d/2} \cdot (1/2^{\Omega(\log n)}) \geq 1 - \binom{\log n}{(\log n)/2} \cdot (1/2^{\Omega(\log n)})$, i.e., high probability, for a sufficiently large constant $\hat{c}$. Using the union bound we obtain this result for all vertices, and thus the maximum neighborhood size is at least halved in each stage, with high probability. Hence throughout the entire execution $\bar{\Delta}$ is an upper bound on the maximum neighborhood size, with high probability.   □

By Lemma 2, within $\log \log n - 1$ stages of Phase 2 the neighborhood size of all vertices becomes $O(1)$, with high probability. In stage $\lfloor \log \log n \rfloor$ of the second phase each active vertex succeeds with a constant probability since $\bar{\Delta} = O(1)$. (See Lemma 1.) Therefore, within $O(\log n)$ iterations of this stage all remaining active vertices succeed, with high probability. Thus we obtain the following result.

Theorem 1. *Procedure Feedback-Broadcasting performs a successful local broadcasting of all vertices, with high probability.*

Next we analyze the running time of the procedure. Each stage of the first phase requires $O(\bar{\Delta})$ time. However, $\bar{\Delta}$ is halved in each stage, and thus the overall running time of the first phase is $O(\bar{\Delta} + \bar{\Delta}/2 + \bar{\Delta}/4 + ...) = O(\bar{\Delta})$. The second phase requires $O(\log n \log \log n)$ time. Hence the overall running time is $O(\bar{\Delta} + \log n \log \log n)$.

Theorem 2. *Local broadcasting in networks with feedback can be performed in $O(\bar{\Delta} + \log n \log \log n)$ time.*

3   Distant Coloring

Our local-broadcasting algorithm produces an $O(\bar{\Delta} + \log n \log \log n)$ time schedule. In other words, a distributed algorithm for networks without interference can be simulated in SINR networks, where each round of the original algorithm is simulated by $O(\bar{\Delta} + \log n \log \log n)$ rounds of the local-broadcasting procedure. This is, however, not optimal, since a schedule of length $O(\bar{\Delta})$ always exists. It is easy to verify that the latter bound is the best possible (up to constant factors). Indeed, given a vertex v, in order to receive the messages of all the $\bar{\Delta}$ vertices at distance at most $\bar{R}$ from v, each of them must transmit in a distinct round.

A schedule of length $O(\bar{\Delta})$ can be obtained by computing a *distance-k coloring*, for a sufficiently large constant k. In this coloring each pair of vertices at (geometrical) distance less than $k \cdot \bar{R}$ from one another are colored by distinct colors. Since the number of vertices in each disk of radius $k \cdot \bar{R}$ is $O(\bar{\Delta})$, a distance-k coloring can always employ $O(\bar{\Delta})$ colors, for any constant k. Unfortunately, it is impossible to obtain such a coloring in the SINR setting (as will be explained shortly), even though it is possible to achieve a *k-hop-coloring*, namely, a coloring in which any pair of vertices within at most k hops from one another are colored by distinct colors. In models without interference a k-hop $O(\bar{\Delta})$-coloring can be computed in $O(\log^* n)$ time for any constant k. This is done by computing an $O(\bar{\Delta})$-coloring of growth-bounded graphs on G^k. Since G is a unit disk graph, G^k is of bounded growth. The running time of the algorithm is $O(\log^* n)$ [11]. Consequently, in SINR networks a k-hop-coloring can be computed within $O(\log^* n \cdot (\bar{\Delta} + \log n \log \log n))$ rounds by performing single-round simulations. (See Theorem 2.) This, however, may increase message size by a factor of $\mathrm{poly}(\bar{\Delta})$ as a consequence of simulating G^k.

Corollary 1. *A k-hop coloring can be computed in $O(\log^* n \cdot (\bar{\Delta} + \log n \log \log n))$ rounds (with high probability) in the SINR setting with uniform power.*

However, a k-hop-coloring does not necessarily produce a feasible SINR schedule. Consider, for instance, three vertices a,b,c, such that $\mathrm{dist}(a,b) = \bar{R}$, and $\mathrm{dist}(a,c) = \mathrm{dist}(b,c) = \bar{R} + \epsilon$, for some $\epsilon > 0$. Then $\varphi(a) = 1$, $\varphi(b) = 2$, $\varphi(c) = 1$ is a proper k-hop coloring for any k, since c cannot receive messages from a and b. On the other hand, by the SINR formula, if a and c transmit simultaneously, they cause interference that prevents b from receiving the message of a. Thus a distance-k coloring is desirable. But it cannot be computed since a and b cannot communicate with c, and cannot make sure they all select distinct color. To solve this problem we propose to use *Steiner points*. In other words, we add some helper vertices that allow to compute a distance-k coloring of the original vertex set. For each original vertex, we add $O(k)$ Steiner vertices in the way illustrated in Figure 1(a). These Steiner vertices have exactly the same status as that of the original vertices of V, i.e., a Steiner vertex is a processor with a transmitter and a receiver. Note that as a result $\bar{\Delta}$ increases only by a multiplicative constant factor of at most 5. Let V' denote the new set of vertices, including vertices of V.

We compute a $2k$-hop-coloring of V' by invoking the algorithm of Corollary 1. We next prove that it results in a *distance-k* coloring of V that employs $O(\bar{\Delta})$ colors.

Lemma 3. *A $2k$-hop coloring of V' is a distance-k coloring of V that employs $O(\bar{\Delta})$ colors.*

Proof. The number of vertices of V' in any disk of radius $\bar{R}$ is at most five times the number of vertices of V in this disk. Consequently, the number of employed colors is $O(5\bar{\Delta}) = O(\bar{\Delta})$. Let $u, v \in V$ be two vertices at distance at most $\bar{R} \cdot k$ from one another. Then there exists a path of at most $2k$ vertices connecting u and v, such that each pair of neighboring vertices on the path are at distance at most $\bar{R}$ from one another. (See Figure 1(b).) Consequently, u and v are colored by distinct colors by the $2k$-hop-coloring algorithm. $\square$

By Lemma 3 and Corollary 1 we obtain the following result.

Theorem 3. *A distance-k coloring of V can be obtained within $O(\log^* n \cdot (\bar{\Delta} + \log n \log \log n))$ rounds, with high probability, using at most $4k \cdot n$ Steiner points.*

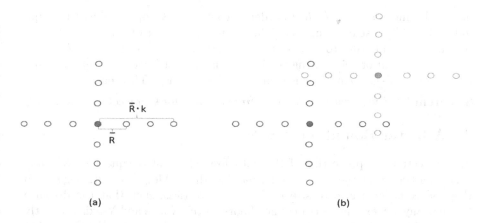

Fig. 1. (a) The vertex in the center is $v \in V$, and it is surrounded by $4k$ Steiner points. (b) If u and v are not too far from one another, there is a path connecting u and v.

Consider all vertices colored by the same color z of a distance-k coloring, for a sufficiently large constant k. Let v be such a vertex. Then the number of vertices in L_i (the ith ring of width $\bar{R}$ around v) whose color is z is $O(i)$. The constant in the O-notation can be made as small as one wishes, by increasing k. As a result, by similar arguments to those in the analysis of Lemma 1 (see also [7]), the interference in the disk of radius $\bar{R}$ of v is sufficiently small to allow a successful broadcast of v. Specifically, for a sufficiently large constant k, no vertices except for v transmit in L_1 and L_2. Hence, the interference I_1 experienced by neighbors of v, i.e., by vertices in L_1 is at most

$$I_1 = \sum_{i=3}^{\infty} P \cdot O(i) \cdot 1/(\bar{R}(i-2))^{\alpha} \leq \sum_{i=3}^{\infty} (1/\bar{R})^{\alpha} \cdot P \cdot O((i-2)^{\alpha-1})$$

$$= (1/\bar{R}^{\alpha}) \cdot P \cdot \sum_{i=1}^{\infty} O(1/i^{\alpha-1}) = (1/\bar{R}^{\alpha}) \cdot P \cdot O((\alpha-1)/(\alpha-2)),$$

where the constant hidden in the O-notation can be made as small as one wishes. In other words, $I_1 \leq \epsilon \cdot P/\bar{R}^{\alpha}$, for an arbitrarily small constant $\epsilon > 0$. This interference is sufficiently small to allow all vertices at distance at most $\bar{R}$ from v to receive the message of v. Consequently, if all vertices of the same color in the k-hop coloring (and only them) transmit simultaneously, they all succeed. Thus Theorem 3 implies the following result.

Theorem 4. *A schedule of length $O(\bar{\Delta})$ can be obtained within $O(\log^* n \cdot (\bar{\Delta} + \log n \log \log n))$ rounds, with high probability, using at most $4k \cdot n$ Steiner points.*

As noted earlier, the schedule length is optimal up to constant factors. Next we show that the number of Steiner points is optimal as well. Consider a vertex set V whose vertices are placed on a line, such that the distance between any pair of neighboring vertices is $k \cdot \bar{R}$. In order to compute a k-hop coloring, the graph induced by V' must be connected. The minimum number of vertices that must be added to V in order to satisfy this requirement is $(n-1)(k-1) = \Omega(k \cdot n)$, since for any pair of neighboring vertices on the line, at least $k-1$ vertices must be placed between them. This is summarized in the next Theorem.

Theorem 5. *The number of Steiner points required for k-hop coloring is $\Omega(kn)$.*

4   A Lower Bound for $\alpha = 2$

In this section we prove that if the path loss exponent α equals 2, then any feasible schedule in the SINR model has length $\Omega(\bar{\Delta} \log n) = \Omega(\Delta \log n)$. To this end consider a grid of size $k \cdot k = n$ of vertices, such that the distance between any vertex and its closest neighbors on the X-axis and Y-axis is exactly one unit. The dimensions of the square containing this grid is $(k-1) \times (k-1)$. Let v be a vertex in a corner of the grid. Suppose that all other vertices $u \in V \setminus \{v\}$ transmit, and let $\bar{R} \geq 1$ be a parameter defining the transmission range, as in Section 2. (Note that $\bar{\Delta} = \Theta(\bar{R}^2)$). Denote by t_u the interference experienced by v as a result of the transmission of u. Then the overall interference experienced by v is at least the interference I_{far} caused by the subset $\bar{V}$ of vertices at distance greater than $\bar{R}$ from v. This interference satisfies

$$I_{far} = \sum_{u \in \bar{V}} t_u = \sum_{u \in \bar{V}} P/d_{uv}^{\alpha} = \sum_{u \in \bar{V}} P/d_{uv}^{2}$$

$$\geq \sum_{i=\bar{R}+1}^{k-1} P \cdot (2i+1)/(2i^2) = \sum_{i=1}^{k-1} P \cdot (2i+1)/(2i^2) - \sum_{i=1}^{\bar{R}} P \cdot (2i+1)/(2i^2).$$

The last inequality follows from the observation that the number of vertices on the boundary of a square of size $(i+1) \times (i+1)$ that do not belong to the inner square of size $i \times i$ is $i + i + 1 = 2i + 1$. On the other hand, each such vertex is at distance at most $\sqrt{2}i$ from v.

Consequently, the interference experienced by v as a result of the transmissions of all other vertices is at least $P \cdot \sum_{i=\bar{R}+1}^{k-1} 1/i$. Whenever $\bar{R} \leq k^{1-\epsilon}$, for an arbitrarily small constant $\epsilon > 0$, we have $P \cdot \sum_{i=\bar{R}+1}^{k-1} 1/i = \Omega(P \cdot \log k - P \cdot \log \bar{R}) = \Omega(P \cdot \log n)$. This is summarized below.

Lemma 4. *Let v be a corner vertex and $\bar{V}$ be the set of vertices at distance greater than $\bar{R}$ from v. If all vertices in $\bar{V}$ transmit, then the interference experienced by v is $\Omega(P \cdot \log n)$.*

Next, Assume for contradiction that there exists a feasible SINR schedule of length $\ell = o(\bar{\Delta} \cdot \log n)$. Then, let $V_1, V_2, ..., V_\ell$ be a partition of $V \setminus \{v\}$, such that the vertices in each V_i, $i \in [\ell]$, can transmit successfully in parallel. Let $j \in [\ell]$ be the index of the set V_j, such that vertices of $V_j \cap \bar{V}$ cause the maximum interference at the corner vertex v. Then, by the Pigeonhole principle,

$$\sum_{u \in V_j \cap \bar{V}} t_u \geq \omega(P/\bar{\Delta}). \tag{1}$$

Note that $\bar{\Delta}$ depends on P linearly. Indeed, increasing P by a multiplicative factor of q results in an increase of the transmission range by $\sqrt{q}$, and thus the number of vertices at distance at most $\sqrt{q} \cdot \bar{R}$ becomes $\Theta(q \cdot \bar{\Delta})$. If we normalize P to be equal to 1 in the case of a transmission of an only-transmitting vertex to a distance of one unit, then $\omega(P/\bar{\Delta}) = \omega(1)$, for any P.

Let $w \in V_j \cap \bar{V}$ be the closest vertex to v. Let $y \in V \setminus V_j$ be a vertex at distance at least $\bar{R} - 1$ and at most $\bar{R}$ from w. See Figure 2. Let $M_w = (V_j \cap \bar{V}) \setminus \{w\}$.

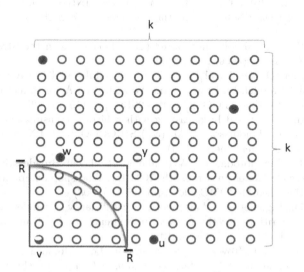

Fig. 2. The set of transmitting vertices $V_j \cap \bar{V}$ is depicted by filled circles.

Note that for any $u \in V_j \cap \bar{V}$ the distance between u and v is at least $1/\sqrt{2}$ the distance between u and w. Thus the interference experienced by w when vertices of $V_j \cap \bar{V}$ transmit is at least $\frac{1}{2} \sum_{u \in M_w} t_u$. The interference experienced by y is at least $\frac{1}{8} \sum_{u \in M_w} t_u$, since the distance between w and each vertex $x \in M_w$ is at least $1/2$ the distance between y and x. Thus, when all vertices of $V_j \cap \bar{V}$ transmit, the SINR formula that determines whether y receives the message of w successfully satisfies (for $\alpha = 2$)

$$\frac{P/d_{wy}^{\alpha}}{N + \sum_{u \in M_w} P/d_{uy}^{\alpha}} \leq \frac{P/(\bar{R}-1)^2}{N + \frac{1}{8} \sum_{u \in M_w} t_u} = \frac{\Theta(P/\bar{\Delta})}{\omega(P/\bar{\Delta})} = o(1) < 1,$$

as $t_w = \Theta(P/\bar{\Delta})$ and $\sum_{u \in M_w} t_u = \omega(P/\bar{\Delta}) - \Theta(P/\bar{\Delta}) = \omega(P/\bar{\Delta})$ by (1).

Hence y fails to receive the message of w, and thus $V_1, V_2, ..., V_\ell$ is not a feasible schedule; contradiction. In summary, we get the following theorem.

Theorem 6. *In settings with loss-path exponent $\alpha = 2$, any feasible SINR schedule with uniform power has length $\Omega(\Delta \log n)$.*

References

1. Barenboim, L., Elkin, M.: Distributed Graph Coloring: Fundamentals and Recent Developments. Morgan & Claypool Synthesis Lectures on Distributed Computing Theory (2013)
2. Barenboim, L., Elkin, M., Pettie, S., Schneider, J.: The locality of distributed symmetry breaking. In: Proc. 53rd Symp. on Foundations of Computer Science (FOCS 2012), pp. 321–330 (2012)
3. Derbel, B., Talbi, E.: Distributed Node Coloring in the SINR Model. In: Proc. 30th IEEE Int. Conf. on Distributed Computing Systems (ICDCS 2010), pp. 708–717 (2010)
4. Fuchs, F., Prutkin, R.: Simple Distributed $(\Delta + 1)$-coloring in the SINR model (2015), http://arxiv.org/abs/1502.02426
5. Fuchs, F., Wagner, D.: On Local Broadcasting Schedules and CONGEST Algorithms in the SINR Model. In: Proc. 9th Int. Workshop on Algorithmic Aspects of Wireless Sensor Networks (ALGOSENSORS 2013), pp. 170–184 (2013)
6. Fuchs, F., Wagner, D.: Local broadcasting with arbitrary transmission power in the SINR model. In: Halldórsson, M.M. (ed.) SIROCCO 2014. LNCS, vol. 8576, pp. 180–193. Springer, Heidelberg (2014)
7. Goussevskaia, O., Moscibroda, T., Wattenhofer, R.: Local Broadcasting in the Physical Interference Model. In: Proc. 5th ACM Int. Workshop on Foundations of Mobile Computing (DialM-POMC 2008), pp. 35–44 (2008)
8. Goussevskaia, O., Pignolet, Y., Wattenhofer, R.: Efficiency of wireless networks: Approximation algorithms for the physical interference model. Foundations and Trends in Networking 4(3), 313–420 (2010)
9. Halldórsson, M., Mitra, P.: Towards Tight Bounds for Local Broadcasting. In: Proc. 8th ACM Int. Workshop on Foundations of Mobile Computing (FOMC 2012), Article No 2 (2012)
10. Jurdzinski, T., Kowalski, D.: Distributed Backbone Structure for Algorithms in the SINR Model of Wireless Networks. In: Proc. 26th Int. Symp. on Distributed Computing (DISC 2012), pp. 106–120 (2012)
11. Schneider, J., Wattenhofer, R.: A Log-Star Distributed Maximal Independent Set Algorithm For Growth Bounded Graphs. In: Proc. 27th ACM Symp. on Principles of Distributed Computing (PODC 2008), pp. 35–44 (2008)
12. Schneider, J., Wattenhofer, R.: Coloring unstructured wireless multi-hop networks. In: Proc. 28th ACM Symp. on Principles of Distributed Computing (PODC 2009), pp. 210–219 (2009)
13. Yu, D., Hua, Q., Wang, Y., Lau, F.: An $O(\log n)$ Distributed Approximation Algorithm for Local Broadcasting in Unstructured Wireless Networks. In: Proc. 8th Int. Conf. on Distributed Computing in Sensor Systems (DCOSS 2012), pp. 132–139 (2012)
14. Yu, D., Wang, Y., Hua, Q., Lau, F.: Distributed Local Broadcasting Algorithms in the Physical Interference Model. In: Proc. 2011 Int. Conf. on Distributed Computing in Sensor Systems (DCOSS 2011), pp. 1–8 (2011)
15. Yu, D., Wang, Y., Hua, Q.-S., Lau, F.C.M.: Distributed $(\Delta + 1)$-Coloring in the Physical Model. In: Erlebach, T., Nikoletseas, S., Orponen, P. (eds.) ALGOSENSORS 2011. LNCS, vol. 7111, pp. 145–160. Springer, Heidelberg (2012)

Byzantine Gathering in Networks*

Sébastien Bouchard[1], Yoann Dieudonné[1], and Bertrand Ducourthial[2]

[1] Laboratoire MIS & Université de Picardie Jules Verne Amiens, France
[2] Heudiasyc, CNRS & Université de Technologie de Compiègne, Compiègne, France

Abstract. This paper investigates an open problem introduced in [14].
Two or more mobile agents start from different nodes of a network and
have to accomplish the task of gathering which consists in getting all
together at the same node at the same time. An adversary chooses the
initial nodes of the agents and assigns a different positive integer (called
label) to each of them. Initially, each agent knows its label but does
not know the labels of the other agents or their positions relative to its
own. Agents move in synchronous rounds and can communicate with
each other only when located at the same node. Up to f of the agents
are Byzantine. A Byzantine agent can choose an arbitrary port when it
moves, can convey arbitrary information to other agents and can change
its label in every round, in particular by forging the label of another
agent or by creating a completely new one. *What is the minimum number $\mathcal{M}$ of good agents that guarantees deterministic gathering of all of
them, with termination?* We provide exact answers to this open problem
by considering the case when the agents initially know the size of the
network and the case when they do not. In the former case, we prove
$\mathcal{M} = f + 1$ while in the latter, we prove $\mathcal{M} = f + 2$. More precisely,
for networks of known size, we design a deterministic algorithm gathering all good agents in any network provided that the number of good
agents is at least $f + 1$. For networks of unknown size, we also design
a deterministic algorithm ensuring the gathering of all good agents in
any network but provided that the number of good agents is at least
$f + 2$. Both of our algorithms are optimal in terms of required number
of good agents, as each of them perfectly matches the respective lower
bound on $\mathcal{M}$ shown in [14], which is of $f + 1$ when the size of the network is known and of $f + 2$ when it is unknown. Perhaps surprisingly,
our results highlight an interesting feature when put in perspective with
known results concerning a relaxed variant of this problem in which the
Byzantine agents cannot change their initial labels. Indeed under this
variant $\mathcal{M} = 1$ for networks of known size and $\mathcal{M} = f + 2$ for networks
of unknown size. Following this perspective, it turns out that when the
size of the network is known, the ability for the Byzantine agents to
change their labels significantly impacts the value of $\mathcal{M}$. However, the
relevance for $\mathcal{M}$ of such an ability completely disappears in the most
general case where the size of the network is unknown, as $\mathcal{M} = f + 2$
regardless of whether Byzantine agents can change their labels or not.

Keywords: deterministic gathering, mobile agent, Byzantine fault.

---

* Partially supported by the European Regional Development Fund (ERDF) and the
Picardy region under Project TOREDY.

C. Scheideler (Ed.): SIROCCO 2015, LNCS 9439, pp. 179–193, 2015.
DOI: 10.1007/978-3-319-25258-2_13

1   Introduction

1.1   Context

Gathering is one of the most fundamental tasks in the field of distributed and mobile systems in the sense that, the ability to gather is in fact a building block to achieve more complex cooperative works. Loosely speaking, the task of gathering consists in ensuring that a group of mobile entities, initially located in different places, ends up meeting at the same place at the same time. These mobile entities, hereinafter called agents, can vary considerably in nature ranging from human beings and robots to animals and software agents. The environment in which the agents are supposed to evolve can vary considerably as well: it may be a terrain, a network modeled as a graph, a three-dimensional space, etc. We can also consider that the sequences of instructions followed by the agents in order to ensure their gathering are either deterministic or randomized.

In this paper, we consider the problem of gathering in a deterministic way in a network modeled as a graph. Thus, the agents initially start from different nodes of the graph and have to meet at the same node by applying deterministic rules. We assume that among the agents, some are Byzantine. A Byzantine agent is an agent subject to unpredictable and arbitrary faults. For instance such an agent may choose to never stop or to never move. It may also convey arbitrary information to the other agents, etc. The case of Byzantine fault is very interesting because it is the worst fault that can occur to agents. As a consequence, gathering in such a context is challenging.

1.2   Model and Problem

The distributed system considered in this paper consists of a group of mobile agents that are initially placed by an adversary at arbitrary but distinct nodes of a network modeled as a finite, connected, undirected graph $G = (V, E)$. We assume that $|V| = n$. In the sequel n is also called the size of the network. Two assumptions are made about the labelling of the two main components of the graph that are nodes and edges. The first assumption is that nodes are anonymous i.e., they do not have any kind of labels or identifiers allowing them to be distinguished from one another. The second assumption is that edges incident to a node v are locally ordered with a fixed port numbering ranging from 0 to $deg(v) - 1$ where $deg(v)$ is the degree of v. Therefore, each edge has exactly two port numbers, one for each of both nodes it links. The port numbering is not supposed to be consistent: a given edge $(u, v) \in E$ may be the i-th edge of u but the j-th edge of v, where $i \neq j$. These two assumptions are not fortuitous. The primary motivation of the first one is that if each node could be identified by a label, gathering would become quite easy to solve as it would be tantamount to explore the graph (via e.g. a breadth-first search) and then meet in the node having the smallest label. While the first assumption is made so as to avoid making the problem trivial, the second assumption is made in order to avoid making the problem impossible to solve. Indeed, in the absence of a way allowing an agent to distinguish locally the edges incident to a node, gathering could be proven as impossible to solve deterministically in view of the fact that some agents could be precluded from traversing some edges and visit some parts of the graph.

An adversary chooses the starting nodes of the agents. The starting nodes are chosen so that there are not two agents sharing initially the same node. At the beginning, an agent has a little knowledge about its surroundings: it does not know either the graph topology, or the number of other agents, or the positions of the others relative to its own. Still regarding agents' knowledge, we will study two scenarios: one in which the agents initially know the parameter n and one in which the agents do not initially know this parameter or even any upper bound on it.

Time is discretized into an infinite sequence of rounds. In each round, every agent, which has been previously woken up (this notion is detailed in the next paragraph), is allowed to stay in place at its current node or to traverse an edge according to a deterministic algorithm. The algorithm is the same for all agents: only the input, whose nature is specified further in the subsection, varies among agents.

Before being woken up, an agent is said to be dormant. A dormant agent may be woken up only in two different ways: either by the adversary that wakes some of the agents at possibly different rounds, or as soon as another agent enters the starting node of the dormant agent. We assume that the adversary wakes up at least one agent. When an agent is woken up in a round r, it is told the degree of its starting node. As mentioned above, in each round $r' \geq r$, the executed algorithm can ask the agent to stay idle or to traverse an edge. In the latter case, this takes the following form: the algorithm ask the agent, located at node u, to traverse the edge having port number i, where $0 \leq i < deg(u) - 1$. Let us denote by $(u, v) \in E$ this traversed edge. In round $r' + 1$, the agents enters node v: it then learns the degree $deg(v)$ as well as the local port number j of (u, v) at node v (recall that in general $i \neq j$). An agent cannot leave any kind of tokens or markers at the nodes it visits or the edges it traverses.

In the beginning, the adversary also assigns a different positive integer (called label) to each agent. Each agent knows its label but does not know the labels of the other agents. When several agents are at the same node in the same round, they see the labels of the other agents and can exchange all the information they currently have. This exchange is done in a "shouting" mode in one round: all the exchanged information becomes common knowledge for agents that are currently at the node. On the other hand when two agents are not at the same node in the same round they cannot see or talk to each other: in particular, two agents traversing simultaneously the same edge but in opposite directions, and thus crossing each other on the same edge, do not notice this fact. In every round, the input of the algorithm executed by an agent a is made up of the label of agent a and the up-to-date memory of what agent a has seen and learnt since its waking up. Note that in the absence of a way of distinguishing the agents, the gathering problem would have no deterministic solution in some graphs. This is especially the case in a ring in which at each node the edge going clockwise has port number 0 and the edge going anti-clockwise has port 1: if all agents are woken up in the same round and start from different nodes, they will always have the same input and will always follow the same deterministic rules leading to a situation where the agents will always be at distinct nodes no matter what they do.

Within the team, it is assumed that up to f of the agents are Byzantine. The parameter f is known to all agents. A Byzantine agent has a high capacity of nuisance: it can choose an arbitrary port when it moves, can convey arbitrary information to other agents and can change its label in every round, in particular by forging the label of another agent or by creating a completely new one. All the agents that are not Byzantine are called good. We consider the task of f-Byzantine gathering which is stated as follows. The adversary wakes up at least one good agent and all good agents must eventually be in the same node in the same round, simultaneously declare termination and stop, provided that there are at most f Byzantine agents. Regarding this task, it is worth mentioning that we cannot require the Byzantine agents to cooperate as they may always refuse to be with some agents. Thus, gathering all good agents with termination is the strongest requirement we can make in such a context.

What is the minimum number $\mathcal{M}$ of good agents that guarantees f-Byzantine gathering?

At first glance, the question might appear as not being really interesting since, after all, the good agents might always be able to gather in some node, regardless of the number of Byzantine agents evolving in the graph. However, this is not the case as pointed out by the study that introduced this question in [14]. More specifically, when this size is initially known to the agents, the authors of this study described a deterministic algorithm gathering all good agents in any network provided that there are at $2f + 1$ of them, and gave a lower bound of $f + 1$ on $\mathcal{M}$ by showing that if the number of good agents is not larger than f, then there are some graphs in which the good agents are not able to gather deterministically with termination. When the size of the network is unknown, they did a similar thing but with different bounds: they gave an algorithm working for a team including at least $4f + 2$ good agents, and showed a lower bound of $f + 2$ on $\mathcal{M}$. However, the question of what the tight bounds are was left as an open problem.

1.3   Our Results

In this paper, we solve this open problem by proving that the lower bounds of $f + 1$ and $f + 2$ on $\mathcal{M}$, shown in [14], are actually also upper bounds respectively when the size of the network is known and when it is unknown. More precisely, we design deterministic algorithms allowing to gather all good agents provided that the number of good agents is at least $f + 1$ when the size of the network is initially known to agents, and at least $f + 2$ when this size is initially unknown.

Perhaps surprisingly, our results highlight an interesting feature when put in perspective with results concerning a relaxed variant of this problem (also introduced in [14]) in which the Byzantine agents cannot change their initial labels. Indeed under this variant $\mathcal{M} = 1$ for networks of known size and $\mathcal{M} = f + 2$ for networks of unknown size[1]. Following this perspective, it turns out that when the size of the network is known, the ability for the Byzantine agents to change their labels significantly impacts the value of $\mathcal{M}$. However, the relevance for $\mathcal{M}$ of such an ability completely disappears in the most general case where the size of the network is unknown, as $\mathcal{M} = f + 2$ regardless of whether Byzantine agents can change their labels or not.

1.4 Related Works

Historically, the first mention of the gathering problem appeared in [28] under the appellation of rendezvous problem. Rendezvous is the term which is usually used when the studied task of gathering is restricted to a team of exactly two agents. From this publication until now, the problem has been extensively studied so that there is henceforth a huge literature about this subject. This is mainly due to the fact that there is a lot of alternatives for the combinations we can make when approaching the problem, e.g., by playing on the environment in which the agents are supposed to evolve, the way of applying the sequences of instructions (i.e., deterministic or randomized) or the ability to leave some traces in the visited locations, etc. Naturally, in this paper we are more interested in the research works that are related to deterministic gathering in networks modeled as graphs. This is why we will mostly dwell on this scenario in the rest of this subsection. However, for the curious reader wishing to consider the matter in greater depth, we invite him to consult [7,1,19] that address the problem in the plane via various scenarios, especially in a system affected by the occurrence of faults or inaccuracies for the last two references. Regarding randomized rendezvous, a good starting point is to go through [2,3,21].

Concerning the context of this paper, the closest work to ours is obviously [14]. Nonetheless, in similar settings but without Byzantine agents, there are some papers that should be cited here. This is in particular the case of [13] in which the author presented a deterministic protocol for solving the rendezvous problem, which guarantees a meeting of the two involved agents after a number of rounds that is polynomial in the size n of the graph, the length l of the shorter of the two labels and the time interval τ between their wake-up times. As an open problem, the authors ask whether it is possible to obtain a polynomial solution to this problem which would be independent of τ. A positive answer to this question was given, independently of each other, in [20] and [29]. While these algorithms ensure rendezvous in polynomial time (i.e., a polynomial number of rounds),

---

[1] The proof that both of these values are enough, under their respective assumptions regarding the knowledge of the network size, relies on algorithms using a mechanism of blacklists that are, informally speaking, lists of labels corresponding to agents having exhibited an "inconsistent" behavior. Of course, in the context of our paper, we cannot use such blacklists as the Byzantine agents can change their labels and in particular steal the identities of good agents.

they also ensure it at polynomial cost since the cost of a rendezvous protocol is the number of edge traversals that are made by the agents until meeting and since each agent can make at most one edge traversal per round. However, it should be noted that despite the fact a polynomial time implies a polynomial cost, the reciprocal is not always true as the agents can have very long waiting periods sometimes interrupted by a movement. Thus these parameters of cost and time are not always linked to each other. This was highlighted in [25] where the authors studied the tradeoffs between cost and time for the deterministic rendezvous problem. More recently, some efforts have been dedicated to analyse the impact on time complexity of rendezvous when in every round the agents are brought with some pieces of information by making a query to some device or some oracle, see, e.g., [11,24]. Along with the works aiming at optimizing the parameters of time and/or cost of rendezvous, some other works have examined the amount of memory that is required to achieve deterministic rendezvous e.g., in [16,17] for tree networks and in [9] for general networks.

All the aforementioned studies that are related to gathering in graphs take place in a synchronous scenario i.e., a scenario in which the agents traverse the edges in synchronous rounds. Some efforts have been also dedicated to the scenario in which the agents move asynchronously: the speed of agents may then vary and is controlled by the adversary. For more details about rendezvous under such a context, the reader is referred to [23,10,15,18] for rendezvous in finite graphs and [4,8] for rendezvous in infinite grids.

Aside from the gathering problem, our work is also in conjunction with the field of fault tolerance via the assumption of Byzantine faults to which some agents are subjected. First introduced in [26], a Byzantine fault is an arbitrary fault occurring in an unpredictable way during the execution of a protocol. Due to its arbitrary nature, such a fault is considered as the worst fault that can occur. Byzantine faults have been extensively studied for "classical" networks i.e., in which the entities are fixed nodes of the graph (cf., e.g., the book [22] or the survey [5]). To a lesser extend, the occurrence of Byzantine faults has been also studied in the context of mobile entities evolving in the plane, cf. [1,12]. Prior to our work, gathering in arbitrary graphs in presence of Byzantine agents was considered only in [14]. As mentioned in the previous section, it is proven in [14] that the minimum number $\mathcal{M}$ of good agents that guarantees f-Byzantine gathering is precisely 1 for networks of known size and $f + 2$ for networks of unknown size, provided that the Byzantine agents cannot lie about their labels. The proof that both of these values are enough, under their respective assumptions regarding the knowledge of the network size, relies on algorithms using a mechanism of blacklists that are, informally speaking, lists of labels corresponding to agents having exhibited an "inconsistent" behavior. Of course, in the context of our paper, we cannot use such blacklists as the Byzantine agents can change their labels and in particular steal the identities of good agents.

2   Preliminaries

Throughout the paper, the number of nodes of a graph is called its size. In this section we present two procedures, that will be used as building blocks in our algorithms. The aim of both of them is graph exploration, i.e., visiting all nodes of the graph by a single agent. The first procedure, based on universal exploration sequences (UXS), is a corollary of the result of Reingold [27]. Given any positive integer N, this procedure allows the agent to traverse all nodes of any graph of size at most N, starting from any node of this graph, using $P(N)$ edge traversals, where P is some polynomial. After entering a node of degree d by some port p, the agent can compute the port q by which it has to exit; more precisely $q = (p + x_i) \mod d$, where x_i is the corresponding term of the UXS of length $P(N)$.

The second procedure [6] needs no assumption on the size of the network but it is performed by an agent using a fixed token placed at a node of the graph. It works in time polynomial in the size of the graph. (It is well known that a terminating exploration even of all anonymous rings of unknown size by a single agent without a token is impossible.) In our applications the roles of the token and of the exploring agent will be played by agents or by groups of agents. At the end of this second procedure, the agent has visited all nodes and determined a BFS tree of the underlying graph.

We call the first procedure $EXPLO(N)$ and the second procedure EST, for *exploration with a stationary token*. We denote by $T(EXPLO(n))$ the execution time of procedure $EXPLO$ with parameter n (note that $T(EXPLO(n)) = P(n) + 1$). We denote by $T(EST(N))$ the maximum time of execution of the procedure EST in a graph of size at most N.

3   Known Graph Size

This section aims at proving the following theorem

Theorem 1. *Deterministic f-Byzantine gathering of k good agents is possible in any graph of known size if, and only if $k \geq f + 1$.*

As mentioned in Subsection 1.2, we know from [14] that:

Theorem 2 ([14]). *Deterministic f-Byzantine gathering of k good agents is not possible in some graph of known size if $k \leq f$.*

Thus, to prove Theorem 1, it is enough to show the following theorem.

Theorem 3. *Deterministic f-Byzantine gathering of k good agents is possible in any graph of known size if $k \geq f + 1$.*

Hence, the rest of this section is devoted to proving Theorem 3. To do so, we show a deterministic algorithm that gathers all good agents in an arbitrary network of known size, provided there are at least $f + 1$ of them.

Before presenting the algorithm, we first give the high level idea which is behind it. Let us assume an ideal situation in which each agent would have as input, besides its label and the network size n, a parameter $\rho = (G^*, L^*)$ corresponding to the initial configuration of the agents in the graph such that:

- G^* represents the n-node graph with all port numbers, in which each node are assigned an identifier belonging to $\{1, \cdots, n\}$. The node identifiers are pairwise distinct. Note that the representation G^* contains more information than there is in the actual graph G as it also includes node identifiers which do not exist in G.
- $L^* = \{(v_1, l_1), (v_2, l_2), \cdots, (v_k, l_k)\}$ where $(v_i, l_i) \in L^*$ iff there is a good agent having label l_i which is initially placed in G at the node having identifier v_i in G^*. Remark that $k \geq f + 1$.

Let us also assume that all the agents in the graph are woken up at the same time by the adversary. In such ideal situation, gathering all good agents can be easily achieved by ensuring that each agent moves towards the node v where the agent having the smallest label is located. Each agent can indeed do that by using the knowledge of $\rho = (G^*, L^*)$ and its own label. Of course, all the good agents do not necessarily reach node v at the same time. However, each agent can compute the remaining time which is required to wait at node v in order to be sure that all good agents are at node v: again this time can be computed using $\rho = (G^*, L^*)$ and the fact that all agents are woken up in the same round. Unfortunately, the agents are not in such ideal situation. First, every agent is not necessarily woken up by the adversary, and for those that are woken by the adversary, this is not necessarily in the same round. Second, the agents do not have configuration ρ as input of the algorithm. In our algorithm we cope with the first constraint by requiring the first action to be a traversal of the entire graph (using procedure $EXPLO(n)$) which allows to wake up all encountered agents that are still dormant. In this way, the agents are "almost synchronized" as the delay between the starting times of any two agents is at most $T(EXPLO(n))$: the waiting time periods can be adjusted regarding this maximum delay. The second constraint i.e., the non-knowledge of ρ, is more complicated to deal with. To handle the lack of information about ρ, agents make successive assumptions about it that are "tested" one by one. More precisely, let $\mathcal{P}$ be the recursively enumerable set of all the configurations $\rho_i = (G_i^*, L_i^*)$ such that G_i^* is a connected n-node graph and $|L_i^*| \geq f + 1$. Let $\Theta = (\rho_1, \rho_2, \rho_3, \cdots)$ be a fixed enumeration of $\mathcal{P}$ (all good agents agree on this enumeration). Each agent proceeds in phases numbered $1, 2, 3, \cdots$. In each phase i, an agent supposes that $\rho = \rho_i$ and, similarly as in the ideal situation, tries to go to the node which is supposed to correspond to node v, where v is the node where the agent having the smallest label is initially located (according to ρ_i). For some reasons detailed in the algorithm (refer to the description of state setup), when $\rho_i \neq \rho$ some agents may be unable to make such a motion. As a consequence, these agents will consider that, rightly, $\rho_i \neq \rho$. On the other hand, whether $\rho_i \neq \rho$ or not, some other good agents may reach a node for which they had no reason to think it is not v (and thus $\rho_i \neq \rho$). The danger here is that when reaching the supposed node v these successful agents could see all the $|L_i^*|$ labels of ρ_i (with the possible "help" of some Byzantine agents). At this point, it may be tempting to consider that gathering is over but this could be wrong especially in the case where $\rho_i \neq \rho$ and some good agents did not reach a supposed node v in phase i. To circumvent this problem, the idea is to get the good agents thinking that $\rho_i = \rho$ to fetch the (possible) others for which $\rho_i \neq \rho$ via a traversal of the entire graph using procedure $EXPLO(n)$ (refer to the description of state tower). To allow this, an agent for which $\rho_i \neq \rho$ will wait a prescribed amount of rounds in order to leave enough time for possible good agents to fetch it (refer to the description of state wait-for-a-tower). For our purposes, it is important to prevent the agents from being fetched any old how by any group, especially those containing only Byzantine agents. Hence our algorithm is designed in such a way that within each phase at most one group, called a *tower* and made up of at least $f + 1$ agents, will be unambiguously recognized as such and be allowed to fetch the other agents via an entire traversal of the graph (this guarantee principally results from the rules that are prescribed in the description of state tower builder). When a tower has finished the execution of procedure $EXPLO(n)$ in some phase i, our algorithm guarantees that all good agents are together and declare gathering is over at the same time (whether the assumed configuration ρ_i corresponds to the real initial configuration or not). On the other hand, in every phase i, if a tower is not created or "vanishes" (because there at not at least $f + 1$ agents inside of it anymore) before the completion of its traversal, no good agent will declare that gathering is over in phase i. In the worst case, the good agents will have to wait until assuming a good hypothesis about the real initial configuration, in order to witness the creation of a tower which will proceed to an entire traversal of the network (and thus declare gathering is over). We now give a detailed description of the algorithm. (Due to the lack of space, the proof of correctness of the algorithm is omitted but will appear in the journal version of the paper).

Algorithm Byz-Known-Size with parameter n (known size of the graph)

The algorithm is made up of two parts. The first part aims at ensuring that all agents are woken up before proceeding to the second part which is actually the heart of the algorithm.

Part 1. As soon as an agent is woken up by the adversary or another agent, it starts proceeding to a traversal of the entire graph and wakes up all encountered agents that are still dormant. This is done using procedure $EXPLO(n)$ where n is the size of the network which is initially known to all agents. Once the execution of $EXPLO(n)$ is accomplished, the agent backtracks to its starting node by traversing all edges traversed in $EXPLO(n)$ in the reverse order and the reverse direction.

Part 2. In this part, the agent works in phases numbered $1, 2, 3, \cdots$. During the execution of each phase, the agent can be in one of the following five states:

setup, tower builder, tower, wait-for-a-tower, failure. Below we describe the actions of an agent A in each of the states as well as the transitions between these states within phase i. We assume that in every round agent A tells the others (sharing the same node as agent A) in which state it is. In some states, the agent will be required to tell more than just its current state: we will mention it in the description of these states. Moreover, in the description of every state X, when we say "agent A transits to state Y", we exactly mean agent A remains in state X until the end of the current round and is in state Y in the following round. Thus, in each round of this part, agent A is always exactly in one state.

At the beginning of phase i, agent A enters state setup.

State setup.

Let ρ_i be the i-th configuration of enumeration Θ (refer to above). If the label l of agent A is not in ρ_i, then it transits to state wait-for-a-tower. Otherwise, let X be the set of the shortest paths in ρ_i leading from the node containing the agent having label l, to the node containing the smallest label of the supposed configuration. Each path belonging to X is represented as the corresponding sequence of port numbers. Let π be the lexicographically smallest path in X (the lexicographic order can be defined using the total order on the port numbers). Agent A follows path π in the real network. If , following path π, agent A has to leave by a port number that does not exist in the node where it currently resides, then it transits to state wait-for-a-tower. In the same way, it also transits to state wait-for-a-tower if, following path π, agent A enters at some point a node by a port number which is not the same as that of path π. Once path π is entirely followed by agent A, it transits to state tower builder.

State tower builder.

When in state tower builder, agent A can be in one of the following three substates: yellow, orange, red. In all of these substates the agent does not make any move: it stays at the same node denoted by v. At the beginning, agent A enters substate yellow. By misuse of language, in the rest of this paper we will sometimes say that an agent "is yellow" instead of "is in substate yellow". We will also use the same kind of shortcut for the two other colors. In addition to its state, we also assume that in every round agent A tells the others in which substate it is.

Substate yellow

Let k be the number of labels in configuration ρ_i. Agent A waits $T(EXPLO(n)) + n$ rounds. If during this waiting period, there are at some point at least k orange agents at node v then agent A transits to substate red. Otherwise, if at the end of this waiting period there are not at least k agents residing at node v such that each of them is either yellow or orange, then agent A transits to state wait-for-a-tower, else it transits to substate orange.

Substate orange

Agent A waits at most $T(EXPLO(n)) + n$ rounds to see the occurrence of one of the following two events. The first event is that there are not at least k agents residing at node v such that each of them is either yellow or orange. The second event is that there are at least k orange agents residing at node v. Note that the two events cannot occur in the same round. If during this waiting period, the first (resp. second) event occurs, then agent A transits to state wait-for-a-tower (resp. substate red). If at the end of the waiting period, none of these events has occurred, then agent A transits to substate wait-for-a-tower.

Substate red

Agent A waits $T(EXPLO(n)) + n$ rounds. If at each round of this waiting period there are at least k red agents at node v, then at the end of the waiting period, agent A transits to state tower. Otherwise, there is a round during the waiting period in which there are not at least k red agents at node v: agent A then transits to state wait-for-a-tower as soon as it notices this fact.

State tower.

Agent A can enter state tower either from state tower builder or state wait-for-a-tower. While in this state, agent A will execute all or part of procedure $EXPLO(n)$. In both cases we assume that, in every round, agent A tells the others the edge traversal number of $EXPLO(n)$ it has just made (in addition to its state). We call this number the index of the agent. Below, we distinguish and detail the two cases.

When agent A enters state tower from state tower builder, it starts executing procedure $EXPLO(n)$. In the first round, its index is 0. Just after making the j-th edge traversal of $EXPLO(n)$, its index is j. Agent A carries out the execution of $EXPLO(n)$ until its term, except if at some round of the execution the following condition is not satisfied, in which case agent A transits to state failure. Here is the condition: the node where agent A is currently located contains a group S of at least $f + 1$ agents in state tower having the same index as agent A. S includes agent A but every agent that is in the same node as agent A is not necessarily in S. If at some point this condition is satisfied and the index of agent A is equal to $P(n)$, which is the total number of edge traversals in $EXPLO(n)$ (refer to Section 2), then agent A declares that gathering is over.

When agent A enters state tower from state wait-for-a-tower, it has just made the s-th edge traversal of $EXPLO(n)$ for some s
(cf. state wait-for-a-tower) and thus, its index is s. Agent A executes the next edge traversals i.e., the $s + 1$-th, $s + 2$-th, $\cdots$, and then its index is successively $s + 1$, $s + 2$, etc. Agent A carries out this execution until the end of procedure $EXPLO(n)$, except if the same condition as above is not fulfilled at some round of the execution of the procedure, in which case agent A also transits to state failure. As in the first case, if at some point the node where agent A is currently located contains a group S of at least $f + 1$ agents in state tower having an index equal to $P(n)$, then agent A declares that gathering is over.

State wait-for-a-tower.

Agent A waits at most $5T(EXPLO(n)) + 4n$ rounds to see the occurrence of the following event: the node where it is currently located contains a group of at least $f + 1$ agents in state tower having the same index t. If during this waiting period, agent A sees such an event, we distinguish two cases. If $t < P(n)$, then it makes the $t + 1$-th edge traversal of procedure $EXPLO(n)$ and transits to state `tower`. If $t = P(n)$, then it declares that gathering is over.

Otherwise, at the end of the waiting period, agent A has not seen such an event, and thus it transits to state `failure`.

State failure. Agent A backtracks to the node where it was located at the beginning of phase i. To do this, agent A traverses in the reverse order and the reverse direction all edges it has traversed in phase i before entering state `failure`. Once at its starting node, agent A waits $10T(EXPLO(n)) + 9n - p$ rounds where p is the number of elapsed rounds between the beginning of phase i and the end of the backtrack it has just made. At the end of the waiting period, phase i is over. In the next round, agent A will start phase $i + 1$.

4   Unknown Graph Size

In this section, we consider the same problem, except we assume that the agents are not initially given the size of the graph. Under this harder scenario, we aim at proving the following theorem.

Theorem 4. *Deterministic f-Byzantine gathering of k good agents is possible in any graph of unknown size if, and only if $k \geq f + 2$.*

As mentioned in Subsection 1.2, we know from [14] that:

Theorem 5 ([14]). *Deterministic f-Byzantine gathering of k good agents is not possible in some graphs of unknown size if $k \leq f + 1$.*

In view of Theorem 5, it is then enough to show the following theorem in order to prove Theorem 4.

Theorem 6. *Deterministic f-Byzantine gathering of k good agents is possible in any graph of unknown size if $k \geq f + 2$.*

Hence, similarly as in Section 3, the rest of this section is devoted to showing a deterministic algorithm that gathers all good agents, but this time in an arbitrary network of unknown size and provided there are at least $f + 2$ good agents.

Before giving the algorithm, which we call *Algorithm Byz-Unknown-Size*, let us provide some intuitive ingredients on which our solution is based.

The algorithm of this section displays a number of similarities with the algorithm of the previous section, but there are also a number of changes to tackle the non-knowledge of the network size. Among the most notable changes, there is firstly the way of enumerating the configurations. Previously, the agents were considering the enumeration $\Theta = (\rho_1, \rho_2, \rho_3, \cdots)$ of $\mathcal{P}$ where $\mathcal{P}$ is the set of every configuration corresponding to a n-node graph in which there are at least $f + 1$ robots with pairwise distinct labels. Now, instead of considering Θ, the agents will consider the enumeration $\Omega = (\phi_1, \phi_2, \phi_3, \cdots)$ of $\mathcal{Q}$ where $\mathcal{Q}$ is the set of all configurations corresponding to a graph of any size (instead of size n only) in which there are at least $f + 2$ agents (instead of at least $f + 1$) with pairwise distinct labels. Note that, as for set $\mathcal{P}$, set $\mathcal{Q}$ is also recursively enumerable. Another change stems from the function performed by a tower, which we also find here. In Algorithm Byz-Known-Size, the role of a tower was to fetch all awaiting good agents (which know that the tested configuration is not good) via procedure $EXPLO(n)$: in the new algorithm, we keep the exact same strategy. However, to be able to use procedure $EXPLO$ with a parameter corresponding to the size of the network, it is necessary, for the good agents that are members of a tower, to know this size. Hence, in our solution, before being considered as a tower and then authorized to make a traversal of the graph, a group of agents will have to learn the size of the graph. To do this, at least each good agent of the group will be required to make a simulation of procedure EST by playing the role of an explorer and using the others as its token. To carry out these simulations, it is also required for the group of agents to contain initially at least $f + 2$ members (explorer + token), even if subsequently it is required for a group of agents forming a tower to contain at least $f + 1$ members. Our algorithm is designed in such a way that if during the simulation of procedure EST by an agent playing the role of an explorer, we have the guarantee there are always at least $f + 1$ agents playing the role of its token, then the explorer will be able to recognize its own token without any ambiguity (and thus will act as if it performed procedure EST with a "genuine" token). Of course, the agents will not always have such a guarantee (especially due to the possible bad behavior of Byzantine agents when testing a wrong configuration) and will not be able to detect in advance whether they will have it or not. Besides, some other problems can arise including, for example, some Byzantine explorer which takes too much time to explore the graph (or worse still, "never finishes" the exploration). However we will show that in all cases, the good agents can never learn an erroneous size of the graph (even with the duplicity of Byzantine agents when testing a wrong configuration). We also show that good agents are assured of learning the size of the network when testing a good configuration at the latest (in particular as the creation of a group of at least $f + 2$ agents and the aforementioned guarantee are ensured when testing a good configuration). As for Algorithm Byz-known-Size, in the worst case the good agents will have to wait until assuming a good hypothesis about the real initial configuration, in order to declare gathering is over. The details of Algorithm Byz-Unknown-Size (sketched above) and its analysis will appear in the journal version of the paper.

5   Conclusion

We provided a deterministic f-Byzantine gathering algorithm for arbitrary connected graphs of known size (resp. unknown size) provided that the number of good agents is at least $f+1$ (resp. $f+2$). By providing these algorithms, we closed the open question of what minimum number of good agents $\mathcal{M}$ is required to solve the problem, as each of our algorithms perfectly matches the corresponding lower bound on $\mathcal{M}$ stated in [14], which is of $f + 1$ when the size of the network is known and of $f + 2$ when it is unknown. Our work also highlighted the fact that the ability for the Byzantine agents to change their labels has no impact in terms of feasibility when the size of the network is initially unknown, since it was proven in [14] that $\mathcal{M}$ is also equal to $f + 2$ when the Byzantine agents do not have this ability.

While we gave algorithms that are optimal in terms of required number of good agents, we did not try to optimize their time complexity. Actually, the time complexity of both our solutions depends on the enumerations of the initial configurations, which clearly makes them exponential in n and the labels of the good agents in the worst case. Hence, the question of whether there is a way to obtain algorithms that are polynomial in n and in the labels of the good agents (with the same bounds on $\mathcal{M}$) remains an open problem.

References

1. Agmon, N., Peleg, D.: Fault-tolerant gathering algorithms for autonomous mobile robots. SIAM J. Comput. 36(1), 56–82 (2006)
2. Alpern, S.: Rendezvous search: A personal perspective. Operations Research 50(5), 772–795 (2002)
3. Alpern, S.: The theory of search games and rendezvous. International Series in Operations Research and Management Science. Kluwer Academic Publishers (2003)
4. Bampas, E., Czyzowicz, J., Gąsieniec, L., Ilcinkas, D., Labourel, A.: Almost optimal asynchronous rendezvous in infinite multidimensional grids. In: Lynch, N.A., Shvartsman, A.A. (eds.) DISC 2010. LNCS, vol. 6343, pp. 297–311. Springer, Heidelberg (2010)
5. Barborak, M., Malek, M.: The consensus problem in fault-tolerant computing. ACM Comput. Surv. 25(2), 171–220 (1993)
6. Chalopin, J., Das, S., Kosowski, A.: Constructing a map of an anonymous graph: Applications of universal sequences. In: Lu, C., Masuzawa, T., Mosbah, M. (eds.) OPODIS 2010. LNCS, vol. 6490, pp. 119–134. Springer, Heidelberg (2010)
7. Cieliebak, M., Flocchini, P., Prencipe, G., Santoro, N.: Distributed computing by mobile robots: Gathering. SIAM J. Comput. 41(4), 829–879 (2012)
8. Collins, A., Czyzowicz, J., Gąsieniec, L., Labourel, A.: Tell me where I am so I can meet you sooner. In: Abramsky, S., Gavoille, C., Kirchner, C., Meyer auf der Heide, F., Spirakis, P.G. (eds.) ICALP 2010. LNCS, vol. 6199, pp. 502–514. Springer, Heidelberg (2010)
9. Czyzowicz, J., Kosowski, A., Pelc, A.: How to meet when you forget: log-space rendezvous in arbitrary graphs. Distributed Computing 25(2), 165–178 (2012)
10. Czyzowicz, J., Pelc, A., Labourel, A.: How to meet asynchronously (almost) everywhere. ACM Transactions on Algorithms 8(4), 37 (2012)
11. Das, S., Dereniowski, D., Kosowski, A., Uznański, P.: Rendezvous of distance-aware mobile agents in unknown graphs. In: Halldórsson, M.M. (ed.) SIROCCO 2014. LNCS, vol. 8576, pp. 295–310. Springer, Heidelberg (2014)
12. Défago, X., Gradinariu, M., Messika, S., Raipin-Parvédy, P.: Fault-tolerant and self-stabilizing mobile robots gathering. In: Dolev, S. (ed.) DISC 2006. LNCS, vol. 4167, pp. 46–60. Springer, Heidelberg (2006)
13. Dessmark, A., Fraigniaud, P., Kowalski, D.R., Pelc, A.: Deterministic rendezvous in graphs. Algorithmica 46(1), 69–96 (2006)
14. Dieudonné, Y., Pelc, A., Peleg, D.: Gathering despite mischief. ACM Transactions on Algorithms 11(1), 1 (2014)

15. Dieudonné, Y., Pelc, A., Villain, V.: How to meet asynchronously at polynomial cost. In: ACM Symposium on Principles of Distributed Computing, PODC 2013, Montreal, QC, Canada, July 22-24, pp. 92–99 (2013)

16. Fraigniaud, P., Pelc, A.: Deterministic rendezvous in trees with little memory. In: Taubenfeld, G. (ed.) DISC 2008. LNCS, vol. 5218, pp. 242–256. Springer, Heidelberg (2008)

17. Fraigniaud, P., Pelc, A.: Delays induce an exponential memory gap for rendezvous in trees. ACM Transactions on Algorithms 9(2), 17 (2013)

18. Guilbault, S., Pelc, A.: Gathering asynchronous oblivious agents with local vision in regular bipartite graphs. Theor. Comput. Sci. 509, 86–96 (2013)

19. Izumi, T., Souissi, S., Katayama, Y., Inuzuka, N., Défago, X., Wada, K., Yamashita, M.: The gathering problem for two oblivious robots with unreliable compasses. SIAM J. Comput. 41(1), 26–46 (2012)

20. Kowalski, D.R., Malinowski, A.: How to meet in anonymous network. Theor. Comput. Sci. 399(1-2), 141–156 (2008)

21. An, H.-C., Krizanc, D., Rajsbaum, S.: Mobile agent rendezvous: A survey. In: Flocchini, P., Gasieniec, L. (eds.) SIROCCO 2006. LNCS, vol. 4056, pp. 1–9. Springer, Heidelberg (2006)

22. Lynch, N.A.: Distributed Algorithms. Morgan Kaufmann (1996)

23. De Marco, G., Gargano, L., Kranakis, E., Krizanc, D., Pelc, A., Vaccaro, U.: Asynchronous deterministic rendezvous in graphs. Theor. Comput. Sci. 355(3), 315–326 (2006)

24. Miller, A., Pelc, A.: Fast rendezvous with advice. In: Algorithms for Sensor Systems - 10th International Symposium on Algorithms and Experiments for Sensor Systems, Wireless Networks and Distributed Robotics, ALGOSENSORS 2014, Wroclaw, Poland, September 12, pp. 75–87 (2014); Revised Selected Papers

25. Miller, A., Pelc, A.: Time versus cost tradeoffs for deterministic rendezvous in networks. In: ACM Symposium on Principles of Distributed Computing, PODC 2014, Paris, France, July 15-18, pp. 282–290 (2014)

26. Pease, M.C., Shostak, R.E., Lamport, L.: Reaching agreement in the presence of faults. J. ACM 27(2), 228–234 (1980)

27. Reingold, O.: Undirected connectivity in log-space. J. ACM 55(4) (2008)

28. Schelling, T.: The Strategy of Conflict. Oxford University Press, Oxford (1960)

29. Ta-Shma, A., Zwick, U.: Deterministic rendezvous, treasure hunts, and strongly universal exploration sequences. ACM Transactions on Algorithms 10(3), 12 (2014)

Signature-Free Asynchronous Byzantine Systems: From Multivalued to Binary Consensus with $t < n/3$, $O(n^2)$ Messages, and Constant Time

Achour Mostéfaoui[1] and Michel Raynal[2,3]

[1] LINA, Université de Nantes, 44322 Nantes Cedex, France
[2] Institut Universitaire de France
[3] IRISA, Université de Rennes 35042 Rennes Cedex, France

Abstract. This paper presents a new algorithm that reduces multivalued consensus to binary consensus in an asynchronous message-passing system made up of n processes where up to t may commit Byzantine failures. This algorithm has the following noteworthy properties: it assumes $t < n/3$ (and is consequently optimal from a resilience point of view), uses $O(n^2)$ messages, has a constant time complexity, and does not use signatures. The design of this reduction algorithm relies on two new all-to-all communication abstractions. The first one allows the non-faulty processes to reduce the number of proposed values to c, where c is a small constant. The second communication abstraction allows each non-faulty process to compute a set of (proposed) values such that, if the set of a non-faulty process contains a single value, then this value belongs to the set of any non-faulty process. Both communication abstractions have an $O(n^2)$ message complexity and a constant time complexity. The reduction of multivalued Byzantine consensus to binary Byzantine consensus is then a simple sequential use of these communication abstractions. To the best of our knowledge, this is the first asynchronous message-passing algorithm that reduces multivalued consensus to binary consensus with $O(n^2)$ messages and constant time complexity (measured with the longest causal chain of messages) in the presence of up to $t < n/3$ Byzantine processes, and without using cryptography techniques. Moreover, this reduction algorithm uses a single instance of the underlying binary consensus, and tolerates message re-ordering by Byzantine processes.

1 Introduction

Consensus in Asynchronous Byzantine Systems. The consensus problem lies at the center of fault-tolerant distributed computing. Assuming that each non-faulty process proposes a value, its formulation is particularly simple, namely, each non-faulty process decides a value (termination), the non-faulty processes decide the same value (agreement), and the decided value is related to the proposed values (validity); the way the decided value is related to the proposed values depends on the failure model. Consensus is binary when only two values can be proposed by the processes, otherwise it is multivalued.

Byzantine failures were introduced in the context of synchronous distributed systems [17,28,31], and then investigated in the context of asynchronous distributed systems [2,19,30]. A process has a *Byzantine* behavior (or commits a Byzantine failure) when it arbitrarily deviates from its intended behavior: it then commits a Byzantine

© Springer International Publishing Switzerland 2015
C. Scheideler (Ed.): SIROCCO 2015, LNCS 9439, pp. 194–208, 2015.
DOI: 10.1007/978-3-319-25258-2_14

failure (otherwise we say it is *non-faulty*). This bad behavior can be intentional (malicious) or simply the result of a transient fault that altered the local state of a process, thereby modifying its behavior in an unpredictable way.

Several validity properties have been considered for Byzantine consensus. This paper considers the following one: a decided value is a value that was proposed by a non-faulty process or a default value denoted $\bot$. Moreover, to prevent trivial or useless solutions, if all the non-faulty processes propose the same value, $\bot$ cannot be decided. As these properties prevent a value proposed only by faulty processes to be decided, such a consensus is called *intrusion-tolerant Byzantine* (ITB) consensus [7,24].

Solving Byzantine Consensus. Let t denote the model upper bound on the number of processes that can have a Byzantine behavior. It is shown in several papers (e.g., see [9,17,28,33]) that Byzantine consensus cannot be solved when $t \geq n/3$, be the system synchronous or asynchronous, or be the algorithm allowed to use random numbers or not.

As far as asynchronous systems are concerned, it is well-known that there is no deterministic consensus algorithm as soon as one process may crash [10], which means that Byzantine consensus cannot be solved either as soon as one process can be faulty. Said another way, the basic asynchronous Byzantine system model has to be enriched with additional computational power. Such an additional power can be obtained by randomization (e.g., see [3,7,13,22,29]), assumption on message delivery schedules (e.g., [5,33]), failure detectors suited to Byzantine systems (e.g., [12,15]), additional –deterministic or probabilistic– synchrony assumptions (e.g., [5,9,20]), or restrictions on the vectors of input values proposed by the processes (e.g., [11,23]). A reduction of atomic broadcast to consensus in the presence of Byzantine processes is presented in [21].

Finally, for multivalued Byzantine consensus, another approach consists in considering a system model enriched with an algorithm solving (for free) binary Byzantine consensus. This reduction approach has been first proposed in the context of synchronous systems [34]. (See [16,18,27] for recent works for such synchronous systems.) Reductions for asynchronous systems where the communication is by message-passing can be found in [6,7,26]. The case where communication is by read/write registers is investigated in [32]. This reduction approach is the approach adopted in this paper to address multivalued Byzantine consensus.

Contributions of the Paper. Considering asynchronous message-passing systems, this paper presents a new reduction from multivalued Byzantine consensus to binary Byzantine consensus, that has the following properties:

- It tolerates up to $t < n/3$ Byzantine processes,
- Its message cost is $O(n^2)$,
- Its time complexity is constant,
- It tolerates message re-ordering by Byzantine processes,
- It does not use cryptography techniques,
- It uses a single instance of the underlying binary Byzantine consensus.

A simple and efficient Byzantine Binary consensus algorithm has recently been proposed in [22]. This algorithm, which is based on Rabin's common coin, is signature-free and round-based, requires $t < n/3$, has an $O(n^2)$ message complexity per round, and its expected number of rounds is constant. It follows that, when the reduction algorithm proposed in this paper is combined with this binary consensus algorithm, we obtain a Byzantine multivalued consensus algorithm that has the five properties listed previously. To our knowledge, this is the first Byzantine multivalued consensus algorithm that is signature-free, optimal with respect to resilience ($t < n/3$), has an $O(n^2)$ expected message complexity, a constant expected time complexity, and tolerates the re-ordering of message deliveries by Byzantine processes.

The design of the proposed reduction algorithm is based on two new communication abstractions, which are *all-to-all* communication abstractions. The first allows the non-faulty processes to reduce the number of values they propose to $k \leq c$ values where c is a known constant. More precisely, $c = 6$ when $t < n/3$ (worst case), $c = 4$ when $n = 4t$, and $c = 3$ when $t < n/4$. The second communication abstraction allows each non-faulty process to compute a set of (proposed) values such that, if the set of a non-faulty process contains a single value, then this value belongs to the set of any non-faulty process. Both communication abstractions have an $O(n^2)$ message complexity and a constant time complexity.

The structure of the resulting Byzantine multivalued consensus algorithm is as follows. It uses the first communication abstraction to reduce the number of proposed values to a constant. Then, it uses sequentially twice the second communication abstraction to provide each non-faulty process with a binary value that constitutes the value it proposes to the underlying binary Byzantine consensus algorithm. Finally, the value decided by a non-faulty process is determined by the output (0 or 1) returned by the underlying binary Byzantine consensus algorithm. Thanks to the communication abstractions, this reduction algorithm is particularly simple (which is a first class design property).

Roadmap. The paper is composed of 6 sections. Section 2 presents the computing model, and defines the multivalued ITB consensus problem. Section 3 defines the first communication abstraction (called RD-broadcast) that reduces the number of proposed values to a constant, presents an algorithm that implements it, and proves it correct. Section 4 defines the second communication abstraction (called MV-broadcast), presents an algorithm that implements it, and proves it correct. Section 5 presents the algorithm reducing multivalued consensus to binary consensus in the presence of Byzantine processes. Due to page limitation, the reader will find all proofs in [25].

2   Computing Model and Intrusion-Tolerant Byzantine Consensus

2.1   Distributed Computing Model

Asynchronous Processes. The system is made up of a finite set Π of $n > 1$ asynchronous sequential processes, namely $\Pi = \{p_1, \ldots, p_n\}$. "Asynchronous" means that each process proceeds at its own pace, which may vary arbitrarily with time, and remains always unknown to the other processes.

Communication Network. The processes communicate by exchanging messages through an asynchronous reliable point-to-point network. "Asynchronous" means that a message that has been sent is eventually received by its destination process, i.e., there is no bound on message transfer delays. "Reliable" means that the network does not lose, duplicate, modify, or create messages. "Point-to-point" means that there is a bi-directional communication channel between each pair of processes. Hence, when a process receives a message, it can identify its sender.

A process p_i sends a message to a process p_j by invoking the primitive "send TAG(m) to p_j", where TAG is the type of the message and m its content. To simplify the presentation, it is assumed that a process can send messages to itself. A process receives a message by executing the primitive "receive()".

The operation broadcast TAG(m) is a macro-operation which stands for "**for each** $j \in \{1, \ldots, n\}$ send TAG(m) to p_j **end for**".This operation is usually called *unreliable* broadcast (if the sender commits a failure in the middle of the **for** loop, it is possible that only an arbitrary subset of processes receives the message).

Failure Model. Up to t processes may exhibit a *Byzantine* behavior. A Byzantine process is a process that behaves arbitrarily: it may crash, fail to send or receive messages, send arbitrary messages, start in an arbitrary state, perform arbitrary state transitions, etc. Hence, a Byzantine process, which is assumed to send a message m to all the processes, can send a message m_1 to some processes, a different message m_2 to another subset of processes, and no message at all to the other processes. Moreover, Byzantine processes can collude to "pollute" the computation. A process that exhibits a Byzantine behavior is also called *faulty*. Otherwise, it is *non-faulty*.

Let us notice that, as each pair of processes is connected by a channel, no Byzantine process can impersonate another process. Byzantine processes can influence the message delivery schedule, but cannot affect network reliability. More generally, the model does not assume a computationally-limited adversary.

Discarding Messages from Byzantine Processes. If, according to its algorithm, a process p_j is assumed to send a single message TAG() to a process p_i, then p_i processes only the first message TAG(v) it receives from p_j. This means that, if p_j is Byzantine and sends several messages TAG(v), TAG(v') where $v' \neq v$, etc., all of them except the first one are discarded by their receivers.

Notation. In the following, this computation model is denoted $\mathcal{BAMP}_{n,t}[\emptyset]$. In the following, this model is restricted with the constraint on $t < n/3$ and is consequently denoted $\mathcal{BAMP}_{n,t}[n > 3t]$.

2.2   Measuring Time Complexity

When computing the time complexity, we consider the longest sequence of messages $m_1, \ldots, m_z$ whose sending are causally related, i.e., for each $x \in [2..z]$, the reception of m_{x-1} is a requirement for the sending of m_x. The time complexity is the length of this longest sequence. Moreover, we implicitly consider that, in each invocation of an all-to-all communication abstraction, the non-faulty processes invoke the abstraction.

2.3 Multivalued Intrusion-Tolerant Byzantine Consensus

Byzantine Consensus. This problem has been informally stated in the Introduction. Assuming that each non-faulty process proposes a value, each of them has to decide on a value in such a way that the following properties are satisfied.

- C-Termination. Every non-faulty process eventually decides on a value, and terminates.
- C-One-shot. A non-faulty process decides at most once.
- C-Agreement. No two non-faulty processes decide on different values.
- C-Obligation (validity). If all the non-faulty processes propose the same value v, then v is decided.

Intrusion-Tolerant Byzantine (ITB) *Consensus.* Byzantine algorithms differ in the validity properties they satisfy. In classical Byzantine consensus, if the non-faulty processes do not propose the same value, they can decide any value (this is captured by the previous C-Obligation property.

As indicated in the Introduction, we are interested here in a more constrained version of the consensus problem in which a value proposed only by faulty processes cannot be decided. This was first investigated in a systematic way in [7,24]. This consensus problem instance is defined by the C-Termination, C-One-shot, C-Agreement, and C-Obligation properties stated above plus the following C-Non-intrusion (validity) property, where $\perp$ is a predefined default value, which cannot be proposed by a process.

- C-Non-intrusion (validity). A value decided by a non-faulty process is a value proposed by a non-faulty process or $\perp$.

The fact that no value proposed only by faulty processes can be decided gives its name (namely *intrusion-tolerant*) to that consensus problem instance[1].

Remark on the Binary Consensus. Interestingly, binary Byzantine consensus (only two values can be proposed by processes) has the following property.

Property 1. The ITB binary consensus problem is such that, if a value v is decided by a non-faulty process, it was proposed by a non-faulty process.
This means that, when considering the ITB binary consensus, $\perp$ can be safely replaced by any of the two possible binary values.

3 The Reducing All-to-All Broadcast Abstraction

3.1 Definition

The *reducing broadcast* abstraction (RD-broadcast) is a one-shot all-to-all communication abstraction, whose aim is to reduce the number of values that are broadcast

---

[1] Directing the non-faulty processes to decide a predefined default value –instead of an arbitrary value, possibly proposed only by faulty processes– in specific circumstances, is close to the notion of an *abortable* object as defined in [14,32] where an operation is allowed to abort in the presence of concurrency. This notion of an abortable object is different from the notion of a query-abortable object introduced in [1].

to a constant. RD-broadcast provides the processes with a single operation denoted RD_broadcast(). This operation has an input parameter, and returns a value. It is assumed that all the non-faulty processes invoke this operation.

When a process p_i invokes RD_broadcast(v_i) we say that it "RD-broadcasts" the value v_i. When a process returns a value v from an invocation of RD_broadcast(), we say that it "RD-delivers" a value (or a value is RB-delivered). The default value denoted $\perp_{rd}$ cannot be RD-broadcast but can be RD-delivered. RD-broadcast is defined by the following properties.

- RD-Termination. Every non-faulty process eventually RD-delivers a value.
- RD-Integrity. No non-faulty process RD-delivers more than one value.
- RD-Justification. The value RD-delivered by a non-faulty process is either a value RD-broadcast by a non-faulty process, or the default value $\perp_{rd}$.
- RD-Obligation. If the non-faulty processes RD-broadcast the same value v, none of them RD-delivers the default value $\perp_{rd}$.
- RD-Reduction. The number of values that are RD-delivered by the non-faulty processes is upper bounded by a constant c.

3.2   An RD-Broadcast Algorithm

An algorithm implementing the RD-broadcast abstraction is described in Figure 1. This algorithm assumes $t < n/3$. The aim of the local variable rd_del_i is to contain the value RD-delivered by p_i; this variable is initialized to "?", a default value that cannot be RD-delivered by non-faulty processes.

When a process p_i invokes RD_broadcast MSG(v_i), it first broadcasts the message INIT(v_i), and then waits until it is allowed to RD-deliver a value (line 1). During this waiting period, p_i receives and processes the messages INIT() or ECHO() sent by the algorithm.

---

let $rd\_pset_i(x)$ **denote** the set of processes from which p_i has received INIT(x) or ECHO(x).

operation RD_broadcast(v_i) **is**
(1)   broadcast INIT(v_i); wait($rd\_del_i \neq$ "?"); return(rd_del_i).

when INIT(v) or ECHO(v) **is received do**
(2)   **if** $(v \neq v_i) \wedge$ (INIT(v) rec. from $(n - 2t)$ different proc.) $\wedge$ (ECHO(v) never broadcast)
(3)   **then** broadcast ECHO(v)
(4)   **end if**;
(5)   **if** $(\exists x : (x \neq v_i) \wedge (|rd\_pset_i(x)| \geq t + 1))$ **then** $rd\_del_i \leftarrow \perp_{rd}$ **end if**;
(6)   **if** $(\exists x : |rd\_pset_i(x)| \geq n - t)$ **then** $rd\_del_i \leftarrow x$ **end if**;
(7)   **let** w be the value such that, $\forall x$ received by p_i: $|rd\_pset_i(w)| \geq |rd\_pset_i(x)|$;
(8)   **if** $(| \cup_x rd\_pset_i(x)| - |rd\_pset_i(w)| \geq t + 1)$ **then** $rd\_del_i \leftarrow \perp_{rd}$ **end if**.

Fig. 1. An algorithm implementing RD-broadcast in $\mathcal{BAMP}_{n,t}[n > 3t]$

The behavior of a process p_i on its server side, i.e. when –while waiting– it receives a message INIT(v) or ECHO(v), is made up of two phases.

- Conditional communication phase (lines 2-4). If the received value v is different from the value v_i it has RD-broadcast, and INIT(v) has been received from "enough" processes (namely $(n - 2t)$), p_i broadcasts the message ECHO(v) if not yet done. Let us notice that, as $n - 2t \geq t + 1$, this means that INIT(v) was broadcast by at least one non-faulty process.

- Try-to-deliver phase (lines 5-8). Then, for any value x it has seen, a process p_i computes first the set $rd\_pset_i(x)$ composed of the processes from which p_i has received a message INIT(x) or ECHO(x). If there is a value x, different from v_i, that has been received from $(t + 1)$ different processes, if p_i is non-faulty, it knows that at least two different values have been RD-broadcast by non-faulty processes (its own value v_i, plus another one). In this case, p_i RD-delivers the default value $\perp_{rd}$ (line 5 and line 1). The RD-delivery of a value by p_i terminates its invocation of the RD-broadcast.

If the predicate of line 5 is not satisfied, p_i checks if there is a value x received from at least $(n - t)$ distinct processes (line 6). Let us notice that, in this case, it is possible that x was RD-broadcast by all correct processes. Hence, p_i RD-delivers this value.

Finally, if p_i has not yet assigned a value to rd_del_i, it computes the value w that, up to now, it received the most often in an INIT$()$ or ECHO$()$ message (line 7). If there are at least $(t + 1)$ different processes that sent INIT$()$ or ECHO$()$ messages with values different from w (this is captured by the predicate of line 8), it is impossible for p_i to have in the future the same value received from $(n - t)$ distinct processes. This claim is trivially true for w, because at least $(t + 1)$ processes sent values different from w. As no value $w' \neq w$ was received more than w, the claim is also true for any such value w'. So, the predicate of line 6 will never be satisfied at p_i, and consequently p_i RD-delivers the default value $\perp_{rd}$.

3.3 Proof of the RD-Broadcast Algorithm

All the proofs assume $t < n/3$.

Lemma 1. *Let nb_echo be the maximal number of different values that a non-faulty process may echo at line 3. We have: $(n/3 > t > n/4) \Rightarrow (nb\_echo \leq 2)$ and $(n/4 \geq t) \Rightarrow (nb\_echo \leq 1)$.*

Lemma 2. *At most c different values can be RD-delivered by the non-faulty processes, where $c = 6$ when $n/3 > t$, $c = 4$ when $n = 4t$, and $c = 3$ when $n/4 > t$.*

Theorem 1. *The algorithm described in Figure 1 implements the RD-broadcast abstraction in the computing model $\mathcal{BAMP}_{n,t}[n > 3t]$.*

Theorem 2. *The number of messages sent by the non-faulty processes is upper bounded by $O(n^2)$. Moreover, in addition to a value sent by a process, a message carries a single bit of control information. The time complexity is $O(1)$.*

3.4 RD-Broadcast vs Byzantine k-Set Agreement

In the k-set agreement problem, each process proposes a value, and at most k different values can be decided by the non-faulty processes. It is shown in [8] that the solvability of k-set agreement in the presence of Byzantine processes depends crucially on the validity properties that are considered.

As the reader can easily check, the specification of the RD-broadcast abstraction defines an instance of the intrusion-based Byzantine c-set agreement problem, where c is the constant defined in Lemma 2. It follows that the algorithm presented in Figure 1 solves this Byzantine k-set agreement instance for any $k \geq c$ in the system model $\mathcal{BAMP}_{n,t}[t < n/3]$. (Let us remind that $t < n/3$ is the lower bound on t to solve Byzantine consensus in a *synchronous* system.)

4 The Multivalued Validated All-to-All Broadcast Abstraction

4.1 Definition

The RD-broadcast abstraction reduces the number of values sent by processes to at most six values (five values RD-broadcast by non-faulty processes, plus the default value denoted $\perp_{rd}$), while keeping the number of messages exchanged by non-faulty processes in $O(n^2)$.

Differently, assuming that each non-faulty process broadcasts a value, and at most k different values are broadcast (where k does not need to be known by the processes), the aim of the one-shot *multivalued validated all-to-all broadcast* abstraction (in short MV-broadcast) is to provide each non-faulty process with an appropriate subset of values (called *validated* values), which can be used to solve multivalued ITB consensus. To that end, the fundamental property of MV-broadcast that is used is the following: if a non-faulty process returns a set with a single value, the set returned by any other non-faulty process contains this value. Moreover, from an efficiency point of view, an important point that has to be satisfied is that the message cost of an MV-broadcast instance has to be $O(kn^2)$.

To MV-broadcast a value v_i, a process p_i invokes the operation MV_broadcast(v_i). This invocation returns to p_i a non-empty set a values, which consists of validated values, plus possibly a default value denoted $\perp_{mv}$. This default value cannot be MV-broadcast by a process. Similarly to RD-broadcast, when a process invokes the operation MV_broadcast(v), we say that it "MV-broadcast v". MV-broadcast is defined by the following properties.

- MV-Obligation. If all the non-faulty processes MV-broadcast the same value v, then no non-faulty process returns a set containing $\perp_{mv}$.
- MV-Justification. If a non-faulty process p_i returns a set including a value $v \neq \perp_{mv}$, there is a non-faulty process p_j that MV-broadcast v.
- MV-Inclusion. Let set_i and set_j be the sets returned by two non-faulty processes p_i and p_j, respectively. $(set_i = \{w\}) \Rightarrow (w \in set_j)$ (let us notice that w can be $\perp_{mv}$).
- MV-Termination. An invocation of MV_broadcast() by a non-faulty process terminates (i.e., returns a non-empty set).

The following property follows directly from the MV-Inclusion property.

- MV-Singleton. Let set_i and set_j be the sets returned by two non-faulty processes p_i and p_j, respectively. $[(set_i = \{v\}) \wedge (set_j = \{w\})] \Rightarrow (v = w)$.

---

let $mv\_pset1_i(x)$ **denote** the set of processes from which p_i has received MV-VAL1(x);
mv_val2_i: set, initially $\emptyset$, of pairs $\langle$process index, value$\rangle$ received in messages MV-VAL2().

operation MV_broadcast MSG(v_i) **is**
(1)   broadcast MV_VAL1(v_i); wait ($\exists v$ such that $|mv\_pset1_i(v)| \geq 2t + 1$);
                              % in the previous wait stateemnt v can be $\perp_{mv}$ %
(2)   broadcast MV_VAL2(v); wait ($|mv\_val2_i| \geq n - t$);
(3)   return ($\{x \mid \langle-, x\rangle \in mv\_val2_i\}$).

when MV_VAL1(y) **is received do**   % y can be $\perp_{mv}$ %
(4)   **if** (($|mv\_pset1_i(y)| \geq t + 1$) $\wedge$ (MV_VAL1() not broadcast))
                              then broadcast MV_VAL1(y) **end if**;
(5)   **let** w be the value such that, $\forall x$ received by p_i: $|mv\_pset1_i(w)| \geq |mv\_pset1_i(x)|$;
(6)   **if** (($| \cup_x mv\_pset1_i(x)| - |mv\_pset1_i(w)| \geq t + 1$)
                              $\wedge$ (MV_VAL1($\perp_{mv}$) not broadcast))
(7)   **then** broadcast MV_VAL1($\perp_{mv}$) **end if**.

when MV_VAL2(x) **is received from** p_j **do**   % x can be $\perp_{mv}$ %
(8)   wait($|mv\_pset1_i(x)| \geq 2t + 1$);
(9) $mv\_val2_i \leftarrow mv\_val2_i \cup \langle j, x\rangle$.

Fig. 2. An algorithm implementing MV-broadcast in $\mathcal{BAMP}_{n,t}[n > 3t]$

4.2   An MV-Broadcast Algorithm

A two-phase algorithm implementing the MV-broadcast abstraction is described in Figure 2. It assumes $t < n/3$, and –as we will see– its message complexity is $O(kn^2)$.

To be *validated*, a value must have been MV-broadcast by at least one non-faulty process. Hence, for a process to locally know whether a value is validated, it needs to receive it from $(t + 1)$ processes.

Each process p_i manages a local variable mv_val2_i, which is a set (initially empty). Its aim is to contain pairs $\langle j, x\rangle$, where j is a process index and x a validated value. The behavior of a non-faulty process p_i is as follows.

- In the first phase (line 1) a process p_i broadcasts its initial value by sending the message MV_VAL1(v_i). It then waits until it knows (a) a validated value v (hence it has received MV_VAL1(v) from at least $(t + 1)$ different processes), (b) and this value v is eventually known by all non-faulty processes. This is captured by the following waiting predicate "the message MV_VAL1(v) has been received from at least $(2t + 1)$ different processes" used at line 1. From then on, p_i will champion this value v for it to belong to the sets returned by the non-faulty processes.

On it server side concerning the reception of a message MV_VAL1(y), a process p_i does the following (line 4). If p_i knows that y is a validated value (i.e., the message MV_VAL1(y) was received from least $(t+1)$ processes), (if not yet done) p_i broadcasts the very same message to help the validated value y to be known by all non-faulty processes.

Then, according to its current knowledge of the global state, p_i checks if there is a possibility that no value at all be present enough to be validated. It there is such a possibility, p_i broadcasts MV_VAL1$(\perp_{mv})$. To that end (as at line 7 of the RD-broadcast algorithm, Figure 1), p_i computes the value w most received from different processes (lines 5). If at least $(t+1)$ processes have broadcast values different from w, p_i broadcasts MV_VAL1$(\perp_{mv})$, if not yet done (lines 6-7); p_i sends the default value because it sees too may different values, and it does not know which ones are from non-faulty processes.

- When it enters the second phase (line 2), a process champions the validated value v it has previously computed with the waiting predicate of line 1. This is done by broadcasting the message MV_VAL2(v). It then waits until the set mv_val2_i contains at least $(n-t)$ pairs $\langle j, x \rangle$, and finally returns the set of values contained in these pairs (line 3). Let us remind that those are validated values.

On its server side, when a process p_i receives a message MV_VAL2(x) from a process p_j, it waits until it has received a message MV_VAL1(x) from at least $(2t+1)$ different processes. This is needed because Byzantine processes can send spurious messages MV_VAL2(x) while they have not validated the value x. More precisely, let us notice that the waiting predicate $(|mv\_pset1_i(x)| \geq 2t+1)$ used by p_i at line 8 is the same as the one used at line 2 by p_j –if it is non-faulty– to champion the value x. Hence, in case p_j is not non-faulty, p_i waits until the same validation predicate $(|mv\_pset1_i(x)| \geq 2t+1)$ becomes true before accepting to process the message MV_VAL2(x) sent by p_j.

Remark. Let us notice that this algorithm is tolerant to message duplication. Moreover, while a non-faulty process is not allowed to MV-broadcast the default value $\perp_{mv}$, a Byzantine process can do it. Let us also remark that $\perp_{mv}$ is the only default value associated with the MV-broadcast abstraction. Hence, for MV-broadcast, $\perp_{rd}$ is a "normal" value, which can be MV-broadcast, as any value different from $\perp_{mv}$.

4.3 Proof of the MV-Broadcast Algorithm

As previously, all the proofs assume $t < n/3$.

Lemma 3. *The waiting predicate $(\exists\, v$ such that $|mv\_pset1_i(v)| \geq 2t+1)$ (used at line 1) is eventually satisfied at any non-faulty process p_i.*

Lemma 4. *The waiting predicate $(|mv\_val2_i| \geq n-t)$ (used at line 2) is eventually satisfied at any non-faulty process p_i.*

Lemma 5. *If all non-faulty processes MV-broadcast the same value v, no non-faulty process returns a set containing $\perp_{mv}$.*

Lemma 6. *If the set returned by a non-faulty process p_i contains a value $v \neq \perp_{mv}$, then v has been MV-broadcast by a non-faulty process.*

Lemma 7. *Let set_i and set_j be the sets returned by two non-faulty processes p_i and p_j, respectively. $(set_i = \{w\}) \Rightarrow (w \in set_j)$.*

Theorem 3. *The algorithm described in Figure 2 implements the MV-broadcast abstraction in the computing model $\mathcal{BAMP}_{n,t}[t < n/3]$.*

Theorem 4. *Let us assume that at most k different values are MV-broadcast by the processes. The number of messages sent by the non-faulty processes is upper bounded by $O(kn^2)$. A message needs to carry a single bit of control information. The time complexity is $O(1)$.*

5 Multivalued Intrusion-Tolerant Byzantine Consensus

The multivalued intrusion-tolerant Byzantine (ITB) consensus problem was defined in Section 2.3. A signature-free algorithm that solves it despite up to $t < n/3$ Byzantine processes is described in this section. This algorithm is such that the expected number of messages exchanged by the non-faulty processes is $O(n^2)$, and its expected time complexity is constant.

5.1 Enriched Computation Model for Multivalued ITB Consensus

In the following, as announced in the introduction, we consider that the additional computational power that allows multivalued ITB consensus to be solved in $\mathcal{BAMP}_{n,t}[t < n/3]$ is an underlying Byzantine binary consensus (BBC) algorithm. Let $\mathcal{BAMP}_{n,t}[t < n/3, \mathrm{BBC}]$ denote the system model $\mathcal{BAMP}_{n,t}[t < n/3]$ enriched with a BBC algorithm. BBC algorithms are described in several papers (e.g., [4,7,13,22,33]).

To obtain a multivalued ITB consensus algorithm with an $O(n^2)$ expected message complexity and a constant expected time complexity, we implicitly consider that the underlying BBC algorithm is the one presented in [22].

5.2 An Efficient Algorithm Solving the Multivalued ITB Consensus Problem

The algorithm is described in Figure 3. The multivalued consensus operation that is built is denoted mv_propose(), while the underlying binary consensus operation it uses is denoted bin_propose(). Extremely simple, this algorithm can be decomposed in four phases. The first three phases are communication phases, while the last phase exploits the result of the previous phases to reduce multivalued Byzantine consensus to BBC.

The second and the third phases are two distinct instances of the MV-broadcast abstraction. Not to confuse them, their corresponding broadcast operations are denoted MV_broadcast$_1$(), and MV_broadcast$_2$(), respectively. Similarly, their default values are denoted $\perp_{mv1}$ and $\perp_{mv2}$. It is assumed that the default values $\perp_{rd}$, $\perp_{mv1}$, $\perp_{mv2}$, and $\perp$ (the consensus default value) are all different. The four phases are as follows, where C_PROP denotes the set of values proposed by the non-faulty processes.

```
operation mv_propose(v_i) is
(1) rd_val_i ← RD_broadcast(v_i);
% ─────────────────────────────────────────────────────────
(2) set1_i ← MV_broadcast_1(rd_val_i);
    %  p_i, p_j non-faulty:  ((|set1_i| = 1) ∧ (|set1_j| = 1)) ⇒ (set1_i = set1_j)  %
(3) if (set1_i = {w}) then aux_i ← w else aux_i ← ⊥ end if;
% ─────────────────────────────────────────────────────────
(4) set2_i ← MV_broadcast_2(aux_i);
    %  p_i, p_j non-faulty:  (set2_i = {w}) ⇒ (w ∈ set2_j)  %
% ─────────────────────────────────────────────────────────
(5) if ((set2_i = {w}) ∧ (w ∉ {⊥_rd, ⊥_mv1, ⊥_mv2, ⊥}))
                    then bp_i ← 1 else bp_i ← 0 end if;
(6) bdec_i ← bin_propose(bp_i);
(7) if (bdec_i = 1) then return(w) such that w ∈ set2_i and w ∉ {⊥_rd, ⊥_mv1, ⊥_mv2, ⊥}
(8)                 else return(⊥)
(9) end if.
```

Fig. 3. An algorithm implementing multivalued ITB consensus in $\mathcal{BAMP}_{n,t}[n > 3t, \mathrm{BBC}]$

- The first phase consists of an RD-broadcast instance. Each non-faulty process p_i invokes RD_broadcast(v_i), where v_i is the value it proposes to consensus, and stores the returned value in its local variable rd_val_i (line 1). Due to properties of the RD-broadcast abstraction, we have

$$rd\_val_i \in RD\_VAL \text{ where } RD\_VAL \subseteq C\_PROP \cup \{\perp_{rd}\},$$

and (due to Lemma 2) $|RD\_VAL| \leq 6$. Moreover, the message cost of this phase is the one of the RD-broadcast, i.e., $O(n^2)$.

- The second phase (lines 2 and 3) consists of the first MV-broadcast instance, namely, a process p_i invokes MV_broadcast$_1(rd\_val_i)$ from which it obtains the non-empty set $set1_i$. Due to the properties of the MV-broadcast abstraction, we have

$$set1_i \subseteq MV\_VAL_1,$$

where $MV\_VAL_1 \subseteq RD\_VAL \cup \{\perp_{mv1}\} \subseteq C\_PROP \cup \{\perp_{rd}, \perp_{mv1}\}$.

Moreover, due to the MV-singleton property, we also have

$$((|set1_i| = 1) \wedge (|set1_j| = 1)) \Rightarrow (set1_i = set1_j).$$

Then, according to the value of $set1_i$, p_i prepares a value aux_i it will broadcast in the second MV-broadcast instance. If $set1_i = \{w\}$, $aux_i = w$, otherwise $aux_i = \perp$ (the consensus default value).

Let $AUX = \cup_{i \in C}\{aux_i\}$, where C denotes the set of non-faulty processes. While preserving the $O(n^2)$ message complexity, the aim of the lines 2 and 3 is to ensure the following property

$$AUX = \{v\} \vee AUX = \{\perp\} \vee AUX = \{v, \perp\}, \text{ where } v \in MV\_VAL_1.$$

Let us notice that, thanks to the MV-Justification property, the set AUX cannot contain a value proposed only by Byzantine processes.

- The third phase (line 4) is a second instance of the MV-broadcast abstraction. The values MV-broadcast by the non-faulty processes are values of the set AUX. So, the set $set2_i$ returned by a non-faulty process p_i is such that

$$set2_i \subseteq MV\_VAL_2 \text{ where } MV\_VAL_2 \subseteq AUX \cup \{\perp_{mv2}\},$$

  and, due to the MV-Inclusion property, the sets returned to any two non-faulty processes p_i and p_j are such that $(set2_i = \{w\}) \Rightarrow (w \in set2_j)$.
- The last phase (lines 5-9) is where the underlying BBC algorithm is exploited. If $set2_i$ contains a single value, that is not a default value, p_i proposes 1 to the underlying BBC algorithm. Otherwise, it proposes 0. Then, according to the value $bdec_i$ returned by the BBC algorithm, there are two cases. If $bdec_i = 1$, p_i return the value of $set2_i$ which is not a default value (line 7). Otherwise, $bdec_i = 0$ and p_i returns the default value $\perp$.

5.3   Proof of the Multivalued ITB Consensus Algorithm and Two Remarks

Theorem 5. *The algorithm described in Figure 3 solves the multivalued ITB consensus problem in the computing model* $\mathcal{BAMP}_{n,t}[t < n/3, \text{BBC}]$.

Theorem 6. *Let us assume an underlying BBC algorithm whose expected message complexity is* $O(n^2)$ *and expected time complexity is constant (e.g., the one presented in [22]). When considering the non-faulty processes, the expected message complexity of the multivalued ITB consensus algorithm described in Figure 3 is* $O(n^2)$*, and its expected time complexity is constant.*

Remark 1. Let us remark that, if we suppress the invocation of the RD-broadcast abstraction, and replace line 1 by the statement *"rd_val_i $\leftarrow v_i$"*, the multivalued ITB consensus remains correct. This modification saves the two communication steps involved in the RD-broadcast, but loses the $O(n^2)$ message complexity, which is now $O(kn^3)$ (this follows from Theorem 4 and the fact that $k \in [1..n]$ is the number of distinct values broadcast by correct processes).

Remark 2. The algorithm of Figure 2 uses two instances of the MV-broadcast abstraction. It is an open problem to know if it is possible to design an algorithm based on a single instance of it.

6   Conclusion

This paper presented an asynchronous message-passing algorithm which reduces multivalued consensus to binary consensus in the presence of up to $t < n/3$ Byzantine processes (n being the total number of processes). This algorithm has the following noteworthy features: its message complexity is $O(n^2)$, its time complexity is $O(1)$, and it does not rely on cryptographic techniques. As far as we know, this is the first consensus reduction owning all these properties, while being optimal with respect to the value of t. This algorithm relies on two new all-to-all communication abstractions. These abstractions consider the values that are broadcast, and not the fact that "this" value was broadcast by "this" process. This simple observation allowed us to design an efficient reduction algorithm. (An n-multiplexing of a one-to-all broadcast abstraction would entail an $O(n^3)$ message complexity.) Interestingly, this reduction algorithm uses a single instance of the Byzantine binary consensus, and tolerates message re-ordering by Byzantine processes.

When combined with the binary Byzantine consensus algorithm presented in [22], we obtain the best algorithm known so far (as far as we know) for multivalued Byzantine consensus in a message-passing asynchronous system (where "best" is with respect the value of t, the message and time complexities, and the absence of limit on the computational power of the adversary).

Acknowledgments. This work has been partially supported by the French ANR project DISPLEXITY devoted to computability and complexity in distributed computing, and the Franco-German ANR project DISCMAT devoted to connections between mathematics and distributed computing.

References

1. Aguilera, M.K., Frolund, S., Hadzilacos, V., Horn, S., Toueg, S.: Abortable and query-abortable objects and their efficient implementation. In: Proc. 26th Annual ACM Symposium on Principles of Distributed Computing (PODC 2007), pp. 23–32 (2007)
2. Attiya, H., Welch, J.: Distributed computing: fundamentals, simulations and advanced topics, 2nd edn., p. 414 pages. Wiley Interscience (2004)
3. Ben-Or, M.: Another advantage of free choice: completely asynchronous agreement protocols. In: Proc. 2nd ACM Symposium on Principles of Distributed Computing (PODC 1983), pp. 27–30. ACM Press (1983)
4. Bracha, G.: Asynchronous Byzantine agreement protocols. Information & Computation 75(2), 130–143 (1987)
5. Bracha, G., Toueg, S.: Asynchronous consensus and broadcast protocols. Journal of the ACM 32(4), 824–840 (1985)
6. Cachin, C., Kursawe, K., Petzold, F., Shoup, V.: Secure and efficient asynchronous broadcast protocols. In: Kilian, J. (ed.) CRYPTO 2001. LNCS, vol. 2139, pp. 524–541. Springer, Heidelberg (2001)
7. Correia, M., Ferreira Neves, N., Verissimo, P.: From consensus to atomic broadcast: time-free Byzantine-resistant protocols without signatures. Computer Journal 49(1), 82–96 (2006)
8. De Prisco, R., Malkhi, D., Reiter, M.: On k-set consensus problems in asynchronous systems. Transactions on Parallel and Distributed Systems 12(1), 7–21 (2001)
9. Dwork, C., Lynch, N., Stockmeyer, L.: Consensus in the presence of partial synchrony. Journal of the ACM 35(2), 288–323 (1988)
10. Fischer, M.J., Lynch, N.A., Paterson, M.S.: Impossibility of distributed consensus with one faulty process. Journal of the ACM 32(2), 374–382 (1985)
11. Friedman, R., Mostéfaoui, A., Rajsbaum, S., Raynal, M.: Distributed agreement problems and their connection with error-correcting codes. IEEE Transactions on Computers 56(7), 865–875 (2007)
12. Friedman, R., Mostéfaoui, A., Raynal, M.: $\diamond P_{mute}$-based consensus for asynchronous Byzantine systems. Parallel Processing Letters 15(1-2), 162–182 (2005)
13. Friedman, R., Mostéfaoui, A., Raynal, M.: Simple and efficient oracle-based consensus protocols for asynchronous Byzantine systems. IEEE Transactions on Dependable and Secure Computing 2(1), 46–56 (2005)

14. Hadzilacos, V., Toueg, S.: On deterministic abortable objects. In: Proc. 32th Annual ACM Symposium on Principles of Distributed Computing (PODC 2013), pp. 4–12 (2013)
15. Kihlstrom, K.P., Moser, L.E., Melliar-Smith, P.M.: Byzantine fault detectors for solving consensus. The Computer Journal 46(1), 16–35 (2003)
16. King, V., Saia, J.: Breaking the $O(n^2)$ bit barrier: scalable Byzantine agreement with an adaptive adversary. In: Proc. 30th ACM Symposium on Principles of Distributed Computing (PODC 2011), pp. 420–429. ACM Press (2011)
17. Lamport, L., Shostack, R., Pease, M.: The Byzantine generals problem. ACM Transactions on Programming Languages and Systems 4(3), 382–401 (1982)
18. Liang, G., Vaidya, N.: Error-free multi-valued consensus with Byzantine failures. In: Proc. 30th ACM Symposium on Principles of Distributed Computing (PODC 2011), pp. 11–20. ACM Press (2011)
19. Lynch, N.A.: Distributed algorithms, 872 pages. Morgan Kaufmann Pub., San Francisco (1996)
20. Martin, J.-P., Alvisi, L.: Fast Byzantine consensus. IEEE Transactions on Dependable and Secure Computing 3(3), 202–215 (2006)
21. Milosevic, Z., Hutle, M., Schiper, A.: On the reduction of atomic broadcast to consensus with Byzantine faults. In: Proc. 30th IEEE Int'l Symposium on Reliable Distributed Systems (SRDS 2011), pp. 235–244. IEEE Computer Press (2011)
22. Mostéfaoui, A., Moumen, H., Raynal, M.: Signature-free asynchronous Byzantine consensus with $t < n/3$ and $O(n^2)$ messages. In: Proc. 33rd Annual ACM Symposium on Principles of Distributed Computing (PODC 2014), pp. 2–9. ACM Press (2014)
23. Mostéfaoui, A., Rajsbaum, S., Raynal, M.: Conditions on input vectors for consensus solvability in asynchronous distributed systems. Journal of the ACM 50(6), 922–954 (2003)
24. Mostéfaoui, A., Raynal, M.: Signature-free broadcast-based intrusion tolerance: never decide a Byzantine value. In: Lu, C., Masuzawa, T., Mosbah, M. (eds.) OPODIS 2010. LNCS, vol. 6490, pp. 143–158. Springer, Heidelberg (2010)
25. Mostéfaoui, A., Raynal, M.: Asynchronous Byzantine systems: from multivalued to binary consensus with $t < n/3$, $O(n^2)$ messages, $O(1)$ time, and no signature. Tech Report 2014, 17 pages, IRISA, Université de Rennes (F) (2015), https://hal.inria.fr/hal-01102496
26. Mostéfaoui, A., Raynal, M., Tronel, F.: From binary consensus to multivalued consensus in asynchronous message-passing systems. Information Processing Letters 73, 207–213 (2000)
27. Patra, A.: Error-free multi-valued broadcast and Byzantine agreement with optimal communication complexity. In: Fernàndez Anta, A., Lipari, G., Roy, M. (eds.) OPODIS 2011. LNCS, vol. 7109, pp. 34–49. Springer, Heidelberg (2011)
28. Pease, M., Shostak, R., Lamport, L.: Reaching agreement in the presence of faults. Journal of the ACM 27, 228–234 (1980)
29. Rabin, M.: Randomized Byzantine generals. In: Proc. 24th IEEE Symposium on Foundations of Computer Science (FOCS 1983), pp. 116–124. IEEE Computer Society Press (1983)
30. Raynal, M.: Communication and agreement abstractions for fault-tolerant asynchronous distributed systems. Morgan & Claypool, 251 pages (2010) ISBN 978-1-60845-293-4
31. Raynal, M.: Fault-tolerant agreement in synchronous message-passing systems, 165 pages. Morgan & Claypool Publishers (2010) ISBN 978-1-60845-525-6
32. Raynal, M.: Concurrent programming: algorithms, principles and foundations, 515 pages. Springer (2013)
33. Toueg, S.: Randomized Byzantine agreement. In: Proc. 3rd Annual ACM Symposium on Principles of Distributed Computing (PODC 1984), pp. 163–178. ACM Press (1984)
34. Turpin, R., Coan, B.A.: Extending binary Byzantine agreement to multivalued Byzantine agreement. Information Processing Letters 18, 73–76 (1984)

A Fast Network-Decomposition Algorithm and Its Applications to Constant-Time Distributed Computation[*]

(Extended Abstract)

Leonid Barenboim[1,**], Michael Elkin[2,***], and Cyril Gavoille[3]

[1] Open University of Israel, Israel
leonidb@openu.ac.il
[2] Ben-Gurion University of the Negev, Israel
elkinm@cs.bgu.ac.il
[3] LaBRI - Universite de Bordeaux, Bordeaux, France
gavoille@labri.fr

Abstract. A partition $(C_1, C_2, ..., C_q)$ of $G = (V, E)$ into clusters of strong (respectively, weak) diameter d, such that the supergraph obtained by contracting each C_i is ℓ-colorable is called a strong (resp., weak) (d, ℓ)-network-decomposition. Network-decompositions were introduced in a seminal paper by Awerbuch, Goldberg, Luby and Plotkin in 1989. Awerbuch et al. showed that strong $(exp\{O(\sqrt{\log n \log \log n})\}, exp\{O(\sqrt{\log n \log \log n})\})$-network-decompositions can be computed in distributed deterministic time $exp\{O(\sqrt{\log n \log \log n})\}$. Even more importantly, they demonstrated that network-decompositions can be used for a great variety of applications in the message-passing model of distributed computing. Much more recently Barenboim (2012) devised a distributed randomized constant-time algorithm for computing strong network decompositions with $d = O(1)$. However, the parameter ℓ in his result is $O(n^{1/2+\epsilon})$.

In this paper we drastically improve the result of Barenboim and devise a distributed randomized constant-time algorithm for computing strong $(O(1), O(n^\epsilon))$-network-decompositions. As a corollary we derive a constant-time randomized $O(n^\epsilon)$-approximation algorithm for the distributed minimum coloring problem. This improves the best previously-known $O(n^{1/2+\epsilon})$ approximation guarantee. We also derive other improved distributed algorithms for a variety of problems.

Most notably, for the extremely well-studied distributed minimum dominating set problem currently there is no known deterministic poly-

---

[*] A full version of this paper with all proofs omitted from the current version due to lack of space is available online [10].

[**] Part of this work has been performed while the author was a postdoctoral fellow at a joint program of the Simons Institute at UC Berkeley and I-Core at Weizmann Institute.

[***] This research has been supported by the Israeli Academy of Science, grant 593/11, and by the Binational Science Foundation, grant 2008390.

C. Scheideler (Ed.): SIROCCO 2015, LNCS 9439, pp. 209–223, 2015.
DOI: 10.1007/978-3-319-25258-2_15

logarithmic -time algorithm. We devise a *deterministic* polylogarithmic-time approximation algorithm for this problem, addressing an open problem of Lenzen and Wattenhofer (2010).

1 Introduction

1.1 Network-Decompositions

In the distributed message-passing model a communication network is represented by an n-vertex graph $G = (V, E)$. The vertices of the graph host processors that communicate over the edges. Each vertex has a unique identity number (ID) consisting of $O(\log n)$ bits. We consider a synchronous setting: computation proceeds in rounds, and each message sent over an edge arrives by the beginning of the next round. The running time of an algorithm is the number of rounds from the beginning until all vertices terminate. Local computation is free.

A *strong* (respectively, *weak*) *diameter* of a cluster $C \subseteq V$ is the maximum distance $\text{dist}_{G(C)}(u, v)$ (resp., $\text{dist}_G(u, v)$) between a pair of vertices $u, v \in C$, measured in the induced subgraph $G(C)$ of C (resp., in G). A partition $(C_1, C_2, ..., C_q)$ of $G = (V, E)$ into clusters of strong (resp., weak) diameter d, such that the supergraph $\mathcal{G} = (\mathcal{V}, \mathcal{E})$, $\mathcal{V} = \{C_1, C_2, ..., C_q\}$, $\mathcal{E} = \{(C_i, C_j) \mid C_i, C_j \in \mathcal{V}, i \neq j, \exists v_i \in C_i, v_j \in C_j, (v_i, v_j) \in E\}$ obtained by contracting each C_i is ℓ-colorable is called a *strong* (resp., *weak*) (d, ℓ)-*network-decomposition*.

Network-decompositions were introduced in a seminal paper by Awerbuch et al. [3]. The authors of this paper showed that strong $(exp\{O(\sqrt{\log n \log\log n})\}, exp\{O(\sqrt{\log n \log\log n})\})$-network-decompositions can be computed in deterministic distributed $exp\{O(\sqrt{\log\log\log n})\}$ time. Even more importantly they demonstrated that many pivotal problems in the distributed message passing model can be efficiently solved if one can efficiently compute (d, ℓ)-network-decompositions with sufficiently small parameters. In particular, this is the case for Maximal Independent Set, Maximal Matching, and $(\Delta + 1)$-Vertex-Coloring.

The result of [3] was improved a few years later by Panconesi and Srinivasan [46] who devised a deterministic algorithm for computing strong $(exp\{O(\sqrt{\log n})\}, exp\{O(\sqrt{\log n})\})$-network-decompositions in $exp\{O(\sqrt{\log n})\}$ time. Awerbuch et al. [1] devised a deterministic algorithm for computing strong $(O(\log n), O(\log n))$-network-decomposition in time $exp\{O(\sqrt{\log n})\}$. Around the same time Linial and Saks [40] devised a randomized algorithm for weak $(O(\log n), O(\log n))$-network-decompositions with $O(\log^2 n)$ time. More generally, the algorithm of Linial and Saks [40] can compute weak $(\lambda, O(n^{1/\lambda} \log n))$-network-decompositions or weak $(O(n^{1/\lambda}), \lambda)$-network-decompositions in time $O(\lambda \cdot n^{1/\lambda} \log n)$.

Observe, however, that all these algorithms [3,46,40] require super-logarithmic time, for all choices of parameters. In ICALP'12 the first-named author of the current paper [5] devised a randomized algorithm for computing strong $(O(1), n^{1/2+\epsilon})$-network-decomposition in $O(1/\epsilon)$ time. Unlike the algorithms of [3,46,40], the algorithm of [5] requires *constant* time. Its drawback however is its very high parameter $\ell = n^{1/2+\epsilon}$. In the current paper we alleviate this drawback, and devise a randomized algorithm for computing strong $(exp\{O(\lambda)\}, n^{1/\lambda})$-network-decomposition in time $exp\{O(\lambda)\}$. In other words, the parameter λ of our new decompositions can be made n^ϵ, for an arbitrarily small constant $\epsilon > 0$, while the running time is still *constant* (specifically, $exp\{O(1/\epsilon)\}$).

1.2   Constant-Time Distributed Algorithms

In their seminal paper titled "What can be computed locally?" [44] Naor and Stockmeyer posed the following question: which distributed tasks can be solved in *constant* time? This question is appealing both from theoretical and practical perspectives. From the latter viewpoint it is justified by the emergence of huge networks. The number of vertices in the latter networks may be so large that even mildest dependence of the running time on n may make the algorithm prohibitively slow.

Naor and Stockmeyer themselves [44] showed that certain types of weak colorings can be computed in constant time. A major breakthrough in the study of distributed constant time algorithms was achieved though a decade after the paper of [44] by Kuhn and Wattenhofer [35]. Specifically, Kuhn and Wattenhofer [35] showed that an $O(\sqrt{k}\Delta^{1/\sqrt{k}} \log \Delta)$-approximate minimum dominating set[1] can be computed in $O(k)$ randomized time. Here $\Delta = \Delta(G)$ is the maximum degree of the input graph G, and k is a positive possibly constant parameter.

An approximation algorithm for another fundamental optimization problem, specifically, for the *minimum coloring* problem, was devised by Barenboim [5] as an application of his aforementioned algorithm for computing network-decompositions. Specifically, it is shown in [5] that an $O(n^{1/2+\epsilon})$-approximation for the minimum coloring problem can be computed in $O(1/\epsilon)$ randomized time. (In the minimum coloring problem one wishes to color the vertices of the graph properly with as few colors as possible.) Observe that since approximating the minimum coloring problem up to a factor of $n^{1-\epsilon}$ is NP-hard [28,25,50], the algorithm of [5] inevitably has to employ very heavy local computations.

In the current paper we employ our improved network-decomposition procedure to come up with a significantly improved constant-time approximation algorithm for the minimum coloring problem. Specifically, our randomized algorithm provides an $O(n^\epsilon)$-approximation for the minimum coloring problem in $exp\{O(1/\epsilon)\}$ time, for an arbitrarily small constant $\epsilon > 0$. We also devise a randomized $O(n^\epsilon)$-approximation algorithm for the *minimum t-spanner* problem with running time $exp\{O(1/\epsilon)\} + O(t)$, for any arbitrarily small constant $\epsilon > 0$. (A subgraph $G' = (V, H)$ of a graph $G = (V, E)$, $H \subseteq E$, is a *t-spanner* of G if for every $u, v \in V$, $\text{dist}_{G'}(u, v) \leq t \cdot \text{dist}_G(u, v)$. In the *minimum t-spanner* problem the objective is to compute a t-spanner of the input graph G with as few edges as possible.)

Ajtai et al. [2] demonstrated that triangle-free n-vertex graphs admit an $O(\sqrt{n}/\sqrt{\log n})$-coloring. Kim [30] showed that this existential bound is tight. We devise a randomized $O(n^{1/2+\epsilon})$-coloring algorithm for triangle-free graphs

---

[1] A subset $U \subseteq V$ in a graph $G = (V, E)$ is a *dominating set* if for every $v \in V \setminus U$ there exists $u \in U$, such that $(u, v) \in E$. In the *minimum dominating set* (henceforth, MDS) problem the goal is to find a minimum-cardinality dominating set of G.

with running time $O(1/\epsilon)$. More generally, we devise a randomized $O(n^{1/k+\epsilon})$-coloring algorithm for graphs of girth greater than $g = 2k, k \geq 2$, with running time $O(1/\epsilon^2)$. Both results apply for any arbitrarily small $\epsilon > 0$, and, in particular, they show that such graph can be colored with a reasonably small number of colors in constant time. Together with our drastically improved constant-time approximation algorithm for the minimum coloring problem, these results significantly expand the set of distributed problems solvable in constant time.

Most our algorithms for constructing network-decompositions use only short messages (i.e., messages of size $O(\log n)$ bits), and employ only polynomially-bounded local computations. Although in general graphs our algorithms for $O(n^{1/\epsilon})$-approximate minimum coloring require large messages, our $O(n^{1/2+\epsilon})$-coloring and $O(n^{1/k+\epsilon})$-coloring algorithms for triangle-free graphs and graphs of large girth employ short messages. Hence the latter coloring algorithms are suitable to serve as building blocks for various tasks. Despite that the number of colors is superconstant, in many tasks it does not affect the overall running time, so the entire task can be performed very quickly. For example, if the colors are used for frequency assignment or code assignment tasks, the running time will not be affected by the number of colors. Instead, the range of frequencies or codes will be affected. However, this is unavoidable in the worst case, in view of the lower bounds on the chromatic number of triangle free graphs and graph of large girth.

1.3   The Minimum Dominating Set Problem

The MDS problem is one of the most fundamental classical problems of distributed graph algorithms. Jia et al. [29] devised the first efficient randomized $O(\log \Delta)$-approximation algorithm for the MDS problem with running time $O(\log n \log \Delta)$. Their result was improved and generalized by Kuhn and Wattenhofer [35] who devised a randomized $O(\sqrt{k}\Delta^{1/\sqrt{k}} \log \Delta)$-approximation algorithm for the problem with time $O(k)$.

The results of [29,35] spectacularly advanced our understanding of the distributed complexity of the MDS problem. However, both these algorithms [29,35] are randomized, and no efficient deterministic distributed algorithms with a nontrivial approximation guarantee for general graphs are currently known. Lenzen and Wattenhofer [38] devised such algorithms for graphs with bounded arboricity. Below we provide a quote from their paper:
"To the best of our knowledge, the deterministic distributed complexity of MDS approximation on general graphs is more or less a blind spot, as so far neither fast (polylogarithmic time) algorithms nor stronger lower bounds are known".

In this paper we address this blind spot and devise a deterministic $O(n^{1/k})$-approximation algorithm for the MDS problem with time $O((\log n)^{k-1})$. Similarly to our approximation algorithms for the minimum coloring and the minimum t-spanner problems, this algorithm is also a consequence of our algorithms for constructing network-decompositions. However, for the MDS we use a deterministic version of these algorithms, while for the minimum coloring and minimum t-spanner problems we use a randomized version. Also, we present a variant of our MDS approximation algorithm that employs only polynomially-bounded local computations, requires $O((\log n)^{k-1})$ time, and provides an $O(n^{1/k} \log \Delta)$ approximation.

1.4   Additional Results

We also use our algorithms for computing network-decompositions for devising algorithms for computing *low-intersecting partitions*. Low-intersecting partitions were introduced by Busch et al. [16] in a paper on universal Steiner trees. A *low-intersecting (α, β, γ)-partition* $\mathcal{P}$ of a graph G is the partition of the vertex set V such that: (1) Every cluster C in $\mathcal{P}$ has strong diameter at most $\alpha \cdot \gamma$. (2) For every vertex $v \in V$, a ball $B_\gamma(v)$ of radius γ around v intersects at most β clusters of $\mathcal{P}$.

Busch et al. showed that given a hierarchy of low-intersecting partitions with certain properties (see [16] for details) one can construct a universal Steiner tree. (See [16] for the definition of universal Steiner tree.) Also, vice versa, given universal Steiner tree they showed that one can construct a low-intersecting partition. They constructed a low-intersecting partition with $\alpha = 4^k, \beta = k \cdot n^{1/k}$, and arbitrary γ.

We devise a distributed randomized algorithm that constructs low-intersecting $((O(\gamma)^k, n^{1/k}, \gamma)$-partitions in time $(O(\gamma))^k \log^{2/3} n$ in general graphs and in $(O(\gamma))^k \cdot exp\{O(\sqrt{\log \log n})\}$ time in graphs of girth $g \geq 6$. This algorithm employs only short messages and polynomially-bounded local computations.

Comparing this result with the algorithm of Busch et al. [16] we note that the partition of [16] has smaller radius. (It is $\gamma \cdot (O(1))^k$ instead of $(O(\gamma))^k$ in our case.) On the other hand, the intersection parameter β of our partitions is smaller. (It is $n^{1/k}$ instead of $k \cdot n^{1/k}$.) In particular, the intersection parameter in the construction of [16] is always $\Omega(\log n)$, while ours can be as small as one wishes. Finally, and perhaps most importantly, the algorithm of [16] is not distributed, and seems inherently sequential.

1.5   Comparison of Our and Previous Techniques

Basically, our algorithms for computing network-decompositions can be viewed as a randomized variant of the deterministic algorithm of Awerbuch et al. [3]. The algorithm of Awerbuch et al. [3] computes iteratively ruling sets for subsets of high-degree vertices in a number of supergraphs. These supergraphs are induced by certain graph partitions which are computed during the algorithm. (A subset $U \subseteq V$ of vertices is called an (α, β)-ruling set if any two distinct vertices $u, u' \in U$ are at distance at least α one from another, and every $v \in V \setminus U$ not in a ruling set has a "ruler" $u \in U$ at distance at most β from v.) As a result of the computation the algorithm of [3] constructs a partition into clusters of diameter at most α, such that the supergraph induced by this partition has arboricity at most β. The algorithm of [3] then colors this partition with $O(\beta)$ colors in time $O(\beta \log n) \cdot O(\alpha)$. (The running time of the algorithm is $O(\beta \log n)$ when running on an ordinary graph. The running time is multiplied by a factor of $O(\alpha)$, because the coloring algorithm is simulated on a supergraph whose vertices are clusters of diameter $O(\alpha)$.) The fact that the running time in the result of [3] is (roughly speaking) the product $\alpha \cdot \beta$ of the parameters of the resulting network-decomposition is the reason that Awerbuch et al [3] made an effort to balance these parameters, and set both of them to be equal to $exp\{O(\sqrt{\log n \log \log n})\}$. The algorithm of Panconesi and Srinivasan [46] is closely related to that of [3] except that it invokes a sophisticated doubly-recursive scheme for computing ruling sets via network-decompositions, and vice versa. This ingenious idea enables [46] to balance the parameters and running time better. Specifically, they are all equal to $2^{O(\sqrt{\log n})}$.

Our algorithm is different from [3,46] in two respects. First, we replace a quite slow (it requires $O(\log n)$ time) deterministic procedure for computing ruling sets by a constant-time randomized one. Note that *generally* computing $(O(1), O(1))$-ruling sets requires $\Omega(\log^* n)$ time [39], but we only need to compute them for *high-degree vertices* of certain supergraphs. This can be easily done in randomized constant time. Second, instead of coloring the resulting partition with $O(\beta)$ colors in $O(\beta \log n) \cdot O(\alpha)$ time, we color it in $O(\beta \cdot n^\epsilon)$ colors in $O(1/\epsilon) \cdot O(\alpha)$ time by a simple randomized procedure, or in $O(\beta^2 \log^{(t)} n)$ colors in $O(t) \cdot O(\alpha)$ time, for a parameter $t > 0$, by a deterministic algorithm Arb-Linial [6]. Hence the number of colors is somewhat greater than in [3,46], but the running time is constant.

The algorithm of Linial and Saks [40] is inherently different from both [3,46] and from our algorithm. It runs for $O(\log n)$ phases, each of which constructs a collection of clusters of diameter $O(\log n)$ at pairwise distance at least 2 which covers at least half of all remaining vertices. The running time of the algorithm of [40], similarly to [3] and [46], is the product of the number of phases and clusters' diameter. Hence the approach of [40] appears to be inherently incapable to give rise to a constant time algorithm.

Our deterministic variant of the network-decomposition procedure is the basis for our deterministic approximation algorithm for MDS. Our deterministic variant is closer to the algorithm of [3] than our randomized one. The main difference between our deterministic variant and the algorithm of [3] is that we use a different much faster coloring procedure for the supergraph induced by the ultimate partition.

1.6   Related Work

Network-decompositions for general graphs were studied in [1,4]. Dubhashi et al. [20] used network decompositions for constructing low-stretch dominating sets. Recently, Kutten et al. [36] extended Linial-Saks network-decompositions to hypergraphs. Many authors [26,34,49] studied network-decompositions for graphs with bounded growth. Distributed approximation algorithms is a vivid research area. See, e.g., [43] and the references therein. Distributed graph coloring is also a very active research area. See a recent monograph [9], and the references therein. Schneider et al. [48] devised a distributed coloring algorithm whose performance depends on the chromatic number of the input graph. However, the algorithm of [48] provides no non-trivial approximation guarantee. Efficient distributed algorithms for constructing sparse undirected spanners can be found in [21,18]. Baswana and Sen [12] devised an approximation algorithm for the minimum t-spanner problem that computes a solution with $O(tn^{1+2/(t+1)})$ expected edges in $O(t^2)$ rounds. For centralized approximation algorithms for the minimum t-spanner problem, see [31,23,12,13].

2   Preliminaries

For a subset $V' \subseteq V$, the graph $G(V')$ denotes the subgraph of G induced by V'. The *degree* of a vertex v in a graph $G = (V, E)$, denoted $\deg_G(v)$, is the number of edges incident on v. A vertex u such that $(u, v) \in E$ is called a *neighbor* of v in G. The *neighborhood* of v in G, denoted $\Gamma_G(v)$, is the set of neighbors of v in G. If the graph G can be understood from context, then we omit the underscript G. For a vertex $v \in V$, the set $v \cup \Gamma(V)$ is denoted by $\Gamma^+(v)$. For a set $W \subseteq V$, we denote by $\Gamma^+(W)$ the set $W \cup \bigcup_{w \in W} \Gamma(w)$. The *distance* between a pair of vertices $u, v \in V$, denoted $\mathrm{dist}_G(u, v)$, is the length of the shortest path between u and v in G. The *diameter* of G is the maximum distance between a pair of vertices in G. The *chromatic number* $\chi(G)$ of a graph G is the minimum number of colors that can be used in a proper coloring of the vertices of G.

3   Network Decomposition

3.1   Procedure Decompose

In this section we devise an algorithm for computing an $(O(1), O(n^\epsilon))$-network-decomposition in $O(1)$ rounds, for an arbitrarily small constant $\epsilon > 0$. More generally, our algorithm computes a $(3^k, O(k \cdot n^{2/k} \cdot \log^2 n))$-network-decomposition Q in $O(3^k \cdot \log^* n)$ rounds, for any positive parameter $k, 1 \le k \le \log n$, along with an $O(k \cdot n^{2/k} \cdot \log^2 n)$-coloring φ of the supergraph induced by Q. (The $\log^* n$ term can be eliminated from the running time at the expense of increasing the number of colors used by φ by a multiplicative factor of $\log^{(t)} n$, for an arbitrarily large constant t. We will later show that the multiplicative factor of k in the second parameter of the network decomposition can also be eliminated without affecting other parameters.) The algorithm is called *Procedure Decompose*. The procedure runs on some supergraph $\hat{G} = (\hat{V}, \hat{E})$ of the original graph G. Each vertex $C \in \hat{V}$ is a cluster (i.e., a subset of vertices) of the original graph $G = (V, E)$, and different clusters are disjoint. Observe that generally it may happen that $V \ne \bigcup_{C \in \hat{V}} C$. The procedure accepts as input the supergraph $\hat{G}$, the number of vertices n of G, the parameter k, and an upper bound s on the number of vertices of the supergraph $\hat{G}$. It also accepts as input two numerical parameters ϵ and t. The parameter $\epsilon > 0$ is a sufficiently small positive constant and $t > 0$ is a sufficiently large integer constant. Initially the supergraph is G itself, with each vertex v forming a singleton cluster $\{v\}$. Hence initially it holds that $n = s$. The procedure is invoked recursively. After each invocation the current supergraph $\hat{G}$ is replaced with a supergraph on fewer vertices, and s is updated accordingly. The parameter n, however, remains unchanged throughout the entire execution.) As a result of an execution of Procedure Decompose every vertex v in $\hat{G}$ is assigned a label $label(v)$. The value of $label(v)$ is equal to the color $\varphi(C_v)$ of the cluster C_v of Q which contains v.

Procedure Decompose partitions the graph $\hat{G}$ into two vertex-disjoint subgraphs with certain helpful properties. Specifically, one of the subgraphs has a sufficiently small maximum degree that allows us to compute a network decomposition in it directly and efficiently. The other subgraph can be partitioned into a sufficiently small number of clusters with bounded diameter. The latter property is used to construct a supergraph whose vertices are formed from the clusters. Since the number of clusters is sufficiently small, the number of vertices of the supergraph is small as well. Then our algorithm proceeds recursively to compute a network decomposition of the new supergraph, using fresh labels that have not been used yet. The recursion continues for k levels. Then each vertex is assigned the label of the supernode it belongs to. (Supernodes of distinct recursion levels may be nested one inside the other. In this case an inner supernode receives the label of an outer supernode. A vertex of the original graph G receives the (same) label of all supernodes it belongs to. Notice that a vertex belongs to exactly one supernode in each recursion level.) This completes the description of the algorithm. Its pseudocode is provided below. (See Algorithm 1.)

The algorithm employs two auxiliary procedures that are described in detail in the full version of this paper [10]. The procedures succeed with high probability, i.e., with probability $1 - 1/n^c$, for an arbitrarily large constant c. The first procedure is called *Procedure Dec-Small*. It accepts a graph G with at most n vertices and maximum degree at most d. Procedure Dec-Small accepts also as input two numerical parameters, ϵ and t, which are relayed to it from Procedure Decompose. Recall that $\epsilon > 0$ is a sufficiently small constant and t is a sufficiently large integer constant. The procedure computes an $O(\min\{d \cdot n^\epsilon, d^2\})$-coloring of G in $O(\log^* n)$ time. (The time is $O(1)$ if $d > n^\epsilon$. Another variant of this procedure computes an $O(d^2 \log^{(t)} n)$-coloring in $O(t)$ time, for an arbitrarily large positive integer t.) Observe that for any integer $p > 0$, a proper p-coloring of a graph G is also a $(0, p)$-network-decomposition of G. (There are p labels, and each cluster consists of a single vertex. Thus the diameter of the decomposition is 0.) Procedure Dec-Small returns a $(0, p)$-network-decomposition S on line 5. It also returns a labeling function $label_S$ for vertices of a subset A. (We will soon describe how this subset is obtained.) The labeling $label_S$ also serves as a proper coloring for the supergraph induced by S.

The second procedure which is invoked by our algorithm is called *Procedure Partition*. This randomized procedure accepts as input an s-vertex supergraph $\hat{G} = (\hat{V}, \hat{E})$ and a parameter $q < \frac{|\hat{V}|}{2c \cdot \log n}$, and partitions $\hat{V}$ into two subsets A and B, such that $\hat{G}(A)$ and $\hat{G}(B)$ have the following properties. The subgraph $\hat{G}(A)$ has maximum degree $O(q \log n)$. The subgraph $\hat{G}(B)$ consists of $O(|V|/q) = O(s/q)$ clusters of diameter at most 2 with respect to $\hat{G}$. The procedure contracts each such cluster into a supernode. Let $\mathcal{B}$ denote the resulting set of supernodes and $\mathcal{G}(\mathcal{B}) = (\mathcal{B}, \mathcal{E}(\mathcal{B}))$ the resulting supergraph. Specifically, the vertex set of $\mathcal{G}(\mathcal{B})$ is $\mathcal{B}$, and its edge set is $\mathcal{E}(\mathcal{B}) = \{(C, C') \mid C, C' \in \mathcal{B}, \exists u \in C, u' \in C'$, such that $(u, u') \in \hat{E}\}$. Procedure Partition returns the subset $A \subseteq \hat{V}$ and the set of supernodes $\mathcal{B}$.

The clusters in B are obtained by computing a dominating set D of B of size $O(|V|/q)$. Each vertex in D becomes a leader of a distinct cluster. Each vertex in $B \setminus D$ selects an arbitrary neighbor in D and joins the cluster of this neighbor. Consequently, in all clusters all vertices are at distance at most 1 from the leader of their cluster. Hence all clusters have diameter at most 2. Initially, each vertex of V joins the set D with probability $1/q$. Then the set B is formed by the vertices of D and their neighbors. Finally, the set A is formed by the remaining vertices, i.e., $A = V \setminus B$. In this stage the procedure returns the set of nodes A and the set of supernodes $\mathcal{B}$ which is obtained from B, and terminates. This completes the description of Procedure Partition.

Algorithm 1. Procedure Decompose($\hat{G}, n, k, s, \epsilon, t$)

1: /* c is an arbitrarily large positive constant */
2: **if** $s \leq 2c \cdot n^{1/k} \log n$ **then**
3:    return Dec-Small($\hat{G}, n, s, \epsilon, t$)
      /* Compute directly a $(0, O(s^2))$-network-decomposition of $\hat{G}$. */
4: **else**
5:    $(A, \mathcal{B}) :=$ Partition($\hat{G}, q := n^{1/k}$)
      /* Partition $\hat{G}$ into A and $\mathcal{B}$. The maximum degree of $\hat{G}(A)$ is $O(n^{1/k} \log n)$.*/
6:    $(S, label_S) :=$ Dec-Small($\hat{G}(A), n, n^{1/k} \log n, \epsilon, t$)
      /* Compute directly a $(0, O(n^{2/k} \cdot \log^2 n))$-network-decomposition of $\hat{G}(A)$. */
7:    $(L, label_L) :=$ Decompose($\mathcal{G}(\mathcal{B}), n, k, \frac{s}{n^{1/k}}$)
      /* A recursive invocation on the supergraph $\mathcal{G}(\mathcal{B})$ that contains at most $\frac{s}{n^{1/k}}$ supernodes. */
8:    **for** each vertex v of $\hat{G}$, **in parallel, do**
9:       **if** $v \in S$ **then**
10:         $label(v) := label_S(v)$
11:      **else if** $v \in L$ **then**
12:         $label(v) := label_L(v) + \Lambda$
            /* $\Lambda = \gamma \cdot \lceil n^{2/k} \cdot \log^2 n \rceil$, where γ is a sufficiently large constant to be determined later. */
13:      **end if**
         /* The labeling function $label$ on $S \cup L$ is defined by: for a cluster $C \in S$ (respectively, $C \in L$) it applies to it the function $label_S()$ (resp., $label_L() + \Lambda$). */
14:   **end for**
15:   return $(S \cup L, label)$
16: **end if**

The recursive invocation of Procedure Decompose on line 7 returns a network decomposition L for the supergraph $\mathcal{G}(B)$. The for-loop (lines 8-14) adds (in parallel) $\Lambda = \gamma \cdot \lceil n^{2/k} \log^2 n \rceil$ to the color of each cluster of the network decomposition L_0 of $\mathcal{G}(B)$, where γ is a sufficiently large constant to be determined later. Since the number of colors used in each recursive level is at most Λ, this loop guarantees that colors used for clusters created on different recursion levels are different. This is because the labeling returned by procedure Dec-Small on line 6 for clusters of S employs the palette $[\Lambda]$ while the labeling computed in lines 11 - 13 for clusters of L employs labels which are greater than Λ. The termination condition of the procedure is the case $s = O(n^{1/k} \log n)$, i.e., when the number s of vertices in the supergraph $\hat{G}$ is already small. At this point the maximum degree of $\hat{G}$ is small as well (at most $s - 1$), and so coloring the supergraph (by Procedure Dec-Small) results in a sufficiently good network decomposition.

Observe that our main algorithm will invoke the procedure on the original graph G. Hence in the first level of the recursion $\hat{G} = G$, and each supernode is actually a node of G. In the second recursion level it is executed on the supernodes of nodes of the original graph G. In the third level it is executed on supernodes of supernodes, etc. Consequently, starting from the second recursion level supernodes have to be simulated using original nodes of the network. To this end each cluster that forms a supernode selects a leader which is used for simulating the supernodes. Moreover, the leader is used to simulate all nested supernodes to which it belongs. Our supernodes are obtained by at most k levels of nesting. In each level of nesting a supernode is a cluster of diameter at most 2 in a graph whose nodes are lower-level supernodes. Hence a simulation of a single round on such a supergraph will require up to 3^{k+1} rounds.

Next we provide several lemmas that will be used for the analysis of the algorithm. We leave the parameters ϵ and t unspecified in all lemmas in this section, because they have no effect on the analysis.

Lemma 31. *Suppose that all invocations of auxiliary procedures of Procedure Decompose have succeeded. Then the invocation computes a $(3^{k-1} - 1, O(k \cdot n^{2/k} \cdot \log^2 n))$-network-decomposition.*

Recall that the auxiliary procedures Dec-Small and Partition succeed with probability $1 - 1/n^c$, for an arbitrarily large constant c. Each of these procedures is invoked at most $k \leq \log n$ times during the execution of Procedure Decompose. Therefore, the probability that all executions of Procedure Dec-Small and Procedure Partition succeed is at least $(1 - 1/n^c)^{2 \log n} \approx 1 - \frac{1}{n^{c/2} \log n}$. Since c is an arbitrarily large constant, all executions of the auxiliary procedures succeed, with high probability. Hence Procedure Decompose computes a $(3^k, O(k \cdot n^{2/k} \cdot \log^2 n))$-network-decomposition, with high probability.

The next lemma analyzes the running time of the algorithm.

Lemma 32. *Let $T_{part}(n, q)$ (respectively, $T_{dec}(n, d)$) denote the running time of Procedure Partition invoked with parameters n and q (resp., Procedure Dec-Small invoked with parameters n and d). We will assume that both these running times are monotone non-decreasing in both parameters. Then the running time of Procedure Decompose is $O(3^k \cdot (T_{part}(n, n^{1/k}) + T_{dec}(n, 2c \cdot n^{1/k} \log n)))$.*

Procedure Dec-Small and Procedure Partition are provided and analyzed in the full version of this paper[10]. Next we state the main results obtained by plugging these procedures into Procedure Decompose. See [10] for the proofs.

Theorem 33. *For any parameter $k, 1 \leq k \leq \log n$, Procedure Decompose computes a $(3^k, O(k \cdot n^{2/k} \cdot \log^2 n))$-network-decomposition along with the corresponding $O(k \cdot n^{2/k} \cdot \log^2 n)$-labeling function in time $O(3^k \cdot \log^* n)$, with high probability. Alternatively, one can also have the second parameter equal to $O(k \cdot n^{2/k} \log n)$ and the running time $O(3^k \cdot k)$.*

It follows that an $(O(1), n^\delta)$-network-decomposition of an arbitrary n-vertex graph along with a proper n^δ-labeling for it can be computed by a randomized algorithm, in $O(1)$ time, with high probability. Additional variants of the algorithm can be found in [10].

4   Applications

We use our network-decomposition techniques to obtain improved algorithms for a variety of problems. The full description of all these applications appear [10]. Due to lack of space we provide here just a few notable results.

4.1   An Approximation Algorithm for the Coloring Problem

The results described in the previous sections (Theroem 33) imply an approximation algorithm for the optimization variant of the coloring problem. A distributed approximation algorithm for the graph coloring problem (based on an $(O(1), O(n^{1/2+\epsilon}))$-network decomposition) was given in [5]. We describe here a generalization of that algorithm which works with any network-decomposition. The generalized algorithm starts by computing a $(3^k - 1, O(n^{1/k} \log n))$-network-decomposition Q with an $O(n^{2/k} \log n)$-labeling $label(\cdot)$ for it. Then in each cluster C the entire induced subgraph $G(C)$ is collected into the leader vertex v_C of C. The leader vertex v_C computes locally the optimum coloring φ_C for C. Finally, v_C broadcasts (a table representation of φ_C) to all vertices of C. Each vertex u that receives this broadcast computes its final color $\psi(u)$ by $\psi(u) = \langle \varphi_C(u), label(u) \rangle$. The running time of this algorithm is the sum of the time required to compute the decomposition Q (i.e., $O(3^k \cdot k^2)$) with the time required for the computation of the colorings φ_C. The latter is dominated by the diameter of Q, times a small constant. The overall running time is therefore $O(3^k \cdot k^2)$. The result is summarized below.

Theorem 41. *For any n-vertex graph $G = (V, E)$ and an integer parameter $k = 1, 2, ...$, an $O(n^{2/k} \log n)$-approximation of the optimal coloring for G can be computed in $O(3^k \cdot k^2)$ time.*

In particular, by setting the parameter k to be an arbitrarily large constant we get a distributed $O(n^\epsilon)$-approximation algorithm for the coloring problem with a *constant* running time, for an arbitrarily small constant $\epsilon > 0$. (The running time is $O(3^{\lceil 1/\epsilon \rceil} \cdot \frac{1}{\epsilon^2})$.) This greatly improves the current state-of-the-art constant-time distributed approximation algorithm for the coloring problem due to [5], which provides an approximation guarantee of $O(n^{1/2+\epsilon})$.

Note that the algorithm in Theorem 41 requires very heavy (exponential in n) local computations and large messages. The heavy computations are inevitable, because unless $NP = P$, the coloring problem cannot be approximated (in polynomial time) up to a ratio of $n^{1-\epsilon}$, for any constant $\epsilon > 0$ [28,25,50]. On the other hand, in triangle-free graphs we can obtain an algorithm with short messages and polynomially-bounded local computation. See [10].

Theorem 42. *An $O(n^{1/2+\epsilon})$-coloring of triangle-free n-vertex graph can be computed in $O(1/\epsilon)$ distributed randomized time, using short messages and polynomially-bounded local computations.*

4.2 An Approximation Algorithm for the Minimum Dominating Set Problem

In this section we employ our network-decomposition algorithm in order to derive approximation algorithms for the minimum dominating set problem. We need the following notion. For positive integer parameters α, β, σ, an (α, β)-network-decomposition Q of a graph $G = (V, E)$ is called σ-*separated* if the clusters of Q can be β-colored in such a way that every pair of clusters $C, C' \in Q$ which are colored by the same color are at distance at least σ from one another, i.e., $\text{dist}_G(C, C') \geq \sigma$. Observe that an ordinary network decomposition is 2-separated.

Suppose that we are given a 3-separated (d, ℓ)-network-decomposition Q of a graph G. For each cluster $C \in Q$, we compute in parallel a dominating set $D \subseteq \Gamma^+(C)$ of C, such that D has minimum cardinality among all dominating sets $D' \subseteq \Gamma^+(C)$ of C. The computation of D is performed by collecting the topology of the clusters and their neighborhoods by the leaders of respective clusters, performing the computation locally using exhaustive search[2], and broadcasting the results to the vertices of the clusters and their neighbors. Since the weak diameter of the clusters is at most d, this requires $O(d)$ rounds. The next lemma show that the resulting set obtained by taking the union of the dominating sets in all clusters constitutes an ℓ-approximate minimum dominating set of the input graph G.

Lemma 43. *For a 3-separated (d, ℓ)-network-decomposition Q, suppose that we have computed a minimum dominating set $D_C \subseteq \Gamma^+(C)$ of C, for each cluster $C \in Q$. Then $|\bigcup\{D_C \mid C \in Q\}| \leq \ell \cdot |MDS(G)|$.*

In the full version of this paper [10] we devise a routine that computes a strong $((O(\log n))^{k-1}, n^{1/k})$-network-decomposition in deterministic time $(O(\log n))^{k-1}$, for any $k = 1, 2, \dots$. Using this network-decomposition in conjunction with Lemma 43 we obtain the following theorem.

Theorem 44. *For an n-vertex graph G, and a positive integer parameter k an $O(n^{1/k})$-approximation for the minimum dominating set problem can be computed in deterministic time $(O(\log n))^{k-1}$.*

---

[2] One can employ polynomial-time local computations instead of exhaustive search in the expense of increasing the approximation ratio by a factor of $O(\log \Delta)$. See the discussion following Theorem 44 .

To avoid heavy local computations by leaders of clusters, we can run a centralized $O(\log \Delta)$-approximation algorithm for the MDS problem in each cluster. (More precisely, since we need a dominating set for C which can use vertices of $\Gamma^+(C)$, we in fact obtain an instance of the Set Cover problem. This instance has left and right degrees bounded by $\Delta + 1$, and thus one can compute an $O(\log \Delta)$-approximate set cover for this instance in centralized polynomial time. As a result the approximation ratio becomes $O(n^{1/k} \log \Delta)$, while the time stays $(O(\log n))^{k-1}$.

In the full version of this paper[10] we employ our network-decompositions for coloring triangle-free graphs. We also show how our network-decomposition algorithm can be employed to obtain low-intersecting partitions. The latter partitions were used in [16] to construct universal Steiner trees. Finally, we devise a distributed approximation algorithm for the minimum t-spanner problem.

Acknowledgments. The authors are grateful to David Peleg for fruitful discussions that helped obtain some of the results in this paper.

References

1. Awerbuch, B., Berger, B., Cowen, L., Peleg, D.: Fast Distributed Network Decompositions and Covers. J. of Parallel and Distr. Computing 39(2), 105–114 (1996)
2. Ajtai, M., Komlos, J., Szemeredi, E.: A note on Ramsey numbers. Journal of Combinatorial Theory, Series A 29, 354–360 (1980)
3. Awerbuch, B., Goldberg, A.V., Luby, M., Plotkin, S.: Network decomposition and locality in distributed computation. In: Proc. of the 30th Annual Symposium on Foundations of Computer Science, pp. 364–369 (1989)
4. Awerbuch, B., Peleg, D.: Sparse partitions. In: Proc. of the 31st IEEE Symp. on Foundations of Computer Science, pp. 503–513 (1990)
5. Barenboim, L.: On the locality of some NP-complete problems. In: Czumaj, A., Mehlhorn, K., Pitts, A., Wattenhofer, R. (eds.) ICALP 2012, Part II. LNCS, vol. 7392, pp. 403–415. Springer, Heidelberg (2012)
6. Barenboim, L., Elkin, M.: Sublogarithmic distributed MIS algorithm for sparse graphs using Nash-Williams decomposition. In: Proc. of the 27th ACM Symp. on Principles of Distributed Computing, pp. 25–34 (2008)
7. Barenboim, L., Elkin, M.: Distributed $(\Delta + 1)$-coloring in linear (in Δ) time. In: Proc. of the 41st ACM Symp. on Theory of Computing, pp. 111–120 (2009)
8. Barenboim, L., Elkin, M.: Deterministic distributed vertex coloring in polylogarithmic time. In: Proc. 29th ACM Symp. on Principles of Distributed Computing, pp. 410–419 (2010)
9. Barenboim, L., Elkin, M.: Distributed Graph Coloring: Fundamentals and Recent Developments. Morgan-Claypool Synthesis Lectures on Distributed Computing Theory (2013)
10. Barenboim, L., Elkin, M., Gavoille, C.: A Fast Network-Decomposition Algorithm and its Applications to Constant-Time Distributed Computation, http://arxiv.org/abs/1505.05697
11. Barenboim, L., Elkin, M., Pettie, S., Schneider, J.: The locality of distributed symmetry breaking. In: Proc. of the 53rd Annual Symposium on Foundations of Computer Science, pp. 321–330 (2012)

12. Baswana, S., Sen, S.: A simple and linear time randomized algorithm for computing sparse spanners in weighted graphs. Random Structures and Algorithms 30(4), 532–563 (2007)

13. Berman, P., Bhattacharyya, A., Makarychev, K., Raskhodnikova, S., Yaroslavtsev, G.: Improved approximation for the directed spanner problem. In: Aceto, L., Henzinger, M., Sgall, J. (eds.) ICALP 2011, Part I. LNCS, vol. 6755, pp. 1–12. Springer, Heidelberg (2011)

14. Bisht, T., Kothapalli, K., Pemmaraju, S.: Super-fast t-ruling sets (Brief Announcement). In: Proc. of the 33th ACM Symposium on Principles of Distributed Computing, pp. 379–381 (2014)

15. Bollobas, B.: Extremal Graph Theory. Dover Publications (2004)

16. Busch, C., Dutta, C., Radhakrishnan, J., Rajaraman, R., Srinivasagopalan, S.: Split and join: strong partitions and universal Steiner trees for graphs. In: Proc. of 53rd Annual IEEE Symp. on Foundations of Computer Science, pp. 81–90 (2012)

17. Cole, R., Vishkin, U.: Deterministic coin tossing with applications to optimal parallel list ranking. Information and Control 70(1), 32–53 (1986)

18. Derbel, B., Gavoille, C., Peleg, D., Viennot, L.: On the locality of distributed sparse spanner construction. In: Proc. of the 27th ACM Symp. on Principles of Distributed Computing, pp. 273–282 (2008)

19. Dinitz, M., Krauthgamer, R.: Directed spanners via flow-based linear programs. In: Proc. of the 43rd ACM Symp. on Theory of Computing, pp. 323–332 (2011)

20. Dubhashi, D., Mei, A., Panconesi, A., Radhakrishnan, J., Srinivasan, A.: Fast distributed algorithms for (weakly) connected dominating sets and linear-size skeletons. Journal of Computer and System Sciences 71(4), 467–479 (2005)

21. Elkin, M.: A near-optimal distributed fully dynamic algorithm for maintaining sparse spanners. In: Proc. of the 26th ACM Symp. on Principles of Distributed Computing, pp. 185–194 (2007)

22. Elkin, M., Peleg, D.: The client-server 2-spanner problem with applications to network design. In: Proc. of the 8th International Colloquium on Structural Information and Communication Complexity, pp. 117–132 (2001)

23. Elkin, M., Peleg, D.: Approximating k-spanner problems for $k \geq 2$. Theoretical Computer Science 337(1-3), 249–277 (2005)

24. Erdős, P., Frankl, P., Füredi, Z.: Families of finite sets in which no set is covered by the union of r others. Israel Journal of Mathematics 51, 79–89 (1985)

25. Feige, U., Kilian, J.: Zero Knowledge and the chromatic number. Journal of Computer and System Sciences 57, 187–199 (1998)

26. Gfeller, B., Vicari, E.: A randomized distributed algorithm for the maximal independent set problem in growth-bounded graphs. In: Proc. of the 26th ACM Symp. on Principles of Distributed Computing, pp. 53–60 (2007)

27. Goldberg, A., Plotkin, S., Shannon, G.: Parallel symmetry-breaking in sparse graphs. SIAM Journal on Discrete Mathematics 1(4), 434–446 (1988)

28. Hastad, J.: Clique is Hard to Approximate Within $n^{1-\epsilon}$. In: Proc. of the 37th Annual Symposium on Foundations of Computer Science, pp. 627–636 (1996)

29. Jia, L., Rajaraman, R., Suel, R.: An efficient distributed algorithm for constructing small dominating sets. In: Proc. of the 20th ACM Symp. on Principles of Distributed Computing, pp. 33–42 (2001)

30. Kim, J.H.: The Ramsey number $R(3,t)$ has order of magnitude $t^2/\log t$. Random Structures and Algorithms 7, 173–207 (1995)

31. Kortsarz, G., Peleg, D.: Generating sparse 2-spanners. Journal of Algorithms 17(2), 222–236 (1994)

32. Kothapalli, K., Pemmaraju, S.: Super-fast 3-ruling sets. In: Proc. of the 32nd IARCS International Conference on Foundations of Software Technology and Theoretical Computer Science, pp. 136–147 (2012)

33. Kuhn, F.: Weak graph colorings: distributed algorithms and applications. In: Proc. of the 21st ACM Symposium on Parallel Algorithms and Architectures, pp. 138–144 (2009)

34. Kuhn, F., Moscibroda, T., Wattenhofer, R.: On the locality of bounded growth. In: Proc. of the 24th ACM Symp. on Principles of Distributed Computing, pp. 60–68 (2005)

35. Kuhn, F., Wattenhofer, R.: Constant-time distributed dominating set approximation. Distributed Computing 17(4), 303–310 (2005)

36. Kutten, S., Nanongkai, D., Pandurangan, G., Robinson, P.: Distributed symmetry breaking in hypergraphs. In: Proc. of the 28th International Symposium on Distributed Computing, pp. 469–483 (2014)

37. Lenzen, C., Oswald, Y., Wattenhofer, R.: What can be approximated locally? case study: dominating sets in planar graphs. In: Proc 20th ACM Symp. on Parallelism in Algorithms and Architectures, pp. 46–54 (2010). See also TIK report number 331, ETH Zurich, 2010

38. Lenzen, C., Wattenhofer, R.: Minimum dominating set approximation in graphs of bounded arboricity. In: Lynch, N.A., Shvartsman, A.A. (eds.) DISC 2010. LNCS, vol. 6343, pp. 510–524. Springer, Heidelberg (2010)

39. Linial, N.: Locality in distributed graph algorithms. SIAM Journal on Computing 21(1), 193–201 (1992)

40. Linial, N., Saks, M.: Low diameter graph decomposition. Combinatorica 13, 441–454 (1993)

41. Luby, M.: A simple parallel algorithm for the maximal independent set problem. SIAM Journal on Computing 15, 1036–1053 (1986)

42. Mitzenmacher, M., Upfal, E.: Probability and Computing: Randomized Algorithms and Probabilistic Analysis. Cambridge University Press (2005)

43. Nanongkai, D.: Distributed approximation algorithms for weighted shortest paths. In: Proc. of the 46th ACM Symp. on Theory of Computing, pp. 565–573 (2014)

44. Naor, M., Stockmeyer, L.: What can be computed locally? In: Proc. 25th ACM Symp. on Theory of Computing, pp. 184–193 (1993)

45. Panconesi, A., Rizzi, R.: Some simple distributed algorithms for sparse networks. Distributed Computing 14(2), 97–100 (2001)

46. Panconesi, A., Srinivasan, A.: On the complexity of distributed network decomposition. Journal of Algorithms 20(2), 581–592 (1995)

47. Saket, R., Sviridenko, M.: New and improved bounds for the minimum set cover problem. In: Gupta, A., Jansen, K., Rolim, J., Servedio, R. (eds.) APPROX/RANDOM 2012. LNCS, vol. 7408, pp. 288–300. Springer, Heidelberg (2012)

48. Schneider, J., Elkin, M., Wattenhofer, R.: Symmetry breaking depending on the chromatic number or the neighborhood growth. Theoretical Computer Science 509, 40–50 (2013)

49. Schneider, J., Wattenhofer, R.: A log-star distributed maximal independent set algorithm for growth bounded graphs. In: Proc. of the 27th ACM Symp. on Principles of Distributed Computing, pp. 35–44 (2008)

50. Zuckerman, D.: Linear Degree Extractors and the Inapproximability of Max Clique and Chromatic Number. Theory of Computing 3(1), 103–128 (2007)

51. http://www.disco.ethz.ch/lectures/ss04/distcomp/lecture/chapter12.pdf

Path-Fault-Tolerant Approximate Shortest-Path Trees⋆

Annalisa D'Andrea[1], Mattia D'Emidio[1], Daniele Frigioni[1],
Stefano Leucci[1], and Guido Proietti[1,2]

[1] Dipartimento di Ingegneria e Scienze dell'Informazione e Matematica,
Università degli Studi dell'Aquila, Via Vetoio, I–67100 L'Aquila, Italy
[2] Istituto di Analisi dei Sistemi ed Informatica "Antonio Ruberti", Consiglio
Nazionale delle Ricerche, Via dei Taurini 19, I–00185 Roma, Italy
{annalisa.dandrea,stefano.leucci}@graduate.univaq.it,
{mattia.demidio,daniele.frigioni,guido.proietti}@univaq.it

Abstract. Let $G = (V, E)$ be an n-nodes non-negatively real-weighted undirected graph. In this paper we show how to enrich a *single-source shortest-path tree* (SPT) of G with a *sparse* set of *auxiliary* edges selected from E, in order to create a structure which tolerates effectively a *path failure* in the SPT. This consists of a simultaneous fault of a set F of at most f adjacent edges along a shortest path emanating from the source, and it is recognized as one of the most frequent disruption in an SPT. We show that, for any integer parameter $k \geq 1$, it is possible to provide a very sparse (i.e., of size $O(kn \cdot f^{1+1/k})$) auxiliary structure that carefully approximates (i.e., within a stretch factor of $(2k − 1)(2|F| + 1)$) the true shortest paths from the source during the lifetime of the failure. Moreover, we show that our construction can be further refined to get a stretch factor of 3 and a size of $O(n \log n)$ for the special case $f = 2$, and that it can be converted into a very efficient *approximate-distance sensitivity oracle*, that allows to quickly (even in optimal time, if $k = 1$) reconstruct the shortest paths (w.r.t. our structure) from the source after a path failure, thus permitting to perform promptly the needed rerouting operations. Our structure compares favorably with previous known solutions, as we discuss in the paper, and moreover it is also very effective in practice, as we assess through a large set of experiments.

1 Introduction

Broadcasting data from a source node to every other node of a network is one of the most basic communication primitives in modern networked applications. Given the widespread diffusion of such applications, in the recent past, there has been an increasing demand for more and more efficient, i.e. scalable and reliable, methods to implement this fundamental feature.

⋆ Research partially supported by the Italian Ministry of University and Research under the Research Grants: 2010N5K7EB PRIN 2010 "ARS TechnoMedia" (Algoritmica per le Reti Sociali Tecno-mediate), and 2012C4E3KT PRIN 2012 "AMANDA" (Algorithms for MAssive and Networked DAta).

C. Scheideler (Ed.): SIROCCO 2015, LNCS 9439, pp. 224–238, 2015.
DOI: 10.1007/978-3-319-25258-2_16

The natural solution is that of modeling the network as a graph (nodes as vertices and links as edges) and building a (fast and compact) structure to be used to transmit the data. In particular, the most common approach of this kind is that of computing a *shortest-path tree* (SPT), rooted at the desired source node, of such graph.

However, the SPT, as any tree-based topology, is prone to unpredictable events that might occur in practice, such as failures of nodes and/or links. Therefore, the use of SPTs might result in a high sensitivity to malfunctioning, which unavoidably causes the undesired effect of disconnecting sets of nodes from the source and thus the interruption of the broadcasting service.

Therefore, a general approach to cope with this scenario is to make the SPT *fault-tolerant* against a given number of simultaneous component failures, by adding to it a set of suitably selected edges from the underlying graph, so that the resulting structure will remain connected w.r.t. the source. In other words, the selected edges can be used to build up alternative paths from the root, each one of them in replacement of a corresponding original shortest path which was affected by the failure. However, if these paths are constrained to be *shortest*, then it can be easily seen that for a non-negatively real weighted and undirected graph of n nodes and m edges, this may require as much as $\Theta(m)$ additional edges, also in the case in which $m = \Theta(n^2)$. In other words, the set-up costs of the strengthened network may become unaffordable.

Thus, a reasonable compromise is that of building *sparse* and fault-tolerant structure which *approximates* the shortest paths from the source, i.e., that contains paths which are guaranteed to be longer than the corresponding shortest paths by at most a given *stretch* factor, for any possible edge/vertex failure that has to be handled. In this way, the obtained structure can be revised as a 2-level communication network: a first *primary* level, i.e., the SPT, which is used when all the components are operational, and an *auxiliary* level which comes into play as soon as a component undergoes a failure.

In this paper, we show that an efficient structure of this sort exists for a prominent class of failures in an SPT, namely those involving a set of adjacent edges along a shortest path emanating from the source of the SPT. Our study is motivated by several applications, such as, for instance, traffic engineering in optical networks or path-congestion management in road-networks, where failures in the above form often affect the SPT [5,11,19]. For this kind of failure, also known as a *path failure*[1], we show that it is possible not only to obtain resilient sparse structures, but also that these can be pre-computed efficiently, and that they can return quickly the auxiliary network level.

1.1 Related Work

In the recent past, many efforts have been dedicated to devising single and multiple edge/vertex fault-tolerant structures. More formally, let r denote a distinguished source vertex of a non-negatively real-weighted and undirected graph

---

[1] Notice that this is a small abuse of nomenclature, since failures we consider are restricted to the path's edges only.

$G = (V(G), E(G))$, with n nodes and m edges. We say that a spanning subgraph H of G is an *Edge/Vertex-fault-tolerant α-Approximate SPT* (in short, α-E/VASPT), with $\alpha > 1$, if it satisfies the following condition: For each edge $e \in E(G)$ (resp., vertex $v \in V(G)$), all the distances from r in the subgraph $H - e$, i.e., H deprived of edge e (resp., the subgraph $H - v$, i.e., H deprived of vertex v and all its incident edges) are α-stretched (i.e., at most α times longer) w.r.t. the corresponding distances in $G - e$ (resp., $G - v$).

An early work on the matter is [20], where the authors showed that by adding at most $n - 1$ edges to the SPT, a 3-EASPT can be obtained. This was shown to be very useful in order to compute a recovery scheme needing only one backup routing table at each node [18]. In [15], the authors showed instead how to build a 1-EASPT in $\widetilde{O}(mn)$ time[2]. Notice that, a 1-EASPT contains *exact* replacement paths from the source, but of course its size might be $\Theta(n^2)$ if G is dense. Then, in [2], Baswana and Khanna devised a 3-VASPT of size $O(n \log n)$. Later on, a significant improvement to this result was provided in [6], where the authors showed the existence of a $(1 + \varepsilon)$-E/VASPT, for any $\varepsilon > 0$, of size $O(\frac{n \log n}{\varepsilon^2})$.

Concerning *unweighted* graphs, in [2] the authors give a $(1 + \varepsilon)$-VABFS (where BFS stands for *breadth-first search tree*) of size $O(\frac{n}{\varepsilon^3} + n \log n)$ (actually, such a size can be easily reduced to $O(\frac{n}{\varepsilon^3})$). Then, Parter and Peleg in [21] present a set of lower and upper bounds to the size of a (α, β)-EABFS, namely a structure for which the length of a path is stretched by at most a factor of α, plus an additive term of β. More precisely, they construct a $(1, 4)$-EABFS of size $O(n^{4/3})$. Moreover, assuming at most $f = O(1)$ edge failures can take place, they show the existence of a $(3(f + 1), (f + 1) \log n)$-EABFS of size $O(fn)$. This was improving onto the general fault-tolerant *spanner* construction given in [9], which, for weighted graphs and for any integer parameter $k \geq 1$, is resilient to up to f edge failures with stretch factor of $2k - 1$ and size $O(f \cdot n^{1+1/k})$.

On the other hand, concerning *approximate-distance sensitivity oracles* (simply α-*oracles* in the following, where α denotes the guaranteed approximation ratio w.r.t. true distances), researchers aimed at computing, with a *low* preprocessing time, a *compact* data structure able to *quickly* answer to some distance query following an edge/vertex failure. The vast literature dates back to the work [23] of Thorup and Zwick, who showed that, for any integer $k \geq 1$, any undirected graph with non-negative edge weights can be preprocessed in $O(km \cdot n^{1/k})$ time to build a $(2k - 1)$-oracle of size $O(k \cdot n^{1+1/k})$, answering in $O(k)$ time to a post-failure distance query, recently reduced to $O(1)$ time in [8]. Due to the long-standing girth conjecture of Erdős [13], this is essentially optimal. Concerning the failure of a set F of at most f edges, in [10] the authors built, for any integer $k \geq 1$, a $(8k - 2)(f + 1)$-oracle of size $O(fk \cdot n^{1+1/k} \log(nW))$, where W is the ratio of the maximum to the minimum edge weight in G, and with a query time of $\widetilde{O}(|F| \cdot \log \log d)$, where d is the actual distance between the queried pair of nodes in $G - F$. As far as *SPT oracles* (i.e., returning distances/paths only from a source node) are concerned, in [2] it is shown how to build in $O(m \log n + n \log^2 n)$ time an SPT oracle of size $O(n \log n)$, that for any

---

[2] The $\widetilde{O}$ notation hides poly-logarithmic factors in n.

single-vertex-failure returns a 3-stretched replacement path in time proportional to the path's size. Finally, for directed graphs with integer positive edge weights bounded by M, in [14] the authors show how to build in $\widetilde{O}(Mn^\omega)$ time and $\Theta(n^2)$ space a randomized single-edge-failure SPT oracle returning *exact* distances in $O(1)$ time, where $\omega < 2.373$ denotes the matrix multiplication exponent.

1.2   Our Results

In this paper, we consider the specific, yet interesting, problem of making a SPT resilient to the failure of any sub-path of size (i.e., number of edges) at most $f \geq 1$ emanating from its source.

More in details, let F be a set of cascading edges of a given SPT, where $0 < |F| \leq f$. We say that a spanning subgraph H of G is a *Path-Fault-Tolerant α-Approximate SPT* (in short, α-PASPT), with $\alpha \geq 1$, if, for each vertex $z \in V(G)$, the following inequality holds: $d_{H-F}(z) \leq \alpha \cdot d_{G-F}(z)$, where $d_{G-F}(z)$ (resp., $d_{H-F}(z)$) denotes the distance from r to z in $G - F$ (resp., $H - F$). For any integer parameter $k \geq 1$, we can provide the following results:

- We give an algorithm for computing, in $O(n \cdot (m + f^2))$ time, a $(2k-1)(2|F|+1)$-PASPT containing $O(kn \cdot f^{1+\frac{1}{k}})$ edges;
- We give an algorithm for computing, in $O(n \cdot (m + f^2))$ time, an oracle of size $O(kn \cdot f^{1+\frac{1}{k}})$ which is able to return: (i) a $(2k-1)(2|F|+1)$-approximate distance in $G - F$ between r and a generic vertex z in $O(k)$ time; (ii) the associated path in $O(k + f + \ell)$ time, where ℓ is the number of its edges; if $k = 1$, this can be further reduced to $O(\ell)$ time.

Concerning the former result, it compares favorably with both the aforementioned general fault-tolerant spanner constructions given in [9], and the unweighted EABFS provided in [21], while concerning instead the latter result, it compares favorably with the fault-tolerant oracle given in [10]. For the sake of fairness, we remind that all these structures were thought to cope with edge failures arbitrarily spread across G, though.

Besides that, we also analyze in detail the special case when at most $f = 2$ failures of cascading edges can occur, for which we are able to achieve a significantly better stretch factor. More precisely, we design: (i) an algorithm for computing, in $O(n \cdot (m + n \log n))$ time, a 3-PASPT containing $O(n \log n)$ edges; (ii) an algorithm for computing, in $O(n \cdot (m + n \log n))$ time, an oracle of size $O(n \log n)$ which is able to return a 3-approximate distance in $G - F$ between r and a generic vertex z in constant time, and the associated path in a time proportional to the number of its edges. Due to space limitations, some of the proofs related to these latter results will be given in the full version of the paper.

Finally, we provide an experimental evaluation of the proposed structures, to assess their performance in practice w.r.t. both size and quality of the stretch.

2   Notation

In what follows, we give our notation for the considered problem. We are given a non-negatively real-weighted, undirected graph $G = (V(G), E(G))$ with $|V(G)| = n$ vertices and $|E(G)| = m$ edges. We denote by $w_G(e)$ or $w_G(u, v)$ the weight of the edge $e = (u, v) \in E(G)$. Given an edge $e = (u, v)$, we denote by $G - e$ or $G - (u, v)$ the graph obtained from G by removing the edge e. Similarly, for a set F of edges, $G - F$ denotes the graph obtained from G by removing the edges in F. Furthermore, given a vertex $v \in V(G)$, we denote by $G - v$ the graph obtained from G by removing vertex v and all its incident edges. Given a graph G, we call $\pi_G(x, y)$ a shortest path between two vertices $x, y \in V(G)$, $d_G(x, y)$ its weighted length (i.e., the distance from x to y in G), $T_G(r)$ a shortest path tree (SPT) of G rooted at a certain distinguished source vertex r. Moreover, we denote by $T_G(r, x)$ the subtree of $T_G(r)$ rooted at vertex x. Whenever the graph G and/or the source vertex r are clear from the context, we might omit them, i.e., we write $\pi(u)$ and $d(u)$ instead of $\pi_G(r, u)$ and $d_G(r, u)$, respectively. When considering an edge (x, y) of an SPT, we assume x and y to be the closest and the furthest endpoints from r, respectively. Furthermore, if P is a path from x to y and Q is a path from y to z, with $x, y, z \in V(G)$, we denote by $P \circ Q$ the path from x to z obtained by concatenating P and Q. We also denote by $w(P)$ the total weight of the edges in P.

For the sake of simplicity we consider only edge weights that are strictly positive. However, our entire analysis also extends to non-negative weights. Throughout the rest of the paper, we assume that, when multiple shortest paths exist, ties are broken in a consistent manner. In particular we fix an SPT $T = T_G(r)$ of G and, given a graph $H \subseteq G$ and $x, y \in V(H)$, whenever we compute the path $\pi_H(x, y)$ and ties arise, we prefer edges in $E(T)$.

A path between any two vertices $u, v \in V(G)$ is said to be an α–approximate shortest path if its length is at most α times the length of the shortest path between u and v in G. For the sake of simplicity, we assume that, if a set of at most f edge failures has to be handled, the original graph is $(f + 1)$–edge connected. Indeed, if this is not the case, we can guarantee the $(f + 1)$–edge connectivity by adding at most $O(nf)$ edges of weight $+\infty$ to G. Notice that this is not actually needed by any of the proposed algorithms.

3   Our PASPT Structure and the Corresponding Oracle

In what follows, we give a high-level description of our algorithm for computing a $(2|F| + 1)$-PASPT, namely H (see Algorithm 1), where $|F| \le f$. We define the level $\ell(v)$ of a vertex $v \in V(G)$ to be the hop-distance between r and v in $T = T_G(r)$, i.e., the number of edges of the unique path from r to v in T. Note that, when a failure of $|F|$ consecutive edges occurs on a shortest path, T will be broken into a forest $\mathcal{C}$ of $|F| + 1$ subtrees. We consider these subtrees as rooted according to T, i.e., each tree T_i is rooted at vertex r_i that minimizes $\ell(r_i)$.

Roughly speaking, the algorithm considers all possible path failures F^* of f vertices by fixing the deepest endpoint v of the failing path. It then reconnects

---

Algorithm 1. Algorithm for building a $(2|F|+1)$-PASPT. Notice that an optional integer parameter $k \geq 1$ is used. By default we set $k = 1$.

    Input : A graph G, $r \in V(G)$, an SPT $T = T_G(r)$, an integer f
    Output: A $(2|F|+1)$-PASPT of G rooted at r

1 $H \leftarrow T = T_G(r)$
2 **foreach** $v \in V(G)$ **do**
3   Let $\langle r = z_0, z_1, \ldots, z_{\ell(v)} \rangle$ be the path from r to v in T
      // F^* contains last $\min\{f, \ell(v)\}$ edges of the path
4   Let $F^* = \{(z_{i-1}, z_i) : i > \ell(v) - \min\{\ell(v), f\}\}$
5   Let $C^* = \{T_1^*, T_2^*, \ldots\}$ be the set of connected components of $T - F^*$

      // Build an auxiliary graph U associated with v
6     $U \leftarrow (\{r_i^* : r_i^* \text{ is the root of } T_i^*\}, \emptyset)$
7   **foreach** $T_i^*, T_j^* \in C^* : T_i^* \neq T_j^*$ **do**
8   Let $E_{i,j} = \{(u,v) \in E(G) \setminus F^* : u \in V(T_i^*), v \in V(T_j^*)\}$
9        $(x', y') \leftarrow \underset{(x,y) \in E_{i,j}}{\arg\min} \{d_T(r_i^*, x) + w_G(x, y) + d_T(y, r_j^*)\}$
         // We say that $(x', y') \in E(G)$ is associated to $(r_i^*, r_j^*) \in E(U)$
10      $E(U) \leftarrow E(U) \cup \{(r_i^*, r_j^*)\}$
11      $w_U(r_i^*, r_j^*) = d_T(r_i^*, x') + w_G(x', y') + d_T(x', r_j^*)$

      // Optional step, executed only if $k \neq 1$. Otherwise, let $U' = U$.
12     $U' \leftarrow$ Compute a $(2k-1)$-spanner of U
13     $E(H) \leftarrow E(H) \cup E(U')$
14 **return** H

---

the resulting $f + 1$ subtrees of $G - F^*$ by selecting at most $O(f^2)$ edges into a graph U, one for each couple of trees T_i^*, T_j^* of the forest $G - F$. These edges are either directly added to the structure H or they are first sparsified into a graph U' by using a suitable multiplicative $(2k-1)$-spanner, so that only $kf^{1+\frac{1}{k}}$ of them are added to H.

In particular, it is known that, given an n-vertex graph and an integer $k \geq 1$, both a $(2k-1)$-spanner and a $(2k-1)$-approximate distance oracle of size $O(kn^{1+\frac{1}{k}})$ can be built in $O(n^2)$ time. The oracle can report an approximate distance between two vertices in $O(k)$ time, and the corresponding approximate shortest path in time proportional to the number of its edges. For further details we refer the reader to [3,4,22]. Recently, it has been shown in [8] that a randomized $(2k-1)$-approximate distance oracle of *expected* size $O(kn^{1+\frac{1}{k}})$ can be built, so that answering a distance query requires only constant time. In what follows, however, we only describe results which are based on deterministic construction and provide a worst case guarantee on the size of the resulting structures.

We start by bounding the running time of Algorithm 1:

Lemma 1. *Algorithm 1 requires $O(n(m + f^2))$ time.*

Proof. Notice that the loop in line 2 considers each vertex of G at most once. We bound the time required by each iteration. For each vertex v a complete auxiliary graph U of $O(f)$ vertices is built. Moreover, the weights of all the edges of U can be computed in $O(m)$ time by scanning all the edges of $E(G) \setminus F^*$ while keeping track, for each pair of vertices $r_i^*, r_j^* \in V(U)$, of the minimum value of the formula in line 9. Finally, the optional spanner construction invoked by line 12 requires $O(f^2)$ time. This concludes the proof.   □

We now bound the size of the returned structure:

Lemma 2. *The structure H returned by Algorithm 1 contains $O(kn \cdot f^{1+\frac{1}{k}})$ edges.*

Proof. At the beginning of the algorithm, H coincides with $T = T_G(r)$, so $|E(H)| = O(n)$. Therefore, we only need to bound the number of edges added to H during the execution of the algorithm. Notice that, for each vertex $v \in V(G)$, Algorithm 1 considers at most $f + 1$ connected components of C^*. For each pair of components, at most one edge is added to U, hence $|E(U)| = O(f^2)$. Either $k = 1$ and $U' = U$ or $k > 1$ and U' is a $(2k - 1)$–spanner of U. In both cases we have $|U'| = O(k|U|^{1+\frac{1}{k}}) = O(kf^{1+\frac{1}{k}})$. As only the edges of U' gets added to H, the claim follows.   □

We now upper-bound the distortion provided by the structure H. For the sake of clarity, we first discuss the case where the step of line 12 of Algorithm 1 is omitted, i.e., we simply set $k = 1$ and $U' = U$. At the end of this section we will argue about the general case.

For each path failure F of $|F| \leq f$ edges, and for each target vertex t, we will consider a suitable path P in $G - F$, whose length is at most $(2|F| + 1)$ times the distance $d_{G-F}(t)$. Then, since P might not be entirely contained in $H - F$, we will show that its length must be an upper bound to the length a path Q in $H - F$ between r an t, and hence to $d_{H-F}(t)$.

We first discuss how P is defined: consider the forest C of the connected components of $T - F$. Let $\pi = \pi_{G-F}(r)$, let $r_0 = r$, and let t_0 be the last vertex of π belonging to T_0. W.l.o.g., we assume $t \notin V(T_0)$, as otherwise we have $d_{H-F}(t) = d_{G-F}(t)$. Moreover, we call (t_0, s_1) the edge following vertex t_0 in π.

Initially, we set $P_0 = \pi_T(s, t_0) \circ (t_0, s_1)$ and $i = 1$. We proceed iteratively: Let T_i be the subtree of C which contains s_i and let t_i be the last vertex of π such that t_i belongs to T_i, i.e., t_i is in the same subtree as s_i (notice that, it may be that $s_i = t_i$). Call r_i the root of T_i. If $t_i = t$ we set $P = P_{i-1} \circ \pi_T(s_i, r_i) \circ \pi_T(r_i, t_i)$, and we are done. Otherwise, let (t_i, s_{i+1}) be the edge following t_i in π. We set $P_i = P_{i-1} \circ \pi_T(s_i, r_i) \circ \pi_T(r_i, t_i) \circ (t_i, s_{i+1})$, we increment i by one, and we repeat the whole procedure. Figure 1 shows an example of such a path P. Let h be the final value of i, at the end of this procedure, so that $t = t_h \in V(T_h)$. Notice that, by construction, the path P does not contain any failed edge. We now argue that the length $w(P)$ of P, is always at most $(2|F| + 1)$ times the distance $d_{G-F}(t)$.

Lemma 3. $d_P(t) \leq (2|F| + 1) \cdot d_{G-F}(t)$, *for every $t \in V(G)$.*

Proof. We proceed by showing, by induction on i, that $d_P(t_i) \leq (2i + 1) \cdot d_{G-F}(t_i)$. The claim follows since $t = t_h$ and $h \leq |F|$.

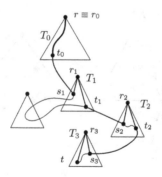

Fig. 1. Example of construction of P. The path P is shown in bold, while the path π is composed of both the light subpaths and of the bold edges with endpoint in different subtrees. In this example P traverses 4 subtrees and hence $h = 3$.

The base case is trivially true, as we have $d_P(t_0) = 1 \cdot d_{G-F}(t_0)$, since t_0 belongs to the same subtree T_0 as r. Now, suppose that the claim is true for $i - 1$. We can prove that it is true also for i by writing:

$$d_P(t_i) = d_P(t_{i-1}) + d_P(t_{i-1}, s_i) + d_P(s_i, r_i) + d_P(r_i, t_i)$$
$$\leq (2i - 1) \cdot d_{G-F}(t_{i-1}) + d_{G-F}(t_{i-1}, s_i) + d_G(s_i, r_i) + d_G(r_i, t_i)$$
$$\leq (2i - 1) \cdot d_{G-F}(t_{i-1}) + d_{G-F}(t_{i-1}, s_i) + d_G(s_i, t_i) + 2d_G(r_i, t_i)$$
$$\leq (2i - 1) \cdot d_{G-F}(t_i) + 2d_G(t_i)) \leq (2i + 1) \cdot d_{G-F}(t_i).$$

□

It remains to show that, even though P might not be entirely contained in $H - F$, its length $w(P)$ is always an upper bound to $d_{H-F}(t)$.

Let v be the deepest endpoint (w.r.t. level) among the endpoints of the edges in F. Moreover, let F^* be the set of failed edges considered by Algorithm 1 when v is examined at line 2, and let U be the the corresponding auxiliary graph. Notice that $F \subseteq F^*$ as F^* always contains $\min\{\ell(v), f\}$ edges. As a consequence, $T_0 \in \mathcal{C}$ contains, in general, several trees in $\mathcal{C}^*$. We let R be the set of the roots of all the subtrees of T_0 which are in $\mathcal{C}_0^*$. Notice that every other tree $T_j \in \mathcal{C}$ such that $T_j \neq T_0$ belongs to $\mathcal{C}^*$ (see Figure 2).

Remember that r_h is the root of the subtree $T_h \in \mathcal{C}^* = T - F^*$ which contains t. Let r'_0 be the root of the last tree $T'_0 \in \mathcal{C}^*$ which is contained in T_0 and is traversed by $\pi_{G-F}(r_h)$. It follows that $r'_0 \in V(P)$. We now construct another path Q, which will be entirely contained in $H - F$. We choose a special vertex $r_0^* \in R$, as follows:

$$r_0^* = \arg\min_{z \in R}\{d_T(z) + d_U(z, r_h)\}. \tag{1}$$

The path Q is composed of three parts, i.e. $Q = Q_1 \circ Q_2 \circ Q_3$. The first one, Q_1, coincides with $\pi_T(r_0^*)$. The second one is obtained by considering the shortest path $\pi_U(r_0^*, r_h)$ and by replacing each edge going from a vertex $r_i^* \in V(U)$ to a vertex $r_j^* \in V(U)$ with the path: $\pi_T(r_i^*, x') \circ (x', y') \circ \pi_T(x', r_j^*)$, where (x', y')

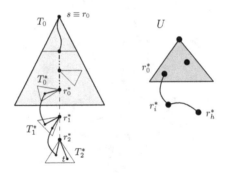

Fig. 2. An example of path Q contained in $H - F$ (left) and of the corresponding edges of U (right). The length of Q is upper-bounded by that of P.

is the edge associated to (r_i^*, r_j^*) by Algorithm 1 when v is considered. Finally, $Q_3 = \pi_T(r_h^*, t)$. In Figure 2, we show an example of how such path Q can be obtained. We now prove that:

Lemma 4. $d_{H-F}(r, t) \leq w(Q) \leq w(P)$

Proof. Notice that the path Q is in H and does not contain any failed edge, hence $d_{H-F}(r, t) \leq w(Q)$ is trivially true.

To prove $w(Q) \leq w(P)$, notice that P can also be decomposed into the three subpaths $P_1 = P[r, r_0']$, $P_2 = P[r_0', r_h]$ and $P_3 = P[r_h, t]$. We have that that $P_3 = Q_3$ and that the endpoints of P_2 coincide with the endpoints of Q_2. By the choice of r_0^*, we must have $w(Q_1) + w(Q_2) \leq w(P_1) + w(P_2)$ as the (weighted length of) path $P_1 \circ P_2$ is considered in equation (1) when $z = r_0'$. This implies that $w(Q) = w(Q_1) + w(Q_2) + w(Q_3) \leq w(P_1) + w(P_2) + w(P_3) = w(P)$. □

By combining Lemma 2 with Lemma 3 and 4, it immediately follows:

Theorem 1. *Algorithm 1 computes, in $O(n(m + f^2))$ time, a $(2|F| + 1)$-PASPT of size $O(nf^2)$, for any $|F| \leq f$.*

We now relax the assumption that $U = U'$. Indeed, if $k \neq 1$, Algorithm 1 computes, in line 12, a $(2k - 1)$-spanner U' of the graph U. In this case, we can construct a path Q' in a similar way as we did for Q, with the exception that we now use the graph U' instead of U. Once we do so, it is easy to prove that a more general version of Lemma 4 holds:

Lemma 5. $d_{H-F}(r, t) \leq (2k - 1)w(Q') \leq (2k - 1)w(P)$

Lemma 5, combined with Lemma 3, immediately implies that $d_{H-F}(r, t) \leq (2k - 1)(2|F| + 1)d_{G-F}(r, t)$. This discussion allows us to show an interesting trade-off between the size of the returned structure and the multiplicative stretch provided, as summarized by the following theorem:

Theorem 2. *Let $k \geq 1$ be an integer. Then, Algorithm 1 can compute, in $O(n(m + f^2))$ time, a $(2k - 1)(2|F| + 1)$-PASPT of size $O(nk \cdot f^{1 + \frac{1}{k}})$.*

Algorithm 2. Algorithm for building an oracle with constant query time.

1 Preprocess $T = T_G(r)$ to answer LCA queries as shown in [16]
2 For each vertex $v \in V(G)$, compute and store its level $\ell(v)$.

3 **foreach** $v \in V(G)$ **do**
4   Let $\langle r = z_0, z_1, \ldots, z_{\ell}(v)$ be the path from r to v in T
5   Build graph U associated with vertex v as in Algorithm 1
6   Compute and store the solution to the all-pairs shortest paths problem on U
7   **foreach** $\eta = 1, \ldots, \min\{f, \ell(v)\}$ **do**
8   **foreach** $r_h : h > \ell(v) - \eta$ **do**
9             $R \leftarrow \{z_i : 0 \le i \le \ell(v) - \eta\}$
10   Let r_0^* be the vertex of R minimizing Equation (1)
11   Store r_0^* with key (v, η, r_i)

Algorithm 3. Algorithm for building an oracle with $O(f)$ query time.

1 Preprocess T to answer LCA queries as shown in [16]
2 For each vertex $v \in V(G)$, compute and store its level $\ell(v)$.

3 **foreach** $v \in V(G)$ **do**
4   Build graph U associated with vertex v as in Algorithm 1
5   Build and store a distance sensitivity oracle of U with stretch $2k - 1$

3.1 Oracle Setting

In what follows, we show how Algorithm 1 can be used to compute an approximate distance oracle of size $O(nf^2)$ (see Algorithm 2). We also show that a smaller-size oracle can be obtained (see Algorithm 3) if we allow for a slightly larger query time.

Theorem 3. *Let F be a path failure of $|F| \le f$ edges and $t \in V(G)$. Algorithm 2 builds, in $O(n(m + f^2))$ time, an oracle of size $O(nf^2)$ which is able to return:*

- *a $(2|F|+1)$-approximate distance in $G - F$ between r and t in constant time;*
- *the associated path in a time proportional to the number of its edges.*

Proof. In order to answer a query we need to find: (i) the root r_0^* of the subtree of C^* which contains t_0, (ii) the root r_h of the subtree of C^* containing t. In order to find r_h, we perform a LCA query on T to find the least common ancestor u between v and t. Either $\ell(v) \ge \ell(u) > \ell(v) - |F|$, in which case $u = r_h$, or $\ell(u) \le \ell(v) - |F|$ which means that t belongs to T_0. As in the latter case we can simply return $d_T(t)$, we focus on the former one. To find r_0^* we look for the vertex associated with the triple $(v, |F|, r_h)$ stored by Algorithm 2 at line 11.

We answer a distance query with the quantity $d_T(r_0^*) + d_{U'}(r_0^*, r_h^*) + d_T(r_h, t)$, which can be computed in constant time by accessing the distances stored in shortest path tree T, plus the solution of the APSP problem on U' computed by Algorithm 2 when vertex v was considered.

To answer a path query we simply construct, and return, the path Q, by expanding the edges of the graph U' into paths which are in $G - F$, as explained before. This clearly takes a time proportional to the number of edges of Q.   □

If we allow for a query time that is proportional to $O(f + k)$, we can reduce the size of the oracle by computing a distance sensitivity oracle (DSO) of U (see Algorithm 3). In this case, we can still find vertex r_h using the LCA query, as shown in the proof of Theorem 3, while vertex r_0^* is guessed among the (up to) f roots of the trees in $G - F^*$ which are contained in T_0. The resulting oracle is summarized by the following:

Theorem 4. *Let F be a path failure of $|F| \leq f$ edges, let $t \in V(G)$ and let $k \geq 1$ be an integer. Algorithm 3 builds, in $O(n(m + f^2))$ time, an oracle of size $O(nkf^{1+\frac{1}{k}})$ which is able to return:*

- *a $(2k-1)(2|F|+1)$-approximate distance in $G-F$ between r and t in $O(f+k)$ time;*
- *the corresponding path in $O(\ell + k + f)$ time, where ℓ is the number of its edges.*

4   Our 3-PASPT Structure for Paths of 2 Edges

In what follows, we provide an algorithm which builds a 3-PASPT (see Algorithm 4) for the special case of at most $f = 2$ cascading edge failures. This structure improves, w.r.t. the quality of the stretch, over the general $(2|F| + 1)$-PASPT of Section 3.

The algorithm starts with a 3-EASPT with $O(n)$ edges [20] and proceeds as follows. As initial building block, it considers a suitable path P in the shortest-path tree $T_G(r)$, and constructs a structure H that is able to handle the failure of a pair of edges $\{e_1, e_2\}$, such that $e_1 \in P$, and guarantees 3-stretched distances from r, for each vertex in G. Then, we make use of the following result of [2]:

Lemma 6 ([2]). *There exists an $O(n)$ time algorithm to compute an ancestor-leaf path Q in $T_G(r)$ whose removal splits $T_G(r)$ into a set of disjoint subtrees $T_G(r, r_1), \ldots, T_G(r, r_j)$ such that, for each $i \leq j$:*

- *$|T_G(r, r_i)| < n/2$ and $V(Q) \cap V(T_G(r, r_i)) = \emptyset$*
- *$T_G(r, r_i)$ is connected to Q through some edge for each $i \leq j$*

This allows us to incrementally add edges to H by considering a set $\mathcal{P}$ of edge-disjoint paths. This set can be obtained by recursively using the path decomposition technique of Lemma 6 on the shortest-path tree $T_G(r)$. We show that, in this way, we are able to build a 3-PASPT of size $O(n \log n)$. Given a path $\pi = \langle s, \ldots, t \rangle$ and a tree T', we denote by $\texttt{FirstLast}(\pi, T')$ the edges of the subpaths of π going (i) from s to the first vertex of π in $V(T')$, and (ii) from the last vertex of π in $V(T')$ to t. If these vertices do not exists, i.e., $V(\pi) \cap V(T') = \emptyset$, then we define $\texttt{FirstLast}(\pi, T') = E(\pi)$. Moreover, we denote by $C(x)$ the edges connecting vertex x to its children in $T_G(r)$. We are able to prove that:

Algorithm 4. Algorithm for building a 3-PASPT for the case of $f = 2$.

Input : A graph G, $r \in V(G)$, an SPT $T = T_G(r)$
Output: A 3-PASPT of G rooted at r

1 $H \leftarrow T_G(r)$
2 $\hat{T} \leftarrow$ compute a 3-EASPT of $T_G(r)$ as shown in [20]
3 $H \leftarrow E(H) \cup E(\hat{T})$
4 Compute a path decomposition $\mathcal{P}$ of $T_G(r)$ by recursively applying Lemma 6
5 **foreach** *Path* $P \in \mathcal{P}$ **do**
6   **foreach** $x \in V(P) : x$ *is not a leaf and* $x \neq r$ **do**
7   Let z be the (unique) child of x in P
8   Let $\hat{e}$ be the edge connecting x and its parent int T

        // Protect vertex x
9          $E(H) \leftarrow E(H) \cup \texttt{FirstLast}(\pi_{G-\hat{e}}(x), T_G(r,z))$
10   **if** $\pi_{G-\hat{e}}(x)$ *contains an edge* e' *in* $C(x)$ **then**
11             $E(H) \leftarrow E(H) \cup \texttt{FirstLast}(\pi_{G-\hat{e}-e'}(x), T_G(r,z))$

        // Protect vertex z
12         $E(H) \leftarrow E(H) \cup E(\pi_{G-\hat{e}}(z))$
13   **foreach** $e' \in \{\pi_{G-\hat{e}}(z) \cap C(x)\}$ **do**
14             $E(H) \leftarrow E(H) \cup E(\pi_{G-\hat{e}-e'}(z))$

        // Protect all the other children of x
15   **foreach** *children* z_i *of* x $z_i \neq z$ **do**
16   Let (u, q) be the first edge of $\pi_{G-\hat{e}-(x,z_i)}(x, z_i)$ with $q \in V(T_G(r,z_i))$
17             $E(H) \leftarrow E(H) \cup \{(u,q)\}$

        // Protect vertices whose paths that do not contain x
18         $T' \leftarrow T_{G-x}(r,)$ with edges oriented towards the leaves
19         $E(H) \leftarrow E(H) \cup \{(x_1, x_2) \in E(T') : x_2 \notin T_G(r,z)\}$
20 **return** H

Theorem 5. *Let F be a path failure of $|F| \leq 2$ edges and $t \in V(G)$. Algorithm 4 computes, in $O(nm + n^2 \log n)$ time, a 3-PASPT of size $O(n \log n)$.*

Notice that the proof of the above theorem will be given in the full version of the paper. Notice also that it is possible to modify Algorithm 4 in order to build an oracle of size $O(n \log n)$ which is able to report, with optimal query time, both a 3-stretched shortest path in $G - F$ and its distance, when F contains two consecutive edges in T. Both the description of the modified algorithm and the proof of the following theorem will be given in the full version of the paper.

Theorem 6. *Let F be a path failure of $|F| \leq 2$ edges and $t \in V(G)$. A modification of Algorithm 4 builds, in $O(nm + n^2 \log n)$ time, an oracle of size $O(n \log n)$ which is able to return:*

- *a 3-approximate distance in $G - F$ between r and t in constant time;*
- *the associated path in a time proportional to the number its edges.*

5   Experimental Study

In this section, we present an experimental study to assess the performance, w.r.t. both the quality of the stretch and the size (in terms of edges), of the proposed structures within SageMath (v. 6.6) under GNU/Linux.

As input to our algorithms, we used weighted undirected graphs belonging to the following graph categories: (i) *Uncorrelated Random Graphs* (ERD): generated by the general *Erdős-Rényi* algorithm [7]; (ii) *Power-law Random Graphs* (BAR): generated by the *Barabási-Albert* algorithm [1]; *Quandrangular Grid Graphs* (GRI): graphs whose topology is induced by a two-dimensional grid formed by squares. For each of the above synthetic graph categories we generated three input graphs of different size and density. We assigned weights to the edges at random, with uniform probability, within $[100, 100\,000]$. We also considered two real-world graphs. In details: (i) a graph (CAI) obtained by parsing the *CAIDA IPv4 topology dataset* [17], which describes a subset of the Internet topology at router level (weights are given by round trip times); (ii) the road graph of Rome (ROM) taken from the 9th Dimacs Challenge Dataset[3] (weights are given by travel times).

Then, for each input graph, we built both the $(2k-1)(2|F|+1)$-PASPT, for which we focused on the basic case of $k = 1$, and the 3-PASPT, as follows: we randomly chose a root vertex, computed the SPT and enriched it by using the corresponding procedures (i.e. Algorithm 1 and 4, resp.). We measured the total number of edges of the resulting structures.

Regarding Algorithm 1, we set $f = 10$, as such a value has already been considered in previous works focused on the effect of path-like disruptions on shortest paths [5,12]. Then, we randomly select path failures of $|F|$ edges to perform on the input graphs, with $|F|$ uniformly chosen at random within the range $[2, f]$. We removed the edges belonging to the path failure from both the original graph and the computed structure. Regarding Algorithm 4, we simply chose at random a pair of edges and removed them from both the original graph and the computed structure.

After the removal, we computed distances, from the root vertex, in both the original graph and the fault tolerant structure, and measured the resulting average stretch. In order to be fair, we considered only those nodes that get disconnected as a consequence of the failures. Our results are summarized in Table 1, where, for each input graph, we report the number of vertices and edges, the average size (number of edges) of the two fault tolerant structures and the corresponding provided average stretch.

First of all, our results show that the quality of the stretch, provided by both the $(2|F|+1)$-PASPT and the 3-PASPT in practice, is always by far better than the estimation given by the worst-case bound (i.e. $2|F|+1$ and 3, resp.). In details, the average stretch is always very close to 1 and does not depend neither on the input size nor on the number of failures. This is probably due to the fact that those cases considered in the worst-case analysis are quite rare.

---

[3] http://www.dis.uniroma1.it/challenge9

Table 1. Average number of edges and stretch factor for both the $(2|F| + 1)$-PASPT and the 3-PASPT.

G	$|V(G)|$	$|E(G)|$	$(2|F| + 1)$-PASPT		3-PASPT	
			#edges	avg stretch	#edges	avg stretch
ERD-1	500	50 000	3 980	1.8015	957	1.0000
ERD-2	1 000	50 000	8 899	1.1360	1 924	1.0000
ERD-3	5 000	50 000	20 198	1.0903	9 501	1.0035
BAR-1	500	1 491	1 366	1.0003	949	1.0041
BAR-2	1 000	2 991	2 765	1.0034	1 871	1.0005
BAR-3	5 000	14 991	13 349	1.0040	9 459	1.0000
GRI-1	500	1 012	1 008	1.0005	868	1.0000
GRI-2	1 000	1 984	1 973	1.0000	1 749	1.0000
GRI-3	5 000	9 940	9 884	1.0000	8 826	1.0000
CAI	5 000	6 328	6 033	1.0000	6 026	1.0000
ROM	3 353	4 831	4 796	1.0000	4 780	1.0000

Similar considerations can be done w.r.t. the number of edges that are added to the SPT by Algorithms 1 and 4. In fact, also in this case, the structures behave better than what the worst-case bound suggests. For instance, the number of edges of the $(2|F| + 1)$-PASPT (the 3-PASPT, resp.) is much smaller than nf^2 ($n \log n$, resp.). In summary, our experiments suggest that the proposed fault tolerant structures might be suitable to be used in practice.

References

1. Albert, R., Barabási, A.-L.: Emergence of scaling in random networks. Science 286, 509–512 (1999)
2. Baswana, S., Khanna, N.: Approximate shortest paths avoiding a failed vertex: Near optimal data structures for undirected unweighted graphs. Algorithmica 66(1), 18–50 (2013)
3. Baswana, S., Sen, S.: Approximate distance oracles for unweighted graphs in õ(n²) time. In: Proc. of 15th ACM-SIAM Symposium on Discrete Algorithms (SODA), pp. 271–280 (2004)
4. Baswana, S., Sen, S.: Approximate distance oracles for unweighted graphs in expected $O(n^2)$ time. ACM Transactions on Algorithms 2(4), 557–577 (2006)
5. Bauer, R., Wagner, D.: Batch dynamic single-source shortest-path algorithms: An experimental study. In: Vahrenhold, J. (ed.) SEA 2009. LNCS, vol. 5526, pp. 51–62. Springer, Heidelberg (2009)
6. Bilò, D., Gualà, L., Leucci, S., Proietti, G.: Fault-tolerant approximate shortest-path trees. In: Schulz, A.S., Wagner, D. (eds.) ESA 2014. LNCS, vol. 8737, pp. 137–148. Springer, Heidelberg (2014)
7. Bollobás, B.: Random Graphs. Cambridge University Press (2001)
8. Chechik, S.: Approximate distance oracles with constant query time. In: Proc. of 46th ACM Symposium on Theory of Computing (STOC), pp. 654–663 (2014)

9. Chechik, S., Langberg, M., Peleg, D., Roditty, L.: Fault-tolerant spanners for general graphs. In: Proc. of 41st ACM Symposium on Theory of Computing (STOC), pp. 435–444. ACM (2009)

10. Chechik, S., Langberg, M., Peleg, D., Roditty, L.: f-sensitivity distance oracles and routing schemes. In: de Berg, M., Meyer, U. (eds.) ESA 2010, Part I. LNCS, vol. 6346, pp. 84–96. Springer, Heidelberg (2010)

11. D'Andrea, A., D'Emidio, M., Frigioni, D., Leucci, S., Proietti, G.: Dynamically maintaining shortest path trees under batches of updates. In: Moscibroda, T., Rescigno, A.A. (eds.) SIROCCO 2013. LNCS, vol. 8179, pp. 286–297. Springer, Heidelberg (2013)

12. D'Andrea, A., D'Emidio, M., Frigioni, D., Leucci, S., Proietti, G.: Experimental evaluation of dynamic shortest path tree algorithms on homogeneous batches. In: Gudmundsson, J., Katajainen, J. (eds.) SEA 2014. LNCS, vol. 8504, pp. 283–294. Springer, Heidelberg (2014)

13. Erdős, P.: Extremal problems in graph theory. In: Theory of Graphs and its Applications, pp. 29–36 (1964)

14. Grandoni, F., Williams, V.V.: Improved distance sensitivity oracles via fast single-source replacement paths. In: Proc. of 53rd IEEE Symposium on Foundations of Computer Science (FOCS), pp. 748–757. IEEE (2012)

15. Gualà, L., Proietti, G.: Exact and approximate truthful mechanisms for the shortest paths tree problem. Algorithmica 49(3), 171–191 (2007)

16. Harel, D., Tarjan, R.E.: Fast algorithms for finding nearest common ancestors. SIAM J. Comput. 13(2), 338–355 (1984)

17. Hyun, Y., Huffaker, B., Andersen, D., Aben, E., Shannon, C., Luckie, M., Claffy, K.C.: The CAIDA IPv4 routed/24 topology dataset. http://www.caida.org/data/active/ipv4_routed_24_topology_dataset.xml

18. Ito, H., Iwama, K., Okabe, Y., Yoshihiro, T.: Polynomial-time computable backup tables for shortest-path routing. In: Proc. of 10th Internaltional Colloquium on Structural Information Complexity (SIROCCO). Proceedings in Informatics, vol. 17, pp. 163–177. Carleton Scientific (2003)

19. Mereu, A., Cherubini, D., Fanni, A., Frangioni, A.: Primary and backup paths optimal design for traffic engineering in hybrid igp/mpls networks. In: Proc. of 7th International Workshop on Design of Reliable Communication Networks (DRCN), pp. 273–280. IEEE (2009)

20. Nardelli, E., Proietti, G., Widmayer, P.: Swapping a failing edge of a single source shortest paths tree is good and fast. Algorithmica 35(1), 56–74 (2003)

21. Parter, M., Peleg, D.: Fault tolerant approximate BFS structures. In: Proc. of 25th ACM-SIAM Symposium on Discrete Algorithms (SODA), pp. 1073–1092. SIAM (2014)

22. Roditty, L., Thorup, M., Zwick, U.: Deterministic constructions of approximate distance oracles and spanners. In: Caires, L., Italiano, G.F., Monteiro, L., Palamidessi, C., Yung, M. (eds.) ICALP 2005. LNCS, vol. 3580, pp. 261–272. Springer, Heidelberg (2005)

23. Thorup, M., Zwick, U.: Approximate distance oracles. Journal of ACM 52(1), 1–24 (2005)

A Faster Computation of All the Best Swap Edges of a Tree Spanner*

Davide Bilò[1], Feliciano Colella[2], Luciano Gualà[3],
Stefano Leucci[4], and Guido Proietti[4,5]

[1] Dipartimento di Scienze Umanistiche e Sociali, Università di Sassari, Italy
[2] Gran Sasso Science Institute, L'Aquila, Italy
[3] Dipartimento di Ingegneria dell'Impresa, Università di Roma "Tor Vergata", Italy
[4] Dipartimento di Ingegneria e Scienze dell'Informazione e Matematica,
Università degli Studi dell'Aquila, Italy
[5] Istituto di Analisi dei Sistemi ed Informatica, CNR, Roma, Italy
`davide.bilo@uniss.it, feliciano.colella@gssi.infn.it,`
`guala@mat.uniroma2.it, {stefano.leucci,guido.proietti}@univaq.it`

Abstract. Given a 2-edge connected, positively real-weighted graph G with n vertices and m edges, a *tree σ-spanner* of G is a spanning tree T in which for every pair of vertices, the ratio of their distance in T over that in G is bounded by σ, the so-called *stretch factor* of T. Tree spanners with provably good stretch factors find applications in communication networks, distributed systems, and network design, but unfortunately –as any tree-based infrastructure– they are highly sensitive to even a single link failure, since this results in a network disconnection. Thus, when such an event occurs, the overall effort that has to be afforded to rebuild an effective tree spanner (i.e., computational costs, set-up of new links, updating of the routing tables, etc.) can be prohibitive. However, if the edge failure is only *transient*, these costs can simply be avoided, by promptly reestablishing the connectivity through a careful selection of a temporary *swap edge*, i.e., an edge in G reconnecting the two subtrees of T induced by the edge failure. According to the tree spanner's nature, a *best swap edge* for a failing edge e is then a swap edge generating a reconnected tree of minimum stretch factor w.r.t. distances in the graph G deprived of edge e. For this problem we provide two efficient linear-space solutions for both the weighted and the unweighted case, running in $O(m^2 \log \alpha(m, n))$ and $O(mn \log n)$ time, respectively. As discussed in the paper, our algorithms also improve on the time complexity of previous results provided for other related settings of the problem.

1 Introduction

Let V be a set of n *sites* that must be reciprocally interconnected, let E be a set of m potential *links* between the sites, and let $w(e)$ be some positive real *cost*

* This work was partially supported by the Research Grant PRIN 2010 "ARS TechnoMedia", funded by the Italian Ministry of Education, University, and Research.

associated with link e. Let $G = (V, E)$ be the corresponding weighted, undirected graph, and assume that G is 2-edge-connected (i.e., to disconnect G we have to remove at least 2 edges). Since we aim to establish an all-to-all communication network in G, we have to design a *connected* spanning subgraph $H = (V, E' \subseteq E)$ of G, which will serve as the actual communication infrastructure. On the one hand, H should be as *sparse* as possible, so that set-up and operational costs will be low, but on the other hand it should be *efficient*, in the sense that it should preserve the structural properties of the underlying graph. For example, if edge costs do now represent *lengths*, then we would like to have a sparse network which will preserve *distances* in G as much as possible. From a theoretical point of view, if we push on the extreme side the sparseness requirement, then H should be a *spanning tree* of G, say T, minimizing the maximum *stretch factor* w.r.t. all the node-to-node distances. Unfortunately, computing such a structure, also known as a *tree σ-spanner*, where σ is exactly the optimal stretch factor, is well-known to be APX-hard. More precisely, if G does not admit a tree 1-spanner, then the problem is not approximable within any constant factor better than 2, unless P=NP [12], while to the best of our knowledge no non-trivial upper bounds are known in terms of approximability, except for the $O(n)$-approximation factor returned by a *Minimum Spanning Tree* (MST) of G. The only positive result is instead the polynomial-time algorithm for computing a tree 1-spanner, that exists only if the input graph admits a unique MST (which coincides with the tree 1-spanner itself) [6]. On the other hand, if G is *unweighted*, then, unless P=NP, the problem is not approximable within an additive term of $o(n)$ [9], and the corresponding decision problem of establishing whether a tree σ-spanner exists is NP-complete for every fixed $\sigma \geq 4$. On the positive side the optimization problem is $O(\log n)$-approximable [9], while establishing whether a tree σ-spanner exists is polynomial-time solvable for $\sigma = 2$ (see again [6]) and it is an open problem for $\sigma = 3$. Furthermore, it is known that constant-stretch tree spanners can be found for several special classes of graphs, like strongly chordal, interval, and permutation graphs (see [5] and the references therein). Finally, for the sake of completeness, we mention that for the related concept of *average* tree σ-spanners, where the focus is on the average stretch w.r.t. all node-to-node distances, it was shown that every graph admits an average tree $O(1)$-spanner [1].

Although near-optimal tree spanners are hard to be found, the use of approximate tree spanners is frequent in the practice. Thus, for them it comes into play a third vital requirement for a communication network, namely its ability to maintain a satisfying level of efficiency also when any of its components (edge or node) fails. However, a tree is not even resilient to a single edge failure! To circumvent this problem, one could associate with each tree edge a *swap edge* that enters into function as soon a failure occurs. This solution is particulary attractive when failures are transient, since switching to the swap tree is rapid and smooth in terms of rerouting processes. Not surprisingly then, this solution has been pursued for several spanning tree structures, and, from an algorithmic point of view, the main question was always the same: once that a suitable criteria for the swapping is fixed, how to compute quickly *all the best possible swap edges* (ABSE problem, in short), one for each tree edge? In this paper, we aim to provide an answer to this question for the tree spanner structure.

1.1   Related Work

The ABSE problem for a tree spanner was introduced in [7]. In that paper, the authors make however the following assumption as far as the swap criteria is concerned: a best swap edge is an edge minimizing the stretch of the swap tree w.r.t. distances in the *original* graph G, and not w.r.t. distances in the fault-free subgraph $G - e = (V, E \setminus \{e\})$. This contrasts with the intuition that a swap tree should be measured in the surviving graph, and not in the original graph, which is in fact the standard assumption in the swap literature. However, as argued by the authors, in this way the swap tree will be built so that the true distances in G are (approximately) preserved, which can be also of interest in terms of quality-of-service. As a matter of fact, in [7] the authors devise two efficient linear-space solutions for both the weighted and the unweighted case, running in $O(m^2 \log n)$ and $O(n^3)$ time, respectively, and using $O(m)$ and $O(n^2)$ space, respectively.

1.2   Our Results

In this paper, we adopt instead the classic approach of measuring the quality of a swap tree in $G - e$, and we present two efficient linear-space solutions for the ABSE tree spanner problem for both the weighted and the unweighted case, and running in $O(m^2 \log \alpha(m, n))$ and $O(mn \log n)$ time, respectively, where α is the inverse of the Ackermann function. For the weighted case, our solution requires an elaboration of the approach proposed in [7]. More precisely, this latter solution would suffer of the fact that once that the stretch factors have to be evaluated in $G - e$, then correspondingly an all-to-all distance problem should be solved in $G - e$, for every e in T. This would lead to an additional $O(mn^2 \log \alpha(m, n))$ time factor, which could be harmful for sparse graphs. We then first show that this step can actually be avoided, and then we devise a faster procedure to lower the $\log n$ factor in [7] to $\log \alpha(m, n)$. Concerning the unweighted case, first of all notice that a closer inspection of the result provided in [7] shows that the same $O(n^3)$ time bound can be obtained also when the stretch factors have to be evaluated in $G - e$. Thus, to improve such a result, we adopt a different approach than that provided in [7], which barely uses the dynamic programming technique given there. In fact, our solution mainly relies on a reduction of our problem to the dynamic maintenance of suitable properties of an auxiliary graph, which can be addressed efficiently by means of a sophisticated data structure, namely the *top tree* [2]. Thus, a comparison with the results one would get by adapting the algorithms given in [7], shows that we are always faster in the weighted case, while for the unweighted case we are better for $m = o\left(\frac{n^2}{\log n}\right)$, i.e., for a large range of graph densities. Conversely, since our algorithms can be used to solve the problems studied in [7] without any additional overhead, we also improve their results under the same conditions.

1.3   Other Related Work

The problem of swapping in spanning trees has received a significant attention from the algorithmic community. There is indeed a line of papers which address ABSE problems starting from different types of spanning trees. Just to mention a few, we recall here the MST, the *minimum diameter spanning tree* (MDST), the *minimum routing-cost spanning tree* (MRCST), and the *single-source shortest-path tree* (SPT). For the MST, a best swap is of course a swap edge minimizing the *cost* of the swap tree, i.e., a swap edge of minimum cost. This problem is also known as the MST *sensitivity analysis* problem, and can be solved in $O(m \log \alpha(m, n))$ time [14]. Concerning the MDST, a best swap is instead an edge minimizing the *diameter* of the swap tree [11,13], and the best solution runs in $O(m \log \alpha(m, n))$ time [4]. Regarding the MRCST, a best swap is clearly an edge minimizing the *all-to-all routing cost* of the swap tree [15], and the fastest solutions for solving this problem has a running time of $O\left(m 2^{O(\alpha(n,n))} \log^2 n\right)$ [3]. Finally, concerning the SPT, the most prominent swap criteria are those aiming to minimize either the maximum or the average distance from the root, and the corresponding ABSE problems can be addressed in $O(m \log \alpha(m, n))$ time [4] and $O(m \alpha(n, n) \log^2 n)$ time [8], respectively.

2   Preliminaries

Let $G = (V, E)$ be an edge-weighted undirected graph with cost function $w : E \to \mathbb{R}^+$. As usual, we denote by n and m the number of nodes and edges of G, respectively. Let T be a spanning tree of a graph G. For any two given nodes x, y in G, we denote by $d_T(x, y)$ and $d_G(x, y)$ the *distances* (i.e., the length of a shortest path) between x and y in T and G, respectively, and we call the *stretch factor* of the pair (x, y) the ratio (≥ 1) between these two distances. Accordingly, the stretch factor of T w.r.t. G is defined as $\sigma(T) = \max_{x,y \in V} \{d_T(x, y)/d_G(x, y)\}$.

For an edge e of T, let $C(e)$ be the set of all edges of $G - e = (V, E \setminus \{e\})$ that crosses the cut induced by the removal of e from T. In the following, we will assume that G is 2-edge-connected, and so for any $e \in E$, $C(e)$ is not empty. Moreover, for any $f \in C(e)$, let $T_{e/f}$ denote the *swap tree* obtained by replacing e with f in T. Then, we say that f is a *best swap edge for e w.r.t. the stretch factor* if $T_{e/f}$ has minimum stretch factor in $G - e$ as compared to any other possible swap tree associated with e. In the rest of the paper we will assume that if $e = (u, v)$ and $f = (u', v')$ then the two vertices u and u' (and therefore also v and v') belong to the same connected component of $T - e$.

In this paper, we are interested in the problem of computing, for every edge e of T, a best swap edge for e that minimizes the stretch factor of the swap tree. We separately consider the case in which G is weighted and G is unweighted.

3   The Weighted Case

Let $e = (u, v)$ be a tree edge and let $f = (u', v') \in C(e)$ be a swap edge for e. Observe that f is a swap edge only for the tree edges along the path from

u' to v' in T. An important property we will use is the following (this is just a rephrasing of Property 1 given in [7]):

Property 1. For any weighted graph $G = (V, E)$ and any pair $x, y \in V$, let $\langle x = v_0, v_1, \dots, v_k = y \rangle$ be the sequence of nodes on a shortest path in G between x and y. Then, for any spanning tree T of G, we have that there exists (at least) an edge of such shortest path, say (v_i, v_{i+1}), such that $d_T(v_i, v_{i+1})/d_G(v_i, v_{i+1}) \geq d_T(x, y)/d_G(x, y)$.

Thus, to evaluate the stretch factor of a swap tree $T_{e/f}$, we can limit our attention to the stretch factor of the pairs (x, y) such that $(x, y) \in E' = E \setminus \{e\}$. In fact,

$$\sigma(T_{e/f}) = \max_{x, y \in V} \left\{ \frac{d_{T_{e/f}}(x, y)}{d_{G-e}(x, y)} \right\} = \max_{(x, y) \in E'} \left\{ \frac{d_{T_{e/f}}(x, y)}{d_{G-e}(x, y)} \right\} \tag{1}$$

$$= \max \left\{ \max_{(x, y) \in E' \setminus C(e)} \frac{d_T(x, y)}{d_{G-e}(x, y)}, \max_{(x, y) \in C(e)} \frac{d_T(x, u') + w(f) + d_T(y, v')}{d_{G-e}(x, y)} \right\}. \tag{2}$$

Notice that the first term is *independent* of f, i.e., it is a lower bound to the stretch factor of any swap tree. Thus, to compare swap edges, we can focus on the evaluation of the second term. In the rest of the paper, a *critical edge* of f will be an edge maximizing the second term of (2), i.e.:

$$\max_{(x, y) \in C(e)} \left\{ \frac{d_T(x, u') + w(f) + d_T(y, v')}{d_{G-e}(x, y)} \right\}. \tag{3}$$

From the above discussions it follows that a best swap edge for e is a swap edge f^* minimizing (3). In the following, for the sake of avoiding technicalities, we assume that critical edges are unique. All the results can be easily extended once this assumption is relaxed. Indeed, if multiple critical edges exist it suffices to select any one of them. We start by proving the following:

Lemma 1. *Let $h = (a, b)$ be the critical edge of f,[1] and assume that, in the swap tree, the stretch factor of (a, b) is larger than the stretch factor of any pair (x, y) such that $(x, y) \in E \setminus C(e)$. Then, h must be a shortest path in $G - e$ between a and b.*

Proof. For the sake of contradiction, assume that h is not a shortest path in $G - e$ between a and b. Let now $\pi_{G-e}(a, b) = \langle a = v_0, v_1, \dots, v_k = b \rangle$ denote a shortest path in $G - e$ between a and b. Notice that such a path contains at least another edge in $C(e)$. Then, from Property 1 we know that in $T_{e/f}$ the stretch factor of (a, b) is less than or equal to the stretch factor of (at least) the endpoints pair of an edge along such a path, say h'. Since h is supposed to be the unique critical edge for f, it follows that h' cannot belong to $C(e)$, and then there must be another edge along $\pi_{G-e}(a, b)$ not belonging to $C(e)$ and whose endpoints pair has a stretch factor greater or equal to that of (a, b), against the assumptions. $\qquad\square$

---

[1] Notice that h may coincide with f.

Under the assumptions of the previous lemma, and for time efficiency reasons that will be clearer later, it follows that the computation of a critical edge of f given in (3) can be safely replaced by the following:

$$\max_{g=(x,y)\in C(e)} \left\{ \frac{d_T(x,u') + w(f) + d_T(y,v')}{w(g)} \right\}. \tag{4}$$

Indeed, since the maximum of (3) is associated with an edge which corresponds to a shortest path between its two endpoints, it follows that increasing the denominator of the fraction associated with the other edges, as it is done in (4), will not affect the computation of the maximum.

The question is now: if the assumption of Lemma 1 does not hold, namely the stretch factor of $T_{e/f}$ is equal to the first term in (2)—and so by definition f is a *best* swap edge—then is it the case that carrying out the evaluation of f by using (4) might eventually affect the correctness of the selection of a best swap edge for e? The answer is no, as proven in the following:

Lemma 2. *Assume that the stretch factor of $T_{e/f}$ is induced by the endpoints pair of an edge in $E' \setminus C(e)$. Then, the measure of the critical edge of f returned by (4) is at most equal to $\sigma(T_{e/f})$.*

Proof. From the assumptions, we have that

$$\sigma(T_{e/f}) \geq \max_{g=(x,y)\in C(e)} \left\{ \frac{d_T(x,u') + w(f) + d_T(y,v')}{d_{G-e}(x,y)} \right\}$$

$$\geq \max_{g=(x,y)\in C(e)} \left\{ \frac{d_T(x,u') + w(f) + d_T(y,v')}{w(g)} \right\}$$

$\square$

In other words, by applying the computation as specified in (4), we either find a critical edge whose measure is larger than the first term of (2), and in this case we know from Lemma 1 that this is exactly the stretch factor of the corresponding swap tree associated with f, or we associate with f an edge which in general is not its critical one according to definition (3), but now the measure of f is not larger than the first term of (2). This means, in both cases when f will be compared against other swap edges for e, the selection process of the best swap edge is guaranteed to be correct.

3.1   A Corresponding Algorithm

From the previous analysis, it follows that an algorithm that, for each edge $e \in E(T)$, computes a best swap edge by finding a swap edge which remains associated with a critical edge as defined by (4), is correct. Then, here comes the reason why we are actually using the modified formula: it does not contain any term depending on e, and so we can apply a classic *sensitivity analysis* approach, which will lead to the following result:

Theorem 1. *The ABSE tree spanner problem on weighted graphs can be solved in time $O(m^2 \log \alpha(m,n))$ and linear space.*

Proof. For each non-tree edge f, we consider the set of tree edges $T(f) = \{e_1, e_2, \ldots, e_k\}$ it covers. Then, we take into consideration all the non-tree edges $E(f) = \{f_1, f_2, \ldots, f_h\}$ covering at least one of the edges in $T(f)$. Hence, after a suitable preprocessing, we assign to each of the edges in $E(f)$ the value specified in (4), and finally we perform in $O(h \log \alpha(h,k)) = O(m \log \alpha(m,n))$ time and $O(m)$ space [14] a corresponding sensitivity analysis (we extract the *maximum* associated with each edge $e_i \in T(f)$). This value is exactly the *measure* of f when it swaps with e_i. This is repeated for all the $\Theta(m)$ non-tree edges, and eventually for each tree-edge e we select the non-tree edge having the *minimum* measure, say f^*. Notice that, as explained before, we are guaranteed this is a best swap edge for e, but its measure is only a lower bound to the actual value of $\sigma(T_{e/f^*})$. Thus, overall, we spend $O(m^2 \log \alpha(m,n))$ time and $O(m)$ space. $\qquad\square$

4   The Unweighted Case

In this section, we provide an algorithm for unweighted graphs having a running time of $O(mn \log n)$ and using $O(m)$ space. A high-level description of the algorithm is the following: For each vertex $z \in V$, it first computes a *candidate* best swap edge for each edge e of T among the non-tree edges incident to z, if any. As we will see, this step can be performed in $O(m \log n)$ time and $O(m)$ space, and it will be repeated n times (once for each vertex $z \in V$). Then, a best swap edge of e is computed by selecting, among the candidate best swap edges of e, a non-tree edge whose corresponding swap tree is of minimum stretch. To optimize space consumption, the algorithm does not explicitly store the set of all the candidate best swap edges but, once a (new) candidate best swap edge of a tree edge e is computed, the algorithm only updates the best swap edge of e found so far in constant time.

Let us now give a detailed description of the algorithm. We fix a vertex $z \in V$ and a tree edge $e = (u, v)$, and we show that the problem of computing the candidate best swap edge of e among the non-tree edges incident to z reduces to a problem instance of the *subset minimum eccentricity problem on trees*. In the *subset minimum eccentricity problem on trees*, we are given a tree T', with a positive real cost $w'(e')$ associated with each edge e' of the tree, and a set $U \subseteq V(T')$, and we are asked to find a vertex of U of minimum *eccentricity*.[2]

The reduction works as follows. Let T_u and T_v be the two subtrees obtained by removing e from T and containing u and v, respectively. Let $\widetilde{V}$ be the subset of vertices of T_v which are incident to some non-tree edge in $C(e)$, let $\widehat{T}$ be the subtree of T_v rooted at v and induced by all the paths from v towards all the vertices of $\widetilde{V}$. If $\widehat{T}$ is not a path, then let r_v be the closest child of v having two or more children in $\widehat{T}$ (possibly, $r_v = v$), otherwise let r_v be equal to (unique)

---

[2] The *eccentricity* of a vertex a of T' is equal to $\max_{b \in V(T')} d_{T'}(a, b)$.

leaf of $\widehat{T}$. Finally, let $\widetilde{T}$ be the subtree of $\widehat{T}$ rooted at r_v, let $E(y)$ be the set of non-tree edges in $C(e)$ incident to y, and let

$$\omega_y = \max_{(x,y)\in E(y)} \{d_T(z,x) + 1\}$$

be the maximum distance from z to y w.r.t. all the trees obtained by swapping e with a non-tree edge incident to y. The instance of the subset minimum eccentricity problem on trees is defined as follows. The tree T' is an unrooted copy of $\widetilde{T}$ augmented by the addition of a new vertex l_y and the edge (y, l_y) for each vertex $y \in \widetilde{V}$; $U = \{y \mid (z,y) \in E(z)\}$ is the set of endvertices in T_v of non-tree edges incident to z (observe that vertex z is not taken into account). Finally, the cost function of a tree edge e' of T' is

$$w'(e') := \begin{cases} \omega_y & \text{if } e' = (y, l_y); \\ 1 & \text{if } e' \in E(\widetilde{T}). \end{cases}$$

Observe that the set of the leaves of T' is $\{l_y \mid y \in \widetilde{V}\}$. The next lemma shows the link between the problem of finding a candidate best swap edge of e and the problem of finding the vertex of U having minimum eccentricity in T'.

Lemma 3. *Let $f = (u', v') \in E(z)$, with $u' = z$. The value of formula (4) computed w.r.t. f is equal to the eccentricity of v' in T'.*

Proof. Indeed, since $u' = z$, (4) can be rewritten as

$$\max_{g=(x,y)\in C(e)} \left\{ \frac{d_T(x,z) + w(f) + d_T(y,v')}{w(g)} \right\}$$

$$= \max_{(x,y)\in C(e)} \{d_T(x,z) + 1 + d_T(y,v')\}$$

$$= \max_{y\in\widetilde{V}} \left\{ \max_{(x,y)\in E(y)} \{d_T(x,z) + 1 + d_T(y,v')\} \right\}$$

$$= \max_{y\in\widetilde{V}} \left\{ d_T(v',y) + \max_{(x,y)\in E(y)} \{d_T(x,z) + 1\} \right\}$$

$$= \max_{y\in\widetilde{V}} \{d_{T'}(v',y) + \omega_y\} = \max_{y\in\widetilde{V}} \{d_{T'}(v',l_y)\} = \max_{b\in V(T')} \{d_{T'}(v',b)\},$$

where the first equality holds because G is unweighted, the second equality holds because $C(e) = \bigcup_{y\in\widetilde{V}} E(y)$, while the last equality holds because the eccentricity of any vertex of T' is given by the length of a path towards some leaf of T', i.e., some l_y. $\square$

The subset minimum eccentricity problem on trees is linear time solvable via a dynamic programming algorithm that computes the vertex eccentricities. Therefore, Lemma 3 already implies an $O(n^2)$ time and $O(m)$ space algorithm for the problem of computing a candidate best swap edge, among the non-tree edges incident to z, of every tree edge, as this problem is equivalent to solving

$n-1$ instances of the subset minimum eccentricity problem on trees (one for each tree edge failure). However, for each vertex z, we can reduce the time complexity of computing a candidate best swap edge of every tree edge to $O(m \log n)$ by exploiting the similarities between instances of the subset minimum eccentricity problem on trees induced by the failure of adjacent tree edges. More precisely, for each vertex z, the algorithm roots T at z, visits the tree edges in preorder, and uses a *top tree* to efficiently generate and solve the corresponding $n - 1$ instances of the subset minimum eccentricity problem on trees.

A *top tree* (see [2]) is a data structure that maintains a dynamic forest on a fixed set of N vertices, some of which are marked, can be initialized as an empty forest in $O(N)$ time and space, and supports each of the following operations in $O(\log N)$ time:

cut($\bar{e}$): if $\bar{e}$ is an edge of the forest, it removes the edge $\bar{e}$ from the forest;

link(a, b, ω): if a and b are vertices of different trees in the forest, it adds the edge (a, b) of cost ω to the forest;

increase-cost(a, b, ω): if (a, b) is an edge of the forest of cost ω', it updates the cost of (a, b) to $\max\{\omega', \omega\}$; otherwise this operation is equivalent to link(a, b, ω);[3]

center(a): it returns a triple $(c, \bar{a}, \bar{b})$, where c is a *center* of the tree,[4] say T'', in the forest that contains a, while $\bar{a}$ and $\bar{b}$ are the two endpoints of a *diametral path* of T'';[5]

node(a, b, k): if a and b are vertices of the same tree, say T'', in the forest and k is a positive integer upper bounded by the hop-distance from a to b in T'', it returns the k-th node along the path from a to b in T'';

closest(a): it returns a marked vertex which is closest to a (w.r.t. forest distances);

dist(a, b): it returns the distance in the forest from a to b.

The algorithm uses the top tree as follows (see Algorithm 1 for the details and Figure 1 for an example). At the beginning of the visit of z the top tree is initialized as an empty forest of $2n$ vertices, where there are two vertices y and l_y for each $y \in V$. Furthermore, the set of marked vertices is $\{y \mid (z, y) \in E(z)\}$.

Let $e = (u, v)$ be the failing tree edge and, w.l.o.g., assume that z is a node of T_u. If e is not incident to z, i.e., $z \neq u$, then let e' be the tree edge incident to u along the path from z to u in T. Finally, let $\widetilde{C}(e) = C(e)$ if $u = z$, and $\widetilde{C}(e) = C(e) \setminus C(e')$ otherwise.

---

[3] The increase-cost operation is not a basic primitive of the top tree, but it can be easily implemented via a cut operation (that has to be modified to return the cost of the removed edge, if any) followed by a link operation.

[4] A tree *center* is a vertex of the tree of minimum eccentricity.

[5] A *diametral path* is a path of the tree of maximum length.

Algorithm 1. Algorithm for ABSE spanner tree problem on unweighted graphs.

Input : a 2-edge-connected graph $G = (V, E)$ and a spanning tree T of G.
Output: $\forall e \in E(T)$, a best swap edge minimizing the stretch factor of the swap tree.

for *every $e \in E(T)$* **do**
  $f_e = \perp$;   // *f_e stores the best swap edge of e*
  $\mu_e = +\infty$;   // *μ_e stores the value of formula (4) w.r.t. f_e*

for *every $z \in V$* **do**
  root T at z; compute the set $\widetilde{C}(e)$ for every $e \in E(T)$;
  initialize the top tree as an empty forest on a set $\mathcal{V}$ of $2n$ vertices, where there are
  two vertices y and l_y for each vertex $y \in V$, and mark all the vertices of the set
  $\{y \in \mathcal{V} \mid (z, y) \in E \setminus E(T)\}$;
  // *We assume that every tree in the top tree is rooted and that the parent of a root*
  vertex is the root vertex itself. The boolean vector B keeps track of the vertices of
  T that are connected to their parents in the underlying structure of the top tree
  (see Figure 1) so as to have a frugal usage of link *operations. The element $r[a]$*
  stores the closest child of a of degree ≥ 2 in the tree that contains a, say T''. If
  such a vertex does not exist, then $r[a]$ stores the parent of the (unique) leaf of T''.
  for *every $a \in V$* **do** $B[a] = 0$; $r[a] = a$;
  for *every $e = (u, v) \in E(T)$ in preorder w.r.t. z* **do**
    // *the instance of the subset min ecc. problem on trees is built*
    for *every $(x, y) \in \widetilde{C}(e)$* **do**
      w.l.o.g., let $y \in V(T_v)$; increase-cost$(y, l_y, d_T(z, x) + 1)$;
      while *$B[y] = 0$ and $y \neq v$* **do**
        let p be the parent of y in T; link$(y, p, 1)$; $B[y] = 1$;
        if p has exactly one child **then** $r[p] = r[y]$;
        if p has exactly two children **then** $r[p] = p$;
        $y = p$;

    cut(e); **if** u has no child **then** $r[u] = u$;
    if u has exactly one child **then** $r[u] = y$, where y is the (unique) child of u;
    // *updating the vector r after the execution of the following instruction is*
    unnecessary as, for each cut$((r[v], p))$*, the corresponding* link$(r[v], p, 1)$ *is*
    executed at the end of the for loop
    if $r[v] \neq v$ **then** cut$((r[v], p))$, where p is the parent of $r[v]$ in T;
    // *the instance of the subset min ecc. problem on trees is solved*
    $(c, \bar{a}, \bar{b}) =$ center$(r[v])$;
    // *a marked vertex s that minimizes the distance from center c, and the*
    corresponding eccentricity value are computed
    $s =$ closest(c); $\mu = \max\left\{\text{dist}(s, \bar{a}), \text{dist}(s, \bar{b})\right\}$;
    // *the second (potential) tree center c' is computed.*
    if dist$(c, \bar{a}) \leq$ dist$(c, \bar{b})$ **then** $c' =$ node$(c, \bar{b}, 1)$; **else** $c' =$ node$(c, \bar{a}, 1)$;
    // *a marked vertex s' that minimizes the distance from vertex c', and the*
    corresponding eccentricity value are computed
    $s' =$ closest(c'); $\mu' = \max\left\{\text{dist}(s', \bar{a}), \text{dist}(s', \bar{b})\right\}$;
    // *the marked vertex s^* of minimum eccentricity, and its corresponding*
    eccentricity value are computed.
    if $\mu \leq \mu'$ **then** $s^* = s$; $\mu^* = \mu$; **else** $s^* = s'$; $\mu^* = \mu'$;
    // *the best swap edge of e is updated*
    if $\mu^* < \mu_e$ **then** $\mu_e = \mu^*$; $f_e = (z, s^*)$;
    if $r[v] \neq v$ **then** link$(r[v], p, 1)$, where p is the parent of $r[v]$ in T;

return $\{f_e \mid e \in E(T)\}$;

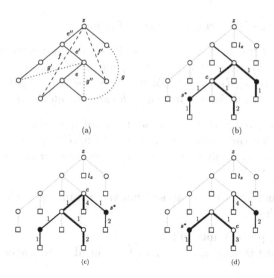

(a)   (b)

(c)   (d)

Fig. 1. An illustration of the execution of Algorithm 1 during the visit of vertex z. The input instance of the ABSE spanner tree problem is shown in (a): solid edges are tree edges, dashed edges are non-tree edges incident to z, and dotted edges are the other non-tree edges. The top trees corresponding to the failure of the tree edges e'', e', and e are shown in (b), (c), and (d), respectively. Square vertices correspond to l_y's vertices, marked vertices are of black color. and white vertices are unmarked, while solid edges denote the underlying top tree structure, and black edges are edges of the top tree. For each failing edge, the corresponding tree in the top tree shows the diametral path (in bold), the tree center (vertex c), and the marked vertex which minimizes the distance from its closest tree center (vertex s^*). The candidate best swap edge of e'' is f which induces a tree having stretch factor equal to 4. The candidate best swap edge of e' is f' which induces a tree having stretch factor equal to 5. Finally, the candidate best swap edge of e is f which induces a tree having stretch factor equal to 5.

For every $(x, y) \in \widetilde{C}(e)$, with $x \in V(T_u)$ and $y \in V(T_v)$, the algorithm first increases the cost of the edge (y, l_y) to $d_T(z, x) + 1$ and then updates the top tree by adding all the missing edges of the path from y to r_v in T, where the cost of each missing edge is 1. Next, the algorithm removes e from the top tree. Finally, it computes a candidate best swap edge of e using a suitable combination of **center**, **node**, **closest**, and **dist** operations according to the following four lemmas which show the relationships among tree center(s), diametral path(s), and vertex eccentricities, as well as an interesting connection between solutions of instances of the subset minimum eccentricity problem on trees generated by the algorithm and tree center(s).

Lemma 4 (folklore). *Any diametral path of a positively edge-weighted tree contains all the tree centers.*

Lemma 5 (folklore). *Any positively edge-weighted tree has either one center or two centers. Furthermore, if the tree has two centers, say c and c', then (c, c') is an edge of the tree.*

The following lemma is a stronger version of Lemma 4.

Lemma 6 ([4]). *Let a be a vertex of a positively edge-weighted tree T'' and let $\bar{a}$ and $\bar{b}$ be the endvertices of a diametral path of T''. The eccentricity of a is equal to $\max\left\{d_{T''}(a,\bar{a}), d_{T''}(a,\bar{b})\right\}$. Furthermore, the longest path in T'' between the one from a to $\bar{a}$ and the one from a to $\bar{b}$ contains all the tree centers.*

Lemma 7. *For a fixed vertex $z \in V$ and a fixed tree edge $e \in E(T)$, an optimal solution of the corresponding instance $\langle T', w', U \rangle$ of the subset minimum eccentricity problem on trees is a vertex of U that minimizes the distance from its closest tree center.*

Proof. By construction, it is easy to see that the instance $\langle T', w', U \rangle$ satisfies the following properties:

(i) edge costs are positive integers;
(ii) all edges of cost strictly greater than 1 are incident to leaves of T', i.e., the vertices l_y's.

We prove the claim by cases according to the number of centers of T'. Thus, according to Lemma 5, we have to distinguish between the following two cases: T' has two centers and T' has exactly one center.

We begin with the case in which T' has two centers, say c and c'. Let $\bar{a}$ and $\bar{b}$ be the endvertices of a diametral path of T'. By Lemma 4, the path from $\bar{a}$ to $\bar{b}$ contains both c and c'. W.l.o.g., we assume that $d_{T'}(\bar{a},c) < d_{T'}(\bar{a},c')$. By Lemma 6, the eccentricities of c and c' are equal to $\max\left\{d_{T'}(c,\bar{a}), d_{T''}(c,\bar{b})\right\} = d_{T'}(\bar{b},c)$ and $\max\left\{d_{T'}(c',\bar{a}), d_{T''}(c',\bar{b})\right\} = d_{T'}(\bar{a},c')$, respectively. Furthermore, as c and c' are both tree centers, we have that $d_{T'}(c,\bar{a}) = d_{T'}(c',\bar{b})$. By Lemma 6, the eccentricity of a vertex $a \in U$ is therefore equal to

$$\max\left\{d_{T'}(a,\bar{a}), d_{T'}(a,\bar{b})\right\} = \min\left\{d_{T'}(a,c) + d_{T'}(c,\bar{b}), d_{T'}(a,c') + d_{T'}(c',\bar{a})\right\}$$
$$= d_{T'}(c,\bar{a}) + \min\left\{d_{T'}(a,c), d_{T'}(a,c')\right\},$$

where the first equality can be proved by case analysis (it clearly holds in the case $d_{T'}(a,c) \leq d_{T'}(a,c')$ as well as in the complementary case $d_{T'}(a,c) > d_{T'}(a,c')$). Since $d_{T'}(c,\bar{a})$ does not depend on a, the above equality implies that a vertex of U having minimum eccentricity is then equal to a vertex of U that minimizes the distance from its closest tree center. Hence, we have proved the claim for the case in which T' has two centers.

Consider now the case in which T' has only one center, say c. Let $\bar{a}$ and $\bar{b}$ be the endvertices of a diametral path of T'. By Lemma 4, the path from $\bar{a}$ to $\bar{b}$ contains c. W.l.o.g., we assume that $d_{T'}(c,\bar{a}) \geq d_{T'}(c,\bar{b})$. We divide the proof into the following two cases: $d_{T'}(c,\bar{a}) = d_{T'}(c,\bar{b})$ and $d_{T'}(c,\bar{a}) > d_{T'}(c,\bar{b})$.

In the former case, i.e., $d_{T'}(c,\bar{a}) = d_{T'}(c,\bar{b})$, by Lemma 6, the eccentricity of a vertex $a \in U$ is upper bounded by

$$\max\left\{d_{T'}(a,\bar{a}), d_{T'}(a,\bar{b})\right\} \leq \max\left\{d_{T'}(a,c) + d_{T'}(c,\bar{a}), d_{T'}(a,c) + d_{T'}(c,\bar{b})\right\}$$
$$= d_{T'}(a,c) + d_{T'}(c,\bar{a})$$

and, since the path in T' from c to $\bar{a}$ is edge-disjoint w.r.t. the path in T' from c to $\bar{b}$ are edge-disjoint, it is lower bounded by

$$\max\left\{d_{T'}(a,\bar{a}), d_{T'}(a,\bar{b})\right\} \geq \min\left\{d_{T'}(a,c) + d_{T'}(c,\bar{a}), d_{T'}(a,c) + d_{T'}(c,\bar{b})\right\}$$
$$= d_{T'}(a,c) + d_{T'}(c,\bar{a}).$$

Therefore, $\max\left\{d_{T'}(a,\bar{a}), d_{T'}(a,\bar{b})\right\} = d_{T'}(a,c) + d_{T'}(c,\bar{a})$. Since $d_{T'}(c,\bar{a})$ does not depend on a, the above equality implies that a vertex of U having minimum eccentricity is then equal to a vertex of U that minimizes the distance from c.

In the latter case, i.e., $d_{T'}(c,\bar{a}) > d_{T'}(c,\bar{b})$, by property (i) we have that $d_{T'}(c,\bar{a}) \geq d_{T'}(c,\bar{b})+1$. As a consequence, and using property (ii), we have that $(c,\bar{a})$ is an edge of T', otherwise the eccentricity value of the vertex immediately following c along the path from c to $\bar{a}$ would be strictly smaller than the eccentricity of c. Therefore, by Lemma 6, the eccentricity of a vertex $a \in U$ is upper bounded by

$$\max\left\{d_{T'}(a,\bar{a}), d_{T'}(a,\bar{b})\right\} \leq \max\left\{d_{T'}(a,c) + d_{T'}(c,\bar{a}), d_{T'}(a,c) + d_{T'}(c,\bar{b})\right\}$$
$$= d_{T'}(a,c) + d_{T'}(c,\bar{a}).$$

Furthermore, since $a \in U$, and vertices of U are internal vertices of T', we have that $a \neq \bar{a}$. As a consequence, the path in T' from a to c does not contain the edge $(\bar{a}, c)$. Therefore, by Lemma 6, the eccentricity of a is lower bounded by

$$\max\left\{d_{T'}(a,\bar{a}), d_{T'}(a,\bar{b})\right\} \geq d_{T'}(a,c) + +d_{T'}(c,\bar{a}).$$

Thus, $\max\left\{d_{T'}(a,\bar{a}), d_{T'}(a,\bar{b})\right\} = d_{T'}(a,c)+d_{T'}(c,\bar{a})$. Once again, as $d_{T'}(c,\bar{a})$ is independent of a, the above equality implies that a vertex of U having minimum eccentricity corresponds to a vertex of U that minimizes the distance from c. This concludes the proof.

$\square$

We can now prove the main theorem.

Theorem 2. *Algorithm 1 solves the ABSE tree spanner problem on unweighted graphs in $O(mn \log n)$ time and $O(m)$ space.*

Proof. For every vertex $z \in V$ and for every tree edge $e \in E(T)$, Algorithm 1 reduces the problem of computing the candidate best swap edge of e, among the non-tree edges incident to z, to an instance of the subset minimum eccentricity problem on trees and then solves the latter instance so as explained in Lemma 7. Therefore, the algorithm correctness follows by the aforementioned lemma and by Lemma 3.

Concerning the time and space complexity of the algorithm, first of all observe that the space consumption is clearly linear in m. To prove the $O(mn \log n)$ time bound, we show that for a fixed vertex $z \in V$, the algorithm runs in $O(m \log n)$ time.

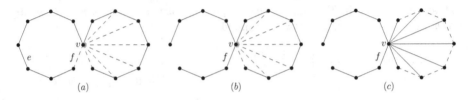

Fig. 2. An example showing that the ratio between the stretch of the best swap tree and the stretch of an optimal tree spanner of the remaining graph can be $\Omega(n)$. The initial graph G has $n = 2k + 1$ vertices and consists of two cycles, each having length k, that share a single vertex v, plus a set of edges joining all the vertices of one of the cycles with v. In (a) an optimal tree spanner with stretch $k - 1$ is shown using solid lines. Picture (b) shows the (unique) best swap tree after the failure of edge e. Notice that the stretch factor of the swap tree in (b) is $k - 1$. Finally, in (c) it is shown an optimal tree spanner of $G - e$ having stretch 2.

Let $z \in V$ be fixed. Observe that all the sets $\widetilde{C}(e)$ can be computed in $O(m)$ time by computing the *least common ancestors* for the endpoints of each non-tree edge (x, y) in constant time (see [10]) and determining the at most two edges along the path from x to y in T which are incident to the least common ancestor of x and y. This implies that each non-tree edge (x, y) is contained in at most two sets among all $\widetilde{C}(e)$'s. The top tree of $2n$ vertices can be initialized in $O(n)$ time (see [2]).

To prove the claim, it remains to show that the overall number of operations performed on the top tree is at most $O(m)$. First of all, observe that the number of cut, center, closest, dist, and node operations is constant for each tree edge $e \in E(T)$. Therefore, the overall number of cut, center, closest, dist, and node operations, for any vertex $z \in V$, is $O(n)$. As there is exactly an increase-cost operation for each non-tree edge $g \in \widetilde{C}(e)$ and since $\sum_{e \in E(T)} |\widetilde{C}(e)| \leq 2m$, the overall number of increase-cost operations, for any $z \in V$, is $O(m)$. Finally, as the top tree contains $2n$ vertices, since the overall number of cut operations is $2(n - 1)$ (at most two cut operations for each tree edge failure), and since a link operation is performed only when an edge is missing, the overall number of link operations, for any vertex $z \in V$, is at most $4n - 3$ as it can never exceed $2n - 1$ (the size of a tree spanning $2n$ vertices) plus the at most $2(n - 1)$ cut operations. □

5   Conclusions

In this paper we provided two efficient linear-space solutions for both the weighted and the unweighted version of the ABSE tree spanner problem, running in $O(m^2 \log \alpha(m, n))$ and $O(mn \log n)$ time, respectively.

Our future research on this problem will follow several directions. First of all, we will work on the extension to the weighted case of the approach we used for the unweighted case, in order to get an improved runtime. Besides that, we will also face the problem of analyzing the *quality* of the swap tree spanner as compared to that of an optimal tree spanner of the graph deprived of the failed edge. To address this task, one should evaluate the ratio between the stretch factors of the two trees. A similar study was performed in [7], where the authors focused on unweighted graphs, and showed that if stretches are measured w.r.t. distances in G (i.e., according to their criteria, but differently from our intent which is that of measuring stretches in the affected graph), and if moreover the initial tree is an *optimal* tree spanner, then this ratio is bounded by 2, and this is tight. Unfortunately, computing an optimal tree spanner is hard, and it is unknown what will be the outcome when the initial tree is suboptimal. On the other hand, when adopting our criteria, we cannot exhibit a similar positive result just right away. Indeed, it is easy to see that in this case the ratio can become unbounded as shown in Figure 2. However, we conjecture that a suitable preprocessing of the initial tree could avoid this pathological behavior. Finally, we also plan to focus our attention on the related problem of handling single node failures.

References

1. Abraham, I., Bartal, Y., Neiman, O.: Embedding metrics into ultrametrics and graphs into spanning trees with constant average distortion. In: Proc. of the 18th ACM-SIAM Symp. on Discrete Algorithms (SODA 2007), pp. 502–511. ACM Press (2007)
2. Alstrup, S., Holm, J., de Lichtenberg, K., Thorup, M.: Maintaining information in fully dynamic trees with top trees. ACM Trans. Algorithms 1(2), 243–264 (2005)
3. Bilò, D., Gualà, L., Proietti, G.: Finding best swap edges minimizing the routing cost of a spanning tree. Algorithmica 68(2), 337–357 (2014)
4. Bilò, D., Gualà, L., Proietti, G.: A faster computation of all the best swap edges of a shortest paths tree. Algorithmica (in press). doi:10.1007/s00453-014-9912-6
5. Brandstädt, A., Chepoi, V., Dragan, F.F.: Distance approximating trees for chordal and dually chordal graphs. Journal of Algorithms 30(1), 166–184 (1999)
6. Cai, L., Corneil, D.G.: Tree spanners. SIAM J. on Disc. Math. 8, 359–387 (1995)
7. Das, S., Gfeller, B., Widmayer, P.: Computing all best swaps for minimum-stretch tree spanners. J. of Graph Algorithms and Applications 14(2), 287–306 (2010)
8. Di Salvo, A., Proietti, G.: Swapping a failing edge of a shortest paths tree by minimizing the average stretch factor. Theor. Comp. Science 383(1), 23–33 (2007)
9. Emek, Y., Peleg, D.: Approximating minimum max-stretch spanning trees on unweighted graphs. SIAM Journal on Computing 38(5), 1761–1781 (2008)
10. Harel, D., Tarjan, R.E.: Fast algorithms for finding nearest common ancestors. SIAM Journal on Computing 13(2), 338–355 (1984)
11. Italiano, G.F., Ramaswami, R.: Maintaining spanning trees of small diameter. Algorithmica 22(3), 275–304 (1998)
12. Liebchen, C., Wünsch, G.: The zoo of tree spanner problems. Discrete Applied Mathematics 156, 569–587 (2008)
13. Nardelli, E., Proietti, G., Widmayer, P.: Finding all the best swaps of a minimum diameter spanning tree under transient edge failures. Journal of Graph Algorithms and Applications 5(5), 39–57 (2001)
14. Pettie, S.: Sensitivity analysis of minimum spanning trees in sub-inverse-Ackermann time. In: Deng, X., Du, D.-Z. (eds.) ISAAC 2005. LNCS, vol. 3827, pp. 964–973. Springer, Heidelberg (2005)
15. Wu, B.Y., Hsiao, C.-Y., Chao, K.-M.: The swap edges of a multiple-sources routing tree. Algorithmica 50(3), 299–311 (2008)

Randomized OBDD-Based Graph Algorithms

Marc Bury*

TU Dortmund, LS2 Informatik, Germany

Abstract. Implicit graph algorithms deal with the characteristic function χ_E of the edge set E of a graph $G = (V, E)$. Encoding the nodes by binary vectors, χ_E can be represented by an Ordered Binary Decision Diagram (OBDD) which is a well known data structure for Boolean functions. OBDD-based graph algorithms solve graph optimization problems by mainly using functional operations and are a heuristic approach to cope with massive graphs. These algorithms heavily rely on a compact representation of the underlying Boolean functions which is why all previously known OBDD-based algorithms are deterministic since random functions are not compressible in general. Here, the first randomized OBDD-based algorithms are presented where random functions with limited independence are used to overcome the large representation size. On the theoretical part, the size of OBDDs representing k-wise independent random functions is investigated and a construction of almost k-wise independent random functions by means of a random OBDD generation is shown. On the algorithmic part, randomization often facilitates the design of simple algorithms which in the context of OBDD-based algorithms means a small number of functional operations and as few input variables of the used Boolean functions as possible. This paper presents a maximal matching algorithm with $O(\log^3 |V|)$ functional operations in expectation using functions with at most $3\log|V|$ variables which is both better than the best known algorithms w.r.t. functional operations and variables. The algorithm may be of independent interest. The experimental evaluation shows that this algorithm outperforms known OBDD-based algorithms for the maximal matching problem.

1 Introduction

In times of Big Data, classical algorithms for optimization problems quickly exceed feasible running times or memory requirements. For instance, the rapid growth of the Internet and social networks results in massive graphs which traditional algorithms cannot process in reasonable time or space. In order to deal with such graphs, implicit (symbolic) algorithms have been investigated where the input graph is represented by the characteristic function χ_E of the edge set and the nodes are encoded by binary numbers. Using *Ordered Binary Decision Diagrams* (OBDDs), which were introduced by Bryant [11] and are a well-known data structure for Boolean functions, to represent χ_E can significantly decrease the space needed to store such graphs. Furthermore, using mainly functional operations, e. g., conjunction, disjunction, and quantification, which are efficiently

---

* Supported by Deutsche Forschungsgemeinschaft, grant BO 2755/1-2.

supported by the OBDD data structure, many optimization problems can be solved on OBDD represented inputs ([16, 17, 19, 34–36, 41]). Implicit algorithms were successfully applied in many areas, e. g., model checking [12], integer linear programming [24] and logic minimization [14]. With one of the first implicit graph algorithms, Hachtel and Somenzi [19] were able to compute a maximum flow on 0-1-networks with up to 10^{36} edges and 10^{27} nodes in reasonable time.

There are two main parameters influencing the actual running time of OBDD-based algorithms: the number of functional operations and the sizes of all intermediate OBDDs used during the computation. But there seems to be a trade-off: The number of operations is an important measure of difficulty [5] but decreasing the number of operations often results in an increase of the number of variables of the used functions. Since the worst case OBDD size of a function $f : \{0,1\}^n \to \{0,1\}$ is $\Theta(2^n/n)$, the number of variables should be as small as possible to decrease the worst case running time. This trade-off was also empirically observed. For instance, an implicit algorithm computing the transitive closure that uses an iterative squaring approach and a polylogarithmic number of operations is often inferior to an implicit sequential algorithm, which needs a linear number of operations in worst case [5, 20]. Another example is the maximal matching algorithm (BP) of Bollig and Pröger [10] that uses only $O(\log^4 N)$ functional operations on functions with at most $6 \log N$ variables while the algorithm (HS) of Hachtel and Somenzi [19] uses $O(N \log N)$ operations in the worst case on function with at most $3 \log N$ variables. However, HS is clearly superior to BP on most instances (see Section 5).

Using randomization in an explicit algorithm often leads to simple and fast algorithms. Here, we propose the first attempt at using randomization to obtain algorithms which have both a small number of variables and a small expected number of functional operations. For this, we want to represent random functions $f_r : \{0,1\}^n \to \{0,1\}$ with $\mathbf{Pr}[f_r(x) = 1] = p$ for every $x \in \{0,1\}^n$ and some fixed probability $0 < p < 1$ by OBDDs where the probability is taken over the random seed r. Using random functions in implicit algorithms is difficult. We need to construct them efficiently but, obviously, if the function values are completely independent (and p is a constant), then the OBDD (and even the more general FBDD or read-once branching program) size of f_r is exponentially large with an overwhelming probability [39]. Thus, we investigate the OBDD size and construction of (almost) k-wise independent random functions where the distribution induced on every k different function values is (almost) uniform. Using the random functions for OBDD-based graph algorithms is challenging in the following sense: Since we have a fixed probability p for every input (which in our case represents a node or an edge of a graph), we often cannot use known randomized algorithms e. g., for maximal matching [1, 26, 21] (a matching M, i. e., a set of edges without a common vertex, is called maximal if M is no proper subset of another matching) because they either do not use limited independence or require distinct event probabilities (depending on the degree of a node). This constraints lead to a new algorithm for the maximal matching problem which can also be used for the similar maximal independent set (MIS) problem where an independent set I is a subset of the nodes such that no two nodes of I are adjacent and any vertex in G is either in I or is adjacent to a node of I.

Related Work. The size of OBDDs representing graphs was investigated for bipartite graphs [32], interval graphs [32, 18], cographs [32] and graphs with bounded tree- and clique-width [28]. Sawitzki [37] showed that the set of problems solved by an implicit algorithm using $O(\log^k N)$ functional operations and functions defined on $O(\log N)$ variables is equal to the complexity class FNC, i.e., the class of all optimization problems that can be efficiently solved in parallel. Implicit algorithms with these properties were designed for instance for topological sorting [41], minimum spanning tree [6], metric TSP approximation [7] and maximal matching [10]. When analyzing implicit algorithms, the actual running time can either be proven for very structured input graphs like [41] did for topological sorting and [8] for maximum matching or the running time is experimentally evaluated like in [19] for maximum flows and in [8, 18] for maximum matching on bipartite graphs or unit interval graphs.

A succinct representation of 2^n random bits, which are k-wise independent, was presented by Alon et al. [1] using $\lfloor k/2 \rfloor n + 1$ independent random bits. This number of random bits is very close to the lower bound of Chor and Goldreich [13]. In order to reduce the number of random bits even further, Naor and Naor [30] introduced the notion of almost k-wise independence where the distribution on every k random bits is "close" to uniform. Constructions of almost k-wise independent random variables are also given in [2] and are using only at most $2(\log n + \log k + \log(1/\varepsilon))$ random bits where ε is a bound on the closeness to the uniform distribution. Looking for a simple representation of almost k-wise independent random variables, Savický [33] presented a Boolean formula of constant depth and polynomial size and used $n \log^2 k \log(1/\varepsilon)$ random bits. In all of these constructions, the running time of computing the i-th random bit with $0 \le i \le 2^n - 1$ depends on k and ε.

Such small probability spaces can be used for a succinct representation of a random string of length 2^n, e.g., in streaming algorithms [3], or for derandomization [1, 26]. The randomized parallel algorithms from [1, 26] compute a maximal independent set of a graph. The computation of a MIS has also been extensively studied in the area of distributed algorithms [4, 25]. An optimal randomized distributed MIS algorithm was presented in [29] where the time and bit complexity (bits per channel) is $O(\log N)$. Using completely independent random bits, Israeli and Itai [21] give a randomized parallel algorithm computing a maximal matching in time $O(\log N)$.

While we are looking for k-wise independent functions with small OBDD size, Kabanets [23] constructed simple Boolean functions which are hard for FBDDs by investigating (almost) $\Theta(n)$-wise independent random functions and showed that the probability tends to 1 as n grows that the size is $\Omega(2^n/n)$.

Our Contribution. In Section 3, we show that the OBDD and FBDD size is at least $2^{\Omega(n + \log(p'))}$ with $p' = 2p(1 - p)$ if the function values of f_r are k-wise independent with $k \ge 4$. We give an efficient construction of OBDDs for 3-wise independent random functions which is based on the known construction of 3-wise independent random variables using BCH-schemes [1]. In Section 4 we investigate a simple construction of a random OBDD which generates almost k-wise independent random functions and has size $O((kn)^2/\varepsilon)$. Reading the actual value of the i-th random bit is just an evaluation of the function on input i which can be done in $O(n)$ time, i.e., it is independent of both k and ε. This construction can be seen as a distribution on graphs with a small OBDD size what enables us to use it as an input distribution for our implicit algorithm in the experimental evaluation. In Section 5 we use pairwise independent random functions to design a simple maximal matching algorithm that uses only $O(\log^3 N)$ functional operations in expectation and functions with at most $3 \log N$ variables which is better than the aforementioned algorithms by Hachtel and Somenzi [19] with $3 \log N$ variables and Bollig and Pröger [10] with $O(\log^4 N)$ operations. This algorithm can easily be extended to the MIS problem and can be implemented as a parallel algorithm using $O(\log N)$ time in expectation or as a distributed algorithm with $O(\log N)$ expected time and bit complexity. To the best of our knowledge, this is the first (explicit or implicit) maximal matching (or independent set) algorithm that does not need any knowledge about the graph (like size or node degrees) as well as uses only pairwise independent random variables. Eventually, we evaluate this algorithm empirically and show that known implicit maximal matching algorithms are outperformed by the randomized algorithm.

2   Preliminaries

All omitted proofs and figures can be found in the full version of this paper.

Binary Decision Diagrams. We denote the set of Boolean functions $f :
\{0,1\}^n \to \{0,1\}$ by B_n. For $x \in \{0,1\}^n$ denote the value of x by $|x| := \sum_{i=0}^{n-1} x_i \cdot 2^i$. Further, for $l \in \mathbb{N}$, we denote by $[l]_2$ the corresponding binary number of l. In his seminal paper [11], Bryant introduced *Ordered Binary Decision Diagrams* (OBDDs), that allow a compact representation of not too few Boolean functions and also supports many functional operations efficiently.

Definition 1 (Ordered Binary Decision Diagram (OBDD))
  Order. A variable order π on the input variables $X = \{x_0, \ldots, x_{n-1}\}$ of a Boolean function $f \in B_n$ is a permutation of the index set $I = \{0, \ldots, n-1\}$.
  Representation. A π-OBDD is a directed, acyclic, and rooted graph G with two sinks labeled by the constants 0 and 1. Each inner node is labeled by an input variable from X and has exactly two outgoing edges labeled by 0 and 1. Each edge (x_i, x_j) has to respect the variable order π, i.e., $\pi^{-1}(i) < \pi^{-1}(j)$.
  Evaluation. An assignment $a \in \{0,1\}^n$ of the variables defines a path from the root to a sink by leaving each x_i-node via the a_i-edge. A π-OBDD G_f represents f iff for every $a \in \{0,1\}^n$ the defined path ends in the sink with label $f(a)$.
  Complexity. The size of a π-OBDD G, denoted by $|G|$, is the number of nodes in G. The π-OBDD size of a function f is the minimum size of a π-OBDD representing f. The OBDD size of f is the minimum π-OBDD size over all variable orders π. The width of G is the maximum number of nodes labeled by the same input variable.

The more general read-once branching programs or *Free Binary Decision Diagrams* (FBDDs) were introduced by Masek [27] where every variable can only be read once on a path from the root to a sink (but the order is not restricted).

A simple function is the inner product $IP_n(x, y) = \bigoplus_{i=0}^{n-1} x_i \wedge y_i$ of two vectors $x, y \in \{0, 1\}^n$. Let π be a variable order where for every $0 \leq i \leq n$, the variables x_i and y_i are consecutive. It is easy to see that the π-OBDD representing IP_n has size $O(n)$ and width 2. Notice that the π-OBDD size is still $O(n)$ if we replace an input vector, e.g., y, by a constant vector $r \in \{0, 1\}^n$.

In the following we describe some important operations on Boolean functions which we will use in this paper (see, e.g., Section 3.3 in [40] for a detailed list). Let f and g be Boolean functions in B_n on the variable set $X = \{x_0, \ldots, x_{n-1}\}$, π a fixed order and let G_f and G_g be π-OBDDs representing f and g, respectively. We denote the subfunction of f where x_j for some $0 \leq j \leq n - 1$ is replaced by a constant $a \in \{0, 1\}$ by $f_{|x_j=a}$.

1. **Negation:** Given G_f, compute a representation for the function $\overline{f} \in B_n$. Time: $O(1)$
2. **Replacement by Constant:** Given G_f, an index $i \in \{0, \ldots, n-1\}$, and a Boolean constant $c_i \in \{0, 1\}$, compute a representation for the subfunction $f_{|x_i=c_i}$. Time: $O(|G_f|)$
3. **Equality Test:** Given G_f and G_g, decide whether f and g are equal. Time: $O(1)$ in most implementations (when using so called *Shared OBDDs*, see [40]), otherwise $O(|G_f| + |G_g|)$.
4. **Synthesis:** Given G_f and G_g and a binary Boolean operation $\otimes \in B_2$, compute a representation for the function $h \in B_n$ defined as $h := f \otimes g$. Time: $O(|G_f| \cdot |G_g|)$
5. **Quantification:** Given G_f, an index $i \in \{1, \ldots, n\}$ and a quantifier $Q \in \{\exists, \forall\}$, compute a representation for the function $h \in B_n$ defined as $h := Qx_i : f$ where $\exists x_i : f := f_{|x_i=0} \vee f_{|x_i=1}$ and $\forall x_i : f := f_{|x_i=0} \wedge f_{|x_i=1}$. Time: see replacement by constant and synthesis

In addition to the operations mentioned above, in implicit graph algorithms (see the next section) the following operation (see, e.g., [36]) is useful to reverse the edges of a given graph. We will use this operation implicitly by writing for instance $f(x, y)$ and $f(y, x)$ in the pseudo code of our algorithm.

Definition 2. *Let $k \in \mathbb{N}$, ρ be a permutation of $\{1, \ldots, k\}$ and $f \in B_{kn}$ with input vectors $x^{(1)}, \ldots, x^{(k)} \in \{0, 1\}^n$. The argument reordering $\mathcal{R}_\rho(f) \in B_{kn}$ with respect to ρ is defined by $\mathcal{R}_\rho(f)(x^{(1)}, \ldots, x^{(k)}) := f(x^{(\rho(1))}, \ldots, x^{(\rho(k))})$.*

This operation can be computed by just renaming the variables and repairing the variable order using $3(k - 1)n$ functional operations (see [9]).

A function f *depends essentially* on a variable x_i iff $f_{|x_i=0} \neq f_{|x_i=1}$. A characterization of minimal π-OBDDs due to Sieling and Wegener [38] can often be used to bound the OBDD size.

Theorem 1 ([38]). *Let $f \in B_n$ and for all $i = 0, \ldots, n-1$ let s_i be the number of different subfunctions which result from replacing all variables $x_{\pi(j)}$ with $0 \leq j \leq i-1$ by constants and which essentially depend on $x_{\pi(i)}$. Then the minimal π-OBDD representing f has s_i nodes labeled by $x_{\pi(i)}$.*

Lower bound techniques for FBDDs are similar but have to take into account that the order can change for different paths. The following property due to Jukna [22] can be used to show good lower bounds for the FBDD size.

Definition 3. *A function $f \in B_n$ with input variables $X = \{x_0, \ldots, x_{n-1}\}$ is called r-mixed if for all $V \subseteq X$ with $|V| = r$ the 2^r assignments to the variables in V lead to different subfunctions.*

Lemma 1 ([22]). *The FBDD size of an r-mixed function is at least $2^r - 1$.*

OBDD-Based Graph Algorithms. Let $G = (V, E)$ be a directed graph with node set $V = \{v_0, \ldots, v_{N-1}\}$ and edge set $E \subseteq V \times V$. Here, an undirected graph is interpreted as a directed symmetric graph. Implicit algorithms work on the characteristic function $\chi_E \in B_{2n}$ of E where $n = \lceil \log N \rceil$ is the number of bits needed to encode a node of V and $\chi_E(x, y) = 1$ if and only if $(v_{|x|}, v_{|y|}) \in E$. Often it is also necessary to store the valid encodings of nodes by the characteristic function χ_V of V. Besides functional operations, OBDD-based algorithms can use $O(\text{polylog}\,|V|)$ additional time, e. g., for constructing OBDDs for a specific function (equality, greater than, inner product, ...).

Small Probability Spaces. A succinct representation of our random function is essential for our randomized implicit algorithm. For this, we have to use random functions with limited independence.

Definition 4 ((Almost) k-wise independence). *Let $X_0, \ldots, X_{m-1}$ be m binary random variables. These variables are called k-wise independent with $k \leq m$ if and only if for all $0 \leq i_1 < \ldots i_k \leq m-1$ and for all $l_1, \ldots, l_k \in \{0, 1\}$ $\mathbf{Pr}\left[X_{i_1} = l_1 \wedge \ldots \wedge X_{i_k} = l_k\right] = 2^{-k}$ and they are called (ε, k)-wise independent iff $\left|\mathbf{Pr}\left[X_{i_1} = l_1 \wedge \ldots \wedge X_{i_k} = l_k\right] - 2^{-k}\right| \leq \varepsilon$.*

If m is a power of 2, the random variables can be seen as function values of a Boolean function.

Definition 5 ((Almost) k-wise Independent Function). *For $r, n \in \mathbb{N}$ let $S = \{0, 1\}^r$ be a sample space and $m = 2^n$. A random function $f : S \to B_n$ maps an element $s \in S$ which is drawn uniformly at random to a Boolean function $f(s)$ which we will denote by f_s. A random function $f : S \to B_n$ is called k-wise $((\varepsilon, k)$-wise$)$ independent if the random variables $X_0(s) := f_s(0^n), \ldots, X_{m-1}(s) := f_s(1^n)$ are k-wise $((\varepsilon, k)$-wise$)$ independent.*

The BCH scheme introduced by Alon et. al [1] is a construction of k-wise independent random variables $X_0, \ldots, X_{2^n-1}$ that only needs $\lfloor k/2 \rfloor n + 1$ independent random bits and works as follows for $k = 3$: Let $r_n \in \{0, 1\}$ be a random

---

Algorithm 1. RandomFunc(x,n)

Input: Variable vector x of length $n \in \mathbb{N}$
Output: 3-wise independent function $f_r(x)$

  Let $r_0, \ldots, r_n$ be $n+1$ independent random bits
  $f_r(x) = \bigoplus_{i=0}^{n-1} (r_i \wedge x_i) \oplus r_n$
  return $f_r(x)$

---

bit, $r^{(j)} \in \{0,1\}^n$ for $1 \le j \le l$ be l uniformly random row vectors, and let the row vector $r = \left[r^{(1)}, \ldots, r^{(l)} \right] \in \{0,1\}^{ln}$ be the concatenation of the vectors. For $0 \le i \le 2^n - 1$ define $X_i = IP_{ln}\left(r, \left[[i]_2, [i^3]_2, \ldots, [i^{2l-1}]_2 \right] \right) \oplus r_n$ where i^{2j-1} for $j = 1, \ldots, l$ is computed in the finite field $GF(2^n)$.

3   OBDD Size of k-wise Independent Random Functions

We start with upper bounds on the OBDD size of 3-wise independent random functions using the BCH scheme. Notice that by means of the BCH scheme it is not possible to construct a pairwise independent function (which is not 3-wise independent) since $X_0 = IP(r, 0^n) = 0$ for every $r \in \{0,1\}^n$.

Theorem 2. *Let $\varepsilon > 0$, $n \in \mathbb{N}$, p be a probability with $0 < p \le 1/2$, and let π be a variable order on the variables $\{x_0, \ldots, x_{n-1}\}$. Define $p(x) := \mathbf{Pr}\left[f_r(x) = 1 \right]$.*

1. *We can construct a π-OBDD representing a 3-wise independent function $f : S \to B_n$ with $S = \{0,1\}^{n+1}$ in time $O(n)$ such that $p(x) = 1/2$ for every $x \in \{0,1\}^n$, and the size of the π-OBDD representing f_r is $O(n)$ with width 2 for every $r \in \{0,1\}^{n+1}$ (see Algorithm 1).*

2. *We can construct a π-OBDD representing a 3-wise independent function $f : S \to B_n$ with $S = \{0,1\}^{t(n+1)}$ where $t = \lceil -\log p - \log \varepsilon \rceil$ in time $O(\frac{n}{p \cdot \varepsilon})$ such that $p \le p(x) \le (1+\varepsilon) \cdot p$ for every $x \in \{0,1\}^n$, and the size of the π-OBDD representing f_r is bounded above by $O(\frac{n}{p \cdot \varepsilon})$ for every $r \in \{0,1\}^{n+1}$.*

Can we also construct small OBDDs for k-wise independent random variables with $k \ge 4$? Unfortunately, this is not possible. In order to show this we need a technical lemma that proves some properties of the subfunctions of a k-wise independent function.

Lemma 2. *Let $f : S \to B_n$ be a k-wise independent random function over a sample space S with $k \ge 4$ and $\mathbf{Pr}\left[f_s(x) = 1 \right] = p$ for all $x \in \{0,1\}^n$. Let π be a fixed variable order. For $s \in S$, $l \in [n]$, and $\alpha \in \{0,1\}^l$ let $f_{s|\alpha}\{0,1\}^{n-l} \to \{0,1\}$ be the subfunction of f_s where the first l variables $x_{\pi(0)}, \ldots, x_{\pi(l-1)}$ are fixed according to α, i.e. $f_{s|\alpha}(z) = f_{s|x_0 = \alpha_0, \ldots, x_{l-1} = \alpha_{l-1}}(z)$. Further, let C_l be the number of collisions of the form $f_{s|\alpha} = f_{s|\alpha'}$ with $\alpha \ne \alpha'$, i.e., $C_l := \left| \{ (\alpha, \alpha') \mid \alpha, \alpha' \in \{0,1\}^l, \alpha \ne \alpha', \text{ and } f_{s|\alpha} = f_{s|\alpha'} \} \right|$ and D_l be the number of different subfunctions $f_{s|\alpha}$. This means $D_l := \left| \{ [f_{s|\alpha}] \mid \alpha \in \{0,1\}^l \} \right|$*

where $[f_{s|\alpha}] := \{\alpha' \in \{0,1\}^l \mid f_{s|\alpha} = f_{s|\alpha'}\}$ is the equivalence class of f_α with respect to function equality. Let $p' = 2p(1-p)$. Then we have $D_l \geq \dfrac{2^l}{\sqrt{2C_l}+1}$ and it holds

$$\mathbf{E}\left[C_l\right] \leq \frac{2^{2l}}{2^{n-l} \cdot p'} \qquad and \qquad \mathbf{E}\left[D_l\right] \geq \frac{2^l}{\sqrt{2 \cdot \mathbf{E}\left[C_l\right]}+1} \geq \frac{1}{\sqrt{\frac{2}{2^n \cdot p'}2^{l/2} + \frac{1}{2^l}}}.$$

Proof. For the sake of simplicity we omit the the index s of f_s and write f_α to denote $f_{s|\alpha}$. Now we fix two different assignments α, α' and define 2^{n-l} random variables $D(z) := D_{\alpha,\alpha'}(z)$ such that $D(z) = 1$ iff $f_\alpha(z) \neq f_{\alpha'}(z)$. Since the function values of the subfunctions are also k-wise independent, we have $\mathbf{E}\left[D(z)\right] = \mathbf{Pr}\left[D(z) = 1\right] = 2p(1-p) := p'$ and $\mathbf{Var}\left[D(z)\right] = \mathbf{E}\left[D(z)^2\right] - \mathbf{E}\left[D(z)\right]^2 = \mathbf{E}\left[D(z)\right] - \mathbf{E}\left[D(z)\right]^2 = p'(1-p')$ for every $z \in \{0,1\}^{n-l}$. Let $D = \sum_z D(z)$. By definition of D, we have $\mathbf{Pr}\left[f_\alpha = f_{\alpha'}\right] = \mathbf{Pr}\left[D = 0\right]$ for a fixed pair (α, α') and the latter term can be bounded from above by the probability that the difference between D and $\mathbf{E}\left[D\right]$ is at least $\mathbf{E}\left[D\right]$, i.e. $\mathbf{Pr}\left[D = 0\right] \leq \mathbf{Pr}\left[|D - \mathbf{E}\left[D\right]| \geq \mathbf{E}\left[D\right]\right]$. Each random variable $D(z)$ depends on two function values, i.e. these variables are $k' = \lfloor k/2 \rfloor$-wise independent. Since $k' \geq 2$, we can use Chebyshev's inequality to bound $\mathbf{Pr}\left[|D - \mathbf{E}\left[D\right]| \geq \mathbf{E}\left[D\right]\right]$ by

$$\frac{\mathbf{Var}\left[D\right]}{\mathbf{E}\left[D\right]^2} = \frac{\sum_z \mathbf{Var}\left[D(z)\right]}{(2^{n-l} \cdot p')^2} = \frac{2^{n-l} \cdot p' \cdot (1-p')}{(2^{n-l} \cdot p')^2} \leq \frac{1}{2^{n-l} \cdot p'}.$$

This implies that $\mathbf{E}\left[C_l\right] \leq \dfrac{\binom{2^l}{2}}{2^{n-l} \cdot p'} \leq \dfrac{2^{2l}}{2^{n-l} \cdot p'}$. Let $M_l = \max_\alpha |[f_\alpha]|$ be the random variable for the size of the largest equivalence class. Since every equivalence class of size $s \geq 2$ causes $\binom{s}{2}$ collisions and no collisions otherwise, we have $C_l \geq \dfrac{M_l(M_l - 1)}{2} \geq \dfrac{(M_l - 1)^2}{2}$ which is equivalent to $\sqrt{2 \cdot C_l} + 1 \geq M_l$. Let $\alpha_1, \dots, \alpha_{D_l} \in \{0,1\}^l$ be representative assignments for the different equivalence classes. Then it holds that $\sum_{i=1}^{D_l} |[f_{\alpha_i}]| = 2^l$. Therefore, we have $M_l \cdot D_l \geq 2^l$ which is equivalent to $D_l \geq 2^l/M_l$ (note that $M_l \geq 1$). This means that we can bound the expected value of D_l by

$$\mathbf{E}\left[D_l\right] \geq \mathbf{E}\left[\frac{2^l}{M_l}\right] \overset{\text{Jensen}}{\geq} \frac{2^l}{\mathbf{E}\left[M_l\right]} \geq \frac{2^l}{\mathbf{E}\left[\sqrt{2 \cdot C_l} + 1\right]}$$

$$\overset{\text{Jensen}}{\geq} \frac{2^l}{\sqrt{2 \cdot \mathbf{E}\left[C_l\right]} + 1} \geq \frac{2^l}{\sqrt{2} \cdot \sqrt{\frac{2^l}{2^{n-l} \cdot p'}} + 1} = \frac{1}{\sqrt{\frac{2}{2^n \cdot p'}2^{l/2} + \frac{1}{2^l}}}.$$

$\square$

Now, we can apply this lemma to show a lower bound on the expected π-OBDD size of a k-wise independent random function.

Theorem 3. *Let* $f : S \to B_n$ *be a k-wise independent random function over a sample space S with* $k \geq 4$ *and* $\mathbf{Pr}\,[f_s(x) = 1] = p$ *for all* $x \in \{0,1\}^n$. *Then, for a fixed variable order* π, *the expected* π-*OBDD size of* f_s *is bounded below by* $\Omega(2^{n/3} \cdot (p')^{(1/3)})$ *with* $p' = 2p(1 - p)$.

Proof. Recall that D_l is the number of different subfunctions $f_{s|\alpha}$ defined as $D_l := \left| \left\{ \left[f_{s|\alpha} \right] \mid \alpha \in \{0,1\}^l \right\} \right|$. For a fixed l, each of the D_l subfunctions needs one node in the OBDD representing f_s (but not necessarily labeled by the variable $x_{\pi(l+1)}$). Therefore, we can bound the expected OBDD size from below by every choice of $\mathbf{E}\,[D_l]$. Thus, we need a lower bound on $\mathbf{E}\,[D_l]$. Due to Lemma 2 we have

$$\mathbf{E}\,[D_l] \geq \frac{1}{\sqrt{\frac{2}{2^n \cdot p'}} 2^{l/2} + \frac{1}{2^l}}. \text{ For } l = 1/3(n - 1 + \log p') \text{ we have } \sqrt{\frac{2}{2^n \cdot p'}} 2^{l/2} = \frac{1}{2^l}$$

which gives us a lower bound of $\Omega(2^{n/3} \cdot (p')^{1/3})$. $\qquad\square$

The next theorem shows that representing k-wise independent random functions with $k \geq 4$ is infeasible even for FBDDs (and with it for OBDDs and all variable orders). The general strategy of the proof of this is similar to the proof as in Wegener's analysis [39] where the OBDD size of completely independent random functions is analyzed: We bound the probability p_l that there is a set of l variables such that the number of different subfunctions deviates too much from the expected value. If $\sum_{l=0}^{n-1} p_l < 1$ holds, then with probability $1 - \sum_{l=0}^{n-1} p_l > 0$ there is no such deviation in any set of l variables. While in [39] the function values are completely independent and, therefore, the calculation can be done more directly and with better estimations, we have to take a detour over the number of subfunctions which are equal (as in Lemma 2 and Theorem 3) and can only use Markov's inequality to calculate the deviation of the expectation.

Theorem 4. *Let* $f : S \to B_n$ *be a k-wise independent random function with* $k \geq 4$ *and* $\mathbf{Pr}\,[f_s(x) = 1] = p$ *for all* $x \in \{0,1\}^n$. *Then, there is an* $s \in S$ *such that* f_s *is an r-mixed function with* $r = \Omega(n + \log(p'))$ *and* $p' = 2p(1 - p)$.

Proof. We prove that there is a function f_s such that for all subsets of r variables, the 2^r assignments of these variables lead to different subfunctions. We start with a sketch of the proof: First, we fix a set of l variables (in other words, a variable order) and prove an upper bound on the probability that the number C_l of collisions $f_{s|\alpha} = f_{s|\alpha}$ deviates by a factor of δ_l from the expectation μ_l. Then we choose δ_l in such a way that the probability is smaller than 1 that there exists a set of l variables where the number of collisions is greater than $\delta_l \mu_l$. Now, we can condition on the event that $C_l \leq \delta_l \mu_l$ for every choice of l variables: By means of Lemma 2 we calculate a value of r such that $D_r > 2^r - 1$. Since D_l is an integer for every l, this implies that $D_r = 2^r$. Thus, all 2^r possible subfunctions are different for all choices of r variables which concludes the proof.

For a fixed set of l variables and a variable order whose set of first l variables coincides with these variables, we know from Lemma 2 that the expected value of C_l is at most $\frac{2^{2l}}{2^{n-l} \cdot p'}$. Due to the dependencies, using Markov's inequality is the best we can do to bound the deviation from the expectation. Thus, for $\delta_l > 1$

we have $\mathbf{Pr}\left[C_l \geq \delta_l \cdot \mathbf{E}\left[C_l\right]\right] \leq \dfrac{1}{\delta_l}$. We have to distinguish $\binom{n}{l}$ possibilities to choose the l variables (and the corresponding variable orders). Let $\delta_l := 2 \cdot \binom{n}{l}$. Then the probability that for all choices of l variables C_l is less than $\delta_l \mathbf{E}\left[C_l\right]$ is bounded below by $1/2$. Now, we condition on the event that $C_l < \delta_l \mathbf{E}\left[C_l\right]$ for all sets of l variables. From Lemma 2 we know that $D_l \geq \dfrac{2^l}{\sqrt{2C_l + 1}}$ which is greater than $\dfrac{2^l}{\sqrt{2\delta_l \mathbf{E}[C_l] + 1}} \geq \dfrac{2^l}{\sqrt{2\delta_l \frac{2^{2l}}{2^n \cdot p'} + 1}}$. Due to the space limitations, we omit the exact calculations but it is possible to choose $l \in \Omega\left(n - \log(1/p')\right)$ such that $D_l > 2^l - 1$ which concludes the proof. $\square$

Due to Lemma 1, the last theorem gives us a lower bound of $2^{\Omega(n + \log(p'))}$ even for FBDDs.

4   Almost k-wise Independent Random Functions

The gap between the OBDD size of 3-wise independent random functions and 4-wise independent random functions is exponentially large. In order to see what kind of random functions have an OBDD size which is in-between these bounds, we show that a construction of a random OBDD of size $O((nk)^2/\varepsilon)$ generates (ε, k)-wise independent functions. The idea is to construct a random OBDD with fixed width w. If w is large enough, the function values of k different inputs are almost uniformly distributed because the paths of the k inputs in the OBDD are likely to be almost independent. For $0 \leq i \leq n - 1$ let layer L_i consists of w nodes labeled by x_i and layer L_n be the two sinks. For all $0 \leq i \leq n - 1$ we choose the $0/1$-successors of every node in layer L_i independently and uniformly at random from the nodes in layer L_{i+1}. Then we pick a random node in layer L_0 as the root of the OBDD.

Theorem 5. *For $w \geq k + nk(k + 1)/\varepsilon$ the above random process generates (ε, k)-wise independent random functions.*

Proof. Let $a_1, \ldots, a_k \in \{0,1\}^n$ be k different inputs and p be the probability that the function values of these inputs are $\alpha_1, \ldots, \alpha_k \in \{0,1\}$. Let $P_1, \ldots, P_k$ the k paths of $a_1, \ldots, a_k$ to the layer L_{n-1}, i.e., the paths end in a node labeled by x_{n-1}. Let D_i be the event that the paths $P_1, \ldots, P_i$ end in different nodes. Since the inputs are different, every P_i has to use an edge which is not used by any other path and, therefore, it holds $\mathbf{Pr}\left[D_i \mid D_{i-1}\right] \geq \left(1 - \frac{i-1}{w}\right)^n$ and with it $\mathbf{Pr}\left[D_k\right] = \prod_{i=2}^{k} \mathbf{Pr}\left[D_i \mid D_{i-1}\right] \geq \prod_{i=2}^{k}\left(1 - \frac{i-1}{w}\right)^n$. We have $\prod_{i=2}^{k}\left(1 - \frac{i-1}{w}\right)^n \geq$ $\prod_{i=2}^{k} e^{-\frac{n}{w/i-1}} \geq 1 - \varepsilon$ for $w \geq k + nk(k + 1)/\varepsilon \geq k + nk(k+1)(1/\ln(\frac{1}{1-\varepsilon}))$. If all paths end in different nodes, then the function values of the k inputs are independent and uniformly distributed, i.e., $p \geq 2^{-k} \cdot \mathbf{Pr}\left[D_k\right] \geq 2^{-k} - \varepsilon$ and $p \leq 1 - (1 - 2^{-k}) \cdot \mathbf{Pr}\left[D_k\right] \leq 2^{-k} + \varepsilon$ which completes the proof. $\square$

5 Randomized Implicit Algorithms

We use the construction of 3-wise independent random functions from the section 3 to design a randomized maximal matching algorithm. Here, the main drawback of our random construction is the missing possibility to use different probabilities for the nodes or at least to do in an efficient way. Randomized algorithms for maximal independent set using pairwise independence like in [1] or [26] or for maximal matching with complete independence [21] choose a node/edge with a probability proportional to the node degree. In order to simulate these selections by our construction, we delete each edge with probability $1/2$, store all isolated edges, and repeat this as long as there are nodes with degree greater than 1. Finally, we add the stored isolated edges to the matching. Algorithm 2 shows the whole randomized implicit maximal matching algorithm. We realize the edge deletions of the inner loop in the following way: We construct two 3-wise independent random functions $f_{r_1}(x), f_{r_2}(y)$ using Algorithm 1 and set $F(x, y) = (x > y) \wedge (f_{r_1}(x) \oplus f_{r_2}(y))$. Since $\mathbf{Pr}_{r_1, r_2}[f_{r_1}(x) \oplus f_{r_2}(y) = 1] = \mathbf{Pr}_{r_1, r_2}[f_{r_1}(x) \neq f_{r_2}(y)] = 1/4 + 1/4 = 1/2$ for inputs $x \leq y$ the function $F(x, y)$ deletes such edges as required. Since we are dealing with undirected graphs, we want $F(x, y) = F(y, x)$ for every (x, y). Therefore, we set $F(x, y) = F(x, y) \vee F(y, x)$ and delete the edges with the operation $\chi_E(x, y) = \chi_E(x, y) \wedge F(x, y)$.

Algorithm 2. Randomized implicit maximal matching algorithm

Input: Graph $\chi_E(x, y)$
Output: Maximal matching $\chi_M(x, y)$

    $\chi_M(x, y) = 0$ // Initial matching
    while $\chi_E(x, y) \not\equiv 0$ **do**
        $\chi_{E'}(x, y) = \chi_E(x, y)$
        $NewEdges(x, y) = 0$
        while $\chi_{E'}(x, y) \not\equiv 0$ **do**
            // Construct 3-wise independent random functions (see Algorithm 1)
            $f_{r_1}(x) = RandomFunc(x, n)$ and $f_{r_2}(y) = RandomFunc(y, n)$
            $F(x, y) = (x > y) \wedge (f_{r_1}(x) \oplus f_{r_2}(y))$
            $F(x, y) = F(x, y) \vee F(y, x)$
            $\chi_{E'}(x, y) = \chi_{E'}(x, y) \wedge F(x, y)$ // Delete edges with probability $1/2$
            $T(x) = \exists z, y : (z \neq y) \wedge \chi_{E'}(x, y) \wedge \chi_{E'}(x, z)$ // Update set of nodes with degree > 1
            // Store isolated edges in NewEdges
            $NewEdges(x, y) = NewEdges(x, y) \vee (\chi_{E'}(x, y) \wedge \overline{T(x)} \wedge \overline{T(y)})$
        end while
        $\chi_M(x, y) = \chi_M(x, y) \vee NewEdges(x, y)$ // Add edges to current matching
        $Matched(x) = \exists y : \chi_M(x, y)$
        $\chi_E(x, y) = \chi_E(x, y) \wedge \overline{Matched(x)} \wedge \overline{Matched(y)}$ // Delete edges incident to matched nodes
    end while
    return $\chi_M(x, y)$

We say that an edge $e \in E'$ (before the inner while-loop) survives iff $e \in E'$ after the inner while-loop of algorithm 2.

Lemma 3. *For every $e = \{u, v\} \in E$ with $deg_E(u) > 1$ or $\deg_E(v) > 1$ before the inner while-loop in algorithm 2 the probability that e survives is at least $\frac{1}{8 \cdot (deg_E(u) + \deg_E(v) - 2)}$.*

Proof. Let $e = \{u, v\} \in E$ be an edge before the inner while-loop and R_e be the number of rounds until edge e is deleted. The random bits in each iteration are 3-wise independent and the iterations themselves are completely independent. Thus, the variables R_e are also 3-wise independent. Denote by $N(e) = \{e' \in E \mid e \cap e' \neq \emptyset\}$ the neighborhood of e, i.e., all edges incident to u or v. Then we have $\mathbf{Pr}\,[e \text{ survives}] = \mathbf{Pr}\,[R_e \text{ is unique maximum in } \{R_{e'} \mid e' \in N(e)\}]$. It is easy to see that $\mathbf{Pr}\,[R_e = i] = \left(\frac{1}{2}\right)^i$ for $i \geq 1$. Let $e' \in N(e)$ and $e' \neq e$ and $z \geq 1$ be fixed. Since the R_e are 3-wise independent, we have $\mathbf{Pr}\,[R_{e'} \geq z \mid R_e = z] = \mathbf{Pr}\,[R_{e'} \geq z] = \sum_{i=z}^{\infty} \left(\frac{1}{2}\right)^i = \left(\frac{1}{2}\right)^{z-1}$. Therefore, the probability that there is an edge $e' \in N(e) \setminus e$ with $R_{e'} \geq z$ is at most $\frac{|N(e)|-1}{2^{z-1}}$, i.e., R_e is unique maximum with probability at least $1 - \frac{|N(e)|-1}{2^{z-1}}$. This is greater than 0 for $z \geq \log(|N(e)| - 1) + 2$. Finally, we have

$$\mathbf{Pr}\,[R_e \text{ is unique maximum}] \geq \left(\frac{1}{2}\right)^{\log(|N(e)|-1)+2} \cdot \left(1 - \frac{|N(e)| - 1}{2^{\log(|N(e)|-1)+1}}\right)$$
$$\geq \frac{1}{8 \cdot (deg_E(u) + deg_E(v) - 2)}$$

$\square$

The number of deleted edges for a matching edge (u, v) that is added to the matching is $deg(u) + deg(v) - 2$ if we do not count the matching edge itself. Thus, the expected number of deleted edges is $\Omega(|E|)$ at the end of the outer loop. This gives us the final result.

Theorem 6. *Let $G = (V, E)$ be a graph with N nodes. Algorithm 2 computes a maximal matching in G. All functions used in algorithm 2 depend on at most $3 \log N$ variables. The expected number of operations is $O(\log^3 N)$.*

Application to the Maximal Independent Set Problem. With a similar idea we are able to design a distributed MIS algorithm: Each node v draws a random bit until this bit is 0. Let r_v be the number of bits drawn by node v. We send r_v to all neighbors and include node v to the independent set iff r_v is a local minimum. The expected number of bits for each channel is 1. A similar analysis as before show that we have an maximal independent set after $O(\log N)$ steps in expectation and the overall expected number of bits per channel is $O(\log N)$.

Experimental Results. All algorithms are implemented in C++ using the BDD framework CUDD 2.5.0[1] by F. Somenzi and were compiled with Visual Studio 2013 in the default 32-bit release configuration. All source files, scripts and random seeds will be publicly available[2]. The experiments were performed on a computer with a 2.5 GHz Intel Core i7 processor and 8 GB main memory running Windows 8.1. The runtime is measured by used processor time in seconds

---

[1] http://vlsi.colorado.edu/~fabio/CUDD/
[2] http://ls2-www.cs.uni-dortmund.de/~gille/

and the space usage of the implicit algorithm is given by the maximum SBDD size which came up during the computation, where an SBDD is a collection of OBDDs which can share nodes. Note that the maximum SBDD size is independent of the used computer system. For our results, we took the mean value over 50 runs on the same graph. Due to the small variance of these values, we only show the mean in the diagrams/tables. We omit the maximal matching algorithm by Bollig and Pröger [10] because the memory limitation was exceeded on every instance presented here.

We choose three types of input instances: First, we used our construction from section 4 as an input distribution in the following way: If the 1-sink is chosen with probability p as a successor of nodes in layer L_{n-1} the expected size of $|f^{-1}(x)|$ is $p \cdot 2^n$. For a fixed $N = 2^{17}$, we used p as a density parameter for our input graph and want to analyze how the density influences the running time of the algorithms. Second, we run the algorithms on some bipartite graphs from a real advertisement application within Google[3] [31]. The motivation was to check whether the randomized algorithm is competitive or even better on instances where the maximal matching algorithm by Hachtel and Somenzi (HS) [19] is running very well. Third, we use non-bipartite graphs from [15]. Since HS is designed for bipartite graphs, a preprocessing step computing a bipartition of these graphs are needed to compute a maximal matching (see, e. g., [10]) while our algorithm also works on general graphs.

We used the following implementation of our algorithm denoted by RM. In order to minimize the running time for the computation of the set of nodes with two or more incident edges, we sparsify the graph at the beginning of the outer while loop by deleting each edge with probability $1/2$ and repeating this D times. Initially, we set $D = \log |E|$ and decrease D by 1 at the end of the outer loop. Asymptotically, the running time does not change since after $O(\log N)$ iterations, i. e., $D = 0$, it does exactly the same as the original algorithm. Initial experiments showed that this is superior to the original algorithm.

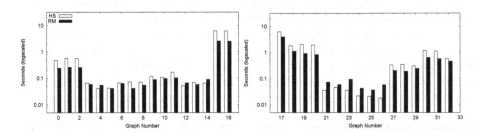

Fig. 1. Running times of HS and RM on the real world instances.

On the random instances the running time and space usage of RM was more or less unaffected by the density of the graph while HS was very slow for small values of p and gets faster with increasing density. For $p \leq 0.2$ RM was much

---

[3] Graph data files can be found at http://www.columbia.edu/~cs2035/bpdata/

Table 1. Running time and space usage of RM on the graphs from [15]

Instance	Nodes	Edges	Time (sec)	Space (SBDD size)
333SP	3712815	22217266	1140.54	66968594
adaptive	6815744	27248640	403.82	22767094
as-Skitter	1696415	22190596	337.53	32020282
hollywood-2009	1139905	113891327	418.36	62253086
roadNet-CA	1971281	5533214	136.18	13177668
roadNet-PA	1090920	3083796	75.26	7633318
roadNet-TX	1393383	3843320	92.62	9125438

faster than HS. In Fig. 1 we see that on the bipartite real world instances RM is similar to HS if the running time is negligibly small but on the largest instances (number 15 to 20) RM is much faster. The graphs from [15] were intentionally chosen to show the potential of RM and indeed do so: It was not possible to run HS on these graphs due to memory limitations whereas RM computed a matching in reasonable time and space (see Table 1). Both graphs from [31] and [15] have very small density and the experiments on the random graphs seem to support the hypothesis that RM is a better choice than HS for such graphs.

References

1. Alon, N., Babai, L., Itai, A.: A fast and simple randomized parallel algorithm for the maximal independent set problem. J. Algorithms 7(4), 567–583 (1986)
2. Alon, N., Goldreich, O., Håstad, J., Peralta, R.: Simple construction of almost k-wise independent random variables. Random Struct. Alg. 3(3), 289–304 (1992)
3. Alon, N., Matias, Y., Szegedy, M.: The space complexity of approximating the frequency moments. J. Comp. and System Sc. 58(1), 137–147 (1999)
4. Awerbuch, B., Goldberg, A.V., Luby, M., Plotkin, S.A.: Network decomposition and locality in distributed computation. In: FOCS, pp. 364–369 (1989)
5. Bloem, R., Gabow, H.N., Somenzi, F.: An algorithm for strongly connected component analysis in $n \log n$ symbolic steps. Formal Meth. in System Design 28(1), 37–56 (2006)
6. Bollig, B.: On symbolic OBDD-based algorithms for the minimum spanning tree problem. Theor. Comput. Sci. 447, 2–12 (2012)
7. Bollig, B., Capelle, M.: Priority functions for the approximation of the metric TSP. Inf. Proc. Letters 113(14-16), 584–591 (2013)
8. Bollig, B., Gillé, M., Pröger, T.: Implicit computation of maximum bipartite matchings by sublinear functional operations. In: Agrawal, M., Cooper, S.B., Li, A. (eds.) TAMC 2012. LNCS, vol. 7287, pp. 473–486. Springer, Heidelberg (2012)
9. Bollig, B., Löbbing, M., Wegener, I.: On the effect of local changes in the variable ordering of ordered decision diagrams. Inf. Proc. Letters 59(5), 233–239 (1996)
10. Bollig, B., Pröger, T.: On efficient implicit OBDD-based algorithms for maximal matchings. Inf. Comput. 239, 29–43 (2014)
11. Bryant, R.E.: Graph-based algorithms for Boolean function manipulation. IEEE Transactions on Computers 35(8), 677–691 (1986)
12. Burch, J.R., Clarke, E.M., McMillan, K.L., Dill, D.L., Hwang, L.J.: Symbolic model checking: 10^{20} states and beyond. Inf. and Comp. 98(2), 142–170 (1992)

13. Chor, B., Goldreich, O.: On the power of two-point based sampling. J. Complexity 5(1), 96–106 (1989)
14. Coudert, O.: Doing two-level logic minimization 100 times faster. In: SODA, pp. 112–121 (1995)
15. Davis, T.A., Hu, Y.: The University of Florida Sparse Matrix Collection. ACM Trans. on Math. Soft. 38(1), 1:1–1:25 (2011)
16. Gentilini, R., Piazza, C., Policriti, A.: Computing strongly connected components in a linear number of symbolic steps. In: SODA, pp. 573–582 (2003)
17. Gentilini, R., Piazza, C., Policriti, A.: Symbolic graphs: Linear solutions to connectivity related problems. Algorithmica 50(1), 120–158 (2008)
18. Gillé, M.: OBDD-based representation of interval graphs. In: Brandstädt, A., Jansen, K., Reischuk, R. (eds.) WG 2013. LNCS, vol. 8165, pp. 286–297. Springer, Heidelberg (2013)
19. Hachtel, G.D., Somenzi, F.: A symbolic algorithms for maximum flow in 0-1 networks. F. Meth. in Sys. Design 10(2/3), 207–219 (1997)
20. Hojati, R., Touati, H., Kurshan, R.P., Brayton, R.K.: Efficient ω-regular language containment. In: Probst, D.K., von Bochmann, G. (eds.) CAV 1992. LNCS, vol. 663, pp. 396–409. Springer, Heidelberg (1993)
21. Israeli, A., Itai, A.: A fast and simple randomized parallel algorithm for maximal matching. Inf. Process. Lett. 22(2), 77–80 (1986)
22. Jukna, S.: Entropy of contact circuits and lower bounds on their complexity. Theor. Comput. Sci. 57, 113–129 (1988)
23. Kabanets, V.: Almost k-wise independence and hard Boolean functions. Theor. Comput. Sci. 297(1-3), 281–295 (2003)
24. Lai, Y., Pedram, M., Vrudhula, S.B.K.: EVBDD-based algorithms for integer linear programming, spectral transformation, and function decomposition. IEEE Trans. on CAD of Int. Circuits and Systems 13(8), 959–975 (1994)
25. Linial, N.: Locality in distributed graph algorithms. SIAM J. Comput. 21(1), 193–201 (1992)
26. Luby, M.: A simple parallel algorithm for the maximal independent set problem. SIAM Journal on Computing 15(4), 1036–1053 (1986)
27. Masek, W.: A fast algorithm for the string editing problem and decision graph complexity. Master's thesis, MIT (1976)
28. Meer, K., Rautenbach, D.: On the OBDD size for graphs of bounded tree- and clique-width. Discrete Mathematics 309(4), 843–851 (2009)
29. Métivier, Y., Robson, J.M., Saheb-Djahromi, N., Zemmari, A.: An optimal bit complexity randomized distributed MIS algorithm. Distributed Computing 23(5-6), 331–340 (2011)
30. Naor, J., Naor, M.: Small-bias probability spaces: Efficient constructions and applications. SIAM J. Comput. 22(4), 838–856 (1993)
31. Negruseri, C.S., Pasoi, M.B., Stanley, B., Stein, C., Strat, C.G.: Solving maximum flow problems on real world bipartite graphs. In: ALENEX, pp. 14–28 (2009)
32. Nunkesser, R., Woelfel, P.: Representation of graphs by OBDDs. Discrete Applied Mathematics 157(2), 247–261 (2009)
33. Savický, P.: Improved Boolean formulas for the Ramsey graphs. Random Struct. Algorithms 6(4), 407–416 (1995)
34. Sawitzki, D.: Implicit flow maximization by iterative squaring. In: Van Emde Boas, P., Pokorný, J., Bieliková, M., Štuller, J. (eds.) SOFSEM 2004. LNCS, vol. 2932, pp. 301–313. Springer, Heidelberg (2004)

35. Sawitzki, D.: The complexity of problems on implicitly represented inputs. In: Wiedermann, J., Tel, G., Pokorný, J., Bieliková, M., Štuller, J. (eds.) SOFSEM 2006. LNCS, vol. 3831, pp. 471–482. Springer, Heidelberg (2006)

36. Sawitzki, D.: Exponential lower bounds on the space complexity of OBDD-based graph algorithms. In: Correa, J.R., Hevia, A., Kiwi, M. (eds.) LATIN 2006. LNCS, vol. 3887, pp. 781–792. Springer, Heidelberg (2006)

37. Sawitzki, D.: Implicit simulation of FNC algorithms. Electronic Colloquium on Computational Complexity (ECCC) 14(028) (2007)

38. Sieling, D., Wegener, I.: NC-algorithms for operations on binary decision diagrams. Parallel Processing Letters 3, 3–12 (1993)

39. Wegener, I.: The size of reduced OBDDs and optimal read-once branching programs for almost all Boolean functions. IEEE Trans. on Comp. 43(11), 1262–1269 (1994)

40. Wegener, I.: Branching programs and binary decision diagrams. In: SIAM Monographs on Discrete Mathematics and Applications (2000)

41. Woelfel, P.: Symbolic topological sorting with OBDDs. J. Disc. Alg. 4, 51–71 (2006)

On Fast and Robust Information Spreading in the Vertex-Congest Model

Keren Censor-Hillel and Tariq Toukan

Technion - Israel Institute of Technology
{ckeren,ttoukan}@cs.technion.ac.il

Abstract. This paper initiates the study of the impact of failures on the fundamental problem of *information spreading* in the Vertex-Congest model, in which in every round, each of the n nodes sends the same $O(\log n)$-bit message to all of its neighbors.

Our contribution to coping with failures is twofold. First, we prove that the randomized algorithm which chooses uniformly at random the next message to forward is slow, requiring $\Omega(n/\sqrt{k})$ rounds on some graphs, which we denote by $G_{n,k}$, where k is the vertex-connectivity.

Second, we design a randomized algorithm that makes dynamic message choices, with probabilities that change over the execution. We prove that for $G_{n,k}$ it requires only a near-optimal number of $O(n \log^3 n/k)$ rounds, despite a rate of $q = O(k/n \log^3 n)$ failures per round. Our technique of choosing probabilities that change according to the execution is of independent interest.

Keywords: distributed computing, information spreading, randomized algorithms, vertex-connectivity, fault tolerance.

1 Introduction

Coping with failures is a cornerstone challenge in the design of distributed algorithms. It is desirable that a distributed system continues to operate correctly despite a reasonable amount of failures, and hence obtaining fault-tolerance has been a fundamental goal in this field. The impact of failures has been studied in various models of computation and for various distributed tasks.

In this paper, we initiate the study of robustness against failures of the task of information spreading in the Vertex-Congest model of computation. Information spreading requires each node of the network to obtain the information of all other nodes. This problem is at the heart of many distributed applications which perform global tasks, and thus is a central issue in distributed computing (see, e.g., [17]). The Vertex-Congest model, where in each round, every node generates an $O(\log n)$-sized packet and sends it to *all* of its neighbours, abstracts the behavior of wireless networks that operate on top of an abstract MAC layer [12] that takes care of collisions.

The time required for achieving information spreading depends on the structure of the communication graph. Even without faults, it is clear that having a

© Springer International Publishing Switzerland 2015
C. Scheideler (Ed.): SIROCCO 2015, LNCS 9439, pp. 270–284, 2015.
DOI: 10.1007/978-3-319-25258-2_19

minimum vertex-cut of size k implies an $\Omega(n/k)$ lower bound for the running time of any algorithm in the above model, and hence our study addresses the k-vertex-connectivity of the graph. The diameter of a graph is a trivial lower bound on the number of rounds required for spreading even without faults, and hence, for k-vertex-connected graphs, $\Omega(n/k)$ is a general lower bound as there exist k-vertex-connected graphs of diameter n/k.

A tempting approach would be to use randomization for choosing which message to forward in each round of communication, in the hope that this would be naturally robust against failures. However, we show that the uniform randomized algorithm is slow on a k-vertex-connected family of graphs, denoted $G_{n,k}$, which consists of n/k cliques of size k that are connected by perfect matchings, requiring $\Omega(n/\sqrt{k})$ rounds.

Instead, this paper presents an algorithm for spreading information in the Vertex-Congest model that uses dynamic probabilities for selecting the messages to be sent in each round. We prove that for $G_{n,k}$, the round complexity of our algorithm is almost optimal and that it is highly robust against node failures.

1.1 Our Contribution

As explained, our first contribution is proving that the intuitive idea of simply choosing at random which message to forward is not efficient. The proof is based on the fact that there is an inverse proportion between the number of received messages in a node and the probability of a message in that node to be chosen and forwarded. The larger the number of messages received in the nodes of a clique, the longer it takes for any newly received message to be forwarded to the nodes of the next clique. The full proof appears in [5, Appendix].

Theorem 1.1. *The uniform random algorithm requires $\Omega(n/\sqrt{k})$ rounds on $G_{n,k}$, in expectation.*

Our main result is an algorithm in which the probabilities for sending messages in each round are not fixed, but rather change dynamically during the execution based on how it evolves. Roughly speaking, the probability of sending a message is set according to the number of times it was received, with the goal of giving higher probabilities for less popular messages. The key intuition behind this approach is that nodes can take responsibility for forwarding messages that they receive few times, while they can assume that messages that have been received many times have already been forwarded throughout the network. This way, we aim to combine qualities of both random and static approaches, obtaining an algorithm that is both fast and robust.

This basic approach alone turns out to be insufficient. It allows each message to be sent fast through multiple paths in the network, but it requires an additional mechanism in order to be robust against failures. Our next step is to augment our algorithm with some additional rounds of communication that allow the paths to change dynamically as the execution unfolds, essentially by-passing faulty nodes. These *shuffle phases* provide fault-tolerance while retaining the efficiency of the algorithm. We consider a strong failure model, in which links are reliable but nodes fail independently with probability q per round and never recover, and prove the following result, which holds with high probability[1].

Theorem 4.3 *Alg. 2 completes full information spreading on $G_{n,k}$ in $O\left(\frac{n}{k}\log^3 n\right)$ rounds, for any node failure probability per round q, $0 \leq q \leq O\left(\frac{k}{n\log^3 n}\right)$, w.h.p.*

While our algorithm is general and does not assume any knowledge of the topology of the network, showing that it is fast and robust for $G_{n,k}$ is important as this graph is basically a k-vertex-connected generalization of a simple path. This constitutes a first step towards understanding this key question. By making minor changes to $G_{n,k}$ we can cover additional graphs with same or similar analysis. We believe that the same approach works for additional families of k-vertex-connected graphs.

1.2   Additional Background and Related Work

One approach for disseminating information that was introduced in [1] and has been intensively studied (e.g. [6, 10, 13, 16]) is *network coding*. Instead of simply relaying the packets they receive, the nodes of a network take several packets and combine them together for transmission. An example is *random linear network coding (RLNC)* presented in [11]. Among its advantages is improving the network's throughput [10]. A conclusion that can be derived from the analysis shown in [9], is that RLNC spreads the information in $\Theta(n/k)$ rounds, w.h.p.

However, network coding requires sending large coefficients, which do not fit within the restriction on the packet size that is imposed in the Vertex-Congest model. An additional disadvantage is derived from the fact that decoding is done by solving a system of linear independent equations of n variables, one variable for each of the original messages. Thus, the decoding process requires the reception of a *sufficient number* of packets by the node, in order to start reproducing the original information. Unfortunately, in most cases, this sufficient number of packets equals the number of original messages, which means that decoding happens only at the end of the process. This issue has supreme importance in applications of broadcasting videos or presentations. For example, when watching online content, one would prefer displaying the downloaded parts of an image immediately on the screen, rather than waiting with an empty screen until the image is fully downloaded.

An almost-optimal algorithm that requires $O(n\log n/k)$ rounds with high probability has been shown in [2]. It is based on a preprocessing stage which constructs vertex-disjoint connected dominating sets (CDSs) which are then used in order to route messages in parallel through all the CDSs. However, this algorithm is non-robust for the following reason. In the basic algorithm the failure of a single node in a CDS suffices to render the entire structure faulty. This sensitivity can be easily fixed by combining $O(\text{polylog}(n))$ CDSs together into

---

[1] We use the phrase "with high probability" (w.h.p.) to indicate that an event happens with probability at least $1 - \frac{1}{n^c}$ for a constant $c \geq 1$.

well-connected components and sending information redundantly over each CDS in the component, incurring a cost of only an $O(\text{polylog}(n))$ factor of slowdown in runtime. Nevertheless, the *construction* itself, of the CDS packings, is highly sensitive to failures. It is an important open problem whether CDS packings can be constructed under faults.

Randomized protocols were designed to overcome similar problems of fault-tolerance in various settings [7, 8], as they are naturally fault-tolerant. The approach taken in this paper, of changing the probabilities of sending messages according to how the execution evolves such that they are inversely proportional to the number of times a message has been received, bears some resemblance and borrows ideas from [4], where a fault-tolerant information spreading algorithm was designed for gossiping, which is a different model of communication. Apart from the high-level intuition, the model of communication and the implementation and analysis are completely different.

1.3 Preliminaries

We assume a network with n nodes that have unique identifiers of $O(\log n)$ bits. Each node u holds one message, denoted m_u. An *information spreading* algorithm distributes the messages of each node in the network to all other nodes.

In the *Vertex-Congest* model, each node knows its neighbours but does not know the global graph topology. The execution proceeds in a sequence of synchronous rounds. In each round, every node generates a packet and sends it to *all* of its neighbours. The packet size is bounded by $O(\log n)$ bits and can encapsulate one message, in addition to some header.

An n-node graph is said to be k-*vertex-connected* if the graph resulting from deleting any (perhaps empty) set of fewer than k vertices remains connected. In this paper we assume that $k = \omega(\log^3 n)$. An equivalent definition [14] is that a graph is k-vertex-connected if for every pair of its vertices it is possible to find k vertex-disjoint paths connecting these vertices.

We consider a strong failure model, in which links are reliable but nodes fail independently with probability q per round and never recover.

2 A Fast Information Spreading Algorithm

In this section, we describe our basic information spreading algorithm. We emphasize that the algorithm does not assume anything about the underlying graph, except for a polynomial bound on its size. In particular, the nodes do not know the vertex-connectivity of the graph, nor any additional information about its topology. Each node u has a set of received messages, whose content at the beginning of round t is denoted $R_u(t)$. We use $cnt_{u,v}(t)$ to denote the number of times a node u has received message m_v by the beginning of round t. Denote by $S_u(t)$ the set of messages sent by node u by the beginning of round t. Define $B_u(t) \equiv R_u(t) - S_u(t)$, the set of messages that are known to node u at the beginning of round t, but not yet sent. We refer to $B_u(t)$ as a logical variable, whose value changes implicitly according to updates in the actual variables

$R_u(t)$ and $S_u(t)$. For every node u, we have that $S_u(0) = \emptyset$, $R_u(0) = \{m_u\}$, $cnt_{u,u}(0) = 1$, and for each $v \neq u$, $cnt_{u,v}(0) = 0$.

We present an algorithm, Alg. 1, that consists of two types of phases: a random phase and ranking phases (see Fig. 3 in [5, Appendix]). Let t_0 be the round number at the beginning of the random phase, and let $\bar{t}_0$ be the round number after the random phase. Let t_p be the round number at the beginning of ranking phase p, and let $\bar{t}_p$ be the round number after ranking phase p, starting from $p = 1$. In this algorithm, it holds that $\bar{t}_p = t_{p+1}$ for every p, and $t_0 = 1$. We will later modify this algorithm in Section 4, where we argue about properties that hold in $\bar{t}_p$ and t_{p+1}, separately. Denote by $\hat{B}_u(t_p)$ the set of node u at time t_p. Unlike $B_u(t)$, $\hat{B}_u(t)$ is an actual variable that does not implicitly change according to $R_u(t)$ and $S_u(t)$. We assign a value to it at the beginning of every phase, that is, $\hat{B}_u(t_p) = B_u(t_p)$, and make sure that its content only gets smaller during a phase. The parameters α and d are constants that are fixed later, at the end of Section 3. The algorithm runs as follows, where in each round every node sends a message and receives messages from all of its neighbors:

(1) Single round (Round 0): This is the first round of the algorithm, where every node u sends the message m_u it has.
(2) Random phase: This is the first phase of the algorithm, which consists of $\tau = \alpha \log n$ rounds. In each round t, every node u picks a message to send from $\hat{B}_u(t_0)$ uniformly at random, and removes it from the set.
(3) Consecutive ranking phases: Each of these phases consists of $\tau' = 8d\tau \log^2 n$ rounds. At the beginning of such a phase, each node uses the Ranking Function (Fig. 1) that defines a probability space over the messages in $\hat{B}_u(t_p)$. In each round, every node u picks a message to send from $\hat{B}_u(t_p)$ according to the probability space, and removes it from the set.

Ranking Function. The ranking function (in Fig. 1) is calculated by each node, and defines a probability space over its messages. Each node u sorts the messages in $\hat{B}_u$ according to their cnt values, smallest to largest, breaking ties arbitrarily. Denote by $rank_m$ the position of the message m within the sorted list, and let $b = |\hat{B}_u|$, be the size of the list. We consider the probability space in which the probability for a message m with $rank_m = r$ to be picked is $\frac{1}{rH_b}$. Namely, the probability is inversely proportional to r. The b-th harmonic number, $H_b = \sum_{i=1}^{b} 1/i$, is a normalization factor (over the whole list of messages). This means that messages in lower positions (lower $rank_m$ values, implying lower cnt values) are more likely to be picked.

Other interesting variants of probability distributions over the messages might work as well. For example, the inverse proportion might be raised to some exponent, and be a function of the cnt values instead of the ranking r. Our ranking function was selected as it is very simple, and fits perfectly in Lemma 2.2. In the algorithm, the probability space used by a node u during a phase is calculated at the start of the phase. In ranking phases, it is defined according to the Ranking function. In the random phase, it is the uniform distribution. Within a phase, the

---

Algorithm 1. for each node u

1: SYNCROUND(m_u)   ▷ Round 0
2: RandomPhase()
3: **loop**
4:     RankingPhase()
5: **end loop**

SyncRound(m)

6: **procedure** SYNCROUND(m)   ▷ A synchronized round
7:     send(m)
8:     $S_u(t) \leftarrow S_u(t) \cup \{m\}$
9:     $R \leftarrow$ received messages
10:    **for all** $m_v \in R$ **do**
11:        $R_u(t) \leftarrow R_u(t) \cup \{m_v\}$
12:        $cnt_{u,v}(t) \leftarrow cnt_{u,v}(t) + 1$
13:    **end for**
14:    $t \leftarrow t + 1$
15: **end procedure**

RandomPhase

16: $\hat{B}_u(t_0) \leftarrow B_u(t)$ ▷ $t = t_0$
17: **loop** τ **times**   ▷ $\tau = \alpha \log n$
18:     $m \leftarrow$ pop message from $\hat{B}_u(t_0)$ uniformly at random
19:     SYNCROUND(m)
20: **end loop**

RankingPhase p

21: $\hat{B}_u(t_p) \leftarrow B_u(t)$ ▷ $t = t_p$
22: $Prob \leftarrow$ RANKINGFUNCTION($\hat{B}_u(t_p)$)
23: **loop** τ' **times**   ▷ $\tau' = 8d\tau \log^2 n$
24:     $m \leftarrow$ pop message from $\hat{B}_u(t_p)$ according to $Prob$
25:     Nullify $Prob[m]$ (update $Prob$ accordingly)
26:     SYNCROUND(m)
27: **end loop**

1: **function** RANKINGFUNCTION(Buffer $\hat{B}_u$)
2:     $mList \leftarrow$ sort $\hat{B}_u$ increasingly according to cnt values
3:     $b \leftarrow$ length($mList$)
4:     **for all** $1 \leq r \leq b$ **do** $Prob[mList[r]] \leftarrow \frac{1}{rH_b}$ **end for**
5:     **return** $Prob$
6: **end function**

Fig. 1. The Ranking Function

only modifications in the probability space of a node are done due to the *non-repetitive* sending policy[2], i.e., the need for nullifying probabilities of messages that are already sent. When a message is sent, the modification can be done, for

---

[2] There is no point in re-sending messages, as all links are reliable.

example, by updating the normalization factor, or alternatively by distributing the probability of the sent message between all other messages (say, proportionally to their current probabilities). Anyhow, this implies that the probability of each message can only get larger during a phase, as long as it is not sent. Namely, the initial probability of a message (at the beginning of a phase) is a lower bound on its probability for the rest of the phase (as long as it is not sent). Probabilities are not defined for messages that were not known at the start of a phase, and were first received during the phase, thus these messages have no chance of being sent until the next phase starts.

The Phase Separation Property. Changes in cnt values during a phase (due to reception of messages) do not affect the probability space of this phase, as it is calculated only at the start of each phase. This implies that messages that are first received by a node after the start of the random phase or a ranking phase have zero probability for being sent during that phase, and can be sent by the node only starting from the next phase, when the probability space is recalculated. We call this *the phase separation property*, and it implies the following:

Proposition 2.1. *At the start of ranking phase p, every message has propagated to a distance of at most $p + 1$.*

The following lemma holds for any node and for a general graph. Its proof appears in [5, Appendix].

Lemma 2.2. *Let m be a message with rank $r \leq 8\tau$ (recall that $\tau = \alpha \log n$), then m is sent during the ranking phase with probability at least $1 - n^{-d}$.*

3 Time Analysis for $G_{n,k}$

Recall that $G_{n,k}$ is the graph that consists of n/k cliques of size k (assume n/k is an integer), with a matching between every two consecutive cliques (see Fig. 2 in [5, Appendix]). Clearly, $G_{n,k}$ is k-vertex-connected.

Additional Definitions. Denote by $\mathcal{C}$ the set of all cliques. Recall the enumeration of the cliques, and denote by C_i clique number i, $i \in \{1, \ldots, \frac{n}{k}\}$. Denote by $C(u)$ the clique that contains node u. A *layer* L is a set of n/k nodes from all distinct cliques that form a path starting in C_1 and ending in $C_{n/k}$. We denote by $\mathcal{L}$ the set of all k layers. The layer $L(u) \in \mathcal{L}$ is the layer that contains node u. Notice that within the same clique, different nodes belong to different layers.

We now analyze the time complexity of the algorithm to spread information over $G_{n,k}$. For simplicity, we analyze the flow of messages from C_j to C_i, where $j \leq i$. The opposite direction of flow and its analysis are symmetric.

Theorem 3.1. *Alg. 1 completes full information spreading on $G_{n,k}$ in $O\left(\frac{n}{k}\log^3 n\right)$ rounds, w.h.p.*

The theorem is directly proved based on Lemma 3.2, as follows.

Lemma 3.2 (Iteration). *For every $i, 1 \leq i \leq \frac{n}{k}$, every node $u \in C_i$, and every node v such that $v \in C_j$ for some $i - p \leq j \leq i$, it holds that $m_v \in R_u(\bar{t}_p)$, w.h.p.*

Proof (Proof of Theorem 3.1). Lemma 3.2 shows that by the end of ranking phase p, w.h.p. each node u knows all messages m_v originating at distance at most p. This implies that full information spreading is completed after n/k phases, since n/k is the diameter of the graph, which proves Theorem 3.1.   □

In the rest of the section we prove Lemma 3.2. The following definition is useful to indicate that a node shares responsibility for disseminating a message.

Definition 3.3 (Fresh message). *A fresh message of a node u at time t, is a message $m_v \in R_u(t)$ for which $cnt_{u,v}(t) < T$, for threshold $T = \frac{1}{2}\tau$.*

General Idea of the Proof. At the end of round 0, every message m_v is disseminated in its own clique $C(v)$. Then, we show that by the end of the random phase, each message m_v is sent w.h.p. by a sufficiently large number of nodes $u \in C(v)$, to become non-fresh in all nodes of the clique $C(v)$. Simultaneously, each of the messages m_v becomes known and fresh in a sufficiently large number of nodes in the neighboring clique.

Then we show that ranking phases shift and preserve this situation. At the beginning of every ranking phase, every fresh message in a node is also fresh in a sufficiently large number of nodes within the same clique. During the phase, all of the fresh messages are sent w.h.p., implying that each one of the messages (i) is disseminated in the clique; (ii) is not fresh in nodes of the clique anymore; and (iii) is fresh in a sufficiently large number of nodes in the neighboring clique.

The combination of properties (ii) and (iii) is the crux of the proof. It guarantees that the process progresses iteratively, as it leads to similar conditions again and again at the beginning of every new ranking phase. This happens because every node can easily distinguish between a new message received from nodes within the clique (becomes non-fresh by the end of the phase), and a new message received from the neighbor in the neighboring clique (stays fresh at the end of the phase, and should be sent during the next phase). We emphasize that all of this is done implicitly, without knowing the structure of the network.

This iterative behavior of the combined properties guarantees that every message propagates one additional clique per phase, until full information spreading completes after $O(n/k)$ phases.

Let t', for $0 \le t' \le \tau-1$, be the time from the first round of the random phase, i.e., $t' = t - t_0$. The following proposition is immediate from the pseudocode:

Proposition 3.4. *At the beginning of the random phase, $\hat{B}_u(t_0)$ for every node $u \in C_i$ contains exactly $k-1$ messages m_v originating at $v \in C_i$, and at most two additional messages, one originating at $v \in C_{i-1} \cap L(u)$, and one originating at $v \in C_{i+1} \cap L(u)$. Thus, it holds that $|\hat{B}_u(t_0+t')| = k+1-t'$, for $i = 2, 3, \cdots, \frac{n}{k}-1$, and $|\hat{B}_u(t_0 + t')| = k - t'$, for $i = 1, \frac{n}{k}$.*

Namely, nodes of inner cliques ($C_i, 1 < i < n/k$) start the random phase with $|\hat{B}_u(t_0)| = k + 1$, while nodes of cliques C_1 and $C_{n/k}$ start the random phase with $|\hat{B}_u(t_0)| = k$.

3.1   Analysis of the Random Phase

The following lemma analyzes the initial random phase, and shows that every message m_v is non-fresh in all nodes of $C(v)$ at the end of the random phase:

Lemma 3.5. *At the end of the random phase, for every message m_v and for all nodes $u \in C(v)$, m_v is non-fresh for u, with probability at least $1 - \frac{1}{n^{\alpha/48-1}}$.*

Proof. Fix v. Message m_v is disseminated in $C(v)$ by the start of the random phase. By Proposition 3.4, for every $u \in C(v)$, it holds that $|\hat{B}_u(t_0 + t')| \le k + 1 - t'$ during the random phase.

Let $\mathbb{1}_{u,v}$, for every $u \in C(v)$, be an indicator variable that indicates whether node u sends m_v during the random phase or not. Then

$$\Pr[\mathbb{1}_{u,v} = 1] \ge 1 - \prod_{t'=0}^{\tau-1} \frac{k - t'}{k + 1 - t'} = 1 - \frac{k + 1 - \tau}{k + 1} \ge \frac{\tau}{(3/2)k} \ .$$

Let $X_v = \sum_{u \in C(v)} \mathbb{1}_{u,v}$, be the number of nodes in $C(v)$ that send m_v during the random phase, i.e., the number of times m_v is received by every node in $C(v)$. Then

$$\mu = E(X_v) = E\left(\sum_{u \in C(v)} \mathbb{1}_{u,v} \right) = \sum_{u \in C(v)} E(\mathbb{1}_{u,v}) \ge \sum_{u \in C(v)} \frac{2\tau}{3k} = \frac{2\tau}{3} \ .$$

Since v is fixed, the indicator variables are independent, as they refer to decisions of distinct nodes. By applying a Chernoff bound [15, Chapter 4], we get

$$\Pr[X_v \le (1 - \delta)\mu] \le \exp\left(-\delta^2 \mu/2\right) \le \exp\left(-\delta^2 \alpha \log n/3\right) < 1/n^{\frac{\alpha\delta^2}{3}} \ .$$

By setting $\delta = \frac{1}{4}$, we get that a message m_v is non-fresh in all nodes $u \in C(v)$ with probability at least $1 - \frac{1}{n^{\alpha/48}}$. By a union bound, this holds for every node v with probability at least $1 - \frac{1}{n^{\alpha/48-1}}$. $\qquad\square$

Definition 3.6. *A pioneer message in node $u \in C_i$ at time t_p (beginning of ranking phase p), is a message $m_v \in R_u(t_p)$ that originated at $v \in C_{i-p-1}$.*

Pioneer Attributes. If a message m_v is a pioneer in node $u \in C_i$ at time t_p, then (i) $v \in L(u)$ (by Proposition 2.1, the message was transmitted over the shortest path), and the following hold at time t_p: (ii) $cnt_{u,v}(t_p) = 1$, and thus m_v is fresh for u, (iii) $m_v \notin R_{u'}(t_p)$ for every $u' \in C_i, u' \ne u$ (by Proposition 2.1), (iv) m_v is disseminated in C_{i-1} (by the node that relayed m_v to its neighbor in C_i), and (v) m_v is fresh in every node $u' \in C_{i-1}$. The following is proved in [5, Appendix].

Lemma 3.7. *With probability at least $1 - 1/n^{\alpha/24-1}$, at the end of the random phase, for every i, the number of pioneer messages that reach C_i is $\le 3\tau$.*

3.2   Analysis of Ranking Phases

After analyzing the single random phase, here we analyze the ranking phases.

Lemma 3.8. *With probability at least $1 - \frac{1}{n^{d-2}}$, every node u that starts ranking phase p with at most 8τ fresh messages, sends all of them during the phase.*

The proof appears in [5, Appendix]. To prove Lemma 3.2, we show a sequence of four inductive properties, that hold for ranking phase p, with probability at least $1 - \left(\frac{2p}{n^{d-2}} + \frac{2}{n^{\alpha/48-1}} \right)$.

Property 1. For every $i, 1 \leq i \leq \frac{n}{k}$, it holds that the number of messages m_v, $v \in C_{i-p-1}$, such that $m_v \in R_u(t_p)$ for some $u \in C_i$ (pioneers), is at most 3τ, and each reaches a distinct node $u \in L(v)$.

Property 2. For every $i, 1 \leq i \leq \frac{n}{k}$, and every node $u \in C_i$, it holds that at time t_p there are at most 4τ *fresh* messages m_v for node u for every one of the two directions of flow (8τ in total). All of them originated at nodes $v \in C_{i-p}$ (similarly, $v \in C_{i+p}$), except for at most one (a pioneer) which originated at $u' \in C_{i-p-1} \cap L(u)$ (similarly, $u' \in C_{i+p+1} \cap L(u)$). All messages $m_v \in R_u(t_p), v \in C_{i-p}$ (similarly, $v \in C_{i+p}$), are fresh.

Property 3. For every $i, 1 \leq i \leq \frac{n}{k}$, and every node $v \in C_{i-p}$, it holds that m_v is fresh for at least T nodes $u \in C_i$ at time t_p. Recall that $T = \tau/2$.

Property 4. For every $i, 1 \leq i \leq \frac{n}{k}$, every node $u \in C_i$, and every node v such that $v \in C_j$ for some $i - p \leq j \leq i$, it holds that $m_v \in R_u(\bar{t}_p)$, and m_v is non-fresh.

We prove the four properties simultaneously by induction on the ranking phase number, p. To prove the base cases, we assume that all events described in Lemma 3.5, Lemma 3.7, and Lemma 3.8 (for $p = 1$) occur. Notice that, by a union bound, the probability for this is at least $1 - \left(\frac{1}{n^{\alpha/24-1}} + \frac{1}{n^{\alpha/48-1}} + \frac{1}{n^{d-2}} \right) \geq 1 - \left(\frac{2}{n^{\alpha/48-1}} + \frac{2}{n^{d-2}} \right)$.

To prove the induction step, we assume that all events described in the four properties for $p-1$, and in Lemma 3.8 for $p-1$ and p, occur. This happens with probability at least $1 - \left(\frac{2}{n^{\alpha/48-1}} + \frac{2(p-1)}{n^{d-2}} + \frac{1}{n^{d-2}} + \frac{1}{n^{d-2}} \right) = 1 - \left(\frac{2}{n^{\alpha/48-1}} + \frac{2p}{n^{d-2}} \right)$.

The complete inductive proof appears in [5, Appendix]. Property 4 guarantees that full information spreading is completed after ranking phase $p = n/k$, with probability at least $1 - \left(\frac{2n/k}{n^{d-2}} + \frac{2}{n^{\alpha/48-1}} \right) \geq 1 - \left(\frac{1}{n^{d-3}} + \frac{1}{n^{\alpha/48-2}} \right) \geq 1 - \frac{1}{n^c}$, for a constant c, by fixing d and α to values $d > c + 3, \alpha > 48c + 96$. This completes the proof of Lemma 3.2, from which Lemma 3.1 follows.

4   Fault Tolerance

Alg. 1 highly depends on the random phase in the following sense. For every node v, consider the set of nodes in neighboring cliques that know message m_v

by the end of the random phase. Then, w.h.p. the algorithm spreads m_v using the layers of nodes in the above set ("carriers"). This means that the paths of a message are fixed very early in the algorithm and do not alternate.

A single failure of a node in each layer (carrier) is sufficient to break down its role. Each message relies on at least T different layers to proceed. Hence, the algorithm is sensitive to failures in which less than T carrier layers are non-faulty.

At the beginning of ranking phase p, consider the case where a message $m_v \in C_{i-p}$ is fresh in $x < T$ nodes in clique C_i, due to failures. The behavior of the algorithm in such case is as follows: During the ranking phase, less than T nodes in the clique send the message, so all other nodes in C_i receive the message less than T times, thus it stays fresh in all of them at the end of ranking phase p. Starting from the next ranking phase, the message m_v propagates regularly over those $x < T$ carriers, but also propagates over all other carriers, with a delay of a phase. This means that every layer becomes responsible for one extra message (in addition to at most 8τ messages), which may still be tolerable. In general, our algorithm can manage a constant number of such occurrences.

We aim to cope with a larger number of failures, so we modify our algorithm to help layers bypass their failing nodes, so they continue operating as carriers.

4.1   Shuffle Phases

We invoke a shuffle phase between every two ranking phases, so phases of the algorithm now proceed as described in Fig. 4 in [5, Appendix]. Roughly speaking, the objective of a shuffle phase, is that nodes of every clique re-divide their responsibilities over messages.

A shuffle phase consists of 8τ rounds. During it, every node sends its fresh messages (and receives fresh messages from all neighbors). Instead of updating the regular *cnt* values, nodes use separate counters, *phasecnt*, to count the number of receptions for each message during the current shuffle phase. Recall that the objective is shuffling the fresh messages between nodes of same clique. Thus, at the end the of the shuffle phase, every node identifies and filters out unwanted messages, which are messages received from neighboring cliques (low *phasecnt* values), and messages that were already non-fresh prior to the start of the shuffle phase. Then it randomly picks 4τ new fresh messages, to start the next ranking phase with.

The important gain from this cooperative division of responsibilities done by the nodes of a clique, is that a node $u \in C_i$ that does not receive new messages from its faulty neighbor $u' \in C_{i-1} \cap L(u)$, can overcome the failure of the carrier layer, and still take part in transmitting relevant messages from one clique to the other, with no delays. The proof of the following appears in [5, Appendix].

Theorem 4.1. *Alg. 2 completes full information spreading on* $G_{n,k}$ *in* $O\left(\frac{n}{k}\log^3 n\right)$ *rounds, w.h.p.*

4.2   Resilience to Faults

Recall that we consider a model of independent failures of nodes, where each node fails at each round with probability q, and never recovers. Let $\tau_e \leq 2\frac{n}{k}\tau' =$

Algorithm 2. for each node u

1: SYNCROUND(m_u)   ▷ Round 0
2: RandomPhase()
3: **loop**
4:     RankingPhase()
5:     ShufflePhase()
6: **end loop**

ShufflePhase p

7: $\hat{B}_u(\bar{t}_p) \leftarrow$ fresh messages in $B_u(t)$ ▷ $t = \bar{t}_p$
8: **for all** $m_v \in \hat{B}_u(\bar{t}_p)$ **do**
9:     $phasecnt_{u,v} \leftarrow 1$
10: **end for**
11: $R \leftarrow \hat{B}_u(\bar{t}_p)$
12: **loop** 8τ **times**
13:     **if** $\hat{B}_u(\bar{t}_p) = \emptyset$ **then**
14:         send own message m_u
15:     **else**
16:         pop and send a fresh message from $\hat{B}_u(\bar{t}_p)$
17:     **end if**
18:     $R' \leftarrow$ receive messages
19:     **for all** $m_v \in R'$ **do**
20:         **if** $m_v \notin R$ **then**
21:             $phasecnt_{u,v} \leftarrow 1$
22:         **else**
23:             $phasecnt_{u,v} \leftarrow phasecnt_{u,v} + 1$
24:         **end if**
25:         $R \leftarrow R \cup \{m_v\}$
26:     **end for**
27:     $t \leftarrow t + 1$
28: **end loop**
29: $R \leftarrow R$ after filtering out unwanted messages.   ▷ Filter out messages m_v with $phasecnt_{u,v} < \hat{c} \cdot T$ ▷ Filter out messages that were non-fresh prior to the start of the phase
30: $R_u(t) \leftarrow R_u(t) \cup R$
31: Select 4τ messages from R randomly, rank them from 1 to 4τ.

$O\left(\frac{n}{k} \log^3 n\right)$ (the round number at the end of ranking phase n/k in Alg. 2). First, we prove the following. The proof appears in [5, Appendix].

Lemma 4.2. *At the end of round τ_e, the number of non-faulty nodes in each clique is at least $(30k/32)$, with probability at least $1 - 1/n^{30}$.*

We show that the algorithm tolerates failures for q, $0 \leq q \leq O\left(\frac{k}{n \log^3 n}\right)$.

Theorem 4.3. *Alg. 2 completes full information spreading on $G_{n,k}$ in $O\left(\frac{n}{k} \log^3 n\right)$ rounds, for any node failure probability per round q, $0 \leq q \leq O\left(\frac{k}{n \log^3 n}\right)$, w.h.p.*

Proof. Fix i, p. Let m_v be a message that is fresh in at least T (non-faulty) nodes in C_{i-1} at the end of shuffle phase $p - 1$. Here we analyze the probability that m_v is *not* shuffled successfully in clique C_i.

An unsuccessful shuffle might occur either because the *phasecnt* values in C_i at the end of shuffle phase p are smaller than the threshold of $T^* = \hat{c}T$, so the message is filtered out (denote this event by A), or because the message was selected by less than T (non-faulty) nodes. By Lemma 3.8, at the beginning of shuffle phase p, the message m_v is supposed to be fresh in at least T nodes in C_i (each of them gets the message from its respective neighbor in C_{i-1}). Of these nodes in C_i, if one does not send m_v during shuffle phase p, then either the node or its neighbor in C_{i-1} (or both) becomes faulty by the end of shuffle phase p. The probability $\hat{q}$ for such a pair of nodes *not* to fail is bounded from below (according to Bernoulli's inequality) by $\hat{q} = ((1 - q)^{\tau_e})^2 \geq (1 - q\tau_e)^2 \geq 1 - 2q\tau_e \geq 1 - 1/16$.

Fix a set of T pairs of nodes $S(m_v) \subseteq C_{i-1} \times C_i$, of those who know message m_v in C_{i-1} at the end of shuffle phase $p - 1$, and their respective neighbors in C_i. There might exist more than T such pairs, but by fixing a set of size T and ignoring the rest, we bound the probability of an unsuccessful shuffle from above, as the ignored nodes can only help and increase the probability of success. A "surviving" pair is a pair of nodes from $S(m_v)$ where both are non-faulty at the end of the shuffle phase, and hence function properly (by sending message m_v) during shuffle phase p. Denote by s, the number of "surviving" pairs. We have:

$$\Pr[A] \leq \sum_{s=0}^{T^*-1} \binom{T}{s} \hat{q}^s (1 - \hat{q})^{T-s} \leq \sum_{s=0}^{T^*-1} \binom{T}{s} (1 - \hat{q})^{T-s} \leq \sum_{s=0}^{T^*-1} \binom{T}{s} \left(\frac{1}{16}\right)^{T-s}.$$

We sum over all $s \in \{0, \ldots, T^* - 1\}$, where the number of "survivors" is lower than the threshold of $\hat{c}T$, which implies that the message m_v is filtered out, improperly, at the end of the shuffle phase due to a low *phasecnt* value.

By setting $0 < \hat{c} \leq \frac{1}{2}$, we get that $\Pr[A] \leq 1/n^{\alpha/3-1}$ (see calculation in [5, Appendix]). Namely, the message is not filtered out with probability at least $1/n^{\alpha/3-1}$. The number of non-faulty nodes in each clique is at least $31k/32$ with probability at least $1 - \frac{1}{n^{30}}$, by Lemma 4.2. An analysis similar to the one in the proof of Lemma 6.3 in [5, Appendix] (with $\delta = 11/15$) gives that, once the message is not filtered out, it is selected by at least T of the non-faulty nodes in C_i with probability at least $1 - 1/n^{11^2\alpha/(15\cdot16)}$. In total, by using a union bound, a message is not shuffled successfully between two consecutive shuffle phases with probability at most $\frac{1}{n^{\alpha/3-1}} + \frac{1}{n^{11^2\alpha/(15\cdot16)}} + \frac{1}{n^{30}} \leq \frac{1}{n^6}$ (for value of α fixed earlier).

We use union bound two more times, for all messages and for all phases, and get an upper bound for the probability that a message is not propagated properly, of $\frac{1}{n^4}$. This proves that the algorithm tolerates failures that occur with probability $0 \leq q \leq \frac{1}{32\tau_e}$ in the given model, with probability at least $1 - \frac{1}{n^4}$. $\square$

5   Discussion

Static-Routes Algorithms. Let ALG be an algorithm that spreads information on k-vertex-connected graphs in $O\left(\frac{n}{k} \cdot \text{polylog}(n)\right)$ rounds, by constructing static routes, and using them to disseminate messages in parallel, each message on a specific route. This makes ALG very sensitive to failures, as a single failure in a route suffices to render the entire route faulty.

However, it can easily be configured so that vertex-disjoint routes are combined into groups of size γ, and every node duplicates its messages and sends them concurrently over these components. Notice that in k-vertex-connected graphs, γ is bounded from above by k. This costs γ slowdown in runtime as a trade-off. Denote this configuration of the algorithm by $ALG(\gamma)$.

We are interested in cases where $\gamma = O(\text{polylog}(n))$, so that the runtime of the algorithm remains $O\left(\frac{n}{k} \cdot \text{polylog}(n)\right)$. Every combination of γ vertex-disjoint routes induces a γ-vertex-connected subgraph, as it stays connected after the removal of any $\gamma - 1$ vertices. Each component functions as long as it stays connected. According to [3, Theorem 1.5], for $\gamma = \Omega(\log^3 n)$, such a component stays connected w.h.p. if its nodes are *sampled* independently with a constant probability. By considering the sampling process imposed by failures, i.e. considering the non-faulty nodes as sampled, then each component stays connected if a constant fraction of its nodes stays non-faulty during the execution, tolerating a constant fraction of nodes that fail. The additional slowdown factor for each message to spread over such a component in the presence of faults can be loosely bounded form above by $O(\gamma)$, as the size of the combined component is $O(\gamma)$ the size of its original routes, (in the worst case a message traverses over all non-faulty nodes of the component). In total, this configuration of the algorithm tolerates the failure of a constant fraction of nodes during its execution, which matches a probability of failure of $q = O\left(\frac{k}{n \cdot \text{polylog}(n)}\right)$ per round, while preserving a time complexity of $O\left(\frac{n}{k} \cdot \text{polylog}(n)\right)$.

The algorithm presented in [2] is static-route, as it constructs CDS packings and routes messages over them. The CDS packings are only fractionally vertex-disjoint, which requires a few modifications to the above analysis. However, despite the above fix, the algorithm remains vulnerable due to the preprocessing stage. Tolerating failures that occur during the preprocessing stage is more complicated, and the construction of CDS packings in the presence of failures is still an open problem.

Summary. In this paper, we show an information spreading algorithm, and prove that it is fast and robust for $G_{n,k}$. The intriguing open question is whether this approach can work for general k-vertex-connected graphs.

To summarize, we find the question of devising a fast and robust information spreading algorithm in the Vertex-Congest model an intriguing open question, and view our result as a first step in this direction. The technique our algorithm leverages, of using probability distributions that change over time according to how the execution unfolds, may have applications in other settings as well.

Acknowledgements. Keren Censor-Hillel is a Shalon Fellow. This research is supported by the Israel Science Foundation (grant number 1696/14). We thank Mohsen Ghaffari, Fabian Kuhn, Yuval Emek and Shmuel Zaks for useful discussions.

References

1. Ahlswede, R., Cai, N., Li, S.Y., Yeung, R.W.: Network information flow. IEEE Transactions on Information Theory 46(4), 1204–1216 (2000)
2. Censor-Hillel, K., Ghaffari, M., Kuhn, F.: Distributed connectivity decomposition. In: Proceedings of the 33rd ACM Symposium on Principles of Distributed Computing, PODC, pp. 156–165 (2014)
3. Censor-Hillel, K., Ghaffari, M., Kuhn, F.: A new perspective on vertex connectivity. In: Proceedings of the Twenty-Fifth Annual ACM-SIAM Symposium on Discrete Algorithms, SODA, pp. 546–561 (2014). http://epubs.siam.org/doi/abs/10.1137/1.9781611973402.41
4. Censor-Hillel, K., Giakkoupis, G.: Fast and robust information spreading (2012) (unpublished manuscript)
5. Censor-Hillel, K., Toukan, T.: On fast and robust information spreading in the vertex-congest model (2015). http://arxiv.org/abs/1507.01181
6. Deb, S., Médard, M., Choute, C.: Algebraic gossip: A network coding approach to optimal multiple rumor mongering. IEEE Transactions on Information Theory 52(6), 2486–2507 (2006)
7. Elsässer, R., Sauerwald, T.: Cover time and broadcast time. In: Proceedings of the 26th International Symposium on Theoretical Aspects of Computer Science, STACS, pp. 373–384 (2009)
8. Feige, U., Peleg, D., Raghavan, P., Upfal, E.: Randomized broadcast in networks. Random Structures & Algorithms 1(4), 447–460 (1990)
9. Haeupler, B.: Analyzing network coding gossip made easy. In: Proceedings of the 43rd Annual ACM Symposium on Theory of Computing, STOC, pp. 293–302 (2011)
10. Ho, T., Koetter, R., Medard, M., Karger, D.R., Effros, M.: The benefits of coding over routing in a randomized setting. In: Proceedings of the IEEE International Symposium on Information Theory, p. 442 (2003)
11. Ho, T., Médard, M., Koetter, R., Karger, D.R., Effros, M., Shi, J., Leong, B.: A random linear network coding approach to multicast. IEEE Transactions on Information Theory 52(10), 4413–4430 (2006)
12. Kuhn, F., Lynch, N., Newport, C.: The abstract MAC layer. Distributed Computing 24(3-4), 187–206 (2011). http://dx.doi.org/10.1007/s00446-010-0118-0
13. Li, S.Y., Yeung, R.W., Cai, N.: Linear network coding. IEEE Transactions on Information Theory 49(2), 371–381 (2003)
14. Menger, K.: Zur allgemeinen kurventheorie. Fundamenta Mathematicae 10(1), 96–115 (1927)
15. Mitzenmacher, M., Upfal, E.: Probability and computing: Randomized algorithms and probabilistic analysis. Cambridge University Press (2005)
16. Mosk-Aoyama, D., Shah, D.: Information dissemination via network coding. In: 2006 IEEE International Symposium on Information Theory, pp. 1748–1752. IEEE (2006)
17. Peleg, D.: Distributed Computing: A Locality-Sensitive Approach. SIAM (2000)

Information Spreading
by Mobile Particles on a Line

Jurek Czyzowicz[1], Evangelos Kranakis[2], Eduardo Pacheco[3],
and Dominik Pająk[4]

[1] Université du Québec en Outaouais, Gatineau, Québec J8X 3X7, Canada
[2] Carleton University, Ottawa, Ontario K1S 5B6, Canada
[3] McGill University, Montreal, Quebec, H3A 0G4, Canada
[4] University of Cambridge, CB3 0FD, UK

Abstract. A set of identical particles is deployed on an infinite line.
Each particle moves freely on the line at arbitrary but constant speed.
When two particles come into contact they bounce acquiring new velocities according to the law of mechanics for elastic collisions. Each particle
initially holds a piece of information. The meeting particles automatically transmit to each other their entire currently possessed information
(i.e., the initial one and the one accumulated by means of previous collisions). Due to the fact that the number of collisions in this setting is
finite [1] communication cannot last forever. This raises some interesting
questions which we address in this paper: Will particle p_j ever obtain the
initial information of p_i? Are colliding particles able to perform broadcasting, convergecast, or gossiping?

We establish necessary and sufficient conditions for any pair of particles to communicate as well as those needed to achieve gossiping, convergecast, and broadcasting. Although these conditions clearly depend
on the initial ordering of the particles along the line, we prove that they
are independent of their starting positions. Further, we show how to efficiently decide whether some of the aforementioned communication primitives can take place. Finally we explain how to compute the necessary
time to carry out all these communication protocols and we describe a
relationship between our problem and an important, longstanding open
question in computational geometry.

Keywords and Phrases: Mobile agents, passive mobility, particles,
communication, broadcasting, convergecast, gossiping, synchronous systems, elastic collisions.

1 Introduction

Mobile agents are autonomous entities that are capable of *sensing*, i.e., ability
to perceive some parameters of the environment, *communication* - ability to
transmit information to other agents, *mobility* - ability to move within their
environment of deployment, and *computation* - ability of processing their data.
Mobile agents usually function in a distributed way, i.e., a collection of mobile

© Springer International Publishing Switzerland 2015
C. Scheideler (Ed.): SIROCCO 2015, LNCS 9439, pp. 285–298, 2015.
DOI: 10.1007/978-3-319-25258-2_20

agents is deployed across a territory and they interact in order to perform a common task.

Frequently, systems of autonomous mobile agents operate in large collections of cheap, tiny, simple entities with very restricted capabilities, mainly due to the limited production cost, size and battery power. These groups of mobile agents, called *swarms*, often perform exploration or monitoring tasks in hard to access or hazardous environments. Other attributes sometimes assumed by mobile agents which need to be produced in massive amounts include anonymity, negligible dimensions, no explicit communication, no common coordinate system, etc. (cf. [2]).

In some research papers it is assumed that the agents are *passively mobile*, i.e., they have no control on their motion, which is exclusively determined by their interaction with the environment (cf. *population protocols* introduced in [3, 4], where extremely resource-limited mobile agents were assumed). In the present paper the agents are represented by mobile particles of equal mass, which are also passively mobile on an infinite line. When two of them collide, their velocities automatically change according to the laws of classical mechanics for elastic collisions. Therefore, two particles exchange their velocities when they meet. We assume that each particle initially possesses a piece of data. During a collision, the interacting particles pass on to each other all the data they have collected until that moment, i.e. their initial data as well as the data accumulated during the previously occurring collisions. Consequently, since the particles have no control on the communication taking place, we can call it *passive communication*. Contrary to the situation in population protocols our process is entirely deterministic and it is possible to compute exactly what the "final state" of the population of particles is.

It is important to note that our particles have absolutely no autonomy, hence it is perhaps inappropriate to call them agents or robots. Although they still possess the capability of mobility, sensing and communication, they are unable to compute or to decide on their future behavior, which is entirely determined by the environment.

We investigate communication protocols in populations of particles. Each particle is initially placed at some position on an infinite line and it is given an initial direction and velocity at which it starts its movement. It was proven in [1] that the number of collisions of elastic particles sliding on an infinite line is finite (although it may be exponential for particles of not necessarily the same mass). Thus, there is a time after which particles stop colliding among themselves, implying that the spreading of information cannot last forever. This raises some fundamental questions which we address in this paper: For two given particles p_i, p_j, will particle p_j eventually obtain the initial information of p_i? What are all the other particles that eventually get the initial information of p_i? Are particles able to perform broadcasting, convergecast, and gossiping?

One may relate our information spreading process to *infection propagation* in the population of particles. When one particle is infected and infection is transmitted by contact, what is the portion of the population which is eventually infected? What is the propagation of infection, i.e., at what moment does the infection reach some given particle? Instead of infection, the process may relate to the progression of some other, positive agent, e.g. curative, detoxifying, fertilizing, etc., whose propagation is made by contact.

As will be seen later, some aspects of our question are related to a fundamental problem in computational geometry. The trajectory of the kth particle corresponds to the k-th level in an arrangement of n lines in the plane. The question had been first asked more than forty years ago and despite a lot of attention devoted to the problem it still remains unsolved.

1.1 Related Work

Despite the simple capabilities of swarms of mobile agents, they have been used to perform tasks, like surveillance and monitoring of hazardous and hard to access environments. In most situations involving such agents, the fundamental research question concerns the feasibility of carrying out a given task [2, 5]. There are numerous papers that study the feasibility of performing tasks by collections of mobile agents such as pattern formation [2, 6–8], gathering [9, 10], and localization [11, 12].

In [3, 4] the authors introduced *population protocols* in order to model wireless sensor networks by extremely limited finite-state computational devices. The agents in population protocols meet at random according to some *interaction graph*. As a result of every meeting the interacting agents change their states correspondingly. Consequently, the agents of population protocols move according to some mobility pattern totally out of their control. This type of movement is called *passive mobility* and is intended to model unstable environments like a flow of water, chemical solutions, wind or unpredictable mobility of agents' carriers (e.g. vehicles or flocks of birds). The convergence of such a process is studied in order to model the eventual behavior of the population of agents. Some recent papers study population protocols in which the interacting agents achieve some specific goals like network construction [13] or determining majority [14]. Clearly, the particles studied in our paper are also subject to passive mobility.

The type of communication allowed for a collection of mobile agents plays an important role in determining the way that agents can interact. Moreover, communication enhances their capabilities and effectiveness. Different models of communication for systems of mobile agents have been studied (cf. [15, 16]) and they can be classified within any of the three following categories: communication via their environment, for instance, via tokens or pebbles that agents are allowed to drop on the environment (e.g. [17, 18]); communication by sensing each other using, for instance, some visibility mechanism (e.g. [7, 9]); and communication by message passing. The main message passing communication problems concern *broadcasting* - when the message of one agent has to reach all other ones, *convergecast* - when the initial information of all agents has to reach one of them and *gossiping* - when each agent has to inform everybody else.

The study of the dynamics of elastic particles sliding on a one dimensional environment has been of great interest in physics for a long time. Much of the work done on this topic has been motivated in order to understand the dynamical properties of gas particles [1, 19–21]. The dynamics emerging from a collection of particles sliding on an infinite line is very rich and not well understood yet [22]. There are, however, some results concerning the total number of elastic collisions of particles of arbitrary masses that move within an infinite line. Sevryuk [1] proved that the number of collisions is upper bounded by $2\left(8n^2(n-1)m_{max}/m_{min}\right)^{n-2}$, where n is the total number of particles and m_{max} and m_{min} are the largest and smallest masses of the particles, respectively. When all particles are of equal mass, at every collision the two particles involved exchange their velocities. So collisions end when all velocities are sorted, in such a setting the number of collisions is upper bounded by n^2. Other results regarding the number of collisions for different dimensions can be found in [23].

A similar model of bouncing particles was considered in [11,12,24], where the authors studied the feasibility of the *localization* task in the cycle and in the segment. The problem of localization consists of each particle determining the starting position (relative to its own position) of all other particles in a finite amount of time. Each particle possessed a clock and a speedometer (hence it was called a robot) but the passive mobility pattern was similar to the one assumed in the present paper.

Some aspects of our paper are related to the problem of computing the kth level of a line arrangement. Given n functions, so that any two of them intersect a bounded number of times, a level of point p is the number of functions strictly above p. Computing the kth level of a set of functions is an extensively studied problem in differential equations and computational geometry and has a close relationship with solving the k-set problem and Davenport-Schinzel sequences [25, 26]. The early papers [27] and [28] showed that the kth level in an arrangement of lines has at most $O(n\sqrt{k})$ and at least $\Omega(n\log k)$ vertices. Major improvements came more than twenty years later when [29] and [30] improved these bounds to $O(nk^{1/3})$ and $n2^{\Omega(\sqrt{\log k})}$. In this paper, we show that computing the trajectory of our kth particle is equivalent to computing the kth level of an arrangement of lines. The number of collisions of this particle equals the combinatorial complexity of the kth level of arrangement.

1.2   Our Results

In Section 3, we establish necessary and sufficient conditions for bouncing particles to *spread* information in a system. Moreover, we give sufficient and necessary conditions for gossiping, convergecast, and broadcasting to take place. We prove that these tasks are independent of the starting positions of the particles. In Section 4, we show how to use the set of velocities of a collection of particles in order to efficiently decide whether or not some of the aforementioned communication primitives take place. Finally, in Section 5, we compute the necessary time to carry out all these communication protocols and we describe a relationship between our problem and the kth level of an arrangement of lines in computational geometry.

Because of the lack of space most proofs have been moved to the Appendix.

2   Preliminaries

Let $\mathcal{S}_n = (\mathcal{H}, \mathcal{V})$ be a collection of n particles $p_1, p_2, \ldots, p_n$ deployed on an infinite line, $\mathcal{L} = (-\infty, \infty)$, with non-zero initial velocities $\mathcal{V} = (v_1, v_2, \ldots, v_n)$, respectively. As in physics, in this paper the notion of velocity includes speed and direction, i.e., $|v_i|$ denotes the initial speed of particle p_i. Let $\mathcal{H} = (h_1, h_2, \ldots, h_n)$ be the initial positions on the line of the particles, we assume that $h_1 < h_2 < \cdots < h_{n-1} < h_n$. Particles update their velocities at the times of their meetings according to the laws of classical mechanics [31] for elastic collisions, since we assume that all particles have equal masses when two of them collide they simply exchange velocities. Similarly as in [1, 19, 22], we assume that no more than two particles may collide (i.e., be at the same position at the same time). We denote by $\nu_i(t)$, the velocity of particle p_i at time t, thus $\nu_i(0) = v_i$.

Each particle p_i initially holds some piece of data (or information) d_i. When two particles collide, they automatically transmit to each other *all the data* that each of them has collected up to that moment.

Fig. 1 shows the diagram of time versus distance to depict the trajectories of a collection of particles and their times of collisions, where the trajectory of a particle r moving with velocity v is depicted by a segment of line of slope $1/v$. As particles never cross each other, their order along the line remains the same forever (besides the meeting points while some of them coincide). Since the particles pairwise exchange velocities, at every moment of time the set of particles $\mathcal{S}_n = (\mathcal{H}, \mathcal{V})$ uses all the velocities $\mathcal{V}$. We can say that in Fig. 1 the lines corresponding to speeds $\mathcal{V}$ form the set of *trains* and that at every moment of time every particle p_i is on some train T_j (and conversely every train contains a particle).

We denote by $\rho(t, i)$ the index of the rightmost particle holding d_i at time t, analogously, $\lambda(t, i)$ denotes the index of the leftmost particle carrying d_i at time t. For every particle p_i we define $M_i = \max\{v_j \in \mathcal{V} | j \leq i\}$ analogously, $m_i = \min\{v_k \in \mathcal{V} | i \leq k\}$. Furthermore, we define $R_i = |\{v_j < M_i | i < j\}|$, and $L_i = |\{v_j > m_i | j < i\}|$.

The *transmission range* of any particle p_i is an interval of particle indices $[a, b] \subseteq [1, n]$ such that for every $j \in [a, b]$, particle p_j receives d_i. A particle p_i *broadcasts* d_i if and only if its transmission range is $[1, n]$. We denote the set of particles of $\mathcal{S}_n$ that perform broadcasting by $\mathcal{B}(\mathcal{S}_n)$. A *convergecast particle* r is a particle that receives every d_i for $1 \leq i \leq n$. *Gossiping* in $\mathcal{S}_n$ takes place if and only if $\mathcal{B}(\mathcal{S}_n) = \{p_1, \ldots, p_n\}$.

3   Transmission Range of Bouncing Particles

Sevryuk [1] proved that the number of collisions in a system of elastic particles is finite. For any collection of particles $\mathcal{S}_n$ there exists a minimal time moment $t^\star$

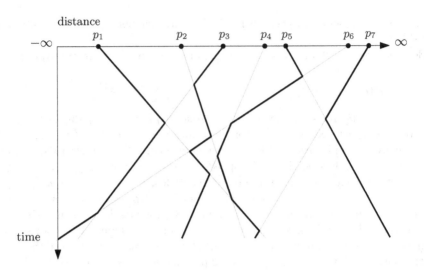

Fig. 1. The trajectories of a collection of seven particles. The thick polylines correspond to the trajectories of p_1, p_3, p_5 and p_7, respectively.

such that for any $t > t^*$ no more collisions take place among the particles of $\mathcal{S}_n$. We call t^* the *expansion time* of $\mathcal{S}_n$. The following lemma follows immediately.

Lemma 1. *After the time of expansion of any collection of particles all the particles are sorted by their velocities.*

Since after the time of expansion no more collisions can take place, any transmission of information among particles must happen before the system expands. The following lemma is a consequence of Lemma 1.

Lemma 2. *For any collection of particles $\mathcal{S}_n$, if particle p_i at some moment of its trajectory is on train T_j for $i \geq j$ $(j > i)$, then p_i received all information d_k for $j \leq k \leq i$ (resp. $i \leq k \leq j$).*

Proof. Let T denote the polyline corresponding to the trajectory of the particle p_i. If p_i enters train T_j at time t then T intersects the line of train T_j. The particle p^* traveling on T_j must received data d_j. However the line of every train T_k for $j < k < i$ must either intersect T or T_j, hence at time t either p^* or p_i contains d_k.   ∎

The initial information d_i is spread through the bounces to successive particles left and right to p_i.

Lemma 3. *The speeds of the rightmost particles $p_{\rho(t,i)}$ holding d_i never decrease in time, i.e., $\nu_{\rho(t_1,i)}(t_1) \leq \nu_{\rho(t_2,i)}(t_2)$ for $t_1 \leq t_2$. Moreover, eventually it acquires velocity M_i, i.e., there exists a time t' such that $\nu_{\rho(t',i)}(t') = M_i$.*

Proof.

It is enough to look at the velocity of the rightmost particle holding d_i at the time moments of its collisions. Consider a collision at time t'' of $p_{\rho(t,i)}$ with its neighbor p which has velocity u just before time t''. If $u > \nu_{\rho(t,i)}(t)$, clearly $\rho(t,i) = \rho(t'',i)$ and after the collision $p_{\rho(t'',i)}$ moves with velocity u. In case of $u < \nu_{\rho(t,i)}(t)$ this can only happen if p is to the right of $p_{\rho(t,i)}$. Thus, at the time of collision p gets d_i and starts moving with velocity $\nu_{\rho(t,i)}(t)$, thus $p_{\rho(t'',i)} = p$. Therefore, the first part of the lemma holds. For the second part of the lemma, let us assume that $\nu_{\rho(m,i)}(m) \neq M_i$ for all m, in particular for $m = t^*$, the time of expansion of the system. Hence, there exists some particle p_j such that $\nu_j(t^*) = M_i$, where $j < \rho(t^*,i)$. This contradicts Lemma 1, since $M_i \geq \nu_{\rho(t^*,i)}(t^*)$. Fig. 2 illustrates this lemma. ∎

Analogously for the leftmost particle we have that the following lemma holds

Lemma 4. *The speeds of the leftmost particles $p_{\lambda(t,i)}$ holding d_i never increase in time, i.e., $\nu_{\lambda(t_1,i)}(t_1) \geq \nu_{\lambda(t_2,i)}(t_2)$ for $t \leq t_2$. Moreover, eventually it acquires velocity m_i, i.e., there exists a time t' such that $\nu_{\lambda(t',i)}(t') = m_i$.*

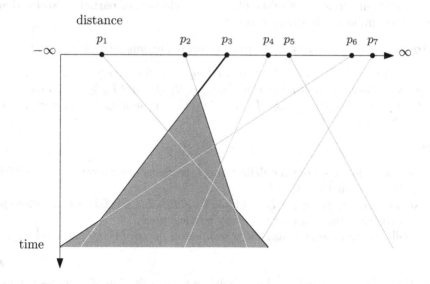

Fig. 2. The bold polylines depict the spread of d_3 among particles.

Notice that $\rho(t^*,i)$ and $\lambda(t^*,i)$ will determine the transmission range of particle p_i since at the time of the expansion of the system no more collisions take place. We denote by $Max_i(t)$ the $\max\{\nu_j(t) | \; j \leq i\}$, i.e., the maximum of the velocities of the particles to the left (including $\nu_i(t)$) of p_i at time t. Analogously, $Min_i(t)$ denotes $\min\{\nu_j(t) | \; j \geq i\}$.

Lemma 5. $p_{\rho(t,j)}$ *(respectively $p_{\lambda(t,j)}$) transfers d_j to its neighbor $p_{\rho(t,j)+1}$ (respectively $p_{\lambda(t,j)-1}$) if and only if $Min_{\rho(t,j)}(t) \leq Max_{\rho(t,j)}(t)$ (respectively $Max_{\lambda(t,j)}(t) \geq Min_{\lambda(t,j)}(t)$).*

Recall that $R_i = |\{v_j < M_i | i < j\}|$ and $L_i = |\{v_j > m_i | j < i\}|$. The following lemma establishes the transmission range of a particle.

Lemma 6. *Information d_i is transferred only to particles $p_{i-L_i}, \ldots, p_{i+R_i}$.*

Proof. Notice that $p_i = p_{\rho(0,i)}$ and let us consider first the transmissions of d_i to successive particles to the right of p_i. Lemma 5 and Lemma 3 guarantee that these changes of successive particles happen exactly R_i times. At time t of the R_ith transmission of d_i to some particle to the right of p_i (which corresponds to the R_ith update of the rightmost particle), all particles p_j such that $j > i+R_i$, it holds that $\nu_j(t) \geq M_i$, thus no more transmission of d_i can take place. Moreover $\rho(t,i) = i + R_i$.

The proof for the transmission of d_i to the left of p_i (eventually reaching p_{i-L_i} is analogous.   ∎

Lemma 6, establishes the transmission range of p_i as $[i - L_i, i + R_i]$. The following corollary establishes the necessary and sufficient conditions for a set of communication primitives to take place in a collection of particles. Notice that they follow immediately from Lemma 6.

Corollary 1 (Communication primitives). *For any particle $p_i \in \mathcal{S}_n$:*

1. *$p_i \in \mathcal{B}(\mathcal{S}_n)$ if and only if $M_i > v_j$ and $m_i < v_k$ for all $j \geq i$ and $k \leq i$;*
2. *p_i is a convergecast particle if and only if $R_1 \geq i$ and $L_n \geq n - i$ and*
3. *Gossiping takes place in $\mathcal{S}_n$ if and only if $v_1 > v_j$ and $v_n < v_k$, for all $j > 1$, and $k < n$.*

Proof.

1. The proof follows from the definition of the transmission range of p_i, which in this case must be $[1, n]$.
2. Notice that $j + R_j \geq j + R_1$ and $j - L_j \leq k - L_n, j < i, k > i$, therefore p_i is in the transmission range of every particle in $\mathcal{S}_n$.
3. It follows from the fact that the transmission range of every particle is $[1, n]$.
    ∎

Notice that the communication primitives for a collection of particles depend on the order of their speeds but not on their initial positions on $\mathcal{L}$.

4   Deciding the Feasibility of Communication

In this section, we show what preprocessing, if any, should be done on the set of velocities of a collection of particles and how to store such information so that we can efficiently decide whether the communication primitives, discussed in the previous section, take place.

Given any collection of particles $\mathcal{S}_n$, we store the velocities of all the particles in table V, such that $\mathtt{V}[\mathtt{i}] = v_i$, and for every particle p_i, we define $M_i' = \max\{v_j \in \mathcal{V}|\ i \le j\}$, $m_i' = \min\{v_k \in \mathcal{V}|\ k \le i\}$. Note that $M_{i+1} \ge M_i$, $m_{i+1} \le m_i$, $M_{i+1}' \ge M_i'$ and $m_{i+1}' \le m_i'$. The next lemma follows immediately from this observation.

Lemma 7. *In $O(n)$ time we can build tables* M, m, M', *and* m' *such that* $\mathtt{M}[\mathtt{i}] = M_i$, $\mathtt{m}[\mathtt{i}] = m_i$, $\mathtt{M}'[\mathtt{i}] = M_i'$, *and* $\mathtt{m}'[\mathtt{i}] = m_i'$.

After constructing these tables, we can use them in order to decide whether any given particle can perform broadcasting. We have the following lemma.

Lemma 8. *For any collection $\mathcal{S}_n$ of particles, there are data structures that allow us to decide whether $p_i \in \mathcal{B}(\mathcal{S}_n)$ in $O(1)$ time.*

Proof. It is sufficient to check the tables introduced in Lemma 7; more specifically we have to check that $\mathtt{M}[\mathtt{i}] > \mathtt{M}'[\mathtt{i}]$ and $\mathtt{m}[\mathtt{i}] < \mathtt{m}'[\mathtt{i}]$. By Corollary 1 we then have that the transmission range of p_i is $[1, n]$.   ∎

By applying Lemma 8 to all particles we immediately obtain the following lemma.

Lemma 9. *For any collection of particles $\mathcal{S}_n$, we can decide in $O(n)$ time whether gossiping takes place.*

The next lemma establishes that the set of particles that are able to perform broadcasting have consecutive indices.

Lemma 10. *Let a and b be the minimum and maximum indices, respectively, of all the particles in $\mathcal{B}(\mathcal{S}_n)$. Then $p_j \in \mathcal{B}(\mathcal{S}_n)$ for every $a \le j \le b$.*

Proof. Notice that for any $a \le j \le b$ we have $a + R_a \le j + R_j$ and $j - L_j \le b - L_b$, therefore the transmission range of p_j is $[1, n]$.   ∎

The next corollary is an immediate consequence of Lemma 8 and Lemma 10.

Corollary 2. *For any collection of particles $\mathcal{S}_n$, we can decide in $O(n)$ time whether broadcast is possible as well as we can determine the range of indices $[a, b]$ of the particles in $\mathcal{B}(\mathcal{S}_n)$.*

The following theorem establishes an interesting result about convergecast in a collection of bouncing particles.

Theorem 1. *For any collection of particles $\mathcal{S}_n$, we can decide in $O(1)$ time whether there exists a convergecast particle.*

Proof. We show that a convergecast particle exists if and only if $v_1 > v_n$. Suppose that $v_1 > v_n$. In such case, the trains T_1 and T_n meet at some time t. Therefore there exists a pair of particles p_a and p_{a+1} on these trains when meeting at time t. By Lemma 2, at time t p_a acquired every $d_i, i \le a$; analogously, at time t p_j collected every $d_k, k \ge j$. Consequently p_a and p_{a+1} are both convergecast particles.

Suppose now that $v_1 \leq v_n$. Since no train intersects T_1 from the left for any time moment t, by Lemma 3 the speed of the rightmost particles $p_{\rho(t,1)}$ holding d_1 are always equal to v_1. Similarly, by Lemma 4 the speed of the leftmost particles $p_{\lambda(t,n)}$ holding d_n are always equal to v_n. As the trajectories of trains T_1 and T_n never intersect, at the expansion time t^* we have $\rho(t^*, 1) < \lambda(t^*, n)$. Consequently, no particle ever acquires d_1 and d_n.

∎

Note that, while deciding existence of a convergecast particle can be determined in constant time, determining which is such a particle cannot.

5 Time of Transmission

In this section we explore the necessary time for the particles to carry out the communication primitives we have studied so far. It turns out that computing the necessary time for the particles to complete their transmission of information is closely related to a well known geometric problem, namely, computing the upper envelope of an arrangement of lines.

The concept of kth level is related to the concepts of k-set and Davenport-Schinzel sequences [25, 26]. The kth level problem stated as in [25] follows: given n univariate linear functions $\mathcal{F} = \{f_1, f_2, \ldots, f_n\}, f_i : \mathbb{R} \to \mathbb{R}$ and a number $k \in \{1, \ldots, n\}$ construct $G : \mathbb{R} \to \mathbb{R}$, where $G(x) =$ the kth smallest of the numbers $f_1(x), \ldots, f_n(x)$. The kth level forms an x-monotone polygonal chain. The complexity of the kth level corresponds to the number of vertices of such a polygonal chain.

When $k = 1, n$, the function G corresponds to the lower envelope and the upper envelope of $\mathcal{F}$, respectively. When the functions define lines in the plane it is known that the upper envelope of $\mathcal{F}$ can be optimally computed in $O(n \log n)$ time and in $O(n)$ time if the lines are sorted (this is because of its duality with computing the convex hull of a set of points).

It is easy to see that our diagram of $time \times distance$ depicting the trajectories of particles gives us an immediate way to relate the trajectories of particles with the kth level problem. Recall that in this diagram, when a particle moves with velocity v its trajectory corresponds to a line of slope $1/v$. Thus, we can define $L = \{l_1, \ldots, l_n\}$ such that l_i corresponds to the equation of the line passing through $(0, h_i)$ with slope $1/v_i$ (recall h_i stands for the initial position of p_i). Fig. 2 depicts the trajectory of p_1 which corresponds to the lower envelope while the trajectory of p_n corresponds to the upper envelope of L. We denote by $lEnv(L)$ and $uEnv(L)$ the lower (left) and the upper (right) envelopes of L, respectively. Consider the rightmost copy of d_i, which is carried at time t by $p_{\rho(t,i)}$. We denote by $traj_i^+(t)$ the trajectory of the rightmost copy of d_i up to time t and by $traj_i^-(t)$ the trajectory of the leftmost copy of d_i up to time t.

Observation 1 *Notice that:*

1. $uEnv(\{l_1, \ldots, l_i\}) = traj_i^+(t)$
2. $lEnv(\{l_i, \ldots, l_n\}) = traj_i^-(t)$

3. p_j gets d_i when $p_{\rho(i,t)} = p_j$ for $j \geq i$ and some t

4. p_j gets d_i when $p_{\lambda(i,t)} = p_j$ for $j \leq i$ and some t

Lemma 11. *For any collection of particles $\mathcal{S}_n$, we can compute in $O(n \log n)$ time the moment at which d_i is transferred to any particle p_j in the transmission range of p_i.*

Proof.

Let $l_1, \ldots, l_n$ be the lines in the plane *time* $\times$ *distance* associated to the initial positions and velocities of the particles and let $uEnv(\{l_1, \ldots, l_i\})$ be the upper envelope of lines $l_1, \ldots, l_i$. For simplicity, let us assume first that $j \in [i+1, i+R_i]$. Notice that there are exactly R_i lines associated to the particles at the right of p_i that intersect $uEnv(\{l_1, \ldots, l_i\})$. Let $l^{(1)} \ldots, l^{(R_i)}$ be such lines sorted in increasing order by the time they intersect $uEnv(\{l_1, \ldots, l_i\})$ (See Fig. 3 for an example). Because of Observation 1, there is a time t at which $p_{\rho(i,t)} = p_j$. This takes place exactly at the intersection of $uEnv(\{l_1, \ldots, l_i\})$ with line $l^{(j-i)}$. Clearly, the time to compute this intersection is dominated by the required time to compute $uEnv(\{l_1, \ldots, l_i\})$.

The proof for the case $j \in [i - L_i, i]$ is analogous. ∎

Fig. 3 illustrates Lemma 11 showing how to compute the time transmission of d_3 to some particle.

It turns out that the time when convergecast takes place can be easily computed. The next theorem states so.

Theorem 2. *For any collection of particles $\mathcal{S}_n$, in $O(1)$ time it is possible to determine the earliest time when convergecast is completed.*

Proof.

By Theorem 1 in $O(1)$ time we can decide if convergecast takes place in $\mathcal{S}_n$. If this is the case, it means that $\mathtt{V}[1] > \mathtt{V}[n]$. Assume that $dist(h_1, h_n) = d$ (the distance between the initial positions of particles p_1 and p_n). Then according to Theorem 1, convergecast takes place at the intersection of trains T_1 and T_n, i.e., at time $\frac{d}{|v_1 - v_n|}$. ∎

The next result follows immediately from Lemma 11.

Theorem 3. *For any collection of bouncing particles, there are $O(n \log n)$ algorithms computing the earliest time t at which:*

1. *The broadcast from particle p_i is completed*
2. *Gossiping in $\mathcal{S}_n$ is carried out*

Proof.

Consider first the broadcast. By of Lemma 8 in $O(1)$ time we decide whether $p_i \in \mathcal{B}(\mathcal{S}_n)$ (after $O(n)$ preprocessing). Lemma 11 guarantees that in $O(n \log n)$ time we can compute the times t_1 and t_2 at which p_1 and p_n received d_i, respectively. Therefore the completion of the broadcasting of d_i by p_i takes place at time $\max\{t_1, t_2\}$.

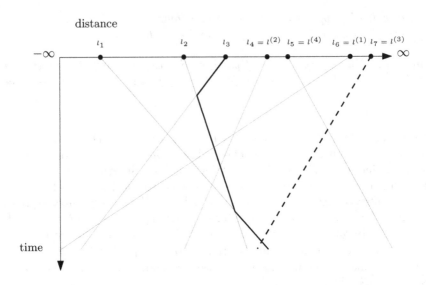

Fig. 3. The fat polyline is the upper envelope of lines l_1, l_2 and l_3. The transmission of d_3 to p_6 takes place at the intersection of $uEnv(\{l_1, l_2, l_3\})$ and $l^{(3)} = l_7$.

For the case of gossiping in $\mathcal{S}_n$, it is completed at the time when p_1 has received d_n and p_n has received d_1. To compute these times we use Lemma 11 again. ∎

Computing the trajectory of the kth particle is equivalent to computing the kth level of L, and the number of collisions of p_k corresponds to the complexity of the kth level of L. The next result follows immediately from the results for the k-level problem (see [29, 30] for further details).

Corollary 3. *In any system of particles $\mathcal{S}_n$, the number of collisions of p_k is upper-bounded by $O(nk^{\frac{1}{3}})$ and it is lower-bounded by $n2^{\Omega(\sqrt{\log k})}$. Moreover, the trajectory of p_k can be computed in $O(nk^{\frac{1}{3}} \log^c n)$ expected time, for some constant c.*

Constructing the trajectories of the first k particles correspond to the construction of all ith levels for $1 \le i \le k$. [32] proved that this can be done in $O(n \log n + nk)$ time and that this is optimal.

6   Conclusions

We analyzed the communication protocols for populations of bouncing particles on a line presenting the propagation of individual information of any particle. We gave efficient algorithms deciding the feasibility of communication between pairs of particles, as well as checking which particles can broadcast, which ones are convergecast particles and whether gossiping is possible. There are several lines of research continuation of the presented model.

One extension involves communication between particles of possibly distinct masses. However, as the crucial concept of train does not easily extend to this model (even if we consider only particles of only two masses or if the initial speeds are the same), a fundamentally different approach is probably needed.

Matching lower and upper bounds on the combinatorial complexity of the trajectory of the kth particle is the obvious open problem of geometrical nature.

The reader may find interesting the following relation of the solution of our problem to a sorting algorithm and its corresponding inversion number. The initial permutation of particle velocities is progressively changing at every moment a bounce is taking place, eventually becoming sorted. Thus, every speed swap corresponds to a bounce and also the total number of bounces corresponds to the number of inversions of the initial permutation.

References

1. Sevryuk, M.: Estimate of the number of collisions of n elastic particles on a line. Theoretical and Mathematical Physics 96(1), 818–826 (1993)
2. Suzuki, I., Yamashita, M.: Distributed anonymous mobile robots: Formation of geometric patterns. SIAM J. Comput. 28(4), 1347–1363 (1999)
3. Angluin, D., Aspnes, J., Diamadi, Z., Fischer, M.J., Peralta, R.: Computation in networks of passively mobile finite-state sensors. Dist. Comp. 18(4), 235–253 (2006)
4. Angluin, D., Aspnes, J., Eisenstat, D.: Stably computable predicates are semilinear. In: PODC, pp. 292–299 (2006)
5. Das, S., Flocchini, P., Santoro, N., Yamashita, M.: On the computational power of oblivious robots: forming a series of geometric patterns. In: PODC, pp. 267–276 (2010)
6. Flocchini, P., Prencipe, G., Santoro, N., Widmayer, P.: Hard tasks for weak robots: The role of common knowledge in pattern formation by autonomous mobile robots. In: Aggarwal, A.K., Pandu Rangan, C. (eds.) ISAAC 1999. LNCS, vol. 1741, pp. 93–102. Springer, Heidelberg (1999)
7. Suzuki, I., Yamashita, M.: Distributed anonymous mobile robots: Formation of geometric patterns. SIAM Journal on Computing 28(4), 1347–1363 (1999)
8. Yamashita, M., Suzuki, I.: Characterizing geometric patterns formable by oblivious anonymous mobile robots. TCS 411(26), 2433–2453 (2010)
9. Cohen, R., Peleg, D.: Local spreading algorithms for autonomous robot systems. TCS 399(1), 71–82 (2008)
10. Flocchini, P., Prencipe, G., Santoro, N., Widmayer, P.: Gathering of asynchronous oblivious robots with limited visibility. In: Ferreira, A., Reichel, H. (eds.) STACS 2001. LNCS, vol. 2010, pp. 247–258. Springer, Heidelberg (2001)
11. Czyzowicz, J., Kranakis, E., Pacheco, E.: Localization for a system of colliding robots. In: Fomin, F.V., Freivalds, R., Kwiatkowska, M., Peleg, D. (eds.) ICALP 2013, Part II. LNCS, vol. 7966, pp. 508–519. Springer, Heidelberg (2013)
12. Czyzowicz, J., Gąsieniec, L., Kosowski, A., Kranakis, E., Ponce, O.M., Pacheco, E.: Position discovery for a system of bouncing robots. In: Aguilera, M.K. (ed.) DISC 2012. LNCS, vol. 7611, pp. 341–355. Springer, Heidelberg (2012)
13. Michail, O., Spirakis, P.G.: Simple and efficient local codes for distributed stable network construction. In: ACM Symposium on Principles of Distributed Computing, PODC 2014, Paris, France, July 15-18, pp. 76–85 (2014)

14. Mertzios, G.B., Nikoletseas, S.E., Raptopoulos, C., Spirakis, P.G.: Determining majority in networks with local interactions and very small local memory. In: Esparza, J., Fraigniaud, P., Husfeldt, T., Koutsoupias, E. (eds.) ICALP 2014, Part I. LNCS, vol. 8572, pp. 871–882. Springer, Heidelberg (2014)
15. Cao, Y.U., Fukunaga, A.S., Kahng, A.B., Meng, F.: Cooperative mobile robotics: Antecedents and directions. In: Proceedings of the 1995 IEEE/RSJ International Conference on Intelligent Robots and Systems 1995. 'Human Robot Interaction and Cooperative Robots, vol. 1, pp. 226–234. IEEE (1995)
16. Dudek, G., Jenkin, M.: Computational principles of mobile robotics. Cambridge University Press (2010)
17. Bender, M.A., Fernández, A., Ron, D., Sahai, A., Vadhan, S.P.: The power of a pebble: Exploring and mapping directed graphs. In: STOC, pp. 269–278 (1998)
18. Czyzowicz, J., Dobrev, S., An, H.-C., Krizanc, D.: The power of tokens: Rendezvous and symmetry detection for two mobile agents in a ring. In: Geffert, V., Karhumäki, J., Bertoni, A., Preneel, B., Návrat, P., Bieliková, M. (eds.) SOFSEM 2008. LNCS, vol. 4910, pp. 234–246. Springer, Heidelberg (2008)
19. Murphy, T.: Dynamics of hard rods in one dimension. Journal of Statistical Physics 74(3), 889–901 (1994)
20. Tonks, L.: The complete equation of state of one, two and three-dimensional gases of hard elastic spheres. Physical Review 50(10), 955 (1936)
21. Wylie, J., Yang, R., Zhang, Q.: Periodic orbits of inelastic particles on a ring. Physical Review E 86, 026601(2) (2012)
22. Cooley, B., Newton, P.: Iterated impact dynamics of n-beads on a ring. SIAM Rev. 47(2), 273–300 (2005)
23. Murphy, T., Cohen, E.: Maximum number of collisions among identical hard spheres. Journal of Statistical Physics 71(5-6), 1063–1080 (1993)
24. Friedetzky, T., Gąsieniec, L., Gorry, T., Martin, R.: Observe and remain silent (Communication-less agent location discovery). In: Rovan, B., Sassone, V., Widmayer, P. (eds.) MFCS 2012. LNCS, vol. 7464, pp. 407–418. Springer, Heidelberg (2012)
25. Chan, T.M.: Remarks on k-level algorithms in the plane. Manuscript, Univ. of Waterloo (1999)
26. Sharir, M., Agarwal, P.K.: Davenport-Schinzel sequences and their geometric applications. Cambridge University Press (1995)
27. Erdös, P., Lovász, L., Simmons, A., Straus, E.G.: Dissection graphs of planar point sets. A Survey of Combinatorial Theory, 139–149 (1973)
28. Lovász, L.: On the number of halving lines. Ann. Univ. Sci. Budapest, Eötvös, Sec. Math. 14, 107–108 (1971)
29. Dey, T.K.: Improved bounds for planar k-sets and related problems. Discrete & Computational Geometry 19(3), 373–382 (1998)
30. Tóth, G.: Point sets with many k-sets. Discrete & Computational Geometry 26(2), 187–194 (2001)
31. Gregory, R.: Classical mechanics. Cambridge University Press (2006)
32. Everett, H., Robert, J.M., Van Kreveld, M.: An optimal algorithm for the ($\leq k$)-levels, with applications to separation and transversal problems. In: Proceedings of the Ninth Annual Symposium on Computational Geometry, ACM, pp. 38–46 (1993)

On Space and Time Complexity
of Loosely-Stabilizing Leader Election*

Taisuke Izumi

Graduate School of Engineering, Nagoya Institute of Technology,
Nagoya, Japan
t-izumi@nitech.ac.jp

Abstract. Loose stabilization is a relaxed notion of self-stabilization, which guarantees algorithms to converge and keep some desired behavior from any initial configuration, but allows the algorithms to drop out of it after a sufficiently long period. In this paper, we investigate the complexity of the loosely-stabilizing leader election problem in the population protocol model under the probabilistic scheduler. The primary contribution is to give lower bounds for the expected length of convergence periods and the memory space. Precisely, for any loosely-stabilizing leader election algorithm with stabilization periods of length $\Omega(\exp(N))$ in expectation, each agent needs $\Omega(\log N)$-bit memory space, and the expected convergence length is $\Omega(Nn)$, where n is the (unknown) number of agents, and N is the upper bound knowledge for n available to the algorithm. We also show the matching upper bounds by proposing a new loosely-stabilizing leader election algorithm, which slightly improves the expected convergence length of the previously known algorithm by Sudo et al. [15] without any degradation of the memory usage or the stabilization length.

1 Introduction

A *passively-mobile* system is a collection of agents that move in a certain region but have no control over how they move. Since the communication range of each agent is limited, two agents can communicate only when they are sufficiently close to each other. A typical example of passively-mobile systems is the network of smart devices attached cars or animals. The *population protocol* is one of the promising models for such a system, which was initiated by Angluin et al. [2]. A population protocol consists of a number of agents, to which some program (algorithm) is deployed. Following the deployed algorithm, each agent changes its state by *pairwise interactions* to other agents (that is, two agents get closer to each other and update their states by exchanging information). The population protocol model is a good abstraction capturing the feature of passively-mobile systems in spite of its mathematical simplicity. Therefore, in the last few years, it has received much attention among the community of the distributed computing.

* This work is supported in part by KAKENHI No. 15H00852 and 25289227.

C. Scheideler (Ed.): SIROCCO 2015, LNCS 9439, pp. 299–312, 2015.
DOI: 10.1007/978-3-319-25258-2_21

In this paper, we consider the self-stabilizing leader election problem on the population protocol model. Self-stabilizing algorithms guarantee that the algorithm eventually satisfies and keeps some desired behavior starting from any initial configuration. Self-stabilizing systems inherently achieve the initialization-freeness and the recoverability from any transient failures, which are very important properties for passively-mobile systems. However the design of self-stabilizing algorithms is often impossible for several problems. Actually, the impossibility of self-stabilizing leader election on the population protocol model has been proved for several different settings [5–7,9–11]. Precisely, to implement the self-stabilizing leader election on the population protocol model, the algorithm needs $\Omega(\log n)$-bit space and the exact knowledge about the total number n of agents in the system [10], or a certain kind of synchrony assumption [8].

Loose stabilization, is one of the approaches to circumvent the impossibility of self-stabilizing solutions [15, 16]. Loose-stabilizing algorithms guarantee that the system reaches some legitimate configuration eventually from any initial configuration, but stability (or closure) property is not strictly ensured. That is, it allows the algorithm to drop out of legitimate configurations after a sufficiently long period of the desired behavior. In the population protocol model under the probabilistic scheduler, Sudo et al. proposed a loosely-stabilizing leader election algorithm only utilizing the knowledge on the upper bound N for n [15], which elects one leader within $O(Nn \log n)$ expected steps, and keeps the elected leader during $\Omega(\exp(N))$ expected steps. Since exponentially long stabilization periods can be regarded as infinite stabilization periods in practice, loose stabilization is so useful as a relaxed concept of self-stabilization.

The main focus of this paper is the complexity issue of loosely-stabilizing leader election algorithms. Currently, it is still open whether we can construct an algorithm more efficient (in the sense of time or space) than the one by Sudo et al. [15] or not. The motivation initiating this study is the existence of a randomized approximated counter using $O(\log \log N)$-bit space [12,14]. Actually, in the original algorithm by Sudo et al., each agent utilizes only $O(1)$-bit memory space and one $O(\log N)$-bit counter to implement the timeout mechanism. Thus, it is a natural idea to reduce the space complexity of that algorithm using small-space approximated counters. However, in this paper, we show that such an approach fails. The primary contribution of this paper is to give lower bounds for the expected length of convergence periods and the memory space of the algorithms with exponentially long stabilization periods, which is precisely stated as follows:

- In any loosely-stabilizing leader election algorithm whose expected stabilization length is $\Omega(\exp(N))$, $\Omega(\log N)$-bit memory space is required to each agent.
- For any loosely-stabilizing leader election algorithm whose expected stabilization length is $\Omega(\exp(N))$, the expected length of convergence periods is $\Omega(Nn)$.

Note that these results hold even for randomized algorithms. Furthermore, we also show that the two bounds above are both tight. That is, we propose a new algorithm for loosely-stabilizing leader election, whose convergence and stabilization lengths are $O(Nn)$ and $\Omega(\exp(N))$ steps while the space for each agent is still $O(\log N)$ bits. This algorithm slightly improves the convergence length of the one by Sudo et al.

The paper is organized as follows: In Section 2 we state the related work. Section 3 provides the notations and definitions used in the paper. The lower bounds are presented in Section 4. Section 5 provides the matching upper bound. Finally the paper is concluded in Section 6.

2 Related Work

The population protocol model is initiated by the two seminal papers by Angluin et al. [2] and Angluin et al. [4], where the main interest was to clarify the class of computable predicates with the inputs distributed over all agents. The leader election problem on the population protocol model is first introduced in the those papers as a subroutine of the predicate computation. Self-stabilizing algorithms for the population protocol model are considered in several papers. The papers by Angluin et al. [5] and Fischer and Jiang [11] are one of the earliest work considering self-stabilizing population protocols. They show several possibility and impossibility results for self-stabilizing leader election in different assumptions. The paper by Beauquier et al. [6] also follows the same direction. The space complexity for the self-stabilizing leader election problem on the population protocol model is obtained by Cai et al. [10]. The problem other than leader election is also considered in several papers [7,9,13].

While most of the results presented in the papers above assumes worst-case schedulers, the complexity analysis under the probabilistic scheduler is also investigated. Angluin et al. [3] shows a fast algorithm for the predicate computation under the assumption that one leader exists at the initial configuration. Very recently, a faster (non-self-stabilizing) leader election algorithm using sublogarithmic space is presented by Alistarh and Gelashvili [1].

The concept of loose stabilization is first introduced by Sudo et al. [15]. Recently, its extension to an arbitrary communication topology is also published [16].

3 Preliminaries

3.1 Agent and Algorithm

A population protocol consists of n agents, which can change their states by interacting with each other. All the pairs cannot necessarily have direct interaction. The possibility of the interaction between two agents is specified by the *interaction graph*. An interaction graph $G = (V, E)$ is a simple directed graph where each vertex, labeled by $v_0, v_1, \ldots v_{n-1}$, corresponds to each agent. The edge $(v_i, v_j) \in E$ implies the possibility of the interaction between agents v_i and

v_j. Throughout this paper, we assume that the interaction graph G is complete, that is, any pair in the system can directly interact with each other.

An algorithm A is a pair $(Q, \delta)^1$, where Q is the set of agent states, and $\delta : Q^2 \times \{0,1\}^* \to Q^2$ is a state transition function. The function takes a pair of agent states and a random-bit sequence, and returns the states of two agents after the interaction. For any $(q_1, q_2) \in Q^2$, if $\delta(q_1, q_2, r)$ is independent of $r \in \{0,1\}^*$, the algorithm is called *deterministic*. Otherwise the algorithm is called *randomized*. Note that all the lower bounds proposed in this paper holds for any randomized algorithm, and the upper bound result is provided by a deterministic algorithm. Given any algorithm A, we define π_A as the transition matrix of A. That is, the row and column of π_A are indexed by the elements in Q^2, and the value $\pi_A((p_1, p_2), (q_1, q_2))$ corresponding to row (p_1, p_2) and column (q_1, q_2) is the transition probability $\Pr[\delta(p_1, p_2, r) = (q_1, q_2)]$.

The space complexity $s(A)$ of an algorithm is defined as the number of bits necessary for representing $|Q|$ distinct states. That is, $s(A) = \lceil \log |Q| \rceil$.

3.2   Execution and Scheduler

A configuration C of n agents is a n-tuple $(q_0, q_1, \ldots, q_{n-1}) \in Q^n$ where i-th entry q_i corresponds to the state of v_i. The state of agent v_i in C is denoted by $C[i]$. We also use the notation $C[i,j] = (C[i], C[j])$ for short. Let $A = (Q, \delta)$ be an arbitrary algorithm. For any two configurations C_0 and C_1, if there exists $(s,t) \in [0, n-1]^2$ satisfying $\pi_A(C_0[s,t], C_1[s,t]) > 0$ and $C_0[j] = C_1[j]$ for any $j \notin \{s,t\}$, we say that C_0 *can go to* C_1 by an interaction. Then the pair (s,t) is called the *activation pair* of the interaction. Note that the activation pair may not be unique. Given two configurations C and C' such that C can go to C', we denote the set of the corresponding activation pairs by $\chi(C, C')$. An *execution* E of A of length l (or with l steps) is a sequence of l configurations $E = C_0, C_1, \ldots, C_l$ such that C_i can go to C_{i+1} for any $i \in [0, l-1]$. The i-th activation in execution E is referred as *step i* of E.

The behavior of an algorithm is determined by the scheduler (and random bits if the algorithm is randomized). Throughout this paper, we assume the *probabilistic scheduler*, that is, at each step, the activation pair is chosen uniformly at random. We denote the execution of A with initial configuration C_0 under the probabilistic scheduler by $E_A(C_0)$. Without ambiguity, we often omit the index A of $E_A(C_0)$.

3.3   Loosely-Stabilizing Leader Election

This subsection provides the formal definition of (α, β)-probabilistic loosely-stabilizing leader election.

A *spec* of $A = (Q, \delta)$ is the set of configurations $\mathcal{L}_A \subseteq Q^n$ and its subset $\mathcal{L}'_A \subseteq \mathcal{L}_A$. The configurations in $\mathcal{L}_A$ and $\mathcal{L}'_A$ are respectively called *legitimate*

---

[1] While the standard definition of the population protocol model is more complex, we adopt a simplified definition because this paper focuses only on the loosely-stabilizing leader election problem.

and *strongly-legitimate* configurations. The leader election problem specifies a set $L \subset Q$ of states called *leader states*. A configuration $C = (q_1, q_2, \ldots, q_n)$ is valid for v_i if $q_i \in L$ and $q_j \notin L$ for any $j \in [0, n-1] \setminus \{i\}$ hold. Any spec of the leader election algorithms must satisfy that any $C \in \mathcal{L}_A$ is valid for some agent $v_i \in V$.

In some execution $E(C_0) = C_0, C_1, \ldots, C_l$, if any configuration appearing in the period $C_k, C_{k+1}, \ldots, C_{k+h}$ is valid for a common $v_i \in V$, we say that $E(C_0)$ is *stabilized* during steps k and $k + h$. We show the formal definition of (α, β)-probabilistic loosely-stabilizing leader election:

Definition 1. *If the legitimate and strongly-legitimate sets $\mathcal{L}_A$ and $\mathcal{L}'_A$ satisfies the following conditions, A solves the (α, β)-probabilistic loosely-stabilizing leader election problem:*

- *For any configuration $C_0 \in Q^n$ and the execution $E(C_0) = C_0, C_1, \ldots, C_k, \ldots$, let $\Gamma(C_0)$ be the random variable representing the minimum value x such that $C_x \in \mathcal{L}'_A$ holds. Then for any $C_0 \in Q^n$, we have $\mathbb{E}[\Gamma(C_0)] \leq \alpha$.*
- *For any configuration $C_0 \in \mathcal{L}'_A$ and the execution $E(C_0) = C_0, C_1, \ldots, C_k, \ldots$, let $\Delta(C_0)$ be the largest value x such that $E(C_0)$ is stabilized during steps zero to x. Then $\mathbb{E}[\Delta(C_0)] \geq \beta$ holds for any $C_0 \in \mathcal{L}'_A$.*

The values α and β are respectively called the *convergence length* and the *stabilization length* of the algorithm. In the following argument, we abbreviate (α, β)-probabilistic loose stabilization by (α, β)-stabilization for short.

3.4 Knowledge on the Number of Agents

In this paper, we assume that each agent knows the upper bound N for the number of agents n. Formally, that knowledge is defined as algorithms taking N as an argument, which is described as $A_N = (Q_N, \delta_N)$. That is, the set Q_N of states and the transition function δ_N can depend on N. For any $N \in \mathbb{N}$, algorithm A_N must work on any system of n agents such that $2 \leq n \leq N$ holds.

4 Complexity Lower Bounds for (α, β)-stabilization

This section provides our new complexity lower bounds for (α, β)-stabilizing leader election with $\beta = \Omega(\exp(N))$. Before the proofs, we first show a useful lemma.

Lemma 1. *Let $A = (Q, \delta)$ be an (α, β)-stabilizing leader election algorithm for some α and β. If there exists a value $l \in \mathbb{N}$ and a function $f(N, n)$ such that for any legitimate configuration $C \in Q^n$ the execution $E(C)$ of length l contains a non-legitimate configuration with probability at least $l/f(N, n)$, then $\beta \leq f(N, n)$ holds.*

Proof. Let $E'(C)$ be the infinite execution starting from C, and C_x be the first non-legitimate configuration in $E'(C)$. Since each period of length l in $E'(C)$ contains a non-legitimate configuration with probability at least $l/f(N,n)$, the expected value $\mathbb{E}[x]$ is bounded by

$$\mathbb{E}[x] \leq l \cdot \sum_{i=1}^{\infty} i \left(1 - \frac{l}{f(N,n)}\right)^{i-1} \left(\frac{l}{f(N,n)}\right) = l \cdot \frac{f(N,n)}{l} = f(N,n).$$

The second equality is obtained from the expectation for the geometric distribution. The lemma is proved. □

4.1   Lower Bound for Space

We first introduce the notion of *execution graphs*.

Definition 2. *Given an algorithm* $A = (Q, \delta)$, (A,n)-*execution graph* $G_{A,n} = (V_{A,n}, E_{A,n}, w_{A,n})$ *is defined as follows:*

- $V_{A,n} = Q^n D$
- $E_{A,n} = \{(C, C') | C \text{ can go to } C'\} D$
- *For any* $(C, C') \in E_{A,n}$, $w_{A,n}((C, C')) = \sum_{(s,t) \in \chi(C,C')} \frac{\pi_A(C[s,t], C'[s,t])}{n(n-1)}$.

Intuitively, the (A,n)-execution graph is the Markov-chain description of the system of n agents running A. We first show a fundamental property of graph $G_{A,n}$.

Lemma 2. *Consider any* (N^c, β)-*stabilizing leader election algorithm* A, *and let* $G'_{A,n} = (V_{A,n}, E'_{A,n})$ *be the graph obtained by removing all the edges with weight less than* $1/(N^{c+1}n(n-1))$ *in* $G_{A,n}$. *In graph* $G'_{A,2}$, *any* $C \in V_A$ *is reachable to some configuration* C' *in* $\mathcal{L}'_A$.

Proof. Suppose for contradiction that some vertex $C \in V_{A,2}$ is unreachable to any vertex in $\mathcal{L}'_A$. Then, $G_{A,2}$ contains a cut $(W, \overline{W})$ ($W \subseteq V_A$) satisfying $C \in W$ and $\mathcal{L}'_A \subseteq \overline{W}$ such that any edge crossing the cut is removed in $G'_{A,2}$. Thus, letting F be the set of edges crossing the cut, any edge in F has weight less than $1/(2N^{c+1})$. Since at least one transition along some edge in F is necessary for the execution $E(C)$ of length N^c to reach a configuration in $\mathcal{L}'_A$, the probability that $E(C)$ does not contain a configuration in $\mathcal{L}'_A$ is bounded by

$$\Pr[E(C) \cap \mathcal{L}'_A = \emptyset] \geq \left(1 - \frac{1}{2N^{c+1}}\right)^{N^c} \geq 1 - \frac{1}{2N}.$$

Then, the expected convergence length of A obviously exceeds N^c, which contradicts the fact that A is (N^c, β)-stabilizing. □

Theorem 1. *Let* $A = (Q, \delta)$ *be a* $(N^c, 2^{dN})$-*stabilizing leader election algorithm for some constant* c *and* d. *Then, we have*

$$s(A) \geq \frac{1}{2} \log \left(\frac{dN}{2(c+3)\log N}\right) = \Omega(\log N).$$

Proof. Suppose for contradiction that $s(A) < \frac{1}{2} \log \left(\frac{dN}{2(c+3) \log N} \right)$ holds. Let $l = |Q^2| = 2^{2s(A)} = dN/(2(c+3) \log N)$ for short. First, we prove that in the system of two agents, the execution of length l starting from any configuration can create a leader with probability at least $(1/(2N^{c+1}))^l$. Since l is the number of vertices in $G'_{A,2}$, from Lemma 2, any configuration C in $G'_{A,2}$ has a path to a node in $\mathcal{L}'_A$ whose length is at most l. Let $P(C) = (p_0, q_0), (p_1, q_1), \ldots, (p_l, q_l)$ be the execution of two-agent configurations corresponding to that path. Note that in the system of two agents, each transition in $P(C)$ occurs with probability at least $1/(2N^{c+1})$ because in the construction of $G'_{A,2}$ we remove all the edges whose transition probability is less than $1/(2N^{c+1})$. Thus, the execution tracing $P(C)$ occurs with probability at least $(1/(2N^{c+1}))^l$. Now we look at the system of n agents ($n \geq 3$), and its legitimate configuration C'. Since $n \geq 3$ holds, C' contains two non-leader agents. Without loss of generality, let $\{v_0, v_1\}$ be the pair of two non-leader agents. We lower bound the probability that $\{v_0, v_1\}$ traces $P(C)$ in the execution $E(C')$ of length l. Under the condition that only v_0 and v_1 are activated in $E(C')$, the probability that v_0 and v_1 trace $P(C)$ is obviously at least $(1/(2N^{c+1}))^l$. Furthermore, the probability satisfying that condition is $(2/n(n-1))^l$. Consequently we can lower bound the probability that v_0 and v_1 trace $P(C)$ in $E(C')$ is at least

$$\left(\frac{2}{n(n-1)} \right)^l \cdot \left(\frac{1}{2N^{c+1}} \right)^l \geq \left(\frac{1}{N^{c+3}} \right)^l \geq 2^{-\frac{dN}{2}} > \frac{l}{2^{dN}}.$$

Since $P(C)$ creates a new leader, the bound above implies that for any legitimate configuration C', the execution $E(C')$ of length l reaches a non-legitimate configuration with probability more than $\frac{l}{2^{dN}}$. From Lemma 1, this contradicts the fact that A is $(N^c, 2^{dN})$-stabilizing. The theorem is proved.   □

4.2 Lower Bound for Convergence Length

This subsection shows that $\alpha = \Omega(Nn)$ holds for any $(\alpha, \Omega(\exp(N)))$-stabilizing leader election algorithm.

Theorem 2. *Let $A = (Q, \delta)$ be any $(g(N, n), \Omega(\exp(N)))$-stabilizing leader election algorithm. Then, we have $g(N, n) = \Omega(Nn)$.*

Proof. Let C be any strongly-legitimate configuration, and X be the indicator random variable which takes one if and only if the execution $E(C)$ of length $2g(N, n)$ contains a non-legitimate configuration. Let v_x be the leader agent in C, and Y be the indicator random variable that $E(C)$ contains no interaction by v_x. Then we have

$$\Pr[Y = 1] = \left(1 - \frac{(n-1)}{n(n-1)} \right)^{2g(N,n)}.$$

Since A is $(g(N, n), \Omega(\exp(N)))$-stabilizing, under the condition of Y, the set V' of $(n-1)$ agents excepting the leader v_x creates a new leader with $g(N, n)$

expected steps. Thus, by applying Markov inequality, V' creates a new leader within $2g(N,n)$ steps with probability at least $1/2$, Consequently we can bound the probability $\Pr[X = 1]$ as follows:

$$\Pr[X = 1] \geq \Pr[X = 1 | Y = 1] \cdot \Pr[Y = 1]$$
$$\geq \frac{1}{2}\left(1 - \frac{(n-1)}{n(n-1)}\right)^{2g(N,n)}$$
$$= e^{-\frac{2g(N,n)}{n}}$$

Applying Lemma 1 with $l = 2g(N,n)$, we obtain $e^{-2g(N,n)/n} \geq g(N,n)/\exp(N)$. That is, $g(N,n) = \Omega(Nn)$ must hold. The theorem is proved. $\square$

5   Matching Upper Bound

In this section, we prove that the time and space lower bounds in Section 4 are tight by proposing an algorithm achieving the matching upper bounds simultaneously. More precisely, we present an algorithm FastLeader, which achieves $(O(Nn), c^N)$-stabilization for some constant $c > 1$. The algorithm is designed in a modular way: We first introduce an algorithm for the $(O(Nn), c^N)$-stabilizing *leader detection* problem, called DetectLeader. The main algorithm follows the idea by Fischer and Jiang [11], which provides a simple self-stabilizing leader election algorithm using leader detector oracles.

5.1   Leader Detector

The objective of leader detection algorithms is to report the existence of leaders in the system. We assume that each agent v_i has a leader flag l_i and an output flag f_i. The agent with $l_i = T$ corresponds to a leader. Since our goal is to construct a loosely-stabilizing leader election algorithm, the leader detection mechanism must be loosely-stabilizing. More precisely, our algorithm satisfies the following two properties:

- In the execution where no leader exists, within $16Nn$ steps in expectation, at least one agent v_i outputs FALSE (i.e., $f_i = F$).
- In the execution where some agent v_i always satisfies $l_i = T$, after the first $24Nn$ steps in expectation, all the agents continue to output TRUE (i.e., $f_i = T$) in the following c^N steps in expectation (c is some constant).

The leader detector can output TRUE even if two agents have the leader flags of TRUE. Algorithm 1 is the pseudocode of DetectLeader. The fundamental idea of the algorithm is a slight but nontrivial modification of the naive timeout mechanism. One of the trivial leader detection mechanisms is the following strategy:

- Each agent v_i prepares a counter variable c_i, which is decreased by one at each interaction, and reset to a large value x (greater than N) if v_i interacts with a leader.

– If the counter variable reaches zero, the agent outputs FALSE. Otherwise, it outputs TRUE.

Unfortunately, this mechanism does not suffice our requirement. To achieve $O(Nn)$-step convergence, we have to set x to $O(N)$. Then, however, with some constant probability, agent may output FALSE despite the existence of a leader: Let $x = \alpha N$, and assume that the system consists of N agents including exactly one leader. In any execution of length $\alpha N n = \alpha N^2$, an agent v_i takes αN steps with some constant probability (because αN is the expected number of interactions performed by v_i). In addition, we can bound the probability that v_i never interacts with the leader by

$$\left(1 - \frac{1}{N(N-1)}\right)^{\alpha N^2} = \Theta(1).$$

Thus, during any execution of length αN^2, some agent provides a wrong output with a not-so-small probability, which violates the requirement of stabilization length $\Omega(c^N)$.

Our key idea for resolving this problem is to use the counter value as a kind of confidence about the existence of leaders. Informally, if some agent v_i has a large counter value close to x, it is expected that v_i interacted with a leader recently, and thus the value of c_i can be regarded as some (weak) witness that a leader actually exists. The first point of the modification is that each agent resets the counter even when it interacts with the one having such a witness. Unfortunately, this modification brings up another problem: Some agent having a witness delays the decrease of the counter value by repetitive resets, which also prevents $O(Nn)$-step convergence in the case of no leader. We avoid this second problem by introducing the "partial reset" of counters: When some agent v_i resets its counter c_i by the interaction with the one having a witness, c_i is set to a moderately large value such that v_i cannot have a witness. That is, we guarantee that only the interactions with leaders can create a witness. If the system has no leader, the number of witnesses decreases, and reaches zero within $O(Nn)$ steps, and then the system follows the standard timeout mechanism afterward.

In details, algorithm DetectLeader takes $16N + 1$ as the maximum counter value, and regards the agents with counter value more than $8N$ as the ones having a witness. We call them *witness agents* (including leader agents). Other agents are called *ordinary agents*. We also refer the counter reset by interactions with a leader and with a witness agent as *full reset* and *partial reset* respectively.

We first prove a fundamental property for the probabilistic scheduler.

Lemma 3. *Given a constant $\alpha > 0$, let E be an execution of steps $\alpha N n$ under the random scheduler, and $\sharp_i(E)$ be the number of the steps in E performed by v_i $(0 \leq i \leq n-1)$. Then, $\alpha N/2 \leq \sharp_i(E) \leq 3\alpha N/2$ holds for all i with probability at least $1 - n2^{-\alpha N/12}$.*

Proof. At each step, an interaction by agent v_i is activated with probability $1/n$. Thus $\sharp_i(E)$ follows the binomial distribution where the success probability

---

Algorithm 1. Algorithm DetectLeader

1:   **At interaction between** (l_0, f_0, c_0) **and** (l_1, f_1, c_1):
2:      **if** $l_0 = T$ **or** $l_1 = T$ **then**
3:         $c_1 \leftarrow 16N + 1$; $c_2 \leftarrow 16N + 1$ /* Full reset */
4:      **endif**
5:      **if** $c_i > 8N$ and $c_{1-i} < 8N$ for some $i \in [0, 1]$ **then** $c_{1-i} \leftarrow 8N$ /* Partial reset */
6:      $c_0 \leftarrow \max\{0, c_0 - 1\}$
7:      $c_1 \leftarrow \max\{0, c_1 - 1\}$
8:      For $i \in [0, 1]$, if $c_i = 0$ then $f_i \leftarrow F$ else $f_i \leftarrow T$ **endif**

---

is $1/n$ and the number of trials is αNn. Let $\mu = \mathbb{E}[\sharp_i(E)] = \alpha N$ for short. By applying Chernoff bound, we can bound the probability $\Pr[|\sharp_i(E) - \mu| \leq \mu] \leq 2e^{-\mu/3 \cdot (1/2)^2} = 2^{-\alpha N/12}$. Using the union bound, we obtain $\Pr[\vee_{i \in [0, n-1]} |\sharp_i(E) - \mu| \leq \mu] \leq n2^{-\alpha N/12}$. The lemma is proved. $\square$

We first show the convergence property of DetectLeader when the system has no leader.

Lemma 4. *In any execution E of length $24Nn$ without leader (i.e., $l_i = F$ holds for any $i \in [0, n-1]$ and any configuration in E), at least one agent v_i satisfies $c_i = 0$ with probability at least $1 - n2^{-4N/3}$.*

Proof. Since there is no leader, no full reset occurs in E. Thus, any agent with counter value greater than or equal to $8N$ never increases the value, and once the counter value of some agent becomes $8N$ or less, it never goes back to the value greater than $8N$. Hence if an agent v_i is activated $8N$ times, its counter value necessarily becomes less than or equal to $8N$. From Lemma 3, during the first $16Nn$ steps of E, all agents have at least $8N$ interactions and any counter value becomes at most $8N$ with probability at least $1 - n2^{-4N/3}$. In the following $8Nn$ steps, no counter value increases. Thus at least one agent is activated more than $8N$ times in those steps. It follows that some agent decreases its counter value to zero and outputs FALSE. $\square$

The next lemma is the core of the convergence property with leaders, and the stabilization property.

Lemma 5. *Assume an execution E of length $4Nn$ with a leader (i.e., there exists at least one x such that $l_x = T$ for any configuration in the execution). Then, with probability at least $1 - 4n2^{-N/12}$, any agent v_i interacts with a witness agent.*

Proof. Let v_x be the leader. First, we prove that after the first $3Nn$ steps more than $(n-1)/2$ agents interact with v_x with high probability. Let X be the set of agents with which v_x interacts in E. Given a set Z of $(n-1)/2$ agents, the probability of $X \cap Z = \emptyset$ is equivalent to the probability that the scheduler does not select $(n-1)$ pairs in $\{(x,y),(y,x)|v_y \in (n-1)/2\}$. Thus it is bounded as follows:

$$\Pr[X \cap Z = \emptyset] = \left(1 - \frac{(n-1)}{n(n-1)}\right)^{3Nn} \leq e^{-3N},$$

When $|X| \leq (n-1)/2$ holds, there exists a set Z of $(n-1)/2$ agents satisfying $X \cap Z = \emptyset$. Thus, using the union bound, the probability $\Pr[|X| \leq (n-1)/2]$ is bounded by

$$\Pr[|X| \leq (n-1)/2] \leq \Pr\left[\bigcup_{Z \subseteq V, |Z|=(n-1)/2} X \cap Z = \emptyset\right]$$
$$\leq \binom{n}{(n-1)/2} e^{-3N}$$
$$\leq (2e)^{n/2} e^{-3N}$$
$$\leq \left(2^{\log_2(2e))-6\log_2 e}\right)^{N/2}$$
$$\leq 2^{-N/2},$$

where we use the bound for the binomial coefficient $\binom{n}{k} \leq (en/k)^k$ and the numeric calculation $\log_2(2e) - 6\log_2 e \leq -1$. From Lemma 3, any agent is activated at most $8Nn$ times with high probability in E, and thus during the last Nn steps (denoted by E') of E the system has at least $n/2$ witness agents. Furthermore, from Lemma 3, with high probability, any agent has at least $N/2$ interactions in E'. Consequently, any agent interacts with a witness agent with probability $1 - 2^{-N/2}$ in E'. Using the union bound, all the agents interact with a witness agent with probability $1 - n2^{-N/2}$. The probability that the scenario above fails is at most $2^{-N/2} + 2n2^{-N/12} + n2^{-N/2} \leq 4n2^{-N/12}$. The lemma is proved.   □

From Lemma 3, the probability that an agent interacts more than $8N$ times in the execution of $4Nn$ steps is exponentially small. Thus we can deduce the corollary from the lemma above:

Corollary 1. *Let C be any execution with a leader, and E_k $(0 \leq k)$ be the period of steps from $4Nnk$ to $4Nn(k+1)$ in $E(C)$. Then, for each $k \geq 1$, $f_i = T$ holds for all agents v_i and any configuration in E_k with probability at least $1 - 4n2^{-N/12}$.*

5.2 Main Algorithm

The main algorithm, called FastLeader, completely follows the algorithm by Fischer and Jiang [11]. Algorithm 2 gives its pseudocode, where local variables l_0 and f_1 are shared with DetectLeader. The algorithm has only two rules: If two leaders interact with each other, one of them are killed, and if the leader detector on some agent v_i reports FALSE, v_i becomes a new leader.

We show the correctness of algorithm FastLeader.

Algorithm 2. Algorithm FastLeader

1:   **At interaction between** (l_0, f_0) **and** (l_1, f_1):
2:      **if** $l_0 = T$ **and** $l_1 = T$ **then** $l_1 \leftarrow F$ **endif**

Theorem 3. *Algorithm* FastLeader *is an* $(O(Nn), \exp(N))$-*stabilizing leader election algorithm.*

Proof. We first consider the case where the initial configuration C_0 has a leader. Algorithm FastLeader never kills all leaders because at least one leader always survives at any interaction. Thus, the execution starting from $E(C_0)$ keeps one leader agent. From Corollary 1, at each period of length $4Nn$ after the first $4Nn$ steps, DetectLeader behaves wrongly with probability at most $4n2^{-N/12}$. It implies that the expected length of the execution where DetectLeader correctly works (after $4Nn$ steps) is $\exp(N)$. Thus the remaining issue we have to show is that the number of leader agents converges to one with $O(Nn)$ expected steps. Now we consider a configuration where k leaders exist. Then, there are $k(k-1)$ interactions decreasing the number of leaders. Thus, the probability that the number of leaders decreases is $k(k-1)/n(n-1)$. That is. the expected number of steps taken to kill one leader is $n(n-1)/k(k-1)$. Since at most n leaders can exist initially, the expected number of steps necessary to elect a single leader is

$$\sum_{i=n}^{2} \frac{n(n-1)}{k(k-1)} = n(n-1) \sum_{i=n}^{2} \left(\frac{1}{k-1} - \frac{1}{k} \right) = O(n^2) = O(Nn).$$

In the case where no leader exists initially, by Lemma 4, at least one leader is generated with $O(Nn)$ expected steps. The following argument is the same as the case where a leader initially exists.   □

6   Concluding Remarks

In this paper, we have shown that for any loosely-stabilizing leader election algorithm with the stabilization period of length $\Omega(\exp(N))$ in expectation, each agent needs $\Omega(\log N)$ memory space, and the expected convergence length must be $\Omega(Nn)$. We have also shown that these bounds are tight by proposing a new algorithm simultaneously achieving both bounds. We conclude this paper with two related open problems:

- The key ingredient of our proof for space complexity is that two agents must create a new leader quickly in the execution isolating them. This strategy does not apply if we assume that the system knows a lower bound N' for n in addition to N. Can we construct an algorithm achieving more compact space with the knowledge on both N' and N?
- All known algorithms, including the ones proposed in this paper, relies on some timeout mechanisms. Thus, the same algorithm seems to work correctly

even in bounded schedulers. Are the loose-stabilization under the probabilistic scheduler and the self-stabilization under bounded (synchronous) schedulers interchangeable?

Acknowledgement. The author thanks Yuichi Sudo and Toshimitsu Masuzawa for their helpful discussion, and Kenji Hata, a former student of the author's group, for his effort on this study.

References

1. Alistarh, D., Gelashvili, R.: Polylogarithmic-time leader election in population protocols. In: Halldórsson, M.M., Iwama, K., Kobayashi, N. (eds.) ICALP 2015, Part II. LNCS, vol. 9135, pp. 479–491. Springer, Heidelberg (2015)
2. Angluin, D., Aspnes, J., Diamadi, Z., Fischer, M., Peralta, R.: Computation in networks of passively mobile finite-state sensors. Distributed Computing 18(4), 235–253 (2006)
3. Angluin, D., Aspnes, J., Eisenstat, D.: Fast computation by population protocols with a leader. Distributed Computing 21(3), 183–199 (2008)
4. Angluin, D., Aspnes, J., Eisenstat, D., Ruppert, E.: The computational power of population protocols. Distributed Computing 20(4), 279–304 (2007)
5. Angluin, D., Aspnes, J., Fischer, M.J., Jiang, H.: Self-stabilizing population protocols. ACM Transactions on Autonomous Adaptive Systtems 3(4), 1–28 (2008)
6. Beauquier, J., Blanchard, P., Burman, J.: Self-stabilizing leader election in population protocols over arbitrary communication graphs. In: Baldoni, R., Nisse, N., van Steen, M. (eds.) OPODIS 2013. LNCS, vol. 8304, pp. 38–52. Springer, Heidelberg (2013)
7. Beauquier, J., Burman, J., Clement, J., Kutten, S.: On utilizing speed in networks of mobile agents. In: Proc. of the 29th ACM SIGACT-SIGOPS Symposium on Principles of Distributed Computing, pp. 305–314 (2010)
8. Beauquier, J., Burman, J., Kutten, S.: Making population protocols self-stabilizing. In: Guerraoui, R., Petit, F. (eds.) SSS 2009. LNCS, vol. 5873, pp. 90–104. Springer, Heidelberg (2009)
9. Beauquier, J., Clement, J., Messika, S., Rosaz, L., Rozoy, B.: Self-stabilizing counting in mobile sensor networks with a base station. In: Pelc, A. (ed.) DISC 2007. LNCS, vol. 4731, pp. 63–76. Springer, Heidelberg (2007)
10. Cai, S., Izumi, T., Wada, K.: How to prove impossibility under global fairness: On space complexity of self-stabilizing leader election on a population protocol model. Theory of Computing Systems 50(3), 433–445 (2012)
11. Fischer, M., Jiang, H.: Self-stabilizing leader election in networks of finite-state anonymous agents. In: Shvartsman, M.M.A.A. (ed.) OPODIS 2006. LNCS, vol. 4305, pp. 395–409. Springer, Heidelberg (2006)
12. Flajolet, P.: Approximate counting: A detailed analysis. BIT 25(1), 113–134 (1985)
13. Izumi, T., Kinpara, K., Izumi, T., Wada, K.: Space-efficient self-stabilizing counting population protocols on mobile sensor networks. Theoretical Computer Science 552, 99–108 (2014)

14. Ogata, M., Yamauchi, Y., Kijima, S., Yamashita, M.: A randomized algorithm for finding frequent elements in streams using o(loglog$N$) space. In: Asano, T., Nakano, S.-i., Okamoto, Y., Watanabe, O. (eds.) ISAAC 2011. LNCS, vol. 7074, pp. 514–523. Springer, Heidelberg (2011)

15. Sudo, Y., Nakamura, J., Yamauchi, Y., Ooshita, F., Kakugawa, H., Masuzawa, T.: Loosely-stabilizing leader election in a population protocol model. Theoretical Computer Science 444, 100–112 (2012)

16. Sudo, Y., Ooshita, F., Kakugawa, H., Masuzawa, T.: Loosely-stabilizing leader election on arbitrary graphs in population protocols. In: Aguilera, M.K., Querzoni, L., Shapiro, M. (eds.) OPODIS 2014. LNCS, vol. 8878, pp. 339–354. Springer, Heidelberg (2014)

Wait-Free Gathering Without Chirality

Quentin Bramas[1,2] and Sébastien Tixeuil[1,2,3]

[1] UPMC Sorbonne Universités, LIP6-CNRS 7606, France
[2] CNRS, LIP6-CNRS 7606, France
[3] Institut Universitaire de France, France

Abstract. We consider the problem of gathering n autonomous robots that evolve in a 2-dimensional Euclidian space at a single location, not known beforehand. We suppose the robots operate in the semi-synchronous Look-Compute-Move model and are anonymous, oblivious, and disoriented.

When robots are capable of strong multiplicity detection (that is, sensing the number of robots at a given location) and the initial configuration is not bivalent, the problem is known to be deterministically solvable for $n > 2$ robots. When an arbitrary number $f < n$ of robots may crash, recent results achieve deterministic gathering of correct robots in the classical model, assuming robots agree on a global common chirality (that is all robots have the same notion of left and right), leaving open the necessity of this assumption.

In this paper, we answer negatively to this question. Our approach is constructive: we present a deterministic gathering algorithm that admits an arbitrary number of crashes and gathers all correct robots even if they do not have a common chirality.

1 Introduction

Networks of mobile robots evolving in a 2-dimensional Euclidian space recently captured the attention of the distributed computing community, as they promise new applications (rescue, exploration, surveillance) in potentially dangerous (and harmful) environments. Since its initial presentation [8], this computing model has grown in popularity and many refinements have been proposed (see [6] for a recent state of the art). From a theoretical point of view, the interest lies in characterizing the exact conditions for solving a particular task.

In the model we consider, robots are anonymous (*i.e.*, indistinguishable from each-other), oblivious (*i.e.*, no persistent memory of the past is available), and disoriented (*i.e.*, they do not agree on a common coordinate system). The robots operate in Look-Compute-Move cycles. In each cycle a robot "Looks" at its surroundings and obtains (in its own coordinate system) a snapshot containing the locations of all robots. Based on this visual information, the robot "Computes" a destination location (still in its own coordinate system) and then "Moves" towards the computed location. Since the robots are identical, they all follow the same deterministic algorithm. The algorithm is oblivious if the computed destination in each cycle depends only on the snapshot obtained in the current

© Springer International Publishing Switzerland 2015
C. Scheideler (Ed.): SIROCCO 2015, LNCS 9439, pp. 313–327, 2015.
DOI: 10.1007/978-3-319-25258-2_22

cycle (and not on the past history of execution). The snapshots obtained by the robots are not consistently oriented in any manner (that is, the robots local coordinate systems do not share a common direction nor a common chirality).

The execution model significantly impacts the solvability of collaborative tasks. Three different levels of synchronization have been considered. The strongest model [8] is the fully synchronized (FSYNC) model where each phase of each cycle is performed simultaneously by all robots. On the other hand, the asynchronous model [6] (ASYNC) allows arbitrary delays between the Look, Compute and Move phases and the movement itself may take an arbitrary amount of time. In this paper, we consider the semi-synchronous (SSYNC) model [8], which lies somewhere between the two extreme models. In the SSYNC model, time is discretized into rounds and in each round an arbitrary yet non-empty subset of the robots are active. The robots that are active in a particular round perform exactly one atomic Look-Compute-Move cycle in that round. It is assumed that the scheduler (seen as an adversary) is fair in the sense that in each execution, every robot is activated infinitely often.

Related Work. The gathering problem is one of the benchmarking tasks in mobile robot networks, and has received a considerable amount of attention (see [6] and references herein). The gathering tasks consist in all robots (considered as dimensionless point in a 2-dimensional Euclidian space) reaching a single point, not known beforehand, in finite time. A foundational result [8,4] shows that in the FSYNC or SSYNC models, no deterministic algorithm can solve gathering for two robots without additional assumptions). This result can be extended [8,4] to the bivalent case, that is when an even number of robots is initially evenly split in exactly two locations. On the other hand, it is possible to solve gathering if $n > 2$ robots start from initially distinct positions if robots are endowed with multiplicity detection: that is, a robot is able to distinguish the two cases where *(i)* a single robots occupies a position, and *(ii)* more than one robot occupies a position.

In hostile environments such as those we envision, robots become likely to fail. So far, three kinds of failures were considered in the context of deterministic gathering [1,3,5,2]:

1. *Transient Faults*: as robots are oblivious (they do not remember their past actions), they naturally are resilient to transient faults that corrupt their memory. However, if the transient fault consequence was to place the robots in some forbidden configuration (*e.g.* a bivalent configuration), some algorithms may not recover. Algorithms can thus be sorted according to the set of admissible initial configurations.

2. *Crash Faults*: when robots may stop executing their algorithm unexpectedly (and correct robots are not able to distinguish a correct robot from a crashed one at first sight), guaranteeing that correct robot still gather in finite time is a challenge. Algorithms can thus be sorted according to the number of admissible crashed robots.

3. *Byzantine Faults*: when robots may have completely (and possibly malicious) behavior, it is known [1] that there exists no deterministic gathering protocol in the SSYNC model even assuming that at most one robot may be Byzantine.

The positive deterministic results so far in a fault tolerant context are as follows. With simple multiplicity detection, and restricting the set of admissible initial configurations to the distinct configurations (that is, the configurations where at most one robot occupies a particular position), gathering is feasible in the SSYNC model with one crash fault [1]. If only transient faults are considered, strong multiplicity detection (that is, being able to sense the exact number of robots at any particular position) permits to extend the set of initial configurations to those that include multiplicity points [5] (however, only the case with an odd number of robots is considered). When a common chirality is available (that is all robots have the same notion of handedness), it becomes possible to tolerate up to $n - 1$ crash faults [3] (that is, the algorithm is wait-free), further expanding the set of initial configurations to those that are not bivalent (so all feasible initial configurations in a deterministic context are considered). When the robots agree on a common direction (*e.g.*, North) it becomes possible to solve gathering without chirality and without any restriction on the set of initial configurations [2]. However, the assumption that a common direction is available trivializes the problem of gathering as it also sufficient to solve gathering starting from a bivalent configuration, but not necessary [7] (in that sense, agreeing on a common direction is a stronger problem than gathering). The various assumptions are summarized in Table 1.

Table 1. Gathering robots in the presence of faults

Reference	Direction	Chirality	Multiplicity	# Crashes	Admissible initial configurations
[1]	No	No	Weak (binary)	$f = 1$	distinct
[5]	No	No	Strong	$f = 0$	n is odd
[3]	No	Yes	Strong	$f = n - 1$	not bivalent
[2]	Yes	No	No	$f = n - 1$	any
This paper	No	No	Strong	$f = n - 1$	not bivalent

So the main open question is whether it is possible to relax the weakest assumptions to date in the "classical" model (that is, without assuming a common direction is available to all robots).

Our Contribution. In this paper, we present a deterministic gathering protocol that can start from any non-bivalent configuration (the largest possible set in the classical model), yet does not require to assume that all robots share a common direction (as in [2]), nor a common chirality (as in [3]). The protocol retains the ability to tolerate up to $n - 1$ crash faults, that it, it is wait-free.

2   Preliminaries

2.1   Model

As previously stated, we consider n robots modeled as dimensionless points evolving in a 2-dimensional Euclidian space. The robots do not share any global knowledge (direction, chirality, etc.), and each robot has its own coordinate system, orientation and unit of length. All robots are anonymous (identical and thus indistinguishable), uniform (they follow the same algorithm), oblivious (they does not remember their previous actions or the previous positions of the other robots) and they do not communicate in an explicit manner.

We assume the semi-synchronous model denoted $SSYNC$. The time is divided into discrete intervals. At each time a robot can be either active or inactive. When active, a robot r makes exactly one Look-Compute-Move cycle. During the Look stage, each active robot r gets the same snapshot containing the location of every other robot with respect to its local coordinate system. With weak multiplicity detection, r knows whether there is "one" or "more than one" robots located at a point. With strong multiplicity detection, r knows the exact number of robots located at a particular point. During the Move stage, robots move towards the computed destination. There exists a constant $\delta > 0$, such that if a destination point is closer than δ, the robot will reach it, otherwise it will move a distance of at least δ toward it.

Let $\mathcal{R} = \{r_1, r_2, \ldots, r_n\}$ denote the set of robots. At any time $\tau \in \mathbb{N}$, $r_i(\tau)$ denotes the robot position at time τ, and $C(\tau)$ denotes the multiset $C(\tau) = \{r_1(\tau), \ldots, r_n(\tau)\}$. In the sequel, we may identify a robot with its position depending on the context. Given a configuration C, $U(C)$ denotes the set of robot positions in C removing multiplicities. $SEC(C)$ denotes the smallest enclosing circle of the set $U(C)$. For $u \in U(C)$, $mult(u)$ denotes the multiplicity of the point u. $\mathcal{R}(P)$ denotes the robots in $\mathcal{R}$ located at P. The predicate GATHERED$(\mathcal{R}, \tau)$ equals $true$ if all non-faulty robots share the same location at time τ.

With $a, b \in \mathbb{R}^2$, let $|a, b|$ denote the Euclidian distance between a and b. Then, $[a, b]$ (resp. $]a, b[$) denote the closed (resp. open) segment line between a and b. $]a, b)$, denote the half line starting at a (excluding a) and passing through b. Since robots do not have a common chirality, $\sphericalangle(u, c, v)$ denote the angle between segment $[c, u]$ and $[c, v]$, in the direction that minimizes it.

The crash fault-tolerance capabilities of a deterministic algorithm come from the following Lemma:

Lemma 1. *[1,3] A gathering algorithm for n robots is tolerant against $f < n$ crashes only if at each configuration, there exists a unique point P such that all robots that are not located at P are instructed to move.*

Definition 1. *Let $C = \{p_1, \ldots, p_n\}$ be a configuration of robots and*

$$c = center(SEC(C))$$

Given a position $p \in U(C)$, the view of p (or, equivalently, the view of the robot(s) located at p), denoted by $\mathcal{V}(p)$, is the multiset of points in C expressed in a polar coordinate system whose origin $(0,0)$ is p and whose point $(1,0)$ is defined as follow. If $(c \neq p)$, then $(1,0) = c$. Otherwise $(1,0)$ is any point $x \neq p \in U(C)$ that maximizes $\mathcal{V}(p)$. The orientation of the polar coordinate system is chosen to maximize $\mathcal{V}(p)$.

Views are compared in a lexicographical manner. Since robots do not share a common chirality, robots may maximize their view in one orientation, while other robots may maximize their view in the other orientation.

Based on the definition of views, we can define an equivalence relation $\backsim_r$ on the set $U(C)$. Two points $u, u' \in U(C)$ are equivalent, denoted $u \backsim_r u'$, if and only if they have the same view with the same orientation.

Definition 2. *The rotational symmetricity of a configuration C, denoted by $sym(C)$, is the cardinality of the largest equivalence class defined by $\backsim_r$.*

Lemma 2. *Let C be a configuration with $k = sym(C) > 1$, and let*

$$c = center(SEC(C))$$

For every $u \in U(C)$, with $c \neq u$, the equivalence class of u is a k-gon with center c and whose corners have the same multiplicity.

Lemma 3. *The perpendicular bisector of two robots that have the same view with different orientation is an axis of symmetry.*

Then a configuration C is symmetric if $sym(C) > 1$. Moreover, if $sym(C) = 1$ and two robots have the same view, with different orientation, then C has an axis of symmetry. Also, if a robot has a unique view (for any choice of chirality, the maximized view is the same), then it is on an axis of symmetry.

2.2   Weber Point

A Weber point minimizes the sum of distances to the robots:

Definition 3. *Given a configuration C, define:*

$$sum(C, r) = \sum_{q \in C} |r, q|$$

$$sum_{\min}(C) = \min_{r \in C} \sum_{q \in C} |r, q|$$

$$W(C) = argmin_{P \in \mathbb{R}^2} sum(C, P)$$

$W(C)$ *is called a Weber point. We say that a point P reduces the sum of distances if $sum(C, P) \leq sum_{\min}(C)$ (note that this does not imply that P is a Weber point).*

A Weber point has the nice property to remain invariant under straight line movement towards it. Even if it is not calculable in general, the Weber point of a symmetric configuration is simply the common center of the regular polygons formed by the equivalency classes of the relation $\backsim_r$. This allows to define a new class of configurations called *quasi regular*, which is the set of configurations obtained from a symmetric configuration after several straight line movements toward the center (see [3] for a formal definition).

Lemma 4. *If $M \in [A, B]$, then $sum(C, M) \leq \max\left(sum(C, A), sum(C, B)\right)$.*

Sketch of proof. The lemma comes from the convexity of the sum of distances.

Lemma 5. *Let $\tau \in \mathbb{N}$ and suppose that, at time τ, all robots move towards a robot that reduces the sum of distances. If $r_0(\tau + 1) \in C(\tau + 1)$ reduces the sum of distances, then either $r_0(\tau)$ reduces the sum of distances, or $r_0(\tau) \in Move(\tau)$.*

Proof. Assume that $r(\tau) \notin Move$, and let r_d be the common destination of all robots. Let $d_r = distance(r, \tau)$ be the distance traveled by Robot r at time τ, and $d = \sum_{r \in \mathcal{R}} d_r$. Then,

$$sum(C, r_d(\tau + 1)) = sum(C, r_d(\tau)) - d$$

By the triangle inequality, we have

$$sum(C, r_0(\tau + 1)) \geq sum(C, r_0(\tau)) - d$$

Since

$$sum_{\min}(C(\tau)) = sum(C, r_d(\tau)) \leq sum(C, r_0(\tau)) \tag{1}$$

we have

$$sum(C, r_d(\tau + 1)) \leq sum(C, r_0(\tau + 1))$$

Finally:

$$sum_{\min}(C(\tau + 1)) \leq sum(C, r_d(\tau + 1)) \leq sum(C, r_0(\tau + 1)) \tag{2}$$

So in order to have equality in Equation 2, we must also have equality in Equation 1. Hence, r_0 reduces the sum of distances at time τ.

3 Robot Configurations

3.1 Classification

We use the same partition of configurations as in [3]. The set of all possible configurations is denoted by $\mathcal{P}$.

- **Bivalent($\mathcal{B}$):**
  $\mathcal{B} = \{C \in \mathcal{P} \mid \forall u \in U(C) \; : \; mult(u) = n/2\}$.

- **Multiple($\mathcal{M}$):**
  $\mathcal{M} = \{C \in \mathcal{P} \mid \exists u \in U(C) : \forall v \neq u \in U(C) : multi(v) < mult(u)\}.$
- **Collinear($\mathcal{L}$)**
  $\mathcal{L} = \{C \in \mathcal{P} \backslash (\mathcal{B} \cup \mathcal{M}) \mid C \text{ is linear}\}.$
  $\mathcal{L}1W = \{C \in \mathcal{L} \mid C \text{ has a unique Weber point}\}.$
  $\mathcal{L}2W = \mathcal{L} \backslash \mathcal{L}1W.$
- **Q-Regular($\mathcal{QR}$)**
  $\mathcal{L} = \{C \in \mathcal{P} \backslash (\mathcal{B} \cup \mathcal{M} \cup \mathcal{L}) \mid C \text{ is quasi regular}\}.$
- **Asymmetric($\mathcal{A}$)**
  $\mathcal{A} = \{C \in \mathcal{P} \backslash (\mathcal{B} \cup \mathcal{M} \cup \mathcal{L} \cup \mathcal{QR}) \mid sym(C) = 1\}.$

3.2   Safe Robots

Definition 4. *Given a configuration C, a robot p is safe if and only if $\forall q \in \mathbb{R}^2 \backslash \{p\}$, the half-line $]p, q)$ contains at most $\lceil n/2 \rceil - 1$ robots of C.*

Safe robots can be used as destination points because they guarantee that the configuration will not become bivalent. We can show the following properties for safe robots (see [3] for a proof of lemma 6 and 7).

Lemma 6. *Any non linear configuration contains a safe robot.*

Lemma 7. *If $C \in \mathcal{B} \cup \mathcal{L}2W$, then C does not have a safe robot.*

Lemma 8. *If the configuration C is not linear, then robots that reduce the sum of distances are safe.*

Proof. Suppose that the robot at point P reduces the sum of distances and is not safe *i.e.* there are $k \geq \lceil \frac{n}{2} \rceil$ robots $r_1, \dots, r_k$ from near to far in a half line L starting at P. We show that $\text{sum}(C(\tau), P) > \text{sum}(C(\tau), r_1)$.
  Firstly, for all $1 \leq i \leq k$, $|P, r_i| = |r_1, r_i| - |r_1, P|$.
  Secondly, for a robot $r \notin L$ $|P, r| \geq |r_1, r| + |r_1, P|$.
  Finally, since the configuration is not linear, the last inequality is strict for at least one robot r'. When we sum the k first equalities and the $n - k$ inequalities, since $k \geq \lceil \frac{n}{2} \rceil$ and one inequality is strict, we have the following contradiction:

$$\sum_{r \in \mathcal{R}} |P, r| > \sum_{r \in \mathcal{R}} |r_1, r|$$

This completes the proof.

Definition 5. *Let $C \in \mathcal{A}$. Consider the subset $S \subset C$ of (safe) robots that reduce the sum of distances. Let $S_1 \subset S$ be the set of robots in S that have a unique view. If S_1 is not empty, safeMin(C) denote the singleton consisting of the robot in S_1 that maximizes the multiplicity and that maximizes the view. Else safeMin(C) denote set of robots in S that maximize the multiplicity and maximize the view.*

From now $\mathcal{A}_1 = \{C \in \mathcal{A} : |\text{safeMin}(C)| = 1\}$ and $\mathcal{A}_2 = \mathcal{A} \backslash \mathcal{A}_1$

3.3   Properties of Configurations

Lemma 9. *Let $C \in \mathcal{A}$, there are exactly one or two robots in safeMin(C). Thus $\mathcal{A}_2 = \{C \in \mathcal{A} : |safeMin(C)| = 2\}$.*

Proof. We just have to show that we cannot have 3 robots located at distinct positions in safeMin(C). By contradiction, suppose $\{r_1, r_2, r_3\} \subset$ safeMin(C).

Since each robot has one chirality, clockwise or counterclockwise, there are two robots, say r_1 and r_2, that have the same chirality. Moreover, since all robots in safeMin(C) maximize the view, r_1 and r_2 must have the same view. So $sym(C) \geq 2$. That is a contradiction with $C \in \mathcal{A}$.

Lemma 10. *If $C \in \mathcal{A}_2$, then there is a unique axis of symmetry and the middle of the two robots in safeMin(C) is on the axis of symmetry.*

Proof. Since $C \in \mathcal{A}_2$, there are two robots in safeMin(C) located at different positions that have the same view with different chirality. By lemma 3, the perpendicular bisector is an axis of symmetry (that contains the middle of the two robots in safeMin(C). Since sym(C) = 1, the axis is unique.

Lemma 11. *Let $C \in \mathcal{A}_2$ and M be the middle of the two robots in safeMin(C), then $sum(C, M) < sum_{min}(C)$*

Proof. Let r_{min} be a robot in safeMin(C). Let r be another robot. If r is located on the axis of symmetry, then $|r, M| < |r, r_{min}|$. Else, let r' be the symmetric robot of r. Since the lines (r, r') and (M, r_{min}) are parallels and M is on the perpendicular bisector of $[r, r']$, we have:

$$|r, M| + |r', M| < |r, r_{min}| + |r', r_{min}|$$

By summing over all robots we have

$$\mathrm{sum}(C, M) < \mathrm{sum}(C, r_{min}) = \mathrm{sum}_{min}(C)$$

Lemma 12. *Let $C \in \mathcal{A}_2$ and P be the middle of the robots in safeMin(C). The configuration obtained after moving some robots toward P is not in $\mathcal{A}_2$ if at least one robot reaches P.*

Proof. Let M be the set of robots moving. Let d be the sum of distances traveled by the robots, Let C' be the configuration after the movements. We have $\mathrm{sum}(C', P) = \mathrm{sum}(C, P) - d$. Moreover, for every robot r:

$$\mathrm{sum}(C', r') \geq \mathrm{sum}(C, r) - d$$

where r' is the position of r after the movement. Also, we have from Lemma 11:

$$\mathrm{sum}(C, r) - d > \mathrm{sum}(C, P) - d = \mathrm{sum}(C', P)$$

So that

$$\mathrm{sum}(C', r') > \mathrm{sum}(C', P)$$

After the movement, a robot is located at P and (strictly) reduce the sum of distances in C' so that, if $C' \in \mathcal{A}$, then $\{P\} = $ safeMin(C') and $C' \in \mathcal{A}_1$.

Lemma 13. *If $C \in \mathcal{A}_2$, then the middle of the robots in safeMin(C) is safe.*

Proof. By Lemma 11, the middle point reduces the sum of distances and by Lemma 8, in a non linear configuration, a point that reduces the sum of distances is safe.

4   The Algorithm

Since the chirality assumption is not crucial in the first four configurations, our algorithm is almost the same as [3] for those classes. The main change appears when the configuration is in $\mathcal{A}$.

Algorithm 1. Fault-Tolerant-Gathering executed by robot r

Let C be the current configuration.

- $C \in \mathcal{L}1\mathcal{W} \cup \mathcal{QR}$: elect the unique Weber point.
- $C \in \mathcal{L}2\mathcal{W}$: Then $U(C)$ is included in a segment $[p_1, p_2]$ where at least one robot is located at each end point. Let c be the middle of $[p_1, p_2]$. If $r \notin \{p_1, p_2\}$ elect c. Otherwise, elect a point e such that $|e, c| = |r, c|$ and $\sphericalangle(r, c, e) = \pi/4$.
- $C \in \mathcal{M}$: Let $e = argmax_{p \in C} mult(p)$. If there is no other robot between r and e, elect e. Otherwise, elect d that verifies $|d, e| = |r, e|$ and

$$\forall v \in \{p \in C \mid \sphericalangle(r, e, p) \neq 0\}, \quad \sphericalangle(r, e, d) < \sphericalangle(r, e, v)/3).$$

- $C \in \mathcal{A}$: If $|safeMin(C)| = 1$, elect the unique robot in safeMin(C), otherwise, there are two robots in safeMin(C), elect the middle point.

Robot r moves toward the elected destination.

---

5   Proof of Correctness

Lemma 14 (see [3]). *Let $C(\tau) \in \mathcal{M} \cup \mathcal{L}1\mathcal{W} \cup \mathcal{QR}$. There exists a time $\tau_c > \tau$ such that $GATHERED(\mathcal{R}, \tau_c) = true$.*

Lemma 15 (see [3]). *Let $C(\tau) \in \mathcal{L}2\mathcal{W}$. There exists a time $\tau_c > \tau$ such that $C(\tau_c) \notin (\mathcal{L}2\mathcal{W} \cup \mathcal{B}) \vee GATHERED(\mathcal{R}, \tau_c) = true$.*

Lemma 16. *Let $C(\tau) \in \mathcal{A}$. There exists a time $\tau_c > \tau$ such that $(C(\tau_c) \in \mathcal{M} \cup \mathcal{L}1\mathcal{W} \cup \mathcal{QR}) \vee (GATHERED(\mathcal{R}, \tau_c) = true))$*

Proof. The lemma follows from Claims $C1$ and $C6$ below. Claims $C2$, $C3$, $C4$ and $C5$ are used to prove $C6$.

$C1 : (C(\tau) \in \mathcal{A}) \Rightarrow (C(\tau + 1) \in \mathcal{M} \cup \mathcal{L}1\mathcal{W} \cup \mathcal{QR} \cup \mathcal{A})$

$C2$: If $C(\tau) \in \mathcal{A}_1$ and $C(\tau + 1) \in \mathcal{A}$, then one of the following assertions holds:

- $C(\tau + 1) = C(\tau)$

- $\text{sum}_{\min}(C(\tau + 1)) \leq \text{sum}_{\min}(C(\tau)) - \delta$

- $\begin{cases} C(\tau + 1) \in \mathcal{A}_1 \\ elected(\tau + 1) = elected(\tau) \\ mult(elected(\tau + 1)) > mult(elected(\tau)) \end{cases}$

$C3: (C(\tau) \in \mathcal{A}_2) \Rightarrow (\text{sum}_{\min}(C(\tau + 1)) \leq \text{sum}_{\min}(C(\tau)))$
$C4: (\forall \tau' \geq \tau : C(\tau') \in \mathcal{A}_1) \Rightarrow (\exists \tau_c \geq \tau : \text{GATHERED}(\mathcal{R}, \tau_c))$
$C5: (\forall \tau' \geq \tau : C(\tau') \in \mathcal{A}_2) \Rightarrow (\exists \tau_c \geq \tau : \text{GATHERED}(\mathcal{R}, \tau_c))$
$C6 : (\forall \tau' \geq \tau : C(\tau') \in \mathcal{A}) \Rightarrow (\exists \tau_c \geq \tau : \text{GATHERED}(\mathcal{R}, \tau_c))$

Proof of C1: If $C(\tau + 1)$ is non linear, then $C(\tau + 1) \notin \mathcal{B} \cup \mathcal{L2W}$. Otherwise, we show that every activated robot elects the same position P, so $C(\tau + 1)$ may be linear only if some robots reach P. Since P is safe at time τ, then P is safe at time $\tau + 1$. Indeed, for every $x \in \mathbb{R}\backslash\{P\}$, the number of robots that are located at $]P, x)$ does not increase (and decreases if some robots reach P). So, there is at least one safe robot located at P. According to Lemma 7, this implies $C(\tau + 1) \notin \mathcal{B} \cup \mathcal{L2W}$.

Proof of C2: Assume that $C(\tau + 1) \neq C(\tau)$ and suppose that:

$$\text{sum}_{\min}(C(\tau + 1)) > \text{sum}_{\min}(C(\tau)) - \delta$$

The destination point P is the same for every robot (it is the unique robot in safeMin($C(\tau)$)) so that $\text{sum}(C(\tau), P) - \text{sum}(C(\tau + 1), P)$ is exactly the sum of distances traveled by moving robots at time τ. Since

$$\text{sum}_{\min}(C(\tau + 1)) = \text{sum}(C(\tau), P)$$

then

$$\text{sum}(C(\tau), P) - \text{sum}(C(\tau + 1), P) < \delta.$$

This is possible if every moving robot reaches its destination. So the multiplicity of P increases at time $\tau + 1$. We have to prove now that $C(\tau + 1) \in \mathcal{A}_1$, and P is the elected destination at time $\tau + 1$.

We know that $C(\tau + 1) \in \mathcal{A}$, so robots in safeMin($C(\tau + 1)$) are safe and reduce the sum of distances. Yet, by Lemma 5, a robot that reduces the sum of distances at time $\tau + 1$ must reduce the sum of distances at time τ or be moving at time τ. In our case, moving robots at time τ are located at P at time $\tau + 1$. Moreover, P already reduces the sum of distances at time τ, so P also reduces the sum of distances at time $\tau + 1$. The multiplicity of other points does not increase because every moving robot reaches P, so P has a strictly greater multiplicity than all other safe robot that reduces the sum of distances. Moreover, P is located on the axis (again, because there is no other safe robot that reduces the sum of distances, with the same multiplicity). Overall, P is safe, reduces the sum of distances, and has a unique view, *i.e.* $C(\tau + 1) \in \mathcal{A}_1$ and $P = elected(\tau + 1)$.

Proof of C3: Let r_1 and r_2 be the two robots in safeMin($C(\tau)$). We have sum($C(\tau), r_1$) = sum($C(\tau), r_2$) = sum$_{min}$($C(\tau)$), and the elected position P is the middle of $[r_1, r_2]$. Let r be another robot. If r is on the axis of symmetry, then the angles $\sphericalangle(P, r_1, r)$ and $\sphericalangle(P, r, r_1)$ are acute, so $|r_1(\tau), r(\tau)| \leq |r_1(\tau{+}1), r(\tau{+}1)|$ (the same inequality holds with r_2). Otherwise, there exists r', the symmetric robot of r with respect to the axis. Let

$$S(\tau) = |r_1(\tau), r(\tau)| + |r_1(\tau), r'(\tau)| + |r_2(\tau), r(\tau)| + |r_2(\tau), r'(\tau)|$$

We want to show that $S(\tau + 1) \leq S(\tau)$.

We decompose the movement of the four robots r_1, r_2, r and r' into two movements: r_1 and r_2 move at time $\tau + \frac{1}{2}$, and r and r' move just after. Note that the way we virtually decompose the movement between τ and $\tau + 1$ does not change the sum at time $\tau + 1$. See Figure 1 to observe two steps.

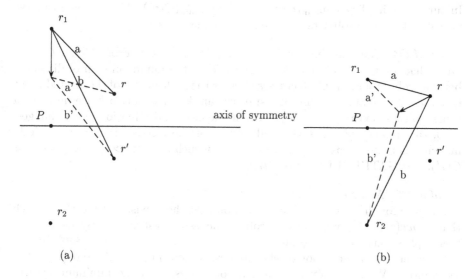

(a)   (b)

Fig. 1. (a) When r_1 moves toward P, we have $a' + b' \leq a + b$. (b) When r moves toward P, we have $a' + b' \leq a + b$

In the first phase, r_1 moves toward P along a segment parallel to the line (r, r'). Since r_1 moves beyond P, which is on the perpendicular bisector of the segment $[r, r']$, we have:

$$|r_1(\tau + \tfrac{1}{2}), r(\tau + \tfrac{1}{2})| + |r_1(\tau + \tfrac{1}{2}), r'(\tau + \tfrac{1}{2})| \leq |r_1(\tau), r(\tau)| + |r_1(\tau), r'(\tau)|$$

The same inequality holds for r_2 and we have $S(\tau + \frac{1}{2}) \leq S(\tau)$.

In the second phase, let T be the triangle formed by $r_1(\tau + \frac{1}{2})$, $r_2(\tau + \frac{1}{2})$, and r. Even upon completion of the movement of r_1 and r_2, we have $P \in [r_1, r_2]$, so r remains inside T at time $\tau + 1$. So:

$$|r_1(\tau+1),r(\tau+1)|+|r_2(\tau+1),r(\tau+1)| \leq |r_1(\tau+\tfrac{1}{2}),r(\tau+\tfrac{1}{2})|+|r_2(\tau+\tfrac{1}{2}),r(\tau+\tfrac{1}{2})|$$

And the same inequality holds with r', so that $S(\tau + 1) \leq S(\tau + \tfrac{1}{2})$. Overall, $S(\tau + 1) \leq S(\tau)$. The same reasoning can be used for every robot, and we get:

$$\text{sum}(C(\tau + 1), r_1(\tau + 1)) + \text{sum}(C(\tau + 1), r_1(\tau + 1))$$
$$\leq \text{sum}(C(\tau), r_1(\tau)) + \text{sum}(C(\tau), r_2(\tau))$$

Which means either

$$\text{sum}(C(\tau + 1), r_1(\tau + 1)) \leq \text{sum}(C(\tau), r_1(\tau))$$

or

$$\text{sum}(C(\tau + 1), r_2(\tau + 1)) \leq \text{sum}(C(\tau), r_2(\tau))$$

In turn, this implies $\text{sum}_{\min}(C(\tau + 1)) \leq \text{sum}_{\min}(C(\tau))$. Also, we can observe that if at least one robot moves then the inequality is strict.

Proof of C4: Assume that $(\forall \tau' \geq \tau \ : \ C(\tau') \in \mathcal{A}_1)$. From Claim $C2$, we know that for $\tau' \geq \tau$, either $C(\tau' + 1) = C(\tau')$, the minimum sum of distances between robots $\text{sum}_{\min}(C)$ decreases by δ, or the multiplicity of the elected point increases. The minimum sum of distances can decrease only a finite number of times, so there exists a time τ_1 such that the elected destination never changes. Also, the number of times the multiplicity of the elected destination strictly increases is finite. Thus, there exists a time τ_2 such that $\forall \tau' > \tau_2, \quad C(\tau' + 1) = C(\tau')$, and GATHERED$(\mathcal{R}, \tau_2) = true$.

Proof of C5: Assume that $(\forall \tau' \geq \tau \ : \ C(\tau') \in \mathcal{A}_2)$.

Suppose for the purpose of contradiction that there exists a time $\tau' > \tau$ such that $Move(\tau') \neq \emptyset$, and a moving robot reaches its destination. By Lemma 12, this implies that $C(\tau' + 1) \notin \mathcal{A}_2$.

So, after time τ, no robot reaches its destination. Let $C \in \mathcal{A}_2$ and $\{r_1, r_2\} = \text{safeMin}(C)$. We now show that if a robot moves, then the minimum sum of distances between robots decreases by a constant factor.

Let ω be a non-null constant and let Ω be the set of points that are at distance at most ω from $[r_1, r_2]$ (see Figure 2).

Case 1: There exists a moving robot inside Ω. Let r be one of the moving robots inside Ω, whose next location $r(t + 1)$ is the closest to I, the middle of $[r_1, r_2]$.

For every $p \in [r_1, r_2]$, we have $\text{sum}(C, p) \leq \text{sum}_{\min}(C)$. Since r is at distance less than ω from $[r_1, r_2]$, we have:

$$\text{sum}(C, r) \leq \text{sum}(C, p) + n\omega \leq \text{sum}_{\min}(C) + n\omega.$$

So, if we choose $\omega = \tfrac{\delta}{2n}$, then $\text{sum}(C, r) \leq \text{sum}_{\min}(C) + \delta/2$. Since r moves towards the axis of symmetry, we also have:

$$\text{sum}(C(t), r(t + 1)) \leq \text{sum}_{\min}(C(t)) + \delta/2$$

Moreover, since the other moving robots do not move closer to I than r, then

$$\forall r' \in Move(t), \quad |r(t+1) - r'(t+1)| \leq |r(t+1) - r'(t)|.$$

Also,

$$0 = |r(t+1) - r(t+1)| \leq |r(t+1) - r(t)| - \delta$$

so that

$$\mathrm{sum}(C(t+1), r(t+1)) \leq \mathrm{sum}(C(t), r(t+1)) - \delta \leq \mathrm{sum}_{\min}(C(t)) - \delta/2$$

Case 2: There exists a moving robot r outside Ω.

Suppose first that r is the only robot to move at time t. Let $\Delta_1 = \mathrm{sum}(C, r_1(t)) - \mathrm{sum}(C, r_1(t+1))$ and $\Delta_2 = \mathrm{sum}(C, r_2(t)) - \mathrm{sum}(C, r_2(t+1))$. Let $\varphi = |r_1 - r_2|$, and d be the distance traveled by r. Let Φ be the diameter of the smallest enclosing circle S of $C(\tau)$. Robots do move outside S, so until the end of the execution, they cannot be at distance more than Φ from one another. We now show that $\Delta_1 + \Delta_2$ is greater than a non-null positive constant.

Let $f : (\varphi, d, \omega, r) \mapsto \Delta_1 + \Delta_2$. We know from $C3$ that, if at least one robot moves, $f > 0$ and for any fixed φ, ω, d, and for r sufficiently far from its destination, $f(\varphi, \omega, d, r) \geq d/2$. So, $r \mapsto f(\varphi, d, \omega, r)$ admits a minimum $\lambda_{\varphi, \omega, d}$. Moreover, f is decreasing with respect to the first variable, and increasing with respect to the second variable, so we obtain:

$$\forall \varphi \in [0, \Phi], \forall d \in [\delta, +\infty[, \forall \omega \in \mathbb{R}^+, \forall r \notin \Omega, \quad f(\varphi, d, \omega, r) \geq \lambda_\omega$$

Thus,

$$\mathrm{sum}(C, r_1(t+1)) + \mathrm{sum}(C, r_2(t+1)) \leq \mathrm{sum}(C, r_1(t)) + \mathrm{sum}(C, r_2(t)) - \lambda_\omega$$

Since the other robots movements do not make $\mathrm{sum}(C, r_1(t+1)) + \mathrm{sum}(C, r_2(t+1))$ increase, then the inequality is true even if other robots move. This implies that either $\mathrm{sum}(C, r_1(t+1)) \leq \mathrm{sum}(C, r_1(t)) - \lambda_\omega/2$, or $\mathrm{sum}(C, r_2(t+1)) \leq \mathrm{sum}(C, r_2(t)) - \lambda_\omega/2$.

Thus we have:

$$\mathrm{sum}_{\min}(C(t+1)) \leq \mathrm{sum}_{\min}(C(t)) - \min\left(\frac{\delta}{2}, \frac{\lambda_{\delta/2n}}{2}\right)$$

So, at each time $\tau' \geq \tau$, if $Move(\tau') \neq \emptyset$, the minimum sum of distances between robots $\mathrm{sum}_{\min}(C(\tau))$ decreases by at least ε, with $\varepsilon = \min\left(\frac{\delta}{2}, \frac{\lambda_{\delta/2n}}{2}\right)$. Thus, there exists a time τ_1 such that robots do not move anymore: $\forall \tau' \geq \tau_1 : C(\tau'+1) = C(\tau')$, and GATHERED$(\mathcal{R}, \tau_1) = true$.

Proof of C6: Assume that $(\forall \tau' \geq \tau : C(\tau') \in \mathcal{A})$. If there exists a time τ_1 and $i \in \{1,2\}$ such that $(\forall \tau' \geq \tau_1 : C(\tau') \in \mathcal{A}_i)$, then we can use claims $C4$ and $C5$ to conclude. Otherwise, there is a strictly increasing sequence $(t_i)_{i \in \mathbb{N}}$ such that:

$$\forall i \in \mathbb{N} \quad C(t_i) \in \mathcal{A}_1 \wedge C(t_i + 1) \in \mathcal{A}_2$$

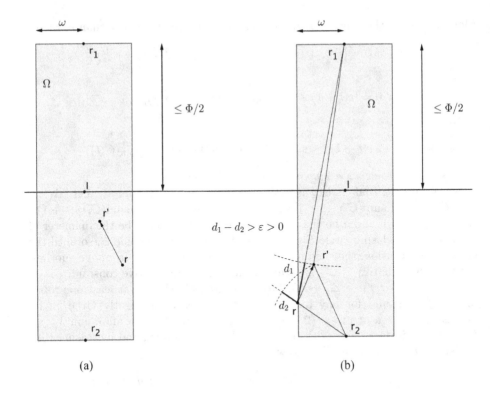

Fig. 2. The two cases of a robot movement. After the movement of r starting *(a)* inside Ω, or *(b)* outside Ω, the minimum sum of the distances decreases by ε.

However, each time this happens, the sum of distances decreases by δ, and it never increases afterward. This is impossible since the sum of distances is a positive number.

Theorem 1. *Algorithm 1 achieves gathering of all correct robots starting from any non-bivalent initial configuration, even if $0 \leq f < n$ robots crashes. The algorithm assumes strong multiplicity detection and makes no common chirality assumption.*

6   Conclusion and Perspectives

We presented a wait-free deterministic gathering protocol that does not require robots to share a common chirality, preserving the most general setting for initial configurations (only bivalent configurations are forbidden). One downside of our approach with respect to Agmon and Peleg solution (that works only for a single crash fault, and only admits distinct configurations as initial possibilities) is that we make use of strong multiplicity, while they use only weak multiplicity. We conjecture that strong multiplicity is required if the admissible set of initial configurations is strictly greater than the set of distinct configurations, and the rest of the hypotheses is preserved (no common direction, no common chirality).

Another challenging open problem would be to extent our results to the ASYNC model without weakening the underlying model with the strong assumption that all robots agree on a common direction (as in [2]).

Acknowledgements. We are grateful to Shantanu Das for helpful comments on a preliminary version of this work. This work was supported in part by LINCS.

References

1. Agmon, N., Peleg, D.: Fault-tolerant gathering algorithms for autonomous mobile robots. SIAM J. Comput. 36(1), 56–82 (2006)
2. Bhagat, S., Gan Chaudhuri, S., Mukhopadhyaya, K.: Fault-tolerant gathering of asynchronous oblivious mobile robots under one-axis agreement. In: Rahman, M.S., Tomita, E. (eds.) WALCOM 2015. LNCS, vol. 8973, pp. 149–160. Springer, Heidelberg (2015)
3. Bouzid, Z., Das, S., Tixeuil, S.: Gathering of mobile robots tolerating multiple crash faults. In: Proceedings of the IEEE International Conference on Distributed Computing Systems (ICDCS 2013), pp. 337–346. IEEE Computer Society Press, Philadelphia (2013)
4. Courtieu, P., Rieg, L., Tixeuil, S., Urbain, X.: Impossibility of gathering, a certification. Information Processing Letters (IPL) 115(3), 447–452 (2015)
5. Dieudonné, Y., Petit, F.: Self-stabilizing gathering with strong multiplicity detection. Theor. Comput. Sci. 428, 47–57 (2012)
6. Flocchini, P., Prencipe, G., Santoro, N.: Distributed computing by oblivious mobile robots. In: Synthesis Lectures on Distributed Computing Theory. Morgan & Claypool Publishers (2012)
7. Izumi, T., Souissi, S., Katayama, Y., Inuzuka, N., Défago, X., Wada, K., Yamashita, M.: The gathering problem for two oblivious robots with unreliable compasses. SIAM J. Comput. 41(1), 26–46 (2012)
8. Suzuki, I., Yamashita, M.: Distributed anonymous mobile robots: Formation of geometric patterns. SIAM J. Comput. 28(4), 1347–1363 (1999)

Treasure Hunt with Advice*

Dennis Komm[1], Rastislav Královič[2], Richard Královič[3], and Jasmin Smula[1]

[1] Department of Computer Science, ETH Zürich, Switzerland
{dennis.komm,jasmin.smula}@inf.ethz.ch
[2] Department of Computer Science, Comenius University, Bratislava, Slovakia
kralovic@dcs.fmph.uniba.sk
[3] Google Zurich, Inc., Switzerland
ri.kralovic@gmail.com

Abstract. The node searching problem (a.k.a. treasure hunt) is a fundamental task performed by mobile agents in a network and can be viewed as an online version of the shortest path problem: an agent starts in a vertex of an unknown weighted undirected graph, and its goal is to reach a given vertex. The cost is the overall distance (measured by the weights of the traversed edges) traversed by the agent. We consider the setting in which the agent sees the identifier of the vertex it is located in, the weights of the incident edges, and also the identifiers of the neighboring vertices. We analyze the problem from the point of view of advice complexity: at the beginning, the agent has a tape with an advice string that gives some a priori information about the input instance. This information has no restricted form; instead, the aim is to study the relationship between the size of this advice and the competitive ratio that can be obtained. We give tight bounds of the form $\Theta(n/r)$ bits of advice for a competitive ratio r (possibly depending on the number of vertices n). In particular, this means that an a priori knowledge of any graph parameter (which would be of size $O(\log n)$) cannot yield a competitive ratio better than $\Omega(n/\log n)$. Moreover, we give a lower bound on the expected competitive ratio of any randomized online algorithm for treasure hunt.

1 Introduction

Problems that involve traversing a graph in a certain way belong to the fundamentals of graph theory [11], and a vast amount of effort has been invested into their study. In such problems, there is an entity (an *agent*) that starts in some vertex of a given graph, and its goal is to explore the graph in some manner such as by traversing all vertices/edges (possibly under some restrictions), finding a (shortest) path to a given vertex, etc. The oldest and most extensive treatment has been the offline case, where the graph and the starting vertex are given as an input and the task is to find the optimal exploration route. The online versions, where the graph is unknown and the agent can only observe its immediate surroundings, have also received significant attention over the last decades [2,18].

---

* Partially funded by the SNF grant 200021-146372 and grant VEGA 1/0979/12.

C. Scheideler (Ed.): SIROCCO 2015, LNCS 9439, pp. 328–341, 2015.
DOI: 10.1007/978-3-319-25258-2_23

In this paper, we consider the following problem, which is called the *treasure hunt* problem. Given is an n-vertex undirected graph with non-negative edge weights. We assume the vertices to have unique identifiers. An agent starts in some vertex, and its goal is to move to a given vertex (the *treasure*); the identity of the treasure vertex is revealed once the agent enters it. The agent can freely move from a vertex to any neighboring vertex, incurring a cost equal to the weight of the traversed edge (an edge can be traversed an arbitrary number of times, each time incurring the same cost). When located at a vertex, the agent can see the identifier of the current vertex, the identifiers of all neighbors, and the weights of the incident edges. This model is known as the *fixed graph scenario* and was introduced by Kalyanasundaram and Pruhs [21]. The agent has (polynomial) memory and can perform an arbitrary computation, but it cannot modify the graph in any way; e. g., it cannot mark any vertices. Our aim is to construct a deterministic algorithm for the agent that minimizes the worst-case competitive ratio (i. e., the ratio of the cost incurred by the algorithm and the optimal cost). First, we observe that a simple greedy agent is $O(n)$-competitive, and that no deterministic agent can achieve a better competitive ratio (even in the unweighted case, where all edges have weight 1). Next, we analyze the problem using the framework of advice complexity.

The advice complexity has been introduced in the context of online problems [4,9,10,20] and distributed problems [16,17]. In both settings, the algorithm has to cope with the lack of information about the instance it is working on (the future input requests, the network topology, etc.).

There are many results that study the impact of some a priori information about the instance on the performance of the algorithm. While the traditional approach is qualitative and analyzes information of a certain type (e. g., the agent may know the size or diameter of the network), the advice complexity approach is quantitative: the agent obtains, at the beginning of the computation, a binary string (the *advice*) that describes the current instance. There is no restriction on the type of the information; it can be viewed as being prepared by a computationally unbounded oracle that knows the algorithm of the agent and the whole input instance (i. e., the network topology, the edge weights, the starting vertex of the agent, and the location of the treasure). The main incentive is to study the relationship between the size of the advice string and the competitive ratio that can be obtained. In recent years, the advice complexity of a large number of online problems was studied; examples include paging [4], the k-server problem [5,10,19,28], and the knapsack problem [6]. Furthermore, it is possible to use special reduction methods to transfer hardness results on the advice complexity from one problem to another [3,7,10].

We show that without any advice, any algorithm is $\Omega(n)$-competitive, even on unweighted graphs, and even using randomization. On the other hand, n bits of advice are sufficient to obtain an optimal solution. Next, we consider unweighted graphs (where all edges have weight 1). We show an asymptotically tight trade-off between the size of the advice and the competitive ratio: for any (not necessarily constant) r, there is an algorithm that uses $O(n/r)$ bits of advice and that achieves a competitive ratio r. On the other hand, any deterministic algorithm that achieves a competitive ratio r needs $\Omega(n/r)$ bits of advice.

Related Work

In 1977, Rosenkrantz et al. [27] proposed the following problem: an agent is initially located in a vertex of an unknown weighted undirected graph, and is able to move from a vertex to any neighboring vertex, inducing a cost equal to the weight of the traversed edge. When located at a vertex, the agent can see the identifier of that vertex, the weights of incident edges, and also the identifiers of its neighbors. Its goal is to traverse all vertices of the graph and induce as little cost as possible. The performance is evaluated in terms of the competitive ratio, i.e., the worst case ratio of the cost induced by the algorithm and the optimum cost. It has been shown that the natural greedy nearest neighbor algorithm is $\Theta(\log_2 n)$-competitive, and a question was posed whether there is any algorithm with a constant competitive ratio. Despite much effort, the question is still open for the general case (although many results are known for special graph classes), the best general lower bound being $5/2 - \varepsilon$ [8]. In the case of directed graphs, the problem is much better understood: the competitive ratio of any (even randomized) algorithm is at least $\Omega(n)$ and $O(n)$-competitiveness is indeed achievable [15].

The same model has been used to study the treasure hunt problem in directed graphs [15]. It has been shown that, on strongly connected weighted graphs, no deterministic or randomized algorithm can have a bounded competitive ratio. For strongly connected unweighted graphs, the best competitive ratio for deterministic algorithms is $\Theta(n^2)$; for randomized algorithms, a lower bound of $\Omega(n)$ is known.

The treasure hunt problem has been studied in unweighted undirected graphs in the framework of advice complexity by Miller and Pelc [24], in the model in which the agent does not see the identifiers of the neighboring vertices (i.e., it only sees the identifier of the vertex it is located in). In particular, in our model, the agent can identify a neighboring vertex as already visited, which can be exploited by the algorithm. The performance of the algorithm in the model of Miller and Pelc was measured in terms of the number of edges e, the competitive ratio r, and the cost of the optimal solution D. The lower and upper bounds $\Omega(D \log_2(\frac{e}{rD}))$ and $O(D \log_2(\frac{e}{r}))$ have been obtained, respectively.

A similar problem of finding a shortest path with a given source and sink was studied for a specific class of graphs by Papadimitriou and Yannakakis [26] and Fiat et al. [14]. The model used in these papers differs from the one studied here in the way in which the vertices are revealed to the agent.

Another related problem is the *rendezvous* problem, where two agents want to meet in an unknown graph [24, 25, 29]. Note that treasure hunt is a special case of rendezvous where the position of one of the agents is fixed.

Finally, note that such problems belong to the class of *search games*, which are further distinguished according to different parameters. Other famous problems from this class include the linear search problem (also known as the cow path problem) [1, 22] and the ANTS (ants nearby treasure search) problem [12, 13].

2   Our Contribution

Let $G = (V, E)$ with $|V| = n$ be an undirected graph with non-negative edge weights $\omega(e)$, $e \in E$ ($\omega(u, v)$, $u, v \in V$, respectively). Let u_0 be the starting vertex of the agent, and let $u_0, u_1, \ldots, u_\ell$ be the shortest path to the treasure vertex u_ℓ. Since the cost of the optimum may be zero, we use the so called non-strict competitive ratio with an additive constant to even out the effects of the special cases with zero (or near-zero) optimum: an algorithm is said to be r-competitive, if there is a constant α such that on any input instance the algorithm incurs cost at most $r \cdot OPT + \alpha$, where OPT is the cost of the optimal solution (i. e., the length of the shortest $u_0 - u_\ell$ path).

As a first observation, let us consider the competitive ratio of a deterministic algorithm without any advice.

Theorem 1. *There is a deterministic algorithm without advice for the treasure hunt problem in weighted graphs that achieves a competitive ratio of $O(n)$.*

Proof. Consider the following simple greedy algorithm on a given graph $G = (V, E)$. The agent works in rounds and maintains a set of vertices that have been visited so far. Let S_i be the set of vertices visited after i rounds, with $S_0 = \{u_0\}$ containing only the starting vertex. In each round i, the agent starts from the vertex u_0, and a new vertex is explored as follows. Let v_i be the vertex from $V \setminus S_{i-1}$ with the shortest distance from u_0. (Note that, since all vertices from S_{i-1} have been already visited, the agent knows the costs of all the outgoing edges; moreover, since also the identifier of the destination is known for an edge, the agent can distinguish which edges lead to vertices in $V \setminus S_{i-1}$.) The agent traverses the shortest path to v_i, which reveals the neighborhood of v_i, and returns back to u_0; the new set is $S_i := S_{i-1} \cup \{v_i\}$. The exploration ends whenever S_i contains the treasure vertex u_ℓ.

Obviously, there are at most n rounds. In each round i, the treasure vertex is located in $V \setminus S_{i-1}$, so the treasure is at least as far as v_i. Hence, in each round, the agent traverses at most twice the optimal cost. $\qquad\square$

The previous theorem is essentially tight, since even when using randomization, and even on unweighted graphs, an asymptotically better competitive ratio cannot be achieved.

Theorem 2. *Any randomized algorithm for the treasure hunt problem on unweighted stars has competitive ratio $\Omega(n)$.*

Proof. Using Yao's principle [30], the expected cost of any randomized algorithm on a worst-case instance from a set of instances $\mathcal{I}$ is at least the expectation of the cost of the best deterministic algorithm over any probability distribution on the instances. Let us take for $\mathcal{I}$ all the instances where the graph is a star with $n-1$ leaves, the agent starts in the root, and the treasure is in some leaf. Consider any deterministic algorithm, and a probability distribution where every leaf is chosen to be the treasure vertex uniformly at random. The algorithm visits the leaves in a fixed order, until it arrives to the leaf with the treasure. Hence, for any i with $1 \leq i \leq n - 1$, the probability that the agent finds the treasure in step $2i - 1$ is $1/(n-1)$. Therefore, the expected cost is

$$\frac{1}{n-1} \sum_{i=1}^{n-1} 2i - 1 \in \Omega(n)$$

as claimed.   □

On the other hand, linear advice is sufficient to achieve optimality, as can be seen from the following theorem.

Theorem 3. *There is an optimal deterministic algorithm with n bits of advice for the treasure hunt problem in weighted graphs.*

Proof. In order to be optimal, the agent must traverse the shortest path $u_0, u_1, \ldots, u_\ell$. The advice is used to query the status of the vertices, i. e., whether they belong to this optimal path or not. Let the agent be located at a vertex u_i. First, it queries, one bit per vertex, the status of all neighbors that have not been queried yet, in the order of increasing identifiers. Since the agent is deterministic, the advice string can be prepared based on the input instance in the correct order. After this step, the agent knows which neighbors of u_i are on the optimal path, and in particular it knows the set of neighbors P that are of the form u_j with $j > i$. Obviously, $u_{i+1} \in P$. Moreover, the weight $w(u_i, u_{i+1})$ is the smallest of all weights $w(u_i, u_j)$, $u_j \in P$, since otherwise $u_0, u_1, \ldots, u_\ell$ would not be shortest. Lastly, there may be several vertices u_j in P with the same weight $w(u_i, u_j) = w(u_i, u_{i+1})$. An arbitrary vertex among them can be chosen (deterministically) since in this case all the edges on the shortest path up to u_j must have zero weight.   □

Again, Theorem 3 cannot be asymptotically improved: advice of linear size is needed even to obtain a constant competitive ratio on unweighted graphs due to Theorem 5.

For the rest of the paper, let us focus on unweighted graphs, i. e., graphs where all edges have weight 1. As our main result, we show an asymptotically tight trade-off between the size of the advice and the best possible competitive ratio. We start by stating an upper bound. To this end, however, we need the following lemma. The lemma basically deals with special vertex separators of size 1 and has been mentioned as folklore by Lipton and Tarjan [23]. We provide the proof for the sake of completeness.

Lemma 1. *Let $G = (V, E)$ be a connected graph with $|V| = n > 6$ vertices. Then there are two sets $C, D \subseteq V$ such that $C \cup D = V$, $|C \cap D| = 1$, both $|C| > n/3$, $|D| > n/3$, and each of them induces a connected subgraph.*

Proof. It is sufficient to prove the lemma for trees, since then it can be applied to a spanning tree of an arbitrary connected graph. Let $G = (V, E)$ be an arbitrary

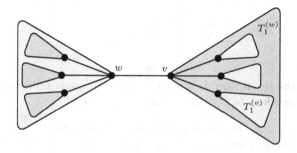

Fig. 1. Schematic drawing of the graph G in the situation described in Lemma 1. The tree $T_1^{(w)}$ contains more than $n/2$ vertices, so the subtree rooted at w must contain less than $n/2$ vertices. The subtree $T_1^{(v)}$ that contains more than $n/2$ vertices must be one of the subtrees within $T_1^{(w)}$.

tree. For any vertex w, let G decompose into n_w trees $T_1^{(w)}, \ldots, T_{n_w}^{(w)}$ when removing w from V. For every $w \in V$ and every i with $1 \le i \le n_w$, denote by $V_i^{(w)}$ the vertex set of $T_i^{(w)}$.

First we prove that there is a vertex $w \in V$ such that, for each i with $1 \le i \le n_w$, we have $|V_i^{(w)}| \le n/2$. Let us assume by contradiction that for each vertex $w \in V$ there exists some index j such that $|V_j^{(w)}| > n/2$. Since for any vertex w there cannot be more than one such subtree, without loss of generality, let $T_1^{(w)}$ be the unique subtree from vertex w with $|V_1^{(w)}| > n/2$. Now consider a vertex w such that $|V_1^{(w)}|$ is minimal among all vertex sets $V_1^{(u)}$ for all $u \in V$, and let v be the (unambiguous) neighbor of w in $T_1^{(w)}$ (see Fig. 1).

From the point of view of v, the subtree $T_i^{(v)}$ that is rooted in w has less than $n/2$ vertices and thus cannot be the subtree $T_1^{(v)}$ with $|V_1^{(v)}| > n/2$. Hence, the subtree of v with more than $n/2$ vertices must be one of the subtrees contained in $T_1^{(w)}$, and therefore $|V_1^{(v)}| < |V_1^{(w)}|$. This is a contradiction to the minimality of $|V_1^{(w)}|$.

Now consider a vertex $w \in V$ such that each $|V_i^{(w)}| \le n/2$ for all $1 \le i \le n_w$. If there is a tree $T_j^{(w)}$ that has $|V_j^{(w)}| \ge n/3$ vertices, then let

$$C := T_j^{(w)} \cup \{w\}$$

and

$$D := (V \setminus C) \cup \{w\}.$$

Else, if all the $T_i^{(w)}$ have $|V_i^{(w)}| < n/3$ vertices, assign the (vertices of) the subtrees $T_i^{(w)}$ one by one greedily to the set C or D that currently contains fewer vertices. When all vertices of all these subtrees are assigned, we additionally add w to both C and D. Hence, in the end, the cardinalities of the two parts differ by at most $n/3$, which means that both sets contain at least

$$\frac{n}{2} - \frac{n/3}{2} + 1 \geq \frac{n}{3} + 1$$

vertices.   □

Now we use the result from Lemma 1 to prove an upper bound on the number of advice bits sufficient to obtain a competitive ratio of r.

Theorem 4. *Let $r := r(n)$ be any function of n such that $18 < r < n$ and r is divisible by 9. There is an r-competitive algorithm for the treasure hunt problem on unweighted graphs that uses $O(n/r)$ bits of advice.*

Proof. Recall that the agent starts in vertex u_0, and that the treasure is located in u_ℓ, so the optimal cost is ℓ. The agent works its way towards u_ℓ in *rounds* that consist of traversing a number of edges, and possibly reading some advice bits. In order to keep track of the advice spent and the distance traversed, two types of accountings are used for the purpose of the analysis: at the beginning, each vertex has a *charge* $9/r$, and the agent has *credit* 0. At some point in time, we may decide to *harvest* some vertices, adding their charge to the agent's credit. We shall make sure that no vertex is harvested twice, and the agent reads only as many advice bits as is its credit. This way, the overall size of the advice is bounded by $9n/r$. The second type of accounting keeps track of the traveled distance: every move of the agent is *booked* to some edge (u_i, u_{i+1}) from the optimal path. In order to bound the competitive ratio, we assert that no edge is booked more than r times.

During each round i, the agent maintains three disjoint sets of vertices. The set H_i contains all vertices that have already been harvested in previous rounds; for each vertex $v \in H_i$, either v or its neighbor has been visited in the previous rounds. The set T_i contains all vertices that have already been visited but not yet harvested. Finally, the set B_i, the *boundary vertices*, are those vertices from $V \setminus (H_i \cup T_i)$ that have a neighbor in T_i. At the beginning, we have $T_1 := \{u_0\}$, $H_1 := \emptyset$, and B_1 contains all the neighbors of u_0.

There are two different kinds of rounds: *traversal* rounds and *advice* rounds. If, in round i, the size of the set $T_i \cup B_i$ is at most $r/3$, the agent traverses these vertices, and books the cost of the traversal to one edge of the optimal path. On the other hand, if $T_i \cup B_i$ is too big for the traversal cost to be amortized, the agent reads one advice bit to narrow down the set of vertices that must be traversed. The rounds are grouped into phases such that each phase starts with a number (possibly zero) of advice rounds and ends with one traversal round.

Let us consider a phase p consisting of rounds $h + 1, \ldots, h + m$. We ensure the following invariants hold at the beginning of each round i.

(a) B_i contains some vertex u_j from the shortest path such that no vertices u_k, $k > j$ are in H_i. Let j^* be the maximum index such that $u_{j^*} \in B_i$; we call u_{j^*} the *distinguished vertex* and $e_i := (u_{j^*}, u_{j^*+1})$ the *distinguished edge* (neither the distinguished vertex, nor the distinguished edge is known to the agent, they are for the purpose of the analysis only).

(b) No costs have been booked to any edge (u_k, u_{k+1}) for $j^* \leq k \leq \ell - 1$.

(c) The agent is located at some vertex $v \in T_{h+1}$.

(d) $T_i \cup B_i$ is connected.

Now let us describe how the algorithm works in greater detail (refer to Fig. 2). The computation starts with round 1 of phase 1. For the sake of simplicity, we say that the preceding (dummy) round was round 0. At the beginning of its computation, the agent is located at u_0. The initial values for the sets are, as we have already mentioned above, $T_1 := \{u_0\}$, $H_1 := \emptyset$, and $B_1 := \{v \in V \mid (u_0, v) \in E\}$. The distinguished vertex is the last vertex from $u_1, \ldots, u_\ell$ that is a neighbor of u_0. It is easy to verify that all invariants hold.

If, in round i, it holds that $|T_i \cup B_i| \geq r/3$, the agent executes an advice round: it internally splits $T_i \cup B_i$ into two parts C_i and D_i using Lemma 1. Then it reads one bit of advice indicating which one of the sets C_i and D_i contains the distinguished vertex u^*. Without loss of generality, let this be C_i. Note that u^* might even be contained in both C_i and D_i, since these sets intersect in one vertex w. If this is the case, the oracle specifies the set C_i to be the one containing u^*.

Then the vertices from $D_i \setminus \{w\}$ are harvested to pay for the advice bit that it just read. Since $D_i \subseteq T_i \cup B_i$, D_i and H_i are disjoint. Hence, the vertices in $D_i \setminus \{w\}$ have not been harvested yet, and Lemma 1 guarantees that both C_i and D_i contain at least $r/9 + 1$ vertices. Thus, the agent gains enough credit by harvesting the vertices from $D_i \setminus \{w\}$ to pay for one advice bit. It sets $H_{i+1} := H_i \cup D_i \setminus \{w\}$, $T_{i+1} := T_i \cap C_i \subseteq T_i$, and $B_{i+1} := C_i \setminus T_i = B_i \cap C_i$. This implies that $u^* \in B_{i+1}$, and the distinguished edge does not change. Invariant (a) is trivially fulfilled. As no costs were booked to any edges in this round, also invariant (b) remains true. Invariant (c) holds since the agent did not move at all in this round. To verify that invariant (d) holds, note that $T_{i+1} \cup B_{i+1} = (T_i \cup B_i) \cap C_i = C_i$, which is connected by Lemma 1.

If, on the other hand, $|T_i \cup B_i| < r/3$, the agent executes a traversal round. Hence, this is the last round (round $h + m$) of phase p. Due to invariant (c) the agent is at some vertex $v \in T_{h+1}$. Before the agent starts to traverse $T_{h+m} \cup B_{h+m}$, it must first enter this set, which incurs cost at most $r/3$ that is booked onto the distinguished edge (which exists due to invariant (a)). The limit of $r/3$ is implied by the fact that T_{h+1} was generated by the traversal round of the previous phase and the properties of traversal rounds shown below.

Now, the agent uses a depth-first search to traverse $T_{h+m} \cup B_{h+m}$, and as soon as it comes across the destination vertex u_ℓ, the algorithm terminates. Otherwise, the traversal incurs cost of less than $2r/3$ that are booked to e_{h+m}. The agent sets $T_{h+m+1} := T_{h+m} \cup B_{h+m}$, since all vertices contained in these sets have been traversed now but have not been harvested yet, and $H_{h+m+1} := H_{h+m}$, as no vertices have been harvested in this round. The agent also computes B_{h+m+1} according to T_{h+m+1} and H_{h+m+1}. From invariant (a) we can conclude that one endpoint of the present distinguished edge e_{h+m} must be $u_{j^*} \in B_{h+m}$ and the other one $u_{j^*+1} \notin (H_{h+m+1} \cup T_{h+m+1})$. Thus, $u_{j^*+1} \in B_{h+m+1}$ is a vertex that guarantees that invariant (a) is fulfilled. However, u_{j^*+1} is not necessarily the distinguished vertex of the next round; this is the last vertex u_k that is

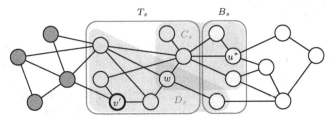

(a). In the preceding round, $s - 1$, the agent traversed the set T_s and ended in some vertex $v' \in T_s$. Now, in round s, the set $T_s \cup B_s$ is too large to be traversed, so the agent splits it into two parts C_s and D_s, which overlap in w.

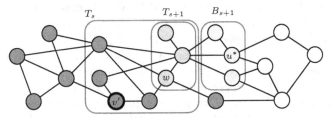

(b). After round s, the agent harvested all vertices from $D_s \setminus \{w\}$ and updated the sets accordingly. The set $T_{s+1} \cup B_{s+1}$ is small enough to be traversed, but the agent is located at the vertex $v' \in T_s \setminus T_{s+1}$, and before starting the traversal, it must first move to some vertex in T_{s+1}.

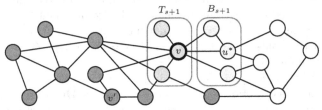

(c). When round $s + 1$ begins, the agent has moved to $v \in T_{s+1}$ and can now start its traversal.

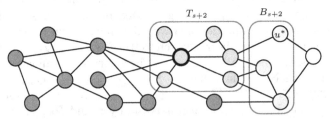

(d). After round $s + 1$, all vertices from $T_{s+1} \cup B_{s+1} = T_{s+2}$ have been traversed. The agent updates the sets T_{s+2} and B_{s+2} and the distinguished vertex u^* accordingly.

Fig. 2. An example of a sequence of rounds performed by the agent.

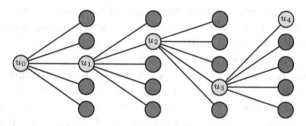

Fig. 3. Example of a pq-tree for $p = 5$, $q = 4$, and $n = 21$. The starting vertex u_0 is the only vertex on level 0. There are q additional levels and p vertices per level. There is one designated vertex per level, colored in gray; u_i is the designated vertex on level i. Each such vertex u_i for $0 \leq i \leq q - 1$ is connected to all vertices on level $i + 1$. The vertex u_q is the destination vertex.

in B_{h+m+1}, for some $k \geq j^* + 1$. Then, the new distinguished edge $e_{h+m+1} = (u_k, u_{k+1})$ is the edge where the optimal path leaves B_{h+m+1} for the last time. Thus, invariants (a) and (b) remain true. At the end of round $h + m$, which is also the beginning of phase $p+1$, the agent is located at some vertex $v \in T_{h+m} = T_{h+m+1}$, making sure that also invariant (c) holds. Invariant (d) holds as well since T_{h+m+1} is connected due this invariant from the previous round and each vertex of B_{h+m+1} is connected to a vertex of T_{h+m+1}.

The traversal stops, whenever the agent enters the treasure vertex u_ℓ for the first time. For every i with $0 \leq i \leq \ell - 1$, every edge $\{u_i, u_{i+1}\}$ is the distinguished edge of at most one phase, and thus cost of at most r are booked to it: not more than $r/3$ for the adjustment move before the traversal of the current search space and at most $2r/3$ for the traversal itself. Hence, the total cost of the agent, adding the additional cost for the last move, is at most $r \cdot \ell + 1$, whereas the cost of an optimal solution is ℓ, and thus the algorithm is r-competitive. Furthermore, each vertex is harvested at most once, amounting for the advice complexity $9n/r$. □

The algorithm from the previous theorem cannot be asymptotically improved, as we show in the next theorem. In the proof we shall use a class of instances called pq-trees (see Fig. 3), which are defined as follows: for given p, q, n, such that $n > pq$, consider q columns (called levels), each containing p vertices. On each level, there is one designated vertex, and every designated vertex on level $1 \leq i < q$ is connected to all vertices of level $i + 1$. Additionally, there is a single vertex on level 0 (root) that is connected to all vertices on level 1, and $\gamma := n - pq - 1$ dummy vertices connected to the root. This way the vertices form a tree with n vertices. The identifiers of the vertices are arbitrary but fixed for a given set of parameters p, q, n. The starting vertex is the root, and the treasure vertex is the designated vertex on level q. There are p^q possible ways to choose the designated vertices $u_1, \ldots, u_q$, and thus $\mathcal{I}$ contains $|\mathcal{I}| = p^q$ different instances, each of them with optimal solution q. We prove the following theorem.

Theorem 5. *Let $r := r(n)$ be any function of n, such that $1 \leq r < n/18$ for each n. Any algorithm for the treasure hunt problem on pq-trees with competitive ratio r needs at least $\Omega(n/r)$ bits of advice.*

Proof. Consider any treasure hunt algorithm $\mathcal{A}$ for pq-trees, with competitive ratio r. This means that there is a constant α, such that on any pq-tree with n vertices, the cost of the algorithm is at most $r \cdot OPT + \alpha$. Let us fix some (large enough) n. We construct a particular pq-tree with n vertices, where the algorithm reads more than $\beta \frac{n}{r}$ bits of advice for some constant $\beta > 0$ that depends on the algorithm and α, but neither on the function r nor on n.

Set $q := \lfloor n/kr \rfloor$ for a suitable constant k specified later, and choose p such that $n = pq + \gamma$ for some $1 \leq \gamma \leq q$. Consider all p^q instances $\mathcal{I}$ with parameters p, q, n. The agent traverses the graph, starting at u_0, until it finally reaches the destination vertex u_q. Until it does, it must visit at least one vertex on each level. The order in which $\mathcal{A}$ traverses the vertices on each level i is fixed for any fixed instance. On any level i with $1 \leq i \leq q$, let a_i be such that $\mathcal{A}$ visits u_i as the a_i-th vertex on level i. Let us assume without loss of generality that $\mathcal{A}$ never visits a leaf twice and never returns to level $i-1$ once it has found u_i. We can identify each instance $I \in \mathcal{I}$ with the characteristic vector $(a_1, a_2, \ldots, a_q)$, with $a_i \in \{1, \ldots, p\}$ for $1 \leq i \leq q$, and the property that the cost of $\mathcal{A}$ on instance I is at least $\sum_{i=1}^{q}(2a_i - 1)$ (we ignore the potential dummy vertices).

Let us call an instance I *good* if $\mathcal{A}$ achieves a competitive ratio of at most r on it, i.e., if the cost of $\mathcal{A}$ on I is at most $qr + \alpha$, which implies

$$\sum_{i=1}^{q} a_i \leq \frac{q(r + 1 + \alpha/q)}{2}.$$

For convenience, let us denote $d := (r + 1 + \alpha/q)/2$. This implies that an instance is good if, for its corresponding characteristic vector $(a_1, \ldots, a_q)$,

$$\sum_{i=1}^{q} a_i \leq qd. \tag{1}$$

For the following argumentation, we interpret each algorithm that uses b bits of advice as a set $\{\mathcal{A}_1, \mathcal{A}_2, \ldots, \mathcal{A}_{2^b}\}$ of deterministic algorithms, as is often done with online algorithms with advice. Since $\mathcal{A}$ reads at most b advice bits and is r-competitive, there is at least one deterministic algorithm $\mathcal{A}_j$ that computes a solution with a competitive ratio of at most r on at least $p^q/2^b$ instances. From now on, let us consider this particular deterministic algorithm $\mathcal{A}_j$, and let us define the set of good instances for $\mathcal{A}_j$ to be $\mathcal{I}^+$. Thus we have

$$|\mathcal{I}^+| \geq \frac{p^q}{2^b}. \tag{2}$$

Now let us bound $|\mathcal{I}^+|$, the number of good instances for $\mathcal{A}_j$, from above. For any good instance, the corresponding characteristic vector must contain at least $q/2$ entries a_i with value at most $2d$; hence, the number of good instances is upperbounded by the number of vectors $(a_1, \ldots, a_q)$, where $a_i \in \{1, \ldots, p\}$, with at least $q/2$ entries with value at most $2d$. To bound this term from above, we make the following considerations. The number of vectors of length $q/2$ with values of at most $2d$ is $(2d)^{q/2}$, the number of vectors of length $q/2$ with values between 1 and p is $p^{q/2}$. The number of possibilities to join two vectors of these two different kinds to construct a vector of length q is $\binom{q}{q/2}$. The same vector of length q might be generated by joining different pairs of vectors of length $q/2$. Nevertheless, these considerations yield an upper bound. The number of characteristic vectors with at least $q/2$ entries with value at most $2d$, and thus also the number of good instances, is therefore

$$|\mathcal{I}^+| \leq (2d)^{\frac{q}{2}} \cdot p^{\frac{q}{2}} \cdot \binom{q}{\frac{q}{2}} . \tag{3}$$

Putting (2) and (3) together yields

$$\frac{p^q}{2^b} \leq (8dp)^{\frac{q}{2}} .$$

We rearrange this inequality to solve it for b and obtain

$$b \geq \frac{q}{2} \cdot \log_2\left(\frac{p}{8d}\right) .$$

If we show that $p/8d > 2$, then

$$b \geq \frac{q}{2} \geq \frac{1}{2}\left(\frac{n}{kr} - 1\right) \geq \frac{1}{2k}\frac{n}{r} - \frac{1}{2} .$$

Since $n/r > 18$, choosing $\beta := 1/(2k) - 1/36$ yields

$$\frac{1}{2k}\frac{n}{r} - \frac{1}{2} > \beta \cdot \frac{n}{r} .$$

It must hold that $\beta > 0$, i. e., $k < 18$. It remains to show that $p/8d > 2$. Recall that $p = (n - \gamma)/q \geq (n - q)/q$. Substituting d, one gets

$$\frac{p}{8d} \geq \frac{n - q}{4q(r + 1) + 4\alpha} ,$$

which is more than 2 if $q < (n - 8\alpha)/(8r + 9)$. Finally, we show that for a suitable choice of k we can obtain

$$q = \left\lfloor \frac{n}{kr} \right\rfloor \leq \frac{n}{kr} < \frac{n - 8\alpha}{8r + 9} .$$

In order for the last inequality to hold, we need to choose k such that

$$k > \frac{n}{n - 8\alpha}\frac{8r + 9}{r} .$$

The second term is always at most 17, and the first term converges to 1 with increasing n. Hence, there is a large enough n (depending on α) such that $k = 17.5$ is suitable. $\qquad\square$

3 Conclusion

We analyzed the treasure hunt problem on undirected graphs in the framework of advice complexity. In particular, we have shown a tight bound of $\Theta(n/r)$ on advice for algorithms with competitive ratio r on unweighted graphs. A natural next step would be to extend this result to weighted graphs.

Furthermore, the upper bound from Theorem 3 can be improved. An anonymous reviewer pointed out a proof that uses a slightly more involved argument and allows to decrease the number of advice bits to $2/3n$.

Acknowledgement. The authors would like to thank Hans-Joachim Böckenhauer and Juraj Hromkovič for very valuable discussions, and an anonymous reviewer.

References

1. Baeza-Yates, R.A., Culberson, J.C., Rawlins, G.J.E.: Searching in the plane. Information and Computation 106(2), 234–252 (1993)
2. Berman, P.: On-line searching and navigation. In: Fiat, A., Woeginger, G.J. (eds.) Online Algorithms 1996. LNCS, vol. 1442, pp. 232–241. Springer, Heidelberg (1998)
3. Böckenhauer, H.-J., Hromkovič, J., Komm, D., Krug, S., Smula, J., Sprock, A.: The string guessing problem as a method to prove lower bounds on the advice complexity. Theoretical Computer Science 554, 95–108 (2014)
4. Böckenhauer, H.-J., Komm, D., Královič, R., Královič, R., Mömke, T.: On the advice complexity of online problems. In: Dong, Y., Du, D.-Z., Ibarra, O. (eds.) ISAAC 2009. LNCS, vol. 5878, pp. 331–340. Springer, Heidelberg (2009)
5. Böckenhauer, H.-J., Komm, D., Královič, R., Královič, R.: On the advice complexity of the k-server problem. In: Aceto, L., Henzinger, M., Sgall, J. (eds.) ICALP 2011, Part I. LNCS, vol. 6755, pp. 207–218. Springer, Heidelberg (2011)
6. Böckenhauer, H.-J., Komm, D., Královič, R., Rossmanith, P.: On the advice complexity of the knapsack problem. In: Fernández-Baca, D. (ed.) LATIN 2012. LNCS, vol. 7256, pp. 61–72. Springer, Heidelberg (2012)
7. Boyar, J., Favrholdt, L.M., Kudahl, C., Mikkelsen, J.W.: Advice complexity for a class of online problems. In: Mayr, E.W., Ollinger, N. (eds.) Proc. of the 32nd International Symposium on Theoretical Aspects of Computer Science (STACS 2015). LIPIcs, vol. 30, pp. 116–129. Schloss Dagstuhl - Leibniz-Zentrum für Informatik (2015)
8. Dobrev, S., Královič, R., Markou, E.: Online graph exploration with advice. In: Even, G., Halldórsson, M.M. (eds.) SIROCCO 2012. LNCS, vol. 7355, pp. 267–278. Springer, Heidelberg (2012)
9. Dobrev, S., Královič, R., Pardubská, D.: Measuring the problem-relevant information in input. RAIRO ITA 43(3), 585–613 (2009)
10. Emek, Y., Fraigniaud, P., Korman, A., Rosén, A.: Online computation with advice. Theoretical Computer Science 412(24), 2642–2656 (2011)
11. Euler, L.: Solutio problematis ad geometriam situs pertinentis. Commentarii Academiae Scientiarum Imperialis Petropolitanae 8, 128–140 (1736)
12. Feinerman, O., Korman, A.: Memory lower bounds for randomized collaborative search and implications for biology. In: Aguilera, M.K. (ed.) DISC 2012. LNCS, vol. 7611, pp. 61–75. Springer, Heidelberg (2012)

13. Feinerman, O., Korman, A., Lotker, Z., Sereni, J.S.: Collaborative search on the plane without communication. In: Kowalski, D., Panconesi, A. (eds.) Proc. of the 31st ACM Symposium on Principles of Distributed Computing (PODC 2012), pp. 77–86 (2012)

14. Fiat, A., Foster, D.P., Karloff, H.J., Rabani, Y., Ravid, Y., Vishwanathan, S.: Competitive algorithms for layered graph traversal. SIAM Journal on Computing 28(2), 447–462 (1998)

15. Förster, K.-T., Wattenhofer, R.: Directed graph exploration. In: Baldoni, R., Flocchini, P., Binoy, R. (eds.) OPODIS 2012. LNCS, vol. 7702, pp. 151–165. Springer, Heidelberg (2012)

16. Fraigniaud, P., Ilcinkas, D., Pelc, A.: Oracle size: A new measure of difficulty for communication tasks. In: Proc. of the 25th Annual ACM symposium on Principles of distributed computing (PODC 2006), pp. 179–187. ACM (2006)

17. Fraigniaud, P., Ilcinkas, D., Pelc, A.: Communication algorithms with advice. Journal of Computer and System Sciences 76(3–4), 222–232 (2010)

18. Ghosh, S.K., Klein, R.: Online algorithms for searching and exploration in the plane. Computer Science Review 4(4), 189–201 (2010)

19. Gupta, S., Kamali, S., López-Ortiz, A.: On advice complexity of the k-server problem under sparse metrics. In: Moscibroda, T., Rescigno, A.A. (eds.) SIROCCO 2013. LNCS, vol. 8179, pp. 55–67. Springer, Heidelberg (2013)

20. Hromkovič, J., Královič, R., Královič, R.: Information complexity of online problems. In: Hliněný, P., Kučera, A. (eds.) MFCS 2010. LNCS, vol. 6281, pp. 24–36. Springer, Heidelberg (2010)

21. Kalyanasundaram, B., Pruhs, K.R.: Constructing competitive tours from local information. Theoretical Computer Science 130(1), 125–138 (1994)

22. Kao, M.-Y., Reif, J.H., Tate, S.R.: Searching in an unknown environment: An optimal randomized algorithm for the cow-path problem. Information and Computation 131(1), 63–79 (1996)

23. Lipton, R.J., Tarjan, R.E.: A separator theorem for planar graphs. SIAM Journal on Applied Mathematics 36(2), 177–189 (1979)

24. Miller, A., Pelc, A.: Tradeoffs between cost and information for rendezvous and treasure hunt. In: Aguilera, M.K., Querzoni, L., Shapiro, M. (eds.) OPODIS 2014. LNCS, vol. 8878, pp. 263–276. Springer, Heidelberg (2014)

25. Miller, A., Pelc, A.: Fast rendezvous with advice. In: Gao, J., Efrat, A., Fekete, S.P., Zhang, Y. (eds.) ALGOSENSORS 2014, LNCS 8847. LNCS, vol. 8847, pp. 75–87. Springer, Heidelberg (2015)

26. Papadimitriou, C.H., Yannakakis, M.: Shortest paths without a map. Theoretical Computer Science 84(1), 127–150 (1991)

27. Rosenkrantz, D.J., Stearns, R.E., Lewis II, P.M.: An analysis of several heuristics for the traveling salesman problem. SIAM Journal on Computing 6(3), 563–581 (1977)

28. Renault, M.P., Rosén, A.: On online algorithms with advice for the k-server problem. In: Solis-Oba, R., Persiano, G. (eds.) WAOA 2011. LNCS, vol. 7164, pp. 198–210. Springer, Heidelberg (2012)

29. Ta-Shma, A., Zwick, U.: Deterministic rendezvous, treasure hunts, and strongly universal exploration sequences. ACM Transactions on Algorithms 10(3), 12:1–12:15 (2012)

30. Yao, A.C.-C.: Probabilistic computations: Toward a unified measure of complexity. In: Proc. of the 18th Annual Symposium on Foundations of Computer Science (FOCS 1977), pp. 222–227 (1977)

Lower Bounds for the Capture Time: Linear, Quadratic, and Beyond

Klaus-Tycho Förster, Rijad Nuridini, Jara Uitto, and Roger Wattenhofer

Computer Engineering and Networks Laboratory,
ETH Zurich, 8092 Zurich, Switzerland
{foklaus,rijadn,juitto,wattenhofer}@ethz.ch

Abstract. In the classical game of Cops and Robbers on graphs, the capture time is defined by the least number of moves needed to catch all robbers with the smallest amount of cops that suffice. While the case of one cop and one robber is well understood, it is an open question how long it takes for multiple cops to catch multiple robbers. We show that capturing $\ell \in \mathcal{O}(n)$ robbers can take $\Omega(\ell \cdot n)$ time, inducing a capture time of up to $\Omega(n^2)$. For the case of one cop, our results are asymptotically optimal. Furthermore, we consider the case of a superlinear amount of robbers, where we show a capture time of $\Omega(n^2 \cdot \log(\ell/n))$.

1 Introduction

This paper brings you back to your childhood, when you played the game of tag with your friends. Particularly interesting is a team version of tag sometimes known as jail, chase, manhunt, smee, or, as in this paper, cops and robbers. In cops and robbers, children are split into two teams, the cops and the robbers, where cops need to touch robbers, in order to jail them. If all children run at approximately the same speed, and the playground is suitably obstructed, the game becomes exciting, and cops usually need to cooperate in order to block possible escape paths of the robbers. Are there playgrounds (graphs) where the cops need a long time to catch all the robbers? This is the central open question we will investigate in this work.

The analytical study of these games on graphs is still relatively young. Breisch [9] first discussed searching for a lost person in a cave in 1967, followed by a formalization by Parsons [22,23] a decade later. The work of Quilliot [25] and Nowakowski and Winkler [21] introduced a game of pursuit-evasion on graphs, today commonly known as *Cops and Robbers*: A cop has to catch a robber, with both alternating in moves along edges. Aigner and Fromme [1] allowed multiple players into the game and showed that in any planar graph, three cops suffice to win. These articles spawned a rich field of interest, with plenty of further work, we refer to the book of Bonato and Nowakowski [7] for an in-depth overview and to [2,8,15] for recent surveys.

There are two central questions in the game of Cops and Robbers: First, how long will these cops need to catch the robbers, i.e., what is the *capture time*?

© Springer International Publishing Switzerland 2015
C. Scheideler (Ed.): SIROCCO 2015, LNCS 9439, pp. 342–356, 2015.
DOI: 10.1007/978-3-319-25258-2_24

Secondly, how many cops are needed to catch the robbers, i.e., what is the *cop number?*

The case of one cop and robber is well understood, with the graphs where one cop suffices being characterized [21,25], and the time needed to capture one robber being at most $n - 4$ in $n \geq 7$ vertex graphs [14]. With multiple cops and robbers, much is still unknown. For the first question, the best current result states that if k cops suffice, then they can capture a single robber in at most n^{k+1} time [3]. Already for one cop this bound is off by a factor of n. For the second question, $\mathcal{O}\left(n / \left(2^{(1-o(1))\sqrt{\log n}}\right)\right)$ cops always suffice [13,18,26] and there are graphs where $\Omega(\sqrt{n})$ cops are needed [24], but it is unclear what the exact bound is. Meyniel conjectured in 1985 that $\mathcal{O}(\sqrt{n})$ cops always suffice [12].

It is an open question what a good capture time lower bound would be for more than a single cop and robber, cf., e.g., [7]. It seems that so far, essentially only the cases of $i)$ the cartesian product of two trees and $ii)$ the d-dimensional hypercube have been successfully investigated for just one robber. For $i)$, the capture time is half the diameter of the graph [19], while for $ii)$, it is $\Theta(d \ln d)$, i.e., polylogarithmic in $n = 2^d$ [6].

After discussing further related work in the following Subsection 1.1 and a formal model in Section 2, we start with the case of one cop and any $\ell \in \mathcal{O}(n)$ robbers in Section 3, and prove that the capture time is $\Theta(\ell \cdot n)$.

In Section 4, we investigate the reversed case of k cops and one robber. As it turns out, the k cops might need $\Omega(n)$ time to capture a single robber, just like in the case of one cop.

Afterwards, we study the case of many cops and many robbers in Section 5, where we show that for k cops and any $\ell \in \mathcal{O}(\sqrt{n/k})$ robbers, the cops need at least $\Omega(\ell \cdot n)$ time to capture all robbers in general graphs. Furthermore, we discuss a superlinear number of robbers and show that the time to capture them all can be as high as $\Omega\left(n^2 \cdot \log(\ell/n)\right)$.

1.1 Further Related Work

The capture time density of a graph can be defined as as the ratio of the capture time to the number of vertices. Bonato et al. extended this notion to infinite graphs, or more precisely to limits of chains of induced subgraphs, and showed that the density can take any value from 0 to 1 for a single cop and robber [5]. It can be tested in polynomial time if fixed $k \in \mathbb{N}$ is the cop number of a graph [3], with these graphs being characterized in [10]. Nonetheless, determining the cop number of a graph is EXPTIME-complete [16].

Many more variants of Cops and Robbers are considered in the literature (e.g., can the cops win if they do not start too far away from the robber [4], applications to compact routing [17], or how to contain worm attacks in networks [11]), we again refer to [2,7,8,15] for an even further overview.

2   Model

The game of Cops and Robbers is a pursuit-evasion game played on undirected graphs $G = (V, E)$ with $|V| = n$ and diameter of D. We denote the set of $k \in \mathbb{N}$ cops as $C = \{p_1, p_2, \ldots, p_k\}$ and the set of $\ell \in \mathbb{N}$ robbers as $R = \{r_1, r_2, \ldots, r_\ell\}$. Throughout this paper, all graphs are assumed to be connected, finite, and, as standard in the literature, reflexive (i.e., each vertex has one self-loop, which is the same as allowing the cops and robbers to stand still).

The game proceeds in rounds, where each round consists of first the cops making a move and then the robbers making a move. In round 0, the cops make a move by placing each cop on a vertex and then the robbers make a move by placing each robber on a vertex. In round 0 and every further round, vertices can be shared by an arbitrary number of cops and robbers. For all subsequent rounds $i \geq 1$, first, a cop move consists of moving each cop along an incident edge, with second, a robber move defined by moving each robber along an incident edge. Both the cops and robbers have perfect information, i.e., they know the whole graph and every previous move played.[1]

A robber is caught if a cop shares its occupied vertex. Once robber r_i is captured, he is removed from the game, i.e., the cops need not to guard the robbers that are already captured. Should at least k cops be needed to catch all robbers on G, then the graph is called k-copwin with a cop number $c(G) = k$.

For $c(G) = k$ and $\ell = 1$, we define the capture time $capt(G, k, 1)$ as the smallest number of moves needed for the k cops to catch the robber, no matter what strategy the robber employs. Note that by this formulation, the capture time $capt(G, k, 1)$ is only defined on graphs where k cops can actually catch a robber. In a similar fashion, $capt(G, k, \ell)$ is the smallest number of moves needed for the k cops to catch all ℓ robbers. Let r_i be the i-th robber to be caught and define $capt(G, k, r_i)$ as the smallest number of moves needed for the k cops to catch the first i robbers.

3   One Cop, Many Robbers

We start with the case of one cop, many robbers. Gavenčiak showed in 2010 that $n - 4$ is the maximum capture time for one cop and one robber if $n \geq 7$ [14].[2] Observe that after catching one robber, the cop could go back to her starting position in at most diameter D moves and then catch the next robber in at most $n - 4$ moves. This gives us an upper bound for the capture time:

Observation 1 *Let G be a 1-copwin graph. Then, $capt(G, 1, \ell) \in \mathcal{O}(\ell \cdot n)$.*

Our next step will be to show a matching lower bound:

Theorem 2. *Let $n \geq 12$. Then, for all $\ell \in \mathcal{O}(n)$, there exists an n-vertex graph G with $capt(G, 1, \ell) \in \Omega(\ell \cdot n)$.*

---

[1] Bonato and Nowakowski also compared it to Pac-Man [7].
[2] We note that it was previously known that $capt(G, 1, 1) \leq n - 3$ for $n \geq 5$ [5].

The combination of both yields that the upper and lower bounds are asymptotically tight, i.e., the capture time of general graphs is $\Theta(\ell \cdot n)$ for $\ell \in \mathcal{O}(n)$.

Before proving Theorem 2, let us start with the example of a tree without branches, i.e., a path: While it might take the cop $D/2 \in \Omega(n)$ time to capture the first robber, all subsequent robbers can be captured in $n-1$ moves, inducing a total capture time of less than $2n$. However, the goal is to force the cop to move a linear number of times for each robber, which is shown in the following proof. Thus, we will construct a graph where, akin to a path, the robber has to move to one end to catch one robber, but then all robbers escape to the other end of the path, inducing a linear capture time for each robber. To ensure that the robbers can escape to the other side, at each end there will be a star with a ray-length of two, cf. Figure 1. Next, we will give a proof for Theorem 2.

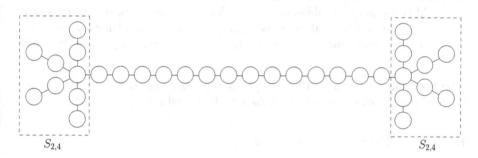

$S_{2,4}$ $S_{2,4}$

Fig. 1. Let $S_{x,y}$ be the star that has y rays of length x, i.e., the common star with n nodes would be $S_{1,n-1}$. In this example, there are two $S_{2,4}$, connected by a path of nodes. Consider a game of one cop and four robbers on this graph with 32 nodes. If in round 0 the four robbers choose a star that is farthest from the cop's initial position, and place themselves at the end of the four rays, then the cop needs at least $10 \geq n/4$ moves to capture the first robber. As the remaining robbers will flee to the other star, the cop needs $19 > n/2$ moves to catch each further robber, inducing a total capture time of 67. This construction can be directly extended to $S_{2,\lfloor n/8 \rfloor}$ for every $\ell \leq n/8$.

Proof (of Theorem 2). We begin by describing the construction of the graph G with capture time $\Omega(\ell \cdot n)$. Let $n \geq 12$ and suppose for now that $\ell \leq n/8$. Let $S_{x,y}$ be the star that has y rays of length x. Create two stars $S_{2,\ell}$, denoted by left star and right star in this proof. Connect the center nodes of both stars by a path of the remaining nodes. Note that this path (including the center of both stars) has a length (of nodes) of least $n/2 \in \Omega(n)$.

When the cop places herself in round 0, she can be either directly in the middle of the path, or closer to the left or the right star center.

Before describing the strategy of the cop, we first describe the strategy of the robbers: In round 0, all the robbers choose a star which has a higher distance to the cop than the other star (or, to break symmetry, the left one if the distance is equal); and place one robber at the end of each the ℓ rays of the star. Until the cop enters one the rays of their star, all robbers stay put. Then, when the cop is one move into the ray from the center of the star, the corresponding robber stays put, but all other robbers move one step towards the center of their star. Should the cop now move back to the center of the star, then all robbers go the end of their rays again. But if the cop moves to the end of the ray to catch a robber, then all the robbers move to the other star and choose pairwise different rays to place themselves at the end. The strategy is iterated until all robbers are caught.

We now describe a lower bound for the number of moves needed by the cop to counter the robbers' strategy: To catch the first robber, no matter where the cop starts, she has to move $\Omega(n)$ times and she has to move to the end of a ray to do so, i.e., $\mathrm{capt}(G, 1, r_1) \in \Omega(n)$. Once a single robber is captured, all other robbers will move to the end of the rays of the other star, and the cop has no possibility to catch them before that. Thus, $\Omega(n)$ moves are needed to capture the second robber, and so on, leading to $\mathrm{capt}(G, 1, r_\ell) = \mathrm{capt}(G, 1, \ell) \in \Omega(\ell \cdot n)$.

Should the amount of robbers be larger than $n/8$, then both stars are created as $S_{2, \lfloor n/8 \rfloor}$, and we ignore all robbers $r_{\lfloor n/8 \rfloor + 1}, \ldots, r_\ell$. Even if they should all be captured in the first round, the capture time for the remaining robbers will still be $\Omega(\ell \cdot n)$.   □

Corollary 3. *For all $n \geq 12$ there exists a n-vertex graph G s.t. the number $\ell \leq n$ of robbers can be chosen with $\mathrm{capt}(G, 1, \ell) \in \Theta(n^2)$.*

4   Many Cops, One Robber

In this section, we turn our attention to the case of many cops and extend our results for an arbitrarily large fixed number of cops:

Theorem 4. *Let $k_0 \geq 2$ be a positive integer. There exists an integer $k \geq k_0$ and a graph $G = (V, E)$ with $|V| = n \in \mathcal{O}(k^2)$ s.t. $c(G) = k$ and $\mathrm{capt}(G, k) \in \Omega(n)$.*

In other words, we claim that the time required to catch the first robber is asymptotically linear in the number of nodes of the graph. Furthermore, we do not only prove the case that the number of cops k is a constant, we propose a stronger claim, which states that the number of nodes n our construction requires is in the order of $\mathcal{O}(k^2)$, i.e., $k \in \mathcal{O}(\sqrt{n})$.

For this effort, we utilize a graph construction by Prałat that shows the existence of graphs with n nodes with a cop number of $\Omega(\sqrt{n})$ [24].

Lemma 5. *[24] Let $c(n)$ denote the maximum of $c(G)$ over all connected graphs with n vertices. Then, $c(n) > \sqrt{n/2} - n^{0.2625}$ for n sufficiently large.*

While Lemma 5 provides us with the existence of graphs with a high cop number, the capture time in these graphs remains low, i.e., a small constant. At first sight, it is not clear if the capture time of a graph can be high in the presence of "many" cops. As a trivial example, the capture time for a single robber is 1 if the number of cops is at least $n/2$, since the initial positions of the cops can be chosen such that each node can be reached within 1 step.

However, finding an equally straightforward bound becomes more elusive when the number of cops is smaller than the size of the smallest dominating set[3] in a graph. For the case of one cop and one robber, it is known that the capture time of any 1-copwin graph is at most linear in the number of nodes. As the next step, we show a more general result showing that a similar bound holds for the case of many cops.

The basic idea is to first fix some integer k_0 and take two copies G_1 and G_2 of a graph G with $\mathcal{O}(k^2)$ nodes with cop number $k \geq k_0$ promised by Lemma 5. Then, we connect these graphs with a long *bridge* (i.e., a path) with $b \in \Omega(n)$ nodes in a similar fashion as in Section 3. The endpoints e_1 and e_2 of the bridge are connected to arbitrary nodes s_1 and s_2 in graphs G_1 and G_2, respectively. Notice that since $k \in \mathcal{O}(\sqrt{n})$, we can choose $b \in \Omega(n)$. We denote the graph constructed in this manner by $\mathcal{D}(G, b)$. See Figure 2 for an illustration.

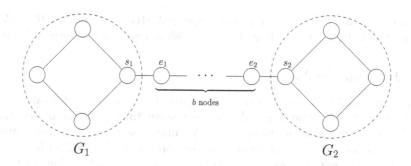

Fig. 2. Construction of the graph $D(G, b)$ for the case of two cops. The entry nodes s_1 and s_2 are chosen arbitrarily and connected by a path of length b. One cop is not sufficient to catch the robber alone in graphs G_1 and G_2.

4.1 Evasion Strategy

The next step is to choose the initial placement for the robber according to the initial placement of the cops in graph $D(G, b) = (V, E)$ for some G and b. Let $X = \{x_1, \ldots, x_k\}$, where $x_i \in V$ for every $1 \leq i \leq k$, be the initial placements of the cops, i.e., cop p_i is initially located in node x_i. In addition, let $d(u, v)$ denote the length of the shortest path between nodes u and v and $d(u, A) = \min\{d(u, v) \mid v \in A\}$, where $A \subseteq V$. We say that G_1 is *away* from the cops if

$$|\{x \in X \mid d(x, G_1) \geq \lfloor b/2 \rfloor\}| \geq \lceil k/2 \rceil ,$$

i.e., half of the cops are at most as close to G_1 as they are to G_2. Being away is defined similarly for G_2. It is easy to verify that either G_1 or G_2 is away from the cops for any k, possibly both.

Intuitively, we aim to locate the robber into one of graphs that is away from the cops, say G_1. By doing this, the fact that the cop number of G_1 is k ensures that the robber has a strategy that allows it to evade any number of cops trying to catch him before every cop has entered G_1. However, before stating the claim

---

[3] A set $D \subseteq V$ is a dominating set for V if every node in $V \setminus D$ has a neighbor in D.

formally, there are some minor technicalities that have to be accounted first. Our construction slightly modifies graph G_1 (and G_2) by adding one additional edge that connects the bridge to G_1 and this can have an effect on the strategy of the robber that allows him to evade up to $k-1$ cops.

Fortunately, there is a simple way to show that this is not an issue. Consider now only the graph G_1 and the strategy $\mathcal{S}$ that the robber has to evade at most $k-1$ cops in G_1 indefinitely. To utilize $\mathcal{S}$ in graph $D(G,b)$, we extend it to the larger graph in the following manner. Consider any configuration of the game, i.e., the placements of the players in which there are $k' \leq k-1$ cops occupying nodes $u_1, \ldots, u_{k'}$ in G_1. Then, the robber selects the move from $\mathcal{S}$ that corresponds to a configuration in which the k' cops occupy the nodes $u_1, \ldots, u_{k'}$ in G_1 and the remaining $k-1-k'$ cops occupy node s_1.

Observation 6 *Let $G = (V, E)$ be a graph with $c(G) = k$ and let $H = (V \cup \{v\}, E \cup \{(u, v)\})$, where $v \notin V$ and u is an arbitrary node in V. Then, $c(H) \geq k$.*

4.2   The Cop Number

We have now gathered the tools that our approach requires to show that capturing one robber with many cops takes at least linear time in our graph construction. However, before going to the claim, we make one more observation about our construction. That is, we show that the cop number of $D(G, b)$ is at most $c(G) + 1$ and at least $c(G)$ for any b. While it might seem that adding the bridge between two graphs with cop number k should not increase the cop number, it is not clear that the robber cannot trick the cops and escape from G_1 to G_2 once all the cops have entered G_1. As a simple example of the aforementioned issue, consider the graph shown in Figure 3. While graph G in the example is a 1-copwin graph, adding an edge between two copies of G breaks this property. The 1-copwin properties are easy to verify by recalling a graph is 1-copwin if and only if it can be reduced to a single vertex by successively removing *corners*, i.e., nodes whose (inclusive) neighborhood is contained in the neighborhood of some other node [1].

We tackle this issue by observing that the cop number in the graph we have constructed is either k or $k+1$, given that the cop number of G is k. It is clearly the case that $c(D(G, b)) \leq k + 1$ since the game, from the perspective of the cops, can be "reduced" to playing it only in G_1 by simply leaving one cop to guard the bridge. Now the cops chasing the robber can use a strategy that does every move with the robber not in G_1 as if the robber was in s_1. If the robber decides to leave G_1, the cop guarding the bridge can capture the robber.

Lemma 7. *Let $G = (V, E)$ be a graph with $c(G) = k \geq 2$. Then, $k \leq c(D(G, b)) \leq k + 1$ for any integer $b > 0$.*

Proof (of Lemma 7). It was shown by Berarducci and Intrigila [3] that if H is an induced subgraph of G and there exists a graph homomorphism from G onto H, which is the identity mapping in H, then $k = c(G) \geq c(H)$. Since $c(G) = k$, G is an induced subgraph of $D(G, b)$ and the homomorphism can be

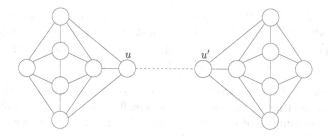

Fig. 3. Two copies of a graph G with cop number 1 connected by an edge denoted by the dashed line. Node u is the unique corner in G. Since adding the dashed edge adds a node into the neighborhood of u that is not in the neighborhood of any other node in G, it follows that u is not a corner after addition of this edge. Due to the symmetry of this example, there are no corners in the resulting construction and therefore, it is not 1-copwin.

found by considering a mapping where G_2 is mapped onto G_1 and every node in the bridge to s_1, it follows that $k \leq c(\mathcal{D}(G,b))$. Given $k + 1$ cops, one of the cops can guard the bridge and force the robber never to exit either G_1 or G_2. Therefore, the remaining k cops can simply apply the strategy promised by the fact that $c(G) = k$ to capture the robber. $\square$

We are now ready to show that the capture time is at least asymptotically linear in the number of the nodes, for an arbitrarily large number of cops.

Proof (of Theorem 4). Let H be a graph with m nodes and a cop number of at least $\sqrt{m/2} - m^{0.2625} \geq k_0$ promised by Lemma 5. Set $G = \mathcal{D}(H,b)$, where $b = m$, and let $c(H) = k'$. Consider now the game with $c(G) = k$ cops, where $k' \leq k \leq k' + 1$ by Lemma 7, one robber, and let G_1 and G_2 be the copies of H in G. Assume without loss of generality that the cops are away from G_1. According to Observation 6, the robber has a strategy that allows him to escape at most $k' - 1$ cops as long as not every cop has entered the subgraph induced by $G_1 \cup \{e_1\}$, where e_1 is the endpoint of the bridge connected to G_1. By definition of the cops being away from G_1, there are at most $\lfloor k/2 \rfloor \geq k' - 1$ cops that are closer to G_1 than to G_2. Thus, there is at least one cop that has to move $\Omega(b) \in \Omega(n)$ times before $G_1 \cup \{e_1\}$ is occupied by at least k' cops for the first time. Since the robber is not captured before this happens and the case for G_2 works analogously, the claim follows. $\square$

5   Many Cops, Many Robbers

Sections 3 and 4 dealt with the case of one cop and one robber, respectively. We now focus on the case of multiple cops and multiple robbers. Our goal is to show that there are k-copwin graphs for arbitrarily large k, s.t. capturing all ℓ robbers with the k cops must take in the order of $\ell \cdot b$ time, with the number of nodes in the graph being in the order of at most $\ell \cdot k^2 + \ell^2 \cdot k + b$. In particular, our goal is establish Theorem 8:

Theorem 8. *Let $k_0 \geq 2$ be a positive integer. There exists an integer $k \geq k_0$, s.t. for all $b \in \mathbb{N}$ and for all $\ell \in \mathbb{N}$ holds: There exists a graph $G_{k,\ell,b} = (V_{k,\ell,b}, E_{k,\ell,b})$ with i) $c(G_{k,\ell,b}) = k$ or $c(G_{k,\ell,b}) = k + 1$ and ii) $|V_{k,\ell,b}| \in \mathcal{O}(\ell \cdot k^2 + \ell^2 \cdot k + b)$, s.t. $capt(G_{k,\ell,b}, c(G_{k,\ell,b}), \ell) \in \Omega(\ell \cdot b)$.*

From Theorem 8, we can directly claim the following corollary, which shows that the capture time is at least asymptotically linear in the number of the nodes times the number of robbers, for an arbitrarily large number of cops and robbers by setting $b \in \Theta(\ell \cdot k^2 + \ell^2 \cdot k)$:

Corollary 9. *Let $k_0 \geq 2$ be a positive integer. There exists an integer $k \geq k_0$ and a graph $G = (V, E)$ with $|V| = n \in \mathcal{O}(\ell \cdot k^2 + \ell^2 \cdot k)$ s.t. $c(G) = k$ and $capt(G, k, \ell) \in \Omega(\ell \cdot n)$.*

We note that the cop number k can be as high as $\mathcal{O}(\sqrt{n/\ell})$, and the amount of robbers as high as $\mathcal{O}(\sqrt{n/k})$ in Corollary 9.

To prove Theorem 8, we cannot use a graph construction similar to the ones in Section 3 or Section 4. In Section 4, the construction relied on the fact that there is just one robber. When connecting the two copies of the graph promised by Lemma 5, the robber can pick the side with less cops – and some cops have to cross the long bride, inducing a linear capture time. However, this construction cannot be coupled with the idea of Section 3, as we cannot rule out a single cop waiting on the bridge: Connecting the graphs promised by Lemma 5 can increase the cop number by one (cf. Figure 3). Then, the robbers cannot escape over the bridge, allowing the cops to capture them in rapid succession. Even if one would add multiple bridges, the extra cop could simulate the behavior of the robbers, capturing at least a fraction of them each time the robbers cross. Thus, we need an improved graph family to establish Theorem 8.

In the following, we first describe the new graph construction (Subsection 5.1) with the desired properties and the strategy of the robbers in these graphs (Subsection 5.2), before we prove Theorem 8 in Subsection 5.3.

5.1   The Graph Construction of $G_{k,\ell,b}$

Given an integer k_0, Lemma 5 promises a graph $G_g = (V_g, E_g)$ with a cop number of $c(G_g) = k \geq k_0$ and at most $|V_g| \in O(k^2)$ nodes. Henceforth, we will refer to these graphs G_g as *gadget graphs* with a diameter of $D(G_g) = D_g$.

The construction idea of this subsection is as follows: We construct a cycle and attach ℓ copies of the gadget graph with long lines to top, bottom, left, and right side of the cycle. A graphical depiction can be found in Figure 4.

We first describe how to attach the copies: Let v_g be a fixed node in G_g. Attach a line of nodes of length $10\ell \cdot D_g$ to v_g. Then, copy the graph G_g with the line ℓ times as $G_{g,1}, \ldots, G_{g,\ell}$, with the other endpoints of the lines called $v_{end,1}, \ldots, v_{end,\ell}$ respectively. Connect $v_{end,i}$ to $v_{end,i+1}$ with a line of length $3 \cdot D_g$ for all $1 \leq i < \ell$, inducing a path from $v_{end,1}$ to $v_{end,\ell}$ consisting of $3(\ell-1) \cdot D_g$ new nodes. Denote this construction as G_g^{top} and copy it three more times as G_g^{bottom}, G_g^{left}, and G_g^{right}.

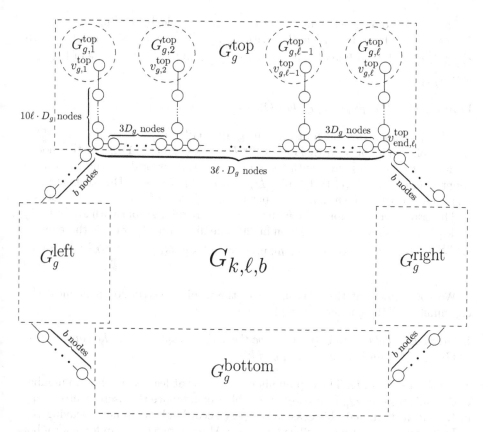

Fig. 4. The graph $G_{k,\ell,b}$ consists of the four subgraphs G_g^{top}, G_g^{bottom}, G_g^{left}, and G_g^{right}, which are connected cyclically by paths of length b. Each of the four subgraphs consists of ℓ copies of the gadget graph G_g with $c(G_g) = k$, which is in turn connected to a path of length $10\ell \cdot D_g$ (with D_g being the diameter of G_g). Initially, the ℓ robbers place themselves all in a subgraph s.t. at least half the cops are at least $b/2$ moves away from the subgraph, one robber in each of the ℓ gadget graphs. As each robber can evade less than k cops in his gadget graph indefinitely, no robber can be caught until k cops enter his gadget graph. I.e., the cops need $\Omega(b)$ moves to capture the first robber. If there are just $G_{k,\ell,b} = k$ cops, the robbers could then all escape to another subgraph, forcing the cops to spend at least $\Omega(b)$ moves for each subsequent robber. However, if there are $c(G_{k,\ell,b}) = k + 1$ cops, the extra cop p_{k+1} could patrol anywhere in the graph, possible blocking the path of the robbers. However, if all robbers always try to move to their respective node v_{end}, but moving back when a cop comes closer than D_g, the extra cop can keep at most one extra robber in check. Then, as soon as k cops enter a gadget graph with a robber, the other robbers can escape to another subgraph: In the top subgraph case, the robbers "left" of p_{k+1} go to G_g^{left}, the robbers "right" of p_{k+1} go to G_g^{right}. As all these robbers always keep a distance of at least D_g to the next cop when escaping to another subgraph, they can position themselves perfectly in their new gadget graph with a diameter of D_g. This can be iterated, enforcing a capture time of $\Omega(\ell \cdot b)$. If $b \in \Omega(n)$ is chosen, then this yield a lower bound of $\Omega(\ell \cdot n)$. We note that three subgraphs would also suffice with a slightly modified strategy.

Lastly, we connect all four structures with a cycle by adding $4 \cdot b$ nodes: Connect $v_{\text{end},\ell}^{\text{top}}$ by a line of length b with $v_{\text{end},1}^{\text{right}}$, $v_{\text{end},\ell}^{\text{right}}$ by a line of length b with $v_{\text{end},1}^{\text{bottom}}$, $v_{\text{end},\ell}^{\text{bottom}}$ by a line of length b with $v_{\text{end},1}^{\text{left}}$, and $v_{\text{end},\ell}^{\text{left}}$ by a line of length b with $v_{\text{end},1}^{\text{top}}$.

Lemma 10. *The graph $G_{k,\ell,b}$ has $O(\ell \cdot k^2 + \ell^2 \cdot k + b)$ nodes.*

Proof (of Lemma 10). Each of the four graphs G_g^{top}, G_g^{bottom}, G_g^{left}, and G_g^{right} consists of ℓ copies of G_g with a line of $10\ell \cdot D_g$ nodes and $3(\ell-1) \cdot D_g$ further nodes connecting them. Together with the $4b$ nodes acting as bridges, the total node count is in $G_{k,\ell,b}$ is $4\left(\ell(\cdot|V_g| + 10\ell \cdot D_g) + 3(\ell-1) \cdot D_g + b\right)$. Due to Lemma 5, $|V_g| \in O(k^2)$, which results in an upper bound of $O(\ell \cdot k^2 + \ell^2 \cdot D_g + b)$ nodes.

The graph construction of Lemma 5 uses $k+1$-regular graphs with $2(k^2+k+1)$ nodes [24]. As shown by, e.g., Moon in [20], the diameter D_g of G_g is therefore in $O(\frac{2(k^2+k+1)}{k+1}) \in O(k)$. Hence, the number of nodes in $G_{k,\ell,b}$ is $O(\ell \cdot k^2 + \ell^2 \cdot k + b)$. $\square$

We now show that the cop number of the whole construction is at most the cop number of the gadget graph plus one:

Lemma 11. *Let G_g with $c(G_g) = k$ be the gadget graph used in the construction of $G_{k,\ell,b}$. The cop number of $G_{k,\ell,b}$ is k or $k+1$.*

Proof (of Lemma 11). The cop number of $G_{k,\ell,b}$ is at least k, as the cop number of the gadget graph G_g is already k: A robber could place themselves into a copy of G_g and just simulate his evasion strategy accordingly, with never leaving G_g.

Furthermore, $k+1$ cops suffice for $G_{k,\ell,b}$: Already two cops can force a robber to place himself into a gadget graph. Then, one cop waits at the exit node v_g, while the remaining k cops capture the robber, simulating their winning strategy from the gadget graph G_g with $c(G_g) = k$. $\square$

5.2 The Robber Strategy

The robber strategy in $G_{k,\ell,b}$ can be summarized as follows: Start in the part of the graph with the fewest cops (each robber in a distinct gadget graph), then try to wait at the end of the line of the current gadget graph, only going back into the gadget graph if a cop comes close. If a cop comes into the current gadget graph, simulate an evasion strategy, which will work for sure until at least k cops enter. Then, as soon as k cops are close to any gadget graph in the subgraph, the other robbers escape to another subgraph without cops, and repeat the initial strategy. If there are $k+1$ cops in the graph, then the cop p_{k+1} may hold back one extra robber from escaping, but all the other robbers can move away from p_{k+1} to another subgraph (possible splitting the robbers into different subgraphs).

We now describe the strategy in detail: After the cops placed themselves, the robbers choose a subgraph G_g^{top}, G_g^{bottom}, G_g^{left}, G_g^{right} to start in that has the most cops being in a distance of at least $b/2$. W.l.o.g., let this subgraph be

G_g^{top}. Note that due to the pigeonhole principle, at most half of the cops can be within a distance of $b/2$ near G_g^{top} and that at least k cops are needed to catch a robber in a gadget graph (cf. Lemma 11). Hence, the remaining cops need at least $b/2 + 10\ell \cdot D_G$ moves to reach any gadget graph in G_g^{top}

Each of the ℓ robbers will place themselves into a pairwise distinct gadget graph $G_{g,1}^{\text{top}}, \ldots, G_{g,\ell}^{\text{top}}$ as follows: Each robber r_i will assume that there are $k-1$ cops in his gadget graph $G_{g,i}^{\text{top}}$, with the missing ones placed all at $v_{g,i}^{\text{top}}$. Then, his placement will be identical as in his evasion strategy for the graph G_g.

Next, as soon as the distance to the nearest (real) cop is larger than D_g, the robber will move towards the node $v_{\text{end},i}^{\text{top}}$, but not surpassing it yet. Should then a cop come closer, then the robber will move back towards $G_{g,i}$, keeping a distance of at least D_g, but move forward again if the cop is further away again. When the robber enters his graph $G_{g,i}^{\text{top}}$ again with a cop close, he resumes simulating the evasion strategy until the distance to the next cop is more than D_g. The distance of D_g is necessary for the robber to assume an arbitrary starting position in the gadget graph again before the first cop enters the gadget graph.

As soon as at least k cops are in a distance of at most D_g to one of the gadget graphs at the top (the robber in this graph now stops moving), there can be at most one other cop, say p_{k+1}, left in the graph. This cop p_{k+1} can be in distance of at most D_g for only one robber (this robber now stops moving as well), allowing all other robbers R' to be at their node $v_{\text{end},i}^{\text{top}}$.

Let $G_{g,j}^{\text{top}}$ be the gadget graph to which the cop p_{k+1} is closest. Due to the construction of $G_{k,\ell,b}$, each robber in R' has now a distance of more than D_g to p_{k+1} (if it exists), and a distance of at least $9\ell \cdot D_g$ to all other cops.

Should $c(G_{k,\ell,b}) = k$, then there is no cop p_{k+1}, and all robbers from R' move to the same of one the other three subgraphs G_g^{bottom}, G_g^{left}, G_g^{right} and place themselves in pairwise distinct gadget graphs, repeating their initial strategy accordingly.

If $c(G_{k,\ell,b}) = k+1$, then the cop p_{k+1} can be closest to the node 1) $v_{\text{end},1}^{\text{top}}$, 2) $v_{\text{end},\ell}^{\text{top}}$, or 3) to some node $v_{\text{end},j}^{\text{top}}$, with $1 < j < \ell$. In the case of 1) (i.e., the cop is at the "left end"), all robbers from R' move to pairwise distinct gadget graphs in G_g^{right}. In the case of 2) (i.e., the cop is at the "right end"), all robbers from R' move to pairwise distinct gadget graphs in G_g^{left}. For the last case of 3), let $R'_<$ be the robbers of R' be at nodes $v_{\text{end},i}^{\text{top}}$ with $i < j$ and $R'_>$ be the robbers of R' be at nodes $v_{\text{end},i}^{\text{top}}$ with $i > j$. All robbers from $R'_<$ move to unoccupied pairwise distinct gadget graphs in G_g^{left}, all robbers from $R'_>$ do the same in G_g^{right}.

Afterwards, the robbers repeat their strategy, adjusted to being in G_g^{top}, G_g^{bottom}, G_g^{left}, G_g^{right} accordingly. Note that the robbers may be split up between all four subgraphs, but that it takes always at least k cops to force them to move to another subgraph.

5.3 A Lower Bound for the Capture Time

In this subsection, we will complete the proof of Theorem 8 by showing a lower bound of $\Omega(\ell \cdot b)$ on the capture time when the robbers use the strategy described in Subsection 5.2 in the graph $G_{k,l,b}$. Essentially, the cops need to move at least b times to capture a constant number of robbers, forcing a lower bound of $\Omega(\ell \cdot b)$.

Proof (of Theorem 8). With Lemma 10 (the size of the graph) and Lemma 11 (the cop number of the graph), all that is left to show of Theorem 8 is a capture time of $capt(G_{k,\ell,b}, c(G_{k,\ell,b}), \ell) \in \Omega(\ell \cdot b)$.

After the cops place themselves initially in the graph $G_{k,l,b}$, the strategy of the robbers will ensure that at least half the cops are at least $b/2$ moves away from each robber. As the robbers are initially in the gadget graphs, at least k cops are required to capture any robber in its gadget graph, requiring at least $b/2$ moves from some cops to capture the first robber.

We begin with the case of $c(G_{k,l,b}) = k$ before discussing $c(G_{k,l,b}) = k+1$. Let w.l.o.g. G_g^{top} be the subgraph where k cops are for the first time within a distance of D_g to a gadget graph. Then, when the other robbers R' in G_g^{top} escape to another subgraph G_g^{bottom}, G_g^{left}, G_g^{right}, these robbers need at most $3\ell \cdot D_g$ moves to exit the subgraph G_g^{top}. However, each of the k cops needs at least $9\ell \cdot D_g$ moves to reach the first node of the type $v_{\text{end}}^{\text{top}}$, ensuring that the other robbers have at least a distance of D_g at all times to these k cops before they enter their new gadget graph to hide in. For every next robber to be captured, these k cops need to move thus to the next gadget graph in another subgraph, enforcing at least b moves for the cops, ensuring a capture time of $\Omega(\ell \cdot b)$.

The case of $c(G_{k,l,b}) = k+1$ is similar, but now there is an additional cop (w.l.o.g. p_{k+1}) that might not need to enter the gadget graphs and is free to move around through the graph, possibly capturing or blocking the other robbers R'. Still, if not at least k cops enter a gadget graph at some point, no robber can be caught, as the robbers can always evade less than k cops in their gadget graph. Consider the move when at least k cops are within distance D_g to a gadget graph. Due to the strategy of the robbers, the remaining cop p_{k+1} can be within distance of D_g to at most one robber in G_g^{top}. Thus, all other robbers, which are at nodes of the type $v_{\text{end}}^{\text{top}}$, can escape to the other subgraphs (G_g^{bottom}, G_g^{left}, G_g^{right}), depending on where p_{k+1} is located – the "ring" structure of $G_{k,\ell,b}$ does not allow p_{k+1} to block the other robbers.

I.e., at most two robbers can be prevented from escaping to another subgraph. When the robbers arrive in the pairwise distinct gadget graphs of their new subgraph G_g^{bottom}, G_g^{left}, G_g^{right}, the initial situation occurs again: The cops need to have moved at least b times to capture again at most two robbers, inducing a total capture time of $\Omega(\ell \cdot b)$.   □

5.4   A Superlinear Number of Robbers

So far, the number of robbers has never exceeded a linear amount, i.e., we did not consider the case of $\omega(n)$ robbers in n-vertex graphs. However, our results can be extended to this case.

We start with the case of one cop and many robbers (cf. Section 3). Fix a number of nodes n for Corollary 3 and let ℓ' be any number of robbers less than $n/8$. If the number of robbers were to be increased to $2\ell'$, then one could always move two robbers as if they were one, with them sharing the same place. Then the capture time would remain the same, as the cop would always capture two robbers at once.

However, after the first two robbers are captured, and all robbers move along the bridge connecting the left and the right star (see Figure 1), the only unoccupied ray of the star can be now be occupied by splitting a pair of robbers into singles. The cop could still capture two robbers in his next catch, but only $\ell'/2$ times in total! After that, all rays would only be occupied by one robber, allowing the cop to capture only one robber at once. Hence, the cop now needs to cross the bridge connecting the two stars at least $\ell'/2 + \ell'$ times.

This concept can be iterated, e.g., for $4\ell'$ robbers, the cop needs to cross the bridge $(1/4 + 1/3 + 1/2 + 1)\ell'$ times. I.e., for $t \cdot \ell'$ robbers, this number increases to $(1/t + \cdots + 1/2 + 1)\ell' \in \Omega(\ell' \log t)$. With the bridge having a length of $\Omega(n)$ nodes, the following corollary holds:

Corollary 12. *For all $n \geq 12$ there exists a 1-copwin graph G s.t. for all numbers $\ell \geq n$ of robbers $capt(G, 1, \ell) \in \Omega\left(n^2 \cdot \log\left(\ell/n\right)\right)$.*

We note that a similar line of thought can be applied to the case of more than one cop, i.e., letting the cops capture multiple robbers at once, and then splitting up the remaining robbers evenly among the gadget graphs.

Acknowledgements. We would like to thank the anonymous reviewers for their helpful comments.

References

1. Aigner, M., Fromme, M.: A Game of Cops and Robbers. Discrete Applied Mathematics 8(1), 1–12 (1984)
2. Alspach, B.: Sweeping and Searching in Graphs: a Brief Survey. Matematiche 59, 5–37 (2006)
3. Berarducci, A., Intrigila, B.: On the Cop Number of a Graph. Advances in Applied Mathematics 14(4), 389–403 (1993)
4. Bonato, A., Chiniforooshan, E.: Pursuit and evasion from a distance: algorithms and bounds. In: Proceedings of the Sixth Workshop on Analytic Algorithmics and Combinatorics (ANALCO), pp. 1–10. SIAM (2009)
5. Bonato, A., Golovach, P.A., Hahn, G., Kratochvíl, J.: The Capture Time of a Graph. Discrete Mathematics 309(18), 5588–5595 (2009)
6. Bonato, A., Gordinowicz, P., Kinnersley, B., Prałat, P.: The Capture Time of the Hypercube. Electr. J. Comb. 20(2), P24 (2013)
7. Bonato, A., Nowakowski, R.J.: The Game of Cops and Robbers on Graphs. Student Mathematical Library, vol. 61. American Mathematical Society, Providence (2011)
8. Bonato, A., Yang, B.: Graph searching and related problems. In: Handbook of Combinatorial Optimization, pp. 1511–1558. Springer, New York (2013)
9. Breisch, R.: An Intuitive Approach to Speleotopology. Southwestern Cavers 6(5), 72–78 (1967)
10. Clarke, N.E., MacGillivray, G.: Characterizations of k-copwin Graphs. Discrete Mathematics 312(8), 1421–1425 (2012)

11. Deo, N., Nikoloski, Z.: The Game of Cops and Robbers on Graphs: a Model for Quarantining Cyber Attacks. Congressus Numerantium, 193–216 (2003)
12. Frankl, P.: Cops and Robbers in Graphs with Large Girth and Cayley Graphs. Discrete Appl. Math. 17(3), 301–305 (1987)
13. Frieze, A.M., Krivelevich, M., Loh, P.-S.: Variations on Cops and Robbers. Journal of Graph Theory 69(4), 383–402 (2012)
14. Gavenciak, T.: Cop-win Graphs with Maximum Capture-time. Discrete Mathematics 310(10–11), 1557–1563 (2010)
15. Hahn, G.: Cops, Robbers and Graphs. Tatra Mt. Math. Publ. 36(163), 163–176 (2007)
16. Kinnersley, W.B.: Cops and Robbers is EXPTIME-complete. J. Comb. Theory, Ser. B 111, 201–220 (2015)
17. Kosowski, A., Li, B., Nisse, N., Suchan, K.: k-Chordal graphs: from cops and robber to compact routing via treewidth. In: Czumaj, A., Mehlhorn, K., Pitts, A., Wattenhofer, R. (eds.) ICALP 2012, Part II. LNCS, vol. 7392, pp. 610–622. Springer, Heidelberg (2012)
18. Lu, L., Peng, X.: On Meyniel's Conjecture of the Cop Number. Journal of Graph Theory 71(2), 192–205 (2012)
19. Mehrabian, A.: The Capture Time of Grids. Discrete Mathematics 311(1), 102–105 (2011)
20. Moon, J.W.: On the Diameter of a Graph. Michigan Math. J. 12(3), 349–351 (1965)
21. Nowakowski, R.J., Winkler, P.: Vertex-to-vertex Pursuit in a Graph. Discrete Mathematics 43(2-3), 235–239 (1983)
22. Parsons, T.D.: Pursuit-evasion in a graph. In: Alavi, Y., Lick, D.R. (eds.) AII 1992. LNCS, vol. 642, pp. 426–441. Springer, Heidelberg (1992)
23. Parsons, T.D.: The search number of a connected graph. In: Proc. 9th Southeast. Conf. on Combinatorics, Graph Theory, and Computing (1978)
24. Pralat, P.: When Does a Random Graph Have a Constant Cop Number. Australasian Journal of Combinatorics 46, 285–296 (2010)
25. Quilliot, A.: Jeux et Pointes Fixes sur les Graphes. Ph.D. thesis, Universite de Paris VI (1978)
26. Scott, A., Sudakov, B.: A Bound for the Cops and Robbers Problem. SIAM J. Discrete Math. 25(3), 1438–1442 (2011)

Collaborative Exploration
by Energy-Constrained Mobile Robots[*]

Shantanu Das[1], Dariusz Dereniowski[2], and Christina Karousatou[1]

[1] LIF, Aix-Marseille University and CNRS, Marseille, France
[2] Faculty of Electronics, Telecommunications and Informatics,
Gdańsk University of Technology, Gdańsk, Poland

Abstract. We study the problem of exploration of a tree by mobile agents (robots) that have limited energy. The energy constraint bounds the number of edges that can be traversed by a single agent. Thus we need a team of agents to completely explore the tree and the objective is to minimize the size of this team. The agents start at a single node, the designated root of the tree and the height of the tree is bounded by the energy bound B. We provide an exploration algorithm without any knowledge about the tree and we compare our algorithm with the optimal offline algorithm that has complete knowledge of the tree. Our algorithm has a competitive ratio of $O(\log B)$, independent of the number of nodes in the tree. We also show that this is the best possible competitive ratio for exploration of unknown trees.

1 Introduction

Overview: Graph exploration is a well studied problem in computer science with a wide range of applications from searching the internet to navigation of robots in unknown environments. The objective is to discover an initially unknown graph by visiting all nodes in a systematic manner starting from a given node of the graph. The problem has been well studied for a single agent exploring a graph [16] or a digraph [1] with the aim of minimizing the exploration time or equivalently the number of edges traversed. Others have studied the problem from the perspective of minimizing the memory needed by the agents for exploration [8,13]. When the nodes of the graph do not have identifiers, the agent may need to mark nodes with a pebble to recognize them and thus, another research direction is to minimize the number of pebbles used for exploration [4].

When the exploration is performed by physical robots, one of the major issues is the energy consumed during the exploration, since each robot may have a limited amount of energy for movement. Surprisingly, most previous studies on exploration have not considered this limitation. Betke et al. [5] and later Awerbuch et al. [2] have studied the problem of exploration with an energy constrained agent. Their solution requires a fuelling station at the starting node and

---

[*] Partially supported by the ANR projects MACARON (anr-13-js02-0002) and AN-COR (anr-14-CE36-0002-01), and by the Polish National Science Center grant DEC-2011/02/A/ST6/00201.

© Springer International Publishing Switzerland 2015
C. Scheideler (Ed.): SIROCCO 2015, LNCS 9439, pp. 357–369, 2015.
DOI: 10.1007/978-3-319-25258-2_25

the agent periodically returns there to refuel. Between two visits to the starting node, the agent can make at most B edge traversals. Thus the diameter of graphs that can be explored is restricted to $B/2$. When refuelling is not allowed, multiple agents may be needed to explore even graphs of restricted diameter. Given a graph G, determining whether a team of k agents, each having an energy constraint of B can explore G is known to be an NP-hard problem, even when the graph G is a tree [12]. When the graph (or the tree) is unknown, there are two possible approaches for online exploration. One approach is to fix the number k of agents and try to bound the amount of energy B required by each agent, as in Dynia et al. [10,11]. In this paper, we take the other approach of fixing the available energy B for each agent and bounding the number of agents used for exploration. Indeed, according to recent trends in robotics [17], it is preferable to use a large number of small robots rather than a few bulky ones and our line of research goes in this direction. In our model, each agent has a limited energy resource without the ability to recharge, thus allowing the agent to traverse at most B edges, and our objective is to limit the total number of such agents used for exploration. We measure the efficiency of the solution in terms of the *competitive ratio* which is defined as the worst case ratio of the cost of the online algorithm for some graph G over the cost of the optimal offline algorithm for the same graph. We restrict ourselves to exploration of trees. The agents start at the designated root of an unknown tree T and they must collectively visit every node of T.

If the height of T i.e. the longest path between the root and a leaf, is greater than B, then T cannot be fully explored, even by an unbounded number of agents.

On the other hand, if the height of the tree is exactly B then each leaf at depth B must be visited by a separate agent. Once the tree is completely explored and known up to depth $B - 1$ then we can send one additional agent to explore each leaf at depth B. Thus it is sufficient to consider algorithms for exploring trees of height at most $B - 1$. Note that the previous results [2,10] for energy-constrained agents were restricted to exploring trees of height at most $B/2$ (or graphs of diameter at most $B/2$).

Related Work: The graph exploration problem has been previously studied with the objective of minimizing the time for exploration. For exploration of undirected graphs by a single agent, the algorithm given by Panaite and Pelc [16] requires $m + O(n)$ time for a graph of m edges and n nodes. For exploration of an unknown tree, exploration in the optimal time of $2(n-1)$ can be achieved by the depth-first search algorithm. Using multiple agents can speedup the exploration and Fraigniaud et al. [12] have presented an algorithm for a team of k mobile agents that explores a tree of height D in $O(D + n/\log k)$ time. They also showed that any algorithm for k-agent exploration of a tree has a $(2 - 1/k)$ overhead over the optimal offline algorithm. While the above results are for small team of agents (where $k \leq \sqrt{n}$), Dereniowski et al. [7] used a large team of agents to reduce the exploration time to $O(D)$ and their solution also works for general graphs where all nodes are within distance D from the starting node. Ortolf et al. [15] gave bounds on the competitive ratio for multiple agent exploration of grid graphs with obstacles. For general graphs, Megow et al. [14] presented a single-agent exploration algorithm having a constant competitive ratio.

The above results do not consider any energy limitation for the agent. For a single, energy constrained agent, the problem of exploration with refuelling, has been studied for grid graphs [5] and also for general graphs [2]. The optimal time algorithm for exploration with refuelling was given by Duncan et al. [9], who also studied exploration under a different type of constraint where the agent is tied to the starting node with a string of fixed length.

For a team of k agents, the problem of exploring a tree using limited energy resources was investigated by Dynia et al. [10] who presented an algorithm that is 8-competitive in terms of the energy consumed by each agent. This was later improved to a competitive ratio of $(4 - 2/k)$ by Dynia et al. [11]. Other problems that have been considered for energy constrained agents (that may not start at the same node) include *broadcast* and *convergecast* [3] as well as *data-delivery* from a source node to a target node in the graph [6]. These are mainly offline solutions where the graph and the starting locations are given as input. The algorithm in the present paper can be seen as an online solution to the problem of data-delivery from the root to the leaves or vice versa, for the special case of colocated agents.

Our Results: We consider the problem of exploration of an unknown tree by a team of mobile agents initially located at the root of the tree. Each agent is equipped with a battery of size B which bounds the total number of edges the agent can traverse during its lifetime. We assume the height of the tree to be at most $B - 1$, and our objective is to find an exploration strategy where every node of the tree is visited by at least one agent, and we wish to minimize the total number of agents used. We study this problem first assuming a global communication model (where agents communicate to each other instantaneously) and provide an algorithm for online exploration, that has a competitive ratio of $O(\log B)$. We then show how to remove the assumption of global communication and achieve the same result in the local communication model, with a constant overhead. Finally we provide a lower bound of $\Omega(\log B)$ on the competitive ratio of any online exploration algorithm for energy-constrained agents, showing that our result is tight. We conclude with some open questions for future research. Due to the space constraint, proofs of some of the lemmas and theorems have been omitted.

2   The Model

The environment to be explored is a rooted tree T. The root r_0 contains an infinite supply of mobile agents, each of which has a limited energy B, allowing it to traverse at most B edges during its lifetime. There is a total order among the agents (i.e. they have distinct identities). The nodes of the tree may be assumed to be anonymous (i.e. we do not require unique identifiers for the nodes of T). Each agent has unlimited memory. When two agents are at the same node, they can freely exchange information. However the agents may not write any information on the nodes of the tree. We call this the *local communication* model. In contrast, in a *global communication* model an agent can communicate instantaneously with any other agent irrespective of their location in the tree.

All agents start at the same time, in the same state. At each time unit, any agent can move to an adjacent node or stay at its current node. Each move costs one unit of time and one unit of energy, while computation and communication between agents are instantaneous and do not consume any energy. The agents cannot exchange their energy resources or recharge their batteries.

The height of the tree (i.e. the distance to the furthest leaf from the designated root r_0) is at most $B - 1$. The size and structure of the tree is initially unknown to the agents. The edges incident at each node are locally ordered with port numbers, allowing the agents to choose edges to visit in a deterministic manner. An exploration strategy for the team of agents is successful if each node of the tree visited by at least one agent. The cost of the exploration strategy is the number of agents which made any non-null moves during the exploration. We denote by OPT the cost of the optimal offline strategy that has complete knowledge of the tree.

For any node $r \in T$ we denote by T_r the subtree of T, rooted at r. Further for any node $v \in T_r$, we define the depth of node v as the length of the path from r to v. We denote by T_r^δ the subtree rooted at r truncated to depth δ from r. We denote by $|T|$, the number of edges in T.

3   Exploration with Global Communication

In this section we describe and analyze a recursive algorithm for tree exploration under the global communication model. The algorithm is called *Global Communication Tree Exploration* (GCTE). The main idea of the algorithm is to explore the tree up until a certain depth and afterwards take advantage of the already known part of the tree to continue the exploration. More specifically, this algorithm proceeds by *levels*. Each level of the algorithm is a set of nodes which are located at a certain depth of the tree. The first level consists of the root r_0. At each level i, agents having energy b_i, expand the explored part of the tree further by increasing its depth by $\varepsilon \cdot b_i$ where ε is a parameter of the algorithm such that $0 < \varepsilon < \frac{1}{4}$. The new frontier of the explored part defines the next level of the algorithm. The algorithm $\mathtt{GCTE}_\varepsilon$ is then recursively called at each node of the newly created level i.

Definition 1. *For $i = 1$, level i of algorithm $\mathtt{GCTE}_\varepsilon$ consists of the root node r_0; the depth d_i of the level i is $d_1 = 0$, the energy b_i at this level is $b_1 = B$. For $i > 1$, level i of $\mathtt{GCTE}_\varepsilon$ consists of all nodes at depth $d_i = d_{i-1} + \lceil \varepsilon \cdot b_{i-1} \rceil$, and $b_i = B - d_i$.*

For any two nodes u and v at the same level, we would like the exploration of the trees T_u and T_v to proceed independently, using disjoint sets of agents. To this end, we allow some overlap between successive levels of the algorithm. More precisely, at each level i, the exploration is extended to the depth of $(\frac{1}{2} + \varepsilon) \cdot b_i$, although the next level still starts at depth $\varepsilon \cdot b_i$ from the current level. This additional extension at each level i allows the algorithm to *look ahead* at the start of the next level $(i+1)$. Thus, at the start of a recursive call to Algorithm $\mathrm{GCTE}_\varepsilon$ at a node r at level $i + 1$, the subtree T_r has been already partially explored to some depth. We show below (c.f. Lemma 1) that the exploration of this partially explored subtree T_r can be done independently to any other subtree at the same level.

Definition 2. *Two partially explored subtrees T_u and T_v, rooted at nodes u and v located at the same depth from r_0, are said to be* independent *if no single agent can visit nodes in the unexplored part of both subtrees.*

Informally, this independence means that disjoint teams of agents can be used for exploring such subtrees during the algorithm. We now formally describe our algorithm $\mathrm{GCTE}_\varepsilon$.

Algorithm $\mathrm{GCTE}_\varepsilon$. An algorithm for tree exploration, $0 < \varepsilon < \frac{1}{4}$

Input: The root r of the tree and an integer b that equals the size of the available energy the agents have.
1: Uncover$(r, \lfloor(\frac{1}{2} + \varepsilon)b\rfloor)$
2: Let $r_1, r_2, \ldots$ be nodes at depth $\lceil \varepsilon \cdot b \rceil$ from r, such that T_{r_i} has some unexplored edges.
3: For each r_i, call Algorithm $\mathrm{GCTE}_\varepsilon$ $(r_i, (b - \lceil \varepsilon \cdot b \rceil))$.

Procedure Uncover(r, δ) with input node r and an integer δ works as follows. During this procedure, the agents explore the unexplored part of subtree T_r rooted at r, using a Depth First Search (DFS) traversal restricted to a depth of δ from r. An agent initially located at the root r_0 arrives at the current root r, having b units of energy and begins to explore the subtree T_r^δ performing DFS. First, this agent goes to the next unexplored node in the DFS traversal. At this node, the agent resumes the DFS traversal. Finally, when the agent has $x(\varepsilon) = \frac{1}{2}(\frac{1}{2} - \varepsilon)b$ units of energy left, it interrupts the exploration (it saves the remaining energy for later use, as explained later in the next section). If the point where the agent is supposed to interrupt the DFS traversal is the middle of an edge, then the agent finishes before traversing this edge. Note that in the global communication model, at any point of exploration, each agent possesses the full knowledge of the part of the tree explored to date and the current locations of all agents. Hence, another agent will arrive at r and continue the DFS exploration by visiting the unexplored node that is supposed to be visited next according to the DFS traversal. This procedure ends when all nodes at depth δ or less have been visited.

Lemma 1. *The subtrees that are created in step 2 of* GCTE$_\varepsilon$ *are pairwise independent. Moreover, for any such subtree T_r rooted at a node r any agent that reaches the unexplored part of T_r cannot return to node r.*

Theorem 1. *For any ε, $0 < \varepsilon < \frac{1}{4}$, Algorithm* GCTE$_\varepsilon$ *called for r_0 and B correctly explores the tree.*

Proof. To prove the correctness of GCTE$_\varepsilon$, we first show that procedure Uncover(r, δ) with $\delta = \lfloor (\frac{1}{2} + \varepsilon)b \rfloor$ correctly explores the subtree rooted at node r up to depth δ. Note that by a simple induction on the distance of r from r_0, any agent that arrives at node r, to execute Uncover(r, δ), has exactly b units of energy. Further any such agent A_j has complete knowledge of the part of subtree T_r already explored by previous agents and thus agent A_j knows the path from r to the next unexplored node v in the DFS traversal of T_r^δ. This node v must be at distance at most δ from r. According to the algorithm, the agent uses

$$l := b - \lceil x(\varepsilon) \rceil = \lfloor b - x(\varepsilon) \rfloor = \lfloor \delta + x(\varepsilon) \rfloor$$

units of energy during the DFS traversal. Since $l \geq \delta$ the agent does succeed in reaching the node v. Hence, each agent used in Uncover visits at least one previously unexplored node in T_r^δ. This implies that eventually all nodes within depth δ in T_r are visited during the DFS exploration. This proves the correctness of procedure Uncover.

In order to complete the proof of the correctness of GCTE$_\varepsilon$, we note that the algorithm makes progress at each level i, that is, level $i + 1$ is at strictly greater depth than level i. Indeed, this follows from $\varepsilon b_i > 0$ for $\varepsilon > 0$, which gives $\lceil \varepsilon b_i \rceil \geq 1$. □

Lemma 2. *The number of levels in Algorithm* GCTE$_\varepsilon$ *is at most* $\log_{(\frac{1}{1-\varepsilon})} B$.

Before proceeding to calculating the cost of Algorithm GCTE$_\varepsilon$, let us make the following useful remark. During the procedure Uncover, each participating agent uses at most δ energy to reach the starting node for its DFS exploration and uses at least $b - \delta - \lceil x(\varepsilon) \rceil = \lfloor \frac{1}{2}(\frac{1}{2} - \varepsilon)b \rfloor$ units of energy to contribute to the DFS exploration of unexplored nodes.

Lemma 3. *Procedure* Uncover(r, δ) *for $r = r_0$ and $\delta = (1/2 + \varepsilon)B$ uses $SOL_r \leq \frac{4}{(\frac{1}{2} - \varepsilon)} \cdot OPT$ agents.*

Theorem 2. *Algorithm* GCTE$_\varepsilon$ *has a competitive ratio of* $\frac{4}{(\frac{1}{2} - \varepsilon)} \cdot \log_{(\frac{1}{1-\varepsilon})} B$.

Proof. Consider a call to GCTE$_\varepsilon(r, b_i)$ at some level $i > 1$, where r is at depth $d_i > 0$ from the global root. Let SOL_r denote the number of agents used by the algorithm to explore edges of the subtree T_r during level i. A DFS exploration walk of T_r that starts and ends at r has length $2 \cdot |T_r|$. As explained before each of the SOL_r agents (except the last one) use at least $\frac{1}{2}(\frac{1}{2} - \varepsilon)b_i$ of their available energy to contribute to the DFS exploration. The last agent may have some available energy after visiting the last unexplored edge in T_r but it does not have enough energy to return to node r (by Lemma 1). Thus if we assume that the last agent attempts to reach the root r with its remaining energy, we can say that the path traversed in total by the agents is at most $2 \cdot |T_r|$. Thus,

$$\frac{1}{2}(\frac{1}{2} - \varepsilon)b_i \cdot SOL_r \leq 2 \cdot |T_r| \implies SOL_r \leq \frac{4}{b_i(\frac{1}{2} - \varepsilon)}|T_r|$$

Furthermore, due to Lemma 1, we know the subtrees at the same level are independent so we can sum up over all subtrees at level i:

$$\sum_{r \in r_1, r_2, \ldots} SOL_r \leq \frac{4}{b_i(\frac{1}{2} - \varepsilon)} \sum_{r \in r_1, r_2, \ldots} |T_r|$$

$$SOL(i) \leq \frac{4}{b_i(\frac{1}{2} - \varepsilon)}|T \setminus T^{d_i}|$$

where $SOL(i)$ denotes the number of agents used by the algorithm at level i. The optimal algorithm uses OPT agents to explore the tree. Any agent that reaches to depth d_i of T has b_i units of energy remaining. Thus, each agent can traverse at most b_i edges below this depth. Hence

$$b_i \cdot OPT \geq |T \setminus T^{d_i}|$$

Combining the above two equations, we have

$$SOL(i) \leq \frac{4}{\frac{1}{2} - \varepsilon}OPT$$

The above bound holds for any level $i > 1$. Moreover, due to Lemma 3, we have exactly the same bound for level $i = 1$ of the algorithm. Since there are at most $\log_{(\frac{1}{1-\varepsilon})} B$ levels in the algorithm (due to Lemma 2), we obtain the total cost SOL of the algorithm,

$$SOL \leq \frac{4}{\frac{1}{2} - \varepsilon} \cdot \log_{(\frac{1}{1-\varepsilon})} B \cdot OPT$$

$\square$

Note that on the termination of algorithm GCTE_ε, each agent that participated in the exploration at level i has at least $x_i(\varepsilon) = \frac{1}{2}(\frac{1}{2} - \varepsilon) \cdot b_i$ units of unused energy. This remaining energy would be used by the algorithm presented in the next section.

4   Exploration with Local Communication

This section is devoted to adaptation of $GCTE_\varepsilon$ for the model with local communication between agents. This is done in two steps. In the first step we introduce an intermediate stage between two models of global and local communication. We call this a *semi-local communication model* and we define it as follows: two agents performing the DFS exploration in Step 1 of an instance of $GCTE_\varepsilon$ can communicate only locally, that is, they can communicate only when present at the same node; on the other hand, the algorithm may call for a new agent that is placed at the root r of a subtree explored by an instance of $GCTE_\varepsilon$. Note that, when an instance of $GCTE_\varepsilon$ calls for a new agent to arrive at the input node r, this agent is initially present at the 'global' root of the entire tree and needs to traverse the path from the global root to r. Thus, in our semi-local communication model this mechanism of calling for agents uses the global communication model. In Section 4.1 we adopt $GCTE_\varepsilon$ so that it operates in the semi-local communication model and we calculate the cost of this modification in terms of the number of agents used. In particular, we prove that with respect to the original algorithm, the total number of agents increases by a constant factor (depending only on ε). Then, in Section 4.2, we add to our algorithm a mechanism for calling for new agents at local roots so that this part is also done via local communication.

4.1   Semi-local Communication Model

We start this section by providing some intuition. We consider an arbitrary execution of $GCTE_\varepsilon(r, b)$ for an input node r and energy level b. Recall Step 1 of $GCTE_\varepsilon$, where the agents, one by one, perform the DFS traversal up to a certain depth of the subtree T_r. Suppose that the agents that perform this traversal are $A_1, \ldots, A_k$ and that they are ordered according to the precedence of their movements, i.e., A_i traverses its path prior to A_{i+1} for each $i \in \{1, \ldots, k-1\}$. For each agent A_i we will add a constant number of $c(\varepsilon)$ additional agents denoted $A_i^1, \ldots, A_i^{c(\varepsilon)}$, where

$$c(\varepsilon) = 2 \cdot \left\lceil \frac{1/2 + \varepsilon}{1/2 - \varepsilon} \right\rceil, \quad 0 < \varepsilon < 1/4. \tag{1}$$

To simplify some statements we sometimes write A_i^0 in place of A_i. The agents $A_i, A_i^1, \ldots, A_i^{c(\varepsilon)}$ are called the *i-th team* for each $i \in \{1, \ldots, k\}$. For the purposes of the analysis we introduce some additional notation that allows us to describe the behavior of agents during this DFS traversal in more details. We denote by brevity

$$x(\varepsilon) = \frac{1}{2}\left(\frac{1}{2} - \varepsilon\right) b. \tag{2}$$

We also say that an agent *heads towards* a node v if in each of the following consecutive time units the agent makes a move that gets it closer to v until either v is reached or the agent runs out of energy. It is said that the i-th team is *successful* if: (i) the agent A_i visited a superset of nodes with respect to its original behavior in Step 1 of GCTE$_\varepsilon$, and (ii) the agent $A_i^{c(\varepsilon)}$ reaches the root r and possesses the information about all moves performed by agents $A_1, \ldots, A_i$.

We now describe the modification of the DFS traversal from Step 1 of GCTE$_\varepsilon$ by describing how A_i and $A_i^1, \ldots, A_i^{c(\varepsilon)}$ operate for each $i \in \{1, \ldots, k\}$.

Behavior of A_i. Recall that in Step 1 of GCTE$_\varepsilon$, each agent A_i, $i \in \{1, \ldots, k\}$, finishes its part of DFS traversal having at least $x(\varepsilon)$ energy left. We now use this energy as follows: the agent heads towards the root r in the next $\lceil x(\varepsilon) \rceil$ time units.

Behavior of A_i^j's. For each $i \in \{1, \ldots, k\}$ and $j \in \{1, \ldots, c(\varepsilon)\}$, the agent A_i^j follows the movements of A_i up to the depth

$$d_j(\varepsilon) = \lfloor j \cdot x(\varepsilon) \rfloor$$

until the completion of the movement of A_i. More precisely, the agent A_i^j mimics each move of A_i from node u to node v if both u and v are within depth (from r) at most $d_j(\varepsilon)$. If, on the other hand, either u or v is at depth greater than $d_j(\varepsilon)$, then A_i^j stays idle in this given time unit. Finally, the agent A_i^j heads towards the root r; we will describe below in which time unit this action is triggered.

Order of Movements. Having described the movements of A_i and A_i^j for each $i \in \{1, \ldots, k\}$ and $j \in \{1, \ldots, c(\varepsilon)\}$, we specify the order of their actions. The agent A_1 starts its movement once all agents of the 1-st team are at r. For each $i \in \{2, \ldots, k\}$, the agent A_i starts its movement once $A_{i-1}^{c(\varepsilon)}$ completed its movement by arriving at r and once all agents of the i-th team are at r. (We will argue later that $A_{i-1}^{c(\varepsilon)}$ indeed returns to the root r.) In other words, once $A_{i-1}^{c(\varepsilon)}$ completes its movement, all agents of the i-th team are called to appear at r. For each $j \in \{1, \ldots, c(\varepsilon)\}$, we only need to describe how they operate once A_i runs out of energy, as their preceding movements are specified above. The agent A_i^j heads towards the root r in time unit in which he occupies the same node as A_i^{j-1} and the latter agent is heading towards r. (Thus, it may happen that for a number of time units both agents will head towards r together.)

In the following we prove that the above actions of agents are valid under the assumption that they have to communicate locally. Considering the order of movements of agents it suffices to argue that each team is successful. We refer to all movements of the agents A_i^j, $i \in \{1, \ldots, k\}$, $j \in \{1, \ldots, c(\varepsilon)\}$, as the *extended DFS traversal* of T.

Lemma 4. *For each $i \in \{1, \ldots, k\}$, the i-th team is successful.*

As a consequence of the above, we obtain the following.

Lemma 5. *The extended DFS traversal correctly explores T_r to a depth of $(\frac{1}{2} + \varepsilon)B$ using $k(c(\varepsilon)+1)$ agents that communicate locally, where k is the number of agents used in the DFS traversal performed in Step 1 of Algorithm GCTE$_\varepsilon$.*

Proof. The fact that the k teams, each of size $c(\varepsilon) + 1$, explore the tree to the required depth follows directly from Lemma 4.   □

4.2   Local Communication between Levels

We start this section with an informal description, also pointing out the obstacles we need to overcome. The mechanism of communication between two consecutive levels will be handled by special agents that we call *managing agents* (see below for a formal definition). A managing agent arrives at a root r for which a call to $\mathrm{GCTE}_\varepsilon$ is performed. This agent is not used for the extended DFS traversal of T_r but will play a crucial role while conducting recursive calls for descendants $r_1, r_2, \ldots$. More precisely, this agent will keep track of which subtrees have been already explored and for which one the recursive call is 'in progress'. By a recursive call, made say for r_i, being in progress we mean that the exploration of T_{r_i} is in progress. Thus, until the exploration of that subtree is completed, the managing agent for T_r is responsible for redirecting all agents arriving at r to this subtree T_{r_i}. Once the exploration of T_{r_i} is completed, the managing agent for T_{r_i} will report this fact to the managing agent for T_r and the latter one may initiate the process of exploration of the next subtree $T_{r_{i+1}}$. Once all subtrees $T_{r_1}, T_{r_2}, \ldots$ are explored the managing agent for T_r returns 'one level up' to report this event to appropriate managing agent.

Observe that the above scheme should be performed in such a way that each subtree T_r 'receives' just enough agents needed for its exploration and not more. This includes one managing agent for the subtree itself, the agents performing the extended DFS traversal of T_r and the agents needed for recursive calls, if any. This is regulated by introducing the agents slowly at the global root so that, within predefined time intervals new agents appear at the global root and are directed gradually by managing agents precisely to the subtree for which the current extended DFS traversal is performed. The time intervals are set up in such a way that if an exploration of a particular subtree is completed then this information has enough time to be carried by the managing agent to the one residing one level up. In this way the flow of agents to a particular subtree is stopped and redirected to the next one supplying the exact amount of agents needed for each of the subtrees. Intuitively, the measurement of time is used indirectly as a communication tool: if a managing agent does not receive for a given amount of time a signal that a recursive call to a subtree is completed, then this means that the exploration of that subtree is not completed and more agents are needed to finish it — hence another agent will be sent to that subtree.

Now we give a detailed description of the modifications to the exploration strategy described in Section 4.1 so that it is valid for agents communicating locally. At the beginning of exploration (i.e., when $\mathrm{GCTE}_\varepsilon$ is called for a tree T), one distinguished agent is selected to be constantly present at the root r_0 of the entire tree T. This agent is called the *managing agent for T*. Similarly, whenever a recursive call of $\mathrm{GCTE}_\varepsilon$ is made for any input node r, the first agent that arrives at r is the managing agent for T_r and it stays at r until the entire subtree T_r is explored.

Extension of Step 1 of $\mathrm{GCTE}_\varepsilon$. Once all $c(\varepsilon) + 1$ members of the i-th team are present at the root r of a subtree for which the extended DFS traversal is performed, the i-th team operates exactly as described in Section 4.1. Recall that the i-th team finishes its work with one of its agents being at the root. The beginning of the operation of the $(i + 1)$-th team is postponed until exactly $c(\varepsilon) + 1$ new agents, each with energy b, appear at r. Then, the $(i + 1)$-th team resumes the extended DFS traversal. We note that the agents forming each team will arrive at r directly from the global root of the tree and this will become clear after description of the extension of Step 3 of $\mathsf{GCTE}_\varepsilon$.

Extension of Step 3 of $\mathsf{GCTE}_\varepsilon$. For this part we need to describe how a recursive call is performed by an instance of $\mathsf{GCTE}_\varepsilon$. This includes two actions: initiating the call and receiving information that a recursive call is completed, i.e., that the exploration of the subtree for which the call was conducted is finished. Suppose that an instance of $\mathsf{GCTE}_\varepsilon$ with input r and b performs a call for a subtree rooted at a node r_i. Recall that the managing agent for T_r, denoted by $A(r)$ is present at r during exploration of T_r. First, $A(r)$ waits until a new agent, denoted by $A(r_i)$, appears at r and after this event this agent is sent to r_i and it becomes the managing agent for T_{r_i}. Then, the algorithm sends each agent arriving at r to the node r_i until the agent $A(r_i)$ returns to r. This completes the recursive call for r_i and $A(r_i)$ stays idle at r indefinitely (and will not play any role in the remaining part of the exploration). Then, the next recursive call, if any, that needs to be done is performed. The information about the current status of each recursive call made by the instance of $\mathsf{GCTE}_\varepsilon(r, b)$, is maintained by $A(r)$, the managing agent for T_r, and once all recursive calls are completed this managing agent returns to the node that is the ancestor of r from which the instance of $\mathsf{GCTE}_\varepsilon(r, b)$ was called.

Distribution of agents at the global root. Note that the above description defines the operation of agents for each instance of $\mathsf{GCTE}_\varepsilon$ except for the managing agent at the global root r_0 for the first call to $\mathsf{GCTE}_\varepsilon$. The managing agent at the global root has all agents at its disposal from the first step and does not need to wait for the arrival of an agent. Therefore we introduce an artificial delay denoted by $d(\varepsilon)$ as defined below. The $d(\varepsilon)$ is an integer and it will be understood that the agents will appear at the global root r in time intervals of $d(\varepsilon)$. This time interval is defined as

$$d(\varepsilon) = (c(\varepsilon) + 2)B. \tag{3}$$

The exploration strategy modified as above is called $\mathsf{LCTE}_\varepsilon$ (*Local Communication Tree Exploration*). We now prove that $\mathsf{LCTE}_\varepsilon$ works correctly in the local communication model.

Lemma 6. *For $0 < \varepsilon < 1/4$, Algorithm $\mathsf{LCTE}_\varepsilon$ correctly explores any tree T using local communication between agents.*

Theorem 3. *Algorithm $\mathsf{LCTE}_\varepsilon$ explores T using at most $O(\log B) \cdot OPT$ agents.*

5 Competitive Ratio of Online Exploration

We now show a lower bound on the competitive ratio of any online exploration algorithm in the local communication model. The following result implies that the competitive ratio of algorithm $\mathsf{LCTE}_\varepsilon$ is asymptotically optimal.

368   S. Das, D. Dereniowski, and C. Karousatou

Theorem 4. *Any online exploration algorithm for exploring a tree of depth $D = B - 1$ has a worst case competitive ratio of at least $\Omega(\log B)$.*

Proof. We consider the family of trees which consist of a line of length $D - 1$ connected to the center of a star with p leaves. Thus all the p leaves of the tree are at distance $D = B - 1$ from the root and there is only one node at distance $D - 1$. An offline algorithm would use exactly p agents for exploring this tree. An online algorithm for exploring this tree can be of two types: We say an algorithm is type-1 if during the algorithm there is no transfer of information from the node at depth $D - 1$ to the root; All other algorithms are of type-2. First notice that if an algorithm of type-1, uses k agents for exploration then k is independent of p, since p remains unknown to the root. Thus, by taking $p > k$, we can make the algorithm fail. So we need to consider only type-2 algorithms where information from the node at depth $D - 1$ is transferred to the root. Any agent visiting this node has at most $B - (D - 1) = 2$ units of energy remaining, so it can return back to depth $D - 3 = B - 4$. Similarly, any agent visiting the node at depth $B - 4$ can return back to depth $B - 8$, and so on. Thus, at least $\Omega(\log B)$ agents are needed to carry the information from the node at depth $D - 1$ back to the root. So any type-2 algorithm would use at least $\Omega(\log B)$ agents. By taking $p = 1$, we get a competitive ratio of $\Omega(\log B)$ for any such algorithm. $\square$

6   Conclusions

We studied the problem of exploring a tree with a team of agents, each of which can traverse at most B edges. We gave matching lower and upper bound of $\Theta(\log B)$ on the competitive ratio of the cost of tree exploration. Unlike previous algorithms for energy constrained agents, the agents in our algorithm do not necessarily return to the root after exploration. This fact allows us to explore trees of larger depth. However there is still a transfer of information from the leaves to the root. Thus the algorithm can be used e.g. to collect information from the leaves of a tree, or to search for a resource and bring it back to the root. Note that the lower bound of $\Omega(\log B)$ on the competitive ratio holds only in the local communication model. An interesting question is whether more efficient algorithms are possible for tree exploration in the global communication model. Another open question is the cost of exploring general graphs or other specific classes of graphs.

References

1. Albers, S., Henzinger, M.R.: Exploring Unknown Environments. SIAM Journal on Computing 29(4), 1164–1188 (2000)
2. Awerbuch, B., Betke, M., Singh, M.: Piecemeal graph learning by a mobile robot. Information and Computation 152, 155–172 (1999)
3. Anaya, J., Chalopin, J., Czyzowicz, J., Labourel, A., Pelc, A., Vaxés, Y.: Convergecast and Broadcast by Power-Aware Mobile Agents. Algorithmica, 1–39 (2014)

4. Bender, M., Fernandez, A., Ron, D., Sahai, A., Vadhan, S.: The power of a pebble: Exploring and mapping directed graphs. In: Proc. 30th ACM Symp. on Theory of Computing (STOC), pp. 269–287 (1998)
5. Betke, M., Rivest, R.L., Singh, M.: Piecemeal learning of an unknown environment. Machine Learning 18(23), 231–254 (1995)
6. Chalopin, J., Das, S., Mihalák, M., Penna, P., Widmayer, P.: Data delivery by energy-constrained mobile agents. In: Flocchini, P., Gao, J., Kranakis, E., der Heide, F.M.a. (eds.) ALGOSENSORS 2013. LNCS, vol. 8243, pp. 111–122. Springer, Heidelberg (2014)
7. Dereniowski, D., Disser, Y., Kosowski, A., Pająk, D., Uznański, P.: Fast collaborative graph exploration. Information and Computation 243, 37–49 (2015)
8. Diks, K., Fraigniaud, P., Kranakis, E., Pelc, A.: Tree exploration with little memory. Journal of Algorithms 51, 38–63 (2004)
9. Duncan, C.A., Kobourov, S.G., Anil Kumar, V.S.: Optimal constrained graph exploration. In: Proc. 12th Annual ACM-SIAM Symposium on Discrete Algorithms (SODA), pp. 807–814 (2001)
10. Dynia, M., Korzeniowski, M., Schindelhauer, C.: Power-aware collective tree exploration. In: Grass, W., Sick, B., Waldschmidt, K. (eds.) ARCS 2006. LNCS, vol. 3894, pp. 341–351. Springer, Heidelberg (2006)
11. Dynia, M., Łopuszański, J., Schindelhauer, C.: Why robots need maps. In: Prencipe, G., Zaks, S. (eds.) SIROCCO 2007. LNCS, vol. 4474, pp. 41–50. Springer, Heidelberg (2007)
12. Fraigniaud, P., Gasieniec, L., Kowalski, D., Pelc, A.: Collective tree exploration. Networks 48(3), 166–177 (2006)
13. Fraigniaud, P., Ilcinkas, D., Peer, G., Pelc, A., Peleg, D.: Graph exploration by a finite automaton. Theoretical Computer Science 345(2-3), 331–344 (2005)
14. Megow, N., Mehlhorn, K., Schweitzer, P.: Online graph exploration: new results on old and new algorithms. In: Aceto, L., Henzinger, M., Sgall, J. (eds.) ICALP 2011, Part II. LNCS, vol. 6756, pp. 478–489. Springer, Heidelberg (2011)
15. Ortolf, C., Schindelhauer, C.: Online multi-robot exploration of grid graphs with rectangular obstacles. In: Proc. 24th ACM Symp. on Parallelism in Algorithms and Architectures (SPAA), pp. 27–36 (2012)
16. Panaite, P., Pelc, A.: Exploring unknown undirected graphs. Journal of Algorithms 33, 281–295 (1999)
17. Rutishauser, S., Correll, N., Martinoli, A.: Collaborative Coverage using a Swarm of Networked Miniature Robots. Robotics and Autonomous Systems 57(5), 517–525 (2009)

Solving the INDUCED SUBGRAPH Problem in the Randomized Multiparty Simultaneous Messages Model*

Jarkko Kari[1], Martin Matamala[2], Ivan Rapaport[2], and Ville Salo[2]

[1] Department of Mathematics and Statistics, University of Turku, Finland
[2] DIM-CMM (UMI 2807 CNRS), Universidad de Chile

Abstract. We study the message size complexity of recognizing, under the broadcast congested clique model, whether a fixed graph H appears in a given graph G as a minor, as a subgraph or as an induced subgraph. The n nodes of the input graph G are the players, and each player only knows the identities of its immediate neighbors. We are mostly interested in the one-round, simultaneous setup where each player sends a message of size $\mathcal{O}(\log n)$ to a referee that should be able then to determine whether H appears in G. We consider randomized protocols where the players have access to a common random sequence. We completely characterize which graphs H admit such a protocol. For the particular case where H is the path of 4 nodes, we present a new notion called twin ordering, which may be of independent interest.

1 Introduction

Yao, in his seminal paper of 1979 [27], not only introduced the two-party communication model but also the much more restricted two-party *simultaneous messages* communication model (SM). The SM model is defined as follows. Alice and Bob wish to evaluate together a function $f : X \times Y \to \{0, 1\}$. Alice receives her input x, Bob receives his input y. Both Alice and Bob send simultaneously a message to a referee, who sees none of the input. The referee then announces the function value $f(x, y)$. Of course, the goal of the game is to minimize the size of the messages. Many results have been obtained in this model and, in particular, clear separations have been proved between the deterministic and the randomized settings [5,9,19].

The extension of the SM model to many players is direct and it is defined as follows. There are n players. These n players wish to evaluate together a function $f : X_1 \times \ldots \times X_n \to \{0, 1\}$. Each player receives an input $x_i \in X_i$. The n players send simultaneously a message to the referee who uses these messages in order to compute the boolean function $f(x_1, \ldots, x_n)$. We call this model the *multiparty simultaneous messages* communication model (MSM).

* This work has been partially supported by CONICYT via Basal in Applied Mathematics (M.M., I.R.), Núcleo Milenio Información y Coordinación en Redes ICM/FIC RC130003 (M.M., I.R.), Fondecyt 1130061 (I.R.) and Fondecyt 3150552 (V.S.).

© Springer International Publishing Switzerland 2015
C. Scheideler (Ed.): SIROCCO 2015, LNCS 9439, pp. 370–384, 2015.
DOI: 10.1007/978-3-319-25258-2_26

The already defined *number-in-hand* multiparty communication model is more general than the MSM model because, in the number-in-hand model, many rounds are allowed and different communication modes can be considered [12,14,17,21,26]. In fact, the MSM model corresponds to the one-round, synchronous, shared-whiteboard number-in-hand model.

The *broadcast congested clique model* is exactly the number-in-hand model but where the joint input, instead of being $(x_1, \ldots, x_n) \in X_1 \times \ldots \times X_n$, is a *graph* [13,16]. This input graph is distributed among the nodes, which are the parties of the communication game. More precisely, in the broadcast congested clique model, the joint input to the n nodes is an undirected n-node graph G, with node v receiving the list of its neighbors in G. Each node broadcasts, in each round, a b-bit message (written on a whiteboard, which is visible to every node).

In this paper we are interested in the simultaneous messages (one-round) broadcast congested clique model SM-BCAST. We assume that the ID of each node is a unique number between 1 and n and that the only information each node has, besides n and its own ID, is the list of IDs of its neighbors in G. These nodes need to send, simultaneously, a b-bit message to the referee allowing him to answer, typically, questions of the form "Does the input graph G belong to the graph class $\mathcal{C}$?".

If there is no restriction on the message size then there is a trivial simultaneous protocol that allows the referee to reconstruct any graph: given an input graph G (with an arbitrary assignment of IDs to each of the n nodes), every node sends the binary vector $x \in \{0,1\}^n$ corresponding to the indicator function of its neighborhood. Clearly, this information determines G completely.

If we restrict the message size then reconstructing G becomes much more difficult. Despite this, in [6] it was proved that if (an upper bound on) the *degeneracy* of G is known in advance, then it is possible to reconstruct G with a one-round protocol of $\mathcal{O}(\log n)$ message size. More precisely,

Proposition 1 (Lemma 2 of [6]). *Let m be a positive number. Then, it is possible to decide deterministically, in the SM-BCAST model, the class of m-degenerate graphs using messages of size $\mathcal{O}(m^2 \log n)$. Moreover, if the degeneracy of G is (upper bounded by) m then G can be completely reconstructed by the referee.*

The degeneracy m of a graph is defined as follows: G is m-*degenerate* if one can remove from G a node r of degree at most m, and then proceed recursively on the resulting graph $G' = G - r$, until obtaining an empty graph; the *degeneracy* of G is the smallest m such that G is m-degenerate. For instance, the degeneracy of trees is 1, and the degeneracy of planar graphs is at most 5. Many other important graph classes have bounded degeneracy and this is the reason why previous result is surprising.

In [1,2,15] the authors introduced a beautiful and powerful technique for *graph sketching*. This technique works both for streaming models and for the SM-BCAST model. It allows the referee to decide whether the input graph is connected when each node sends one message of size $\mathcal{O}(\log^3 n)$. The protocol for generating the messages is randomized.

Some negative results for the SM-BCAST model have also been obtained. In [7] the authors prove that it is impossible to decide whether the input graph G has diameter at most 3 or whether G has a triangle unless the messages sent by the nodes are all of size $\Omega(n)$, even if randomness is allowed. Deciding whether the input graph G contains a cycle requires at least one node to write a message of length at least $\lceil \log d \rceil - 1$, where d is the maximum degree of G [4].

It should be pointed out that negative results in the general broadcast congested clique model yield negative results in the SM-BCAST model. In fact, if one can prove that any solution for some problem in the broadcast congested clique model allowing messages of size at most b needs at least r rounds, then one can conclude that any solution of the same problem in the SM-BCAST model needs messages of size at least $\Omega(rb)$.

Let H be a fixed graph. The question we address in this paper is the following: "Does H appear in the input graph G?" In graph theory, the word "appear" has at least three interpretations: H may appear as a *minor* of G, as a *subgraph* of G or as an *induced subgraph* of G.

1.1 Minors

An interesting application of Proposition 1 is related to the problem of detecting the presence of particular minors in the input graph G. The study of graph classes defined by graph minors is one of the most important branches of graph theory, culminating in the Robertson–Seymour theorem [22], also known as the Graph Minor Theorem, which states that every minor-closed family of graphs is defined by a *finite* set of forbidden minors. Many classes of graphs are minor-closed, and have known characterizations in terms of minors. For example, the famous theorem of Kuratowski states that planar graphs are exactly those not containing K_5 or $K_{3,3}$ as minors.

Let H be a fixed graph. We say that H is a minor of G if H can be extracted from G by deleting edges, deleting nodes and contracting edges. We say that G is H-minor free if G does not have H as a minor. H-minor free graphs have bounded degeneracy [18,24,25]. This fact, together with Proposition 1, allows us to conclude that, in the SM-BCAST model, it is possible to decide deterministically whether H is a minor of G using messages of size $\mathcal{O}(\log n)$. Moreover, if G is H-minor free then G can be completely reconstructed by the referee. This implies that for *every* minor-closed class $\mathcal{C}$, there must be an $\mathcal{O}(\log n)$ message size deterministic protocol that decides class $\mathcal{C}$ (and that even reconstructs the input graphs belonging to the class). Unfortunately, for many minor-closed classes we have not discovered the corresponding finite set of forbidden minors yet, and therefore, we can only conclude the *existence* of such protocol (as it occurs with the existence of polynomial time algorithms for recognizing minor-closed classes in the sequential, classical setting).

1.2 Subgraphs

We say that G *contains* H if H is a (not necessarily induced) subgraph of G. The problem H-SUBGRAPH consists in deciding whether H is a subgraph of G. Proposition 1 can also be used for tackling H-SUBGRAPH. In fact, Drucker, Kuhn and Oshman [13] made the following remark: the degeneracy of graphs which do not contain H as a subgraph (H-subgraph free graphs) can be upper bounded in terms of the Turán number $ex(n, H)$, the maximal number of edges of an n-node graph which does not contain a subgraph isomorphic to H. More precisely, they showed that the degeneracy of H-subgraph free graphs with n nodes is at most $4ex(n, H)/n$. This gives the following result.

Proposition 2 ([13]). *Let H be a fixed graph. Then, the problem H-SUBGRAPH can be solved in the* SM-BCAST *model with a $\mathcal{O}(ex(n, H)^2 \log n/n^2)$ message size deterministic protocol.* ⋆

The previous proposition gives some interesting upper bounds. For instance, if H is a tree or a forest then $ex(n, H) = \Theta(n)$ [3]. Therefore, in this case, H-SUBGRAPH can be solved with messages of size $\mathcal{O}(\log n)$. It is also known that $ex(n, C_{2\ell}) = \Theta(n^{1+1/\ell})$, where $C_{2\ell}$ is the even length cycle of length 2ℓ [8]. In other words, $C_{2\ell}$-SUBGRAPH can be solved with messages of size $\mathcal{O}(n^{2/\ell} \log n)$.

The authors in [13] obtained interesting lower bounds which also depend on the Turán number. For instance, consider the ℓ-node cycle C_ℓ. They show that if $\ell \geq 4$, then any protocol that solves C_ℓ-SUBGRAPH needs at least $\Omega(ex(n, C_\ell)/(nb))$ rounds, where b is the message size each node can broadcast in each round. This yields a lower bound of $\Omega(ex(n, C_\ell)/n)$ message size for the SM-BCAST model. Considering that $ex(n, C_\ell) = \Theta(n^2)$ if $\ell > 3$ is odd and that $ex(n, C_\ell) = \Theta(n^{1+2/\ell})$ if ℓ is even we conclude the following: any randomized protocol which solves C_ℓ-SUBGRAPH in the SM-BCAST model uses messages of size at least $\Omega(n)$ if ℓ is odd and $\Omega(n^{2/\ell})$ if ℓ is even.

In the case of triangles C_3 they obtain $\Omega(n/(e^{\mathcal{O}(\sqrt{\log n})}b))$ rounds as a lower bound for the deterministic case. This yields a lower bound of $\Omega(n/e^{\mathcal{O}(\sqrt{\log n})})$ for any deterministic protocol that solves C_3-SUBGRAPH in the SM-BCAST model. In Corollary 5, we state a similar result more generally: if H contains a cycle, then messages of polynomial size are needed in the problem H-SUBGRAPH.

1.3 Induced Subgraphs

An *induced subgraph* of a graph $G = (V, E)$ is a graph $G' = (V', E')$ with $V' \subseteq V$ and such that $vw \in E'$ if and only if $vw \in E$. In other words, the edges of the induced graph G' are all those whose endpoints are both in V'. A class of graphs $\mathcal{G}$ is said to be *hereditary* if every induced subgraph of every member of $\mathcal{G}$ is also in $\mathcal{G}$.

⋆ In [13] the authors say that the message size is $\mathcal{O}(ex(n, H) \log n/n)$. But this bound is an optimistic interpretation of the upper bound of Proposition 1, because instead of considering m^2 they consider m. The conclusions they obtain do not depend on this issue.

A graph G is H-free if H is not an induced subgraph of G. It is easy to show that a graph class $\mathcal{G}$ is hereditary if and only if $\mathcal{G}$ is defined by a (finite or infinite) set $\mathcal{H}$ of forbidden graphs. More precisely, $\mathcal{G}$ is hereditary if and only if for some $\mathcal{H}$, $\mathcal{G} = \{G \mid G \text{ is } H\text{-free, for every } H \in \mathcal{H}\}$.

There is no analog of the Graph Minor Theorem for induced subgraphs, and many classes of graphs have an infinite minimal set of forbidden induced subgraphs. For instance, the class of bipartite graphs is hereditary and the (minimal) set of forbidden induced subgraphs is the set of odd cycles. There are, however, many interesting classes of graphs defined by finite families of forbidden induced subgraphs. For example, the class P_4-free, the class of graphs without an induced copy of the 4-node path P_4, is the class of *cographs*. It arises often in algorithmic graph theory, and also plays a major role in this article.

Problem H-Induced Subgraph consists in deciding whether H is an induced subgraph of G. This problem has not yet been addressed in the congested clique model (with the exception of H being a clique, because K_k is an induced subgraph of G if and only if it is a subgraph of G). This work intends to initiate this research line.

1.4 Notation

In this work a "graph" is always a "simple undirected labeled graph". In particular, the nodes of the n-node input graph $G = (V, E)$ are labeled by their IDs. The open neighborhood of a node $v \in V$ is denoted by $N_G(v)$ and corresponds to the set of nodes which are adjacent to v. The closed neigborhood is $N_G[v] = N_G(v) \cup \{v\}$.

Let H be a graph. The number of nodes of H is denoted by $|H|$. Its complement $\overline{H}$ is the graph with the same set of nodes $V(H)$ but such that $e \in E(\overline{H}) \iff e \notin E(H)$. We write $H_1 \cong H_2$ when the two graphs are isomorphic. Let v be a node of H. We denote by $H - v$ the graph with $|H| - 1$ nodes where, besides removing v, we also remove all the edges incident to v. Similarly, we denote by $H - e$ the graph obtained by removing the edge e from H.

The path of k nodes is denoted by P_k, the cycle of k nodes is denoted by C_k, the clique of k nodes is denoted by K_k. The disjoint union of H_1 and H_2 is denoted by $H_1 + H_2$. The disjoint union of t isomorphic graphs is denoted by tH (where each of the t graphs is isomorphic to H).

A *deterministic protocol* $\mathcal{P}$ in the SM-BCAST model describes the mechanisms of the nodes (for generating the messages) and the mechanism of the referee (for retrieving the final result) that correctly computes the output on all inputs. An ϵ-*error randomized protocol* $\mathcal{P}$ for some problem is a protocol in which every node and the referee are allowed to use a *public* sequence of random bits, and for every input the referee outputs the correct answer with probability at least $1 - \epsilon$. The *cost* of a protocol $\mathcal{P}$, denoted $C(\mathcal{P})$, is the length of the longest message sent to the referee. The *deterministic message size complexity*, denoted $C(f)$, is the minimum cost of any deterministic protocol computing f. Analogously, we denote as $C_\epsilon(f)$ the message size complexity for ϵ-error (public) randomized protocols.

1.5   Our Results

We study the message size complexity of the problem of determining whether a fixed graph H "appears" in a given graph G, mostly under the one-round SM-BCAST model. In particular, we are interested in finding out which graphs H admit (deterministic or randomized) solutions with message size that is logarithmic in n, the number of nodes of the input graph G. Note that a $\log(n)$-size message allows one to identify a node in G, so each node can broadcast the identities of a bounded number of nodes.

As already discussed in Section 1.1, for any graph H, a logarithmic message size is enough to determine – even deterministically – if H is a minor of an arbitrary input graph G.

By Section 1.2, the same is true for the problem of determining whether H is a subgraph of a given G when H is a forest. In other words, if H is a forest, then H-SUBGRAPH can be decided by a deterministic protocol with simultaneous messages of logarithmic size. On the other hand, in Section 2 we prove that if H is not a forest then any protocol (even randomized) requires polynomial size messages. These results are summarized in Corollary 5.

Our results of Section 3 concern the appearance of H as an induced subgraph in G (with $|V(H)| \geq 3$, because otherwise the problem is trivial). Corollary 6 (together with Comment 1) states that polynomial message size is required to solve H-INDUCED SUBGRAPH – even with a randomized protocol – for all H except for $H \in \{P_1 + P_2, P_3, P_4\}$. These are exactly the graphs of order at least three that both themselves and their complements are without cycles. We then provide a randomized protocol with logarithmic message size for the case $H = P_3$ (equivalently $H = P_1 + P_2$) in Proposition 7. Note that P_3-INDUCED SUBGRAPH is equivalent to asking if a graph G is a disjoint union of cliques.

Our most involved result is the one of Section 4, where we provide a randomized protocol with logarithmic message size for problem P_4-INDUCED SUBGRAPH (Proposition 10). For doing this we give a characterization of P_4-free graphs (or cographs) based on the notion of twin ordering. This characterization of cographs is, to the best of our knowledge, a new one.

We are not aware of deterministic one-round solutions for P_3- and P_4-INDUCED SUBGRAPH problems, so these remain open. However, the problems can be solved with logarithmic message size in two rounds (Proposition 8) and in $2(h-1)$ rounds (Proposition 11), respectively, where h bounds the cograph level to be checked.

Every connected cograph has diameter 2. Proposition 10 tells us that cographs can be recognized, in the SM-BCAST model, with a randomized $\mathcal{O}(\log n)$ message size protocol with $1/n^c$ error. It is interesting to point out that, from the paper of Holzer and Pinsker [16], one can conclude that for deciding whether a graph has diameter 2, the size of the messages must be $\Omega(n)$, even if randomness is allowed.

2   Lower Bounds for Detecting Subgraphs and Induced Subgraphs

As mentioned in Section 1.2, it follows from [13] that any randomized ϵ-error protocol that solves the problem C_ℓ-SUBGRAPH in the SM-BCAST model uses messages of size $\Omega(n)$ for $\ell > 3$ odd and $\Omega(n^{2/\ell})$ for ℓ even

The following two propositions generalize these results from cycles C_ℓ to arbitrary graphs H that contain a cycle. Our proofs work also in the case $H = C_3$ of a triangle, and the same proofs provide the lower bounds also for the H-INDUCED SUBGRAPH problem.

The proofs are reductions from the INDEX problem. Consider the INDEX function in the two players SM model: the first player, say Alice, has as input an N-bit boolean vector x and the second player, Bob, has an integer $q \in [1, N]$. Then $\text{INDEX}(x, q) = x_q$, the q'th coordinate of Alice's vector. We will use the fact that for any $\epsilon < 1/2$, any public coin randomized protocol for INDEX requires $\Omega(N)$ bits (see, e.g., [19] for a proof).

Let H and G be two disjoint graphs, let ab be an edge of H and let r, t be two nodes of G. We denote by $G_{rt} \oplus H_{ab}$ the graph obtained from G and $H - ab$ by identifying nodes r and a, and nodes t and b. Then, the set of nodes of $G_{rt} \oplus H_{ab}$ is $V(G) \cup V(H) \setminus \{a, b\}$, where we still denote by r (resp. t), the new node obtained under the identification of r and a (resp. t and b). We call $\tilde{G}$ the subgraph of $G_{rt} \oplus H_{ab}$ induced by the set of nodes $V(G)$; we have $\tilde{G} \cong G$. We call $\tilde{H}$ the subgraph of $G_{rt} \oplus H_{ab}$ induced by the set $U := (V(H) \cup \{r, t\}) \setminus \{a, b\}$. We notice that $\tilde{H} \cong H$ if and only if rt is an edge of G.

A cycle in $G_{rt} \oplus H_{ab}$ is called a *crossing* cycle if it contains nodes from $V(G) \setminus \{r, t\}$ and from $V(H) \setminus \{a, b\}$. Then, the length of a crossing cycle is at least the distance in $H - ab$ between a and b, which we denote by k_H, plus the distance in $G - rt$ between r and t, which we denote by k_G.

Let $\mathcal{P}$ be a protocol for a graph problem. We denote by $\mathcal{P}(|G|, v, N_G(v))$ the message generated in the protocol by node v having neighborhood $N_G(v)$ in a graph with $|G|$ nodes.

We first consider the case that H contains a cycle of odd length.

Proposition 3. *Let H be a non-bipartite graph. Any randomized ϵ-error protocol that solves H-INDUCED SUBGRAPH or H–SUBGRAPH uses messages of size $\Omega(n)$.*

Proof. Let $N \in \mathbb{N}$ be even. Consider the following instance of the INDEX problem. Alice receives the indicator vector of a set $X \subseteq V_L \times V_R$ where V_L and V_R are two disjoint sets of cardinality $|V_L| = |V_R| = \frac{N}{2}$. Bob receives a couple (p, q). The question the referee needs to answer is whether $(p, q) \in X$. We already know that any ϵ-error randomized protocol that solves this problem needs Alice to send $\Omega(N^2)$ bits.

Suppose that there exists a randomized ϵ-error protocol $\mathcal{P}$ that solves H-INDUCED SUBGRAPH or H-SUBGRAPH using messages of size $c(n)$. We are going to use $\mathcal{P}$ to solve INDEX.

Consider the N-node graph $G = (V_L \cup V_R, E)$ with $E = X$. Let a, b be two nodes of H such that the edge ab lies in a shortest odd cycle (it must exist, H is non-bipartite), and let k be the length of this cycle. Let $i \in V_L$ and $j \in V_R$. Then, any odd-length crossing cycle of $G_{ij} \oplus H_{ab}$ has length at least $k + 2$: paths in G between i and j have odd length, and the shortest even-length path in $H - ab$ between a and b has length $k - 1$. Hence, any cycle of length k in $G_{ij} \oplus H_{ab}$ is either included in G or in $\tilde{H}$. But as k is odd and G is bipartite, any cycle of length k belongs to $\tilde{H}$. In conclusion, H is an (induced) subgraph of $G_{ij} \oplus H_{ab}$ if and only if ij is an edge of G: if H is a subgraph of $G_{ij} \oplus H_{ab}$, then ij is an edge of G as otherwise $\tilde{H}$ has fewer cycles of length k than H and, conversely, if ij is an edge of G then $\tilde{H} \cong H$ is an induced subgraph of $G_{ij} \oplus H_{ab}$.

Alice can take advantage of the previous fact in order to generate a message from her input X. She generates N messages, one for each node in G.

For each $i \in V_L$ she generates the message (M_a^i, M^i), where

- $M_a^i = \mathcal{P}(|G| + |H| - 2, i, N_G(i) \cup N_H(a) \setminus \{b\})$ is the message node i would send in the graph $G_{ij'} \oplus H_{ab}$ with $j' \in V_R$ arbitrary.
- $M^i = \mathcal{P}(|G| + |H| - 2, i, N_G(i))$ is the message node i would send in the graph $G_{i'j'} \oplus H_{ab}$ with $i' \in V_L \setminus \{i\}$, $j' \in V_R$, both arbitrary.

For each $j \in V_R$ she generates the message (M_b^j, M^j), where

- $M_b^j = \mathcal{P}(|G| + |H| - 2, j, N_G(j) \cup N_H(b) \setminus \{a\})$ is the message node j would send in the graph $G_{i'j} \oplus H_{ab}$ with $i' \in V_L$ arbitrary.
- $M^j = \mathcal{P}(|G| + |H| - 2, j, N_G(j))$ is the message node j would send in the graph $G_{i'j'} \oplus H_{ab}$ with $i' \in V_L$, $j' \in V_R \setminus \{j\}$, both arbitrary.

Suppose that Bob sends (p, q) to the referee. How can the referee decide whether $(p, q) \in X$? He simply simulates protocol $\mathcal{P}$ considering for node p the message M_a^p, for node q the message M_b^q and for every other node r the message M^r (recall that H is fixed, known by the referee).

The size of the message sent by Alice is $\mathcal{O}(Nc(N + |H| - 2))$. Therefore, since the randomized complexity of INDEX is $\Omega(N^2)$ and $|H|$ is constant, we conclude that $c(N) = \Omega(N)$. $\square$

Another reduction in the same style provides a lower bound in the case of bipartite H containing cycles.

Proposition 4. *Let H be a bipartite graph containing a cycle. Any randomized ϵ-error protocol that solves H-INDUCED SUBGRAPH or H–SUBGRAPH uses messages of size $\Omega(n^{2/k})$ where k is the (even) length of the shortest cycle in H.*

Proof. Let $N = ex(n, C_k)$ for some n, and let $G = (V, E)$ be a graph with n nodes and N edges which does not contain a subgraph isomorphic to C_k. Recall that $ex(n, C_k) = \Theta(n^{1+2/k})$ for even k.

Consider the following instance of the INDEX problem. Alice receives a vector $X \in \{0, 1\}^N$ and Bob receives a natural number $p \in [1, N]$. The question the referee needs to answer is whether $X_p = 1$. We already know that any ϵ-error randomized protocol that solves this problem needs Alice to send $\Omega(N)$ bits.

Suppose that there exists a randomized ϵ-error protocol $\mathcal{P}$ that solves either H-INDUCED SUBGRAPH or H-SUBGRAPH using messages of size $c(n)$. We are going to use $\mathcal{P}$ to solve INDEX.

Let $e_1, e_2, \ldots, e_N$ be an enumeration of the edges of G, and consider the subgraph $G' = (V, E')$ of G with $e_i \in E' \iff X_i = 1$. Let a, b be two nodes of H such that the edge ab lies in a shortest cycle (that is, on a cycle of length k).

For any $e_i = (r, t)$, $1 \leq i \leq N$, consider the graph $G'_{rt} \oplus H_{ab}$. Then any crossing cycle of $G'_{rt} \oplus H_{ab}$ has length at least $k+1$: $k_G \geq 2$ and $k_H = k-1$. By definition, G' has no cycle of length k. Hence, any cycle of length k in $G'_{rt} \oplus H_{ab}$ must appear in the subgraph $\tilde{H}$ of $G'_{rt} \oplus H_{ab}$. Therefore, if H is a subgraph of $G'_{rt} \oplus H_{ab}$ then rt is an edge of G' as otherwise $\tilde{H}$ has fewer cycles of length k than H. And of course, conversely, presence of edge rt in G' means that $\tilde{H} \cong H$ is an induced subgraph of $G'_{rt} \oplus H_{ab}$.

Alice can take advantage of the previous fact in order to generate a message from her input X. She generates n messages, one for each node in G'. For each $i \in V$ she generates the message (M^i, M^i_a, M^i_b), where

- $M^i_a = \mathcal{P}(n + |H| - 2, i, N_{G'}(i) \cup (N_H(a) \setminus \{b\}))$ is the message node i would send in the graph $G'_{ij} \oplus H_{ab}$ with $j \in V \setminus \{i\}$ arbitrary.
- $M^i_b = \mathcal{P}(n + |H| - 2, i, N_{G'}(i) \cup (N_H(b) \setminus \{a\}))$ is the message node i would send in the graph $G'_{ij} \oplus H_{ba}$ with $j \in V \setminus \{i\}$ arbitrary.
- $M^i = \mathcal{P}(n + |H| - 2, i, N_{G'}(i))$ is the message node i would send in the graph $G'_{i'j'} \oplus H_{ab}$ with $i', j' \in V \setminus \{i\}$, both arbitrary.

Suppose that Bob sends p to the referee. How can the referee decide whether $X_p = 1$? If $e_p = (y, z)$ he simply simulates the protocol $\mathcal{P}$ considering for node y the message M^y_a, for node z the message M^z_b and for every other node i the message M^i (recall that H is fixed, known by the referee).

The size of the message sent by Alice is $\mathcal{O}(nc(n + |H| - 2))$. Therefore, since the randomized complexity of INDEX is $\Omega(N)$ we conclude that $c(n + |H| - 2)$ is $\Omega(N/n)$. We have $N = ex(n, C_k) = \Omega(n^{1+2/k})$, which proves the claim.   □

Combining Propositions 3 and 4 with the observations of Section 1.2, we obtain the following:

Corollary 5 *If H is a forest, then the problem H-SUBGRAPH can be decided by a deterministic protocol with simultaneous messages of logarithmic size. If H contains a cycle, then a randomized protocol with simultaneous messages for H-SUBGRAPH requires messages of polynomial size.*

3   The Problem H-INDUCED SUBGRAPH

Lemma 1. *Let H be a fixed graph. The problems H-INDUCED SUBGRAPH and $\overline{H}$-INDUCED SUBGRAPH are equivalent. More precisely, there exists a protocol*

*with message size b for solving H-INDUCED SUBGRAPH if and only if there exists
a protocol with message size b for solving $\overline{H}$-INDUCED SUBGRAPH.*

Proof. Let H be a fixed graph. Suppose that we have a protocol for solving
$\overline{H}$-INDUCED SUBGRAPH. We can use this protocol for solving H-INDUCED SUB-
GRAPH as follows. Let G be the input graph. Note that H is an induced subgraph
of G if and only if $\overline{H}$ is an induced subgraph of $\overline{G}$. Therefore, every node v can
consider the nodes that are not its neighbors and apply the protocol for detect-
ing $\overline{H}$ with this new, complementary neighborhood. Of course, if there is enough
information for reconstructing $\overline{G}$ (when the answer is positive) then there is
enough information for reconstructing G.   □

Corollary 6 *Let H be an arbitrary graph with at least 3 nodes. If $H \notin \{P_1 +
P_2, P_3, 2P_2, C_4, P_4\}$ then any randomized ϵ-error protocol that solves H-INDUCED
SUBGRAPH uses messages of size $\Omega(n)$.*

Proof. This follows directly from Proposition 3 and Lemma 1 because, the graphs
listed above, are the only graphs with at least 3 nodes which are bipartite both
themselves and their complements.   □

Comment 1 *Notice that $\overline{P_1 + P_2} = P_3$, $\overline{2P_2} = C_4$ and $\overline{P_4} = P_4$. Therefore,
in order to understand completely problem H-INDUCED SUBGRAPH, the only
problems we need to study are P_3-INDUCED SUBGRAPH, C_4-INDUCED SUBGRAPH
and P_4-INDUCED SUBGRAPH. The case $H = C_4$ in Proposition 4 directly provides
an $\Omega(n^{1/2})$ lower bound on the message size for any randomized ϵ-error protocol
that solves C_4-INDUCED SUBGRAPH. Therefore, the only two cases for which we
do not know the message size complexity yet are $H = P_3$ and $H = P_4$.*

3.1   The Problem P_3-INDUCED SUBGRAPH

Notice that a graph is P_3-free if and only if it is the disjoint union of cliques.
There is a classical randomized "fingerprint" technique for testing whether two
vectors are equal. We are going to use this technique for solving P_3-INDUCED
SUBGRAPH. It works as follows. Let $n^{c+3} < p \leq 2n^{c+3}$ be a prime number. A
value $t \in \mathbb{Z}_p$ is chosen uniformly at random using $\mathcal{O}(\log(n))$ public random bits.
Given an n-bits vector $a = (a_1, \ldots, a_n)$, consider the polynomial $P_a = a_1 +
a_2X + a_3X^2 + \ldots a_nX^{n-1}$ in $\mathbb{Z}_p[X]$ and let $FP(a, t) = P_a(t)$. The value $FP(a, t)$
is sometimes called the "fingerprint" of vector a. Clearly two equal vectors have
equal fingerprints, and, more important, for any two different vectors a and b, the
probability that $FP(a, t) = FP(b, t)$ is at most $1/n^{c+2}$ (because the polynomial
$P_a - P_b$ has at most n roots and t was chosen uniformly at random, thus the
probability that t is a root of $P_a - P_b$ is at most $1/n^{c+2}$, see [20]).

Proposition 7. *For any constant $c > 0$, P_3-INDUCED SUBGRAPH can be solved
with a randomized $\mathcal{O}(\log n)$ message size protocol with $1/n^c$ error.*

Proof. Let $x_i \in \{0,1\}^n$ be the input vector of node i, i.e., the characteristic function of its closed neighborhood $N[i] = N(i) \cup \{i\}$. A protocol for P_3-INDUCED SUBGRAPH consists in each node sending two numbers: its degree d_i and its fingerprint $m_i = FP(x_i, t)$. Let $l(m)$ be the number of nodes that send the same fingerprint m. The referee concludes that the input graph G is the disjoint union of cliques (and therefore P_3 is not an induced subgraph of G) if and only if all the nodes with the same fingerprint m have degree $l(m) - 1$. It is not difficult to realize that the previous protocol fails if and only if there are at least two nodes i, j with different neighborhoods such that $FP(m_i, t) = FP(m_j, t)$. For each fixed pair of nodes this probability is at most $1/n^{c+2}$, so altogether the probability of a wrong answer is at most $1/n^c$. □

3.2 A Deterministic Protocol for P_3-INDUCED SUBGRAPH

Recall that when more than one round is allowed the messages, instead of being sent to a referee, are written on a shared whiteboard.

Proposition 8. *There exists a $\mathcal{O}(\log n)$ message size deterministic two-round protocol for solving P_3-INDUCED SUBGRAPH.*

Proof. Let G be the input graph. Our protocol does the following. In the first round each node v writes on the whiteboard its own ID together with the minimum ID of its closed neighborhood $M_v = \min\{\text{ID}(u) \mid u \in N_G[v]\}$. In the second round each node v writes only one bit. It writes the bit 1 if and only if for all $u \in N_G[v]$ $M_u = M_v$ and for all $u \notin N_G[v]$ $M_u \neq M_v$.

Obviously, every node writes a 1 in the second round if and only if G is a disjoint union of cliques. If G is indeed a disjoint union of cliques then, with the information written on the whiteboard, it is possible to reconstruct it. □

Open Problem 1. *Is it possible to solve deterministically, in the SM-BCAST model, the problem P_3-INDUCED SUBGRAPH using messages of size $\mathcal{O}(\log n)$?*

4 The Problem P_4-INDUCED SUBGRAPH

Let G_1 and G_2 be two disjoint graphs. The join operation $G_1 \star G_2$ consists in connecting all the nodes of G_1 with all the nodes of G_2. Formally, it is defined as follows: $G_1 \star G_2 = \overline{(\overline{G_1} + \overline{G_2})}$. The class of *cographs* is defined recursively. First, an isolated node K_1 is a cograph. Second, $G \neq K_1$ is a cograph if and only if G is the join or the union of two disjoint cographs [11,23].

In this paper we provide a new characterization of cographs based on a new notion we introduce here that we call *twin ordering*. Two nodes u and v of a graph G are called twins if $N_G(u) \setminus \{v\} = N_G(v) \setminus \{u\}$.

A twin ordering of an n-node graph is an ordering $v_1, \ldots, v_n$ such that for each $j \geq 2$, the vertex v_j has a twin in the graph induced by $\{v_1, \ldots, v_j\}$.

Proposition 9. *For a graph G the following are equivalent.*

1. *G is a cograph.*
2. *Every non trivial induced subgraph of G has a pair of twins.*

3. G is P_4-free.

4. G has a twin ordering.

Proof. The equivalence between the first three characterizations was proved in [11] and [23]. It is clear that the second implies the fourth. Moreover, not only there exists a twin ordering, but one can find it by repeatedly picking an arbitrary node having a twin and removing such node. This follows from the assumption that every non trivial induced subgraph has a pair of twins.

We prove that if G has a twin ordering, then it is P_4-free. Take any subset of nodes $U = \{v_t, v_l, v_k, v_j\}$, with $t < l < k < j$. For the sake of contradiction, let us assume that the graph induced by U is P_4. Among the choices for U, pick one with j as small as possible. From hypothesis, there is a $i < j$ such that v_i and v_j are twins in the graph induced by $\{v_1, \ldots, v_j\}$. Since P_4 has no pairs of twins we get that $v_i \notin U$. But this is a contradiction with the choice of U because the graph induced by $\{v_t, v_l, v_k, v_i\}$ is P_4 and $\max\{t, l, k, i\} < j$. $\square$

Let $G = (V, E)$ be an n-node graph with $V \subseteq \mathbb{N}$. The *canonical ordering* of G is the twin ordering of G defined as follows. Instead of picking an arbitrary node having a twin we select the lexicographically first pair of twins. Then we choose, among these two nodes, the smaller one. The process continues, starting by removing v_n, until no further twins appear. So, a canonical ordering of an arbitrary graph is of the form $v_k, \ldots, v_n$ and $k = 1$ if and only if G is a cograph.

Let p be a prime and let $\phi = (\phi_w)_{w \in V}$ be a linearly independent family of polynomials in $\mathbb{Z}_p[X]$. Let $q = (q_w)_{w \in V}$ and $\bar{q} = (\bar{q}_w)_{w \in V}$ be defined by $q_w = \sum_{w' \in N_G(w)} \phi_{w'}$ and $\bar{q}_w = q_w + \phi_w$, for each $w \in V$. We also define $\alpha_{u,v} = \phi_u - \phi_v$, $\beta_{u,v} = q_u - q_v$ and $\gamma_{u,v} = \bar{q}_u - \bar{q}_v$, for each $u, v \in V$. The *derivated polynomials* of family ϕ are the following polynomials: $(\alpha_{u,v})_{u,v \in V}, (\beta_{u,v})_{u,v \in V}, (\gamma_{u,v})_{u,v \in V}$.

Let u and v be twins. We associate to $G - v$ the polynomials $(\phi'_w)_{w \in V \setminus \{v\}}$ by $\phi'_w = \phi_w$, when $w \neq u$, and $\phi'_u = \phi_u + \phi_v$. By using this construction, starting with $\phi_u(x) = x^u$, and following the canonical ordering $v_k, \ldots v_n$, we obtain polynomials ϕ^i_u, for each $k \leq i \leq n$ and each u in the graph $G - \{v_n, \ldots, v_{i+1}\}$. We call these polynomials the *basic* polynomials of G. The *canonical family of polynomials* of G is the union of basic polynomials and their derivated. This family has at most $n \times 3n^2 = 3n^3$ polynomials.

We say that a vector $m = ((a_w, b_w))_{w \in V} \in (\mathbb{Z}_p)^{2n}$ is *valid for G at $t \in \mathbb{Z}_p$* if there is a linearly independent family of polynomials $(\phi_w)_{w \in V}$ in $\mathbb{Z}_p[X]$ such that $a_w = \phi_w(t)$ and $b_w = q_w(t)$, for each $w \in V$.

Lemma 2. *Let $m = ((a_w, b_w))_{w \in V} \in (\mathbb{Z}_p)^{2n}$ be valid for G at t. Let u, v be twins in G such that $a_u \neq a_v$. Then, the following vector $m' = ((a'_w, b'_w))_{w \in V \setminus \{v\}} \in (\mathbb{Z}_p)^{2n-2}$ is valid for $G - v$ at t. For each $w \in V \setminus \{u, v\}$: $a'_w = a_w$ and $b'_w = b_w$. For $w = u$: $a'_u = a_u + a_v$ and $b'_u = b_u - a_v \delta_{uv}$, where $\delta_{uv} \in \{0, 1\}$ and $\delta_{uv} = 1$ if and only if $a_u + b_u = a_v + b_v$.*

Proof. Let $(\phi_w)_{w \in V}$ be a linearly independent family of polynomials associated to m. Since u and v are twins and $a_u \neq a_v$, we have that $a_u + b_u = a_v + b_v$ if and only if u and v are adjacent. Hence, $\delta_{uv} = 1$ if and only if u and v are adjacent.

Let $(\phi'_w)_{w \in V \setminus \{v\}}$ be given by $\phi'_w = \phi_w$ for each $w \neq u$, and $\phi'_u = \phi_u + \phi_v$. Clearly, this family is linearly independent.

For $w \neq u$ we have that $a'_w = a_w = \phi_w(t) = \phi'_w(t)$. Moreover, $b'_w = b_w$ and $b_w = q_w(t)$. Since u and v are twins either both are in $N_G(w)$ or none. In both cases we have that $b'_w = q'_w(t)$. By definition, $a'_u = a_u + a_v = \phi_u(t) + \phi_v(t) = \phi'_u(t)$. Also, by definition, $b'_u = b_u - \delta_{uv} a_v$. We know that $b_u = q_u(t) = \delta_{uv}\phi_v(t) + q'_u(t)$. Hence, $b'_u = q'_u(t)$. □

Proposition 10. *For any constant $c > 0$, P_4-INDUCED SUBGRAPH can be solved with a randomized $\mathcal{O}(\log n)$ message size protocol with $1/n^c$ error.*

Proof. Let $G = (V, E)$ be an n-node graph. Let p be prime with $3n^{c+4} \leq p \leq 6n^{c+4}$. The protocol applied to G is the following. Each node sends to the referee the message m_w such that $m = (m_w)_{w \in V}$ is valid for G at t, where t is picked uniformly at random in $\mathbb{Z}_p$. Each node computes such a message by defining $\phi_w(x) = x^w$.

On input $m \in (\mathbb{Z}_p)^{2n}$ the referee iterates at most $n - 1$ times trying to build the canonical ordering $\{v_1, \ldots, v_n\}$. In iteration i he starts with a graph G^i and a vector $m^i \in (\mathbb{Z}_p)^{2(n-i+1)}$ (with $G^1 = G$ and $m^1 = m$). He determines if there is a pair of nodes u and v in G^i such that $a^i_u \neq a^i_v$ and either $b^i_u = b^i_v$ or $a^i_u + b^i_u = a^i_v + b^i_v$ He selects, among all these, the lexicographically first pair of nodes. If no such pair exists, then he rejects. Otherwise, he sets $G^{i+1} = G^i - v$, setting $v_{n-i+1} = v$ (w.l.o.g, we assume that $v < u$). Then he computes m^{i+1} from m^i according to Lemma 2. If the referee reaches iteration $n - 1$ he accepts.

What is the probability that the referee does not construct the canonical ordering of G? This could happen only when the chosen t is a zero of at least one member of the canonical family of G. As this family has at most $3n^3$ polynomials, this occurs with probability at most $3n^3(n/p) \leq 1/n^c$. □

4.1   A Deterministic Protocol for P_4-INDUCED SUBGRAPH

The definition of cographs by closure operations comes with the following natural hierarchy, which we call the *bottom-up hierarchy*, and which will be needed in the proof of Proposition 11. First, $\Sigma_0 = \Pi_0 = \{K_1\}$. Second, for $i \geq 0$, Σ_{i+1} is the set of disjoint unions of graphs in Π_i and Π_{i+1} is the set of joins of graphs in Σ_i. A graph G is a cograph if and only if $G \in \Sigma_i$ for some i. Notice that Σ_1 corresponds exactly to the class of disjoint unions of isolated nodes $K_1 + \ldots + K_1$. On the other hand, Π_1 corresponds to the class of all cliques. More precisely, $\Pi_1 = \{K_n\}_{n>0}$, where K_n is the n-node clique.

Notice that $G \in \Sigma_i \iff \overline{G} \in \Pi_i$. We can prove this by induction. This is obviously true for $i = 1$. Assume now that $G = G_1 + G_2 \in \Sigma_i$ for some $i > 1$. The result follows from the induction hypothesis because $\overline{G} = \overline{G_1 + G_2} = \overline{G_1} \star \overline{G_2}$.

Σ_2 is exactly the class of P_3-free graphs because, as we saw previously, P_3-free graphs are exactly the disjoint unions of cliques. Since $\overline{P_3} = P_1 + P_2$ and considering the previous observation, we conclude that Π_2 is the class of $(P_1 + P_2)$-free graphs.

Let G be a cograph. In [10] the authors define the *height* of G as the minimum i such that $G \in \Sigma_i \cup \Pi_i$. We do not know if there is a one-round deterministic protocol. However, any fixed level of the bottom-up hierarchy has a deterministic protocol with a bounded number of rounds:

Proposition 11. *Let $h > 0$ be a fixed positive integer. Then, there exists a $2(h-1)$-round protocol for the classes Σ_h and Π_h with messages of size $\log n$.*

Proof. We prove the existence/correctness/complexity of such protocols by induction on h. For $h = 2$ we use the two-round protocol of Proposition 8.

For the general case, first note that if the distance between two nodes is finite and strictly larger than 2, then P_4 is an induced subgraph of G. Let us now describe the protocol for Σ_h when $h > 2$ (the one for Π_h is symmetric). In the first round of the protocol, every node v writes on the whiteboard the minimum ID of the nodes in its closed neighborhood $N_G[v] = N_G(v) \cup \{v\}$. In the second round, every node v writes the minimum ID among the IDs written by the nodes in its closed neighborhood.

If the graph G is indeed P_4-free, then after these two rounds, every node knows the partition $G = G_1 + G_2 + \cdots + G_k$ of G into its connected components. In the third round, every node includes in its message the verification that this partition is correct: if some node in G_i is connected to a node in G_j with $i \neq j$ then, its message will state this fact. If this happens, then protocol concludes that G is not in Σ_h: in this case G is not even a cograph, because it contains an induced path of length 3, that is, a copy of P_4.

Assuming G is a cograph, every node knows its partition into connected components after the second round. Thus, in the third round, we can start performing the protocol for Π_{h-1} separately in each of the connected components G_i. If some G_i is not in Π_{h-1}, then $G \notin \Sigma_h$. If all of these graphs are in Π_{h-1}, the recursively called protocols reconstruct the graphs G_i, and our protocol for Σ_h reconstructs G as their disjoint union. $\square$

Open Problem 2. *Is it possible to solve deterministically, in the* SM-BCAST *model, the problem P_4-INDUCED SUBGRAPH using messages of size $\mathcal{O}(\log n)$?*

References

1. Ahn, K.J., Guha, S., McGregor, A.: Analyzing graph structure via linear measurements. In: Proc. of the 23rd Annual ACM-SIAM Symp. on Discrete Algorithms (SODA 2012), pp. 459–467 (2012)
2. Ahn, K.J., Guha, S., McGregor, A.: Graph sketches: Sparsification, spanners, and subgraphs. In: Proc. of PODS 2012, pp. 5–14 (2012)
3. Ajtai, M., Komlós, J., Simonovits, M., Szemerédi, E.: The exact solution of the Erdos-T. Sós conjecture for (large) trees (in preparation)
4. Arfaoui, H., Fraigniaud, P., Ilcinkas, D., Mathieu, F.: Distributedly testing cycle-freeness. In: Kratsch, D., Todinca, I. (eds.) WG 2014. LNCS, vol. 8747, pp. 15–28. Springer, Heidelberg (2014)
5. Babai, L., Kimmel, P.G.: Randomized simultaneous messages: Solution of a problem of Yao in communication complexity. In: Proc. of the 12th Annual IEEE Conference on Computational Complexity, pp. 239–246 (1997)

6. Becker, F., Matamala, M., Nisse, N., Rapaport, I., Suchan, K., Todinca, I.: Adding a referee to an interconnection network: What can (not) be computed in one round. In: Proc. of IPDPS 2011, pp. 508–514 (2011)
7. Becker, F., Montealegre, P., Rapaport, I., Todinca, I.: The simultaneous number-in-hand communication model for networks: Private coins, public coins and determinism. In: Halldórsson, M.M. (ed.) SIROCCO 2014. LNCS, vol. 8576, pp. 83–95. Springer, Heidelberg (2014)
8. Bondy, J.A., Simonovits, M.: Cycles of even length in graphs. Journal of Combinatorial Theory, Series B 16(2), 97–105 (1974)
9. Chakrabarti, A., Shi, Y., Wirth, A., Yao, A.: Informational complexity and the direct sum problem for simultaneous message complexity. In: Proc. of FOCS 2001, pp. 270–278 (2001)
10. Chudnovsky, M., Scott, A., Seymour, P.: Excluding pairs of graphs. Journal of Combinatorial Theory, Series B 106, 15–29 (2014)
11. Corneil, D.G., Lerchs, H., Burlingham, L.S.: Complement reducible graphs. Discrete Applied Mathematics 3(3), 163–174 (1981)
12. Drucker, A., Kuhn, F., Oshman, R.: The communication complexity of distributed task allocation. In: Proc. of PODC 2012, pp. 67–76 (2012)
13. Drucker, A., Kuhn, F., Oshman, R.: On the power of the congested clique model. In: Proc. of PODC 2014, pp. 367–376 (2014)
14. Gronemeier, A.: Asymptotically optimal lower bounds on the NIH-multi-party information complexity of the AND-function and disjointness. In: Proc. of STACS 2009, pp. 505–516 (2009)
15. Guha, S., McGregor, A., Tench, D.: Vertex and hyperedge connectivity in dynamic graph streams. In: Proc. of PODS 2015, pp. 241–247 (2015)
16. Holzer, S., Pinsker, N.: Approximation of distances and shortest paths in the broadcast congest clique. CoRR, abs/1412.3445 (2014)
17. Jayram, T.S.: Hellinger strikes back: A note on the multi-party information complexity of AND. In: Dinur, I., Jansen, K., Naor, J., Rolim, J. (eds.) APPROX and RANDOM 2009. LNCS, vol. 5687, pp. 562–573. Springer, Heidelberg (2009)
18. Kostochka, A.V.: Lower bound of the hadwiger number of graphs by their average degree. Combinatorica 4(4), 307–316 (1984)
19. Kremer, I., Nisan, N., Ron, D.: On randomized one-round communication complexity. Computational Complexity 8(1), 21–49 (1999)
20. Kushilevitz, E.: Communication complexity. Adv. Computers 44, 331–360 (1997)
21. Phillips, J.M., Verbin, E., Zhang, Q.: Lower bounds for number-in-hand multiparty communication complexity, made easy. In: Proc. of the 23rd Annual ACM-SIAM Symp. on Discrete Algorithms (SODA 2012), pp. 486–501 (2012)
22. Robertson, N., Seymour, P.D.: Graph minors XX: Wagner's conjecture. Journal of Combinatorial Theory, Series B 92(2), 325–357 (2004)
23. Seinsche, D.: On a property of the class of n-colorable graphs. Journal of Combinatorial Theory, Series B 16(2), 191–193 (1974)
24. Thomason, A.: An extremal function for contractions of graphs. In: Mathematical Proc. of the Cambridge Philosophical Society, vol. 95, pp. 261–265 (1984)
25. Thomason, A.: The extremal function for complete minors. Journal of Combinatorial Theory, Series B 81(2), 318–338 (2001)
26. Woodruff, D.P., Zhang, Q.: When distributed computation is communication expensive. In: Afek, Y. (ed.) DISC 2013. LNCS, vol. 8205, pp. 16–30. Springer, Heidelberg (2013)
27. Yao, A.: Some complexity questions related to distributive computing (preliminary report). In: Proc. of STOC 1979, pp. 209–213 (1979)

A Separation of n-consensus and $(n + 1)$-consensus Based on Process Scheduling

Carole Delporte-Gallet[1], Hugues Fauconnier[1], and Sam Toueg[2]

[1] Université Paris Diderot, France
[2] University of Toronto, ON, Canada

Abstract. A fundamental research theme in distributed computing is the comparison of systems in terms of their ability to solve basic problems such as consensus that cannot be solved in completely asynchronous systems. In particular, in a seminal work [12], Herlihy compares shared-memory systems in terms of the shared objects that they have: he proved that there are shared objects that are powerful enough to solve consensus for n processes, but are too weak to solve consensus for $n + 1$ processes; such objects are placed at level n of a *wait-free hierarchy*.

As in [12], we compare shared-memory systems with respect to their ability to solve consensus for n processes. But instead of comparing systems defined by the shared objects that they have, we compare read-write systems defined by the set of *process schedules* that can occur in these systems. Defining systems this way can capture many types of systems, e.g., systems whose synchrony ranges from fully synchronous to completely asynchronous, several systems with failure detectors, and "obstruction-free" systems. In this paper, we consider read-write systems defined in terms of sets of process schedules, and investigate the following fundamental question: Is there a system of $n+1$ processes such that consensus can be solved for every subset of n processes in the system, but consensus cannot be solved for the $n + 1$ processes of the system? We show that the answer to the above question is "yes", and so these systems can be classified into hierarchy akin to Herlihy's hierarchy.

1   Motivation and Related Work

A fundamental research theme in distributed computing is the comparison of systems in terms of their ability to solve basic problems such as consensus or k-set agreement that cannot be solved in completely asynchronous systems [10,9,4,14,18,17,11,19,8,1,3]. In particular, in a seminal work [12], Herlihy compares shared-memory systems in terms of the shared objects that they have: he proved that there are shared objects that are powerful enough to solve consensus for n processes, but are too weak to solve consensus for $n + 1$ processes; such objects are placed at level n of a *wait-free hierarchy*. The importance of this hierarchy comes from Herlihy's universality result: intuitively, every object at level n of this hierarchy can be used to implement *any* object shared by n processes in a wait-free manner.

© Springer International Publishing Switzerland 2015
C. Scheideler (Ed.): SIROCCO 2015, LNCS 9439, pp. 385–398, 2015.
DOI: 10.1007/978-3-319-25258-2_27

As in [12], in this paper we compare shared-memory systems with respect to their ability to solve consensus for n processes. But instead of comparing systems defined by the shared objects that they have, we compare systems (with shared registers) defined by the *process schedules* that they allow, as we explain below.

First, note that several types of read-write shared-memory systems, e.g., *asynchronous*, *partially synchronous* and *synchronous* systems, can be defined by the set of process schedules that may occur in these systems.[1] For example, a completely asynchronous system is one where every process schedule can occur. Similarly, a partially synchronous system is one where the process schedules satisfy some timeliness or "fairness" conditions [2,16,1] which effectively define the set of process schedules that may occur. Perfectly synchronous systems can also be defined by the set of process schedules that can occur. Furthermore, several systems with *failure detectors* [5] are equivalent to systems defined in terms of process schedules: for several well-known failure detectors $\mathcal{D}$ (including $P, \Diamond P, S$ and $\Diamond S$) an asynchronous system augmented with $\mathcal{D}$ is *equivalent* to a system where schedules satisfy some fairness conditions [16].[2] Finally, *obstruction-free* algorithms work in systems with a specific set of process schedules, namely, schedules where some process eventually executes solo [13].

Thus, shared-memory systems defined in terms of process schedules capture a large set of systems, e.g., systems whose synchrony ranges from fully synchronous to completely asynchronous, several systems with failure detectors, and "obstruction-free" systems. In this paper, we consider such systems and investigate the following natural question:

> Is there a system of $n + 1$ processes such that consensus can be solved for every subset of n processes in the system, but consensus cannot be solved for the $n + 1$ processes of the system?

If this is true for every n, it would imply that these systems can be classified into hierarchy akin to Herlihy's hierarchy.

The answer to the above question is not obvious. In [7] it is shown that if a failure detector $\mathcal{D}$ is powerful enough to solve consensus for every subset of n processes in a system of $n+1$ processes, then it is powerful enough to solve consensus for all the $n+1$ processes in the system. Since [16] shows that several systems with failure detectors are equivalent to systems defined in terms of process schedules, it is tempting to conjecture that the answer is "no".

In this paper we show that the answer to the above question is "yes". More precisely, we prove that for every $n \geq 1$, there is a read-write shared-memory system S of $n + 1$ processes such that: (a) consensus can be solved for every subset of n processes of S, and (b) consensus cannot be solved for the $n + 1$ processes of S.

In fact we prove the following slightly stronger result. For any finite set of processes P, and all $n \geq 1$, there is a system of P such that: (a) consensus can be solved for *every*

---

[1] Intuitively, a process schedule specifies the order in which processes take steps.
[2] The results in [16] were for message-passing systems, but similar results hold for read-write shared-memory systems.

set of n or fewer processes of P, and (b) consensus cannot be solved for *any* set of $n+1$ or more processes of P.

Roadmap. In Section 2, we describe the systems under consideration, namely, read-write shared-memory systems defined in terms of their schedules. In Section 3, we recall the definition of consensus and explain what it means for an algorithm to solve this problem in a given system. In Section 4, we give our main results: we prove that there is a schedule-based read-write system such that: (1) consensus can be solved for every subset of processes of size n of this system (Section 5), and (2) consensus cannot be solved for any subset of processes of size $n + 1$ (Section 6). Some brief remarks conclude the paper in Section 7.

2 Model

We consider shared-memory systems of processes with SWMR multivalued atomic registers. Processes proceed by executing atomic steps: in each step, a single process can read or write a single register. In the following, P is a finite set of processes, and a process in P is denoted as p_j for some $j \in \mathbb{N}$.

2.1 Process Schedules

A *schedule* σ of a finite set of processes P is a finite or infinite sequence where each element of the sequence is a process in P, e.g., $\sigma = p_2 p_4 p_3 p_1 p_2 p_5 p_4 p_3 p_4 p_2 p_5 p_3 p_3 p_3 p_5 \cdots$ is a schedule of $P = \{p_1, p_2, p_3, p_4, p_5\}$. Each occurrence of a process p in a schedule σ is called a *step of p (in σ)*. We consider systems with process crashes. A process is *correct in a schedule σ* if it occurs infinitely often in σ, otherwise it is *faulty (or crashes) in σ*.

Roughly speaking, a schedule σ is *k-solo*, if σ has a process that runs solo for at least k consecutive steps infinitely often. More precisely, *a schedule σ of a set of processes P is k-solo* if σ is finite or there is a process $p \in P$ such that σ contains an infinite number of subsequences consisting of k or more consecutive steps of p (and only of p). For example, the schedule $\sigma = p_1 p_2 p_3 p_1 p_2 p_3 p_1 p_2 p_3 \cdots$ of processes $P = \{p_1, p_2, p_3\}$ is not 2-*solo*, while the schedule $\sigma = p_1 p_2 p_3 p_3 p_1 p_2 p_3 p_3 p_1 p_2 p_3 p_3 \cdots$ is 2-*solo*. Note that if a schedule is *k-solo* then it is also $(k - 1)$-*solo*, and every schedule is trivially 1-*solo*.

2.2 Systems and Subsystems

Intuitively, a system of a set of processes P is defined by the set Σ of schedules of P that can occur in this system. More precisely, *a system of P* is a set Σ of schedules of P. We say that *σ is a schedule of system Σ* if $\sigma \in \Sigma$. Moreover, *Σ' is a subsystem of Σ* if $\Sigma' \subseteq \Sigma$.

In the following, P is a finite set of processes and Q is a subset of P. If σ is a schedule of P, we denote by $\sigma(Q)$ the subsequence of σ obtained by keeping only the steps of the processes that are in Q; e.g., if $\sigma = p_2p_4p_3p_1p_2p_5p_4p_3p_4p_2p_5p_3p_3p_3p_5$ and $Q = \{p_2, p_5\}$ then $\sigma(Q) = p_2p_2p_5p_2p_5p_5$. Note that $\sigma(Q)$ is a schedule of Q. If Σ is a set of schedules of P (i.e., Σ is system of P) we denote by $\Sigma(Q)$ the set of schedules obtained by keeping only the steps of the processes that are in Q in the schedules of Σ; more precisely: $\Sigma(Q) = \{\sigma' \mid \exists \sigma \in \Sigma \text{ such that } \sigma' = \sigma(Q)\}$. Note that all the schedules of $\Sigma(Q)$ are schedules of Q.

We now define some systems that are central to our results. Let P be a finite set of processes.

- $\Sigma_P = \{\sigma \mid \sigma \text{ is a schedule of } P\}$; this is *the asynchronous system of P* (because it contains *all* the possible schedules of P).
- $\Sigma_{P,n}^k = \{\sigma \mid \sigma \text{ is a schedule of } P \text{ and for all } Q \subseteq P \text{ such that } |Q| \leq n: \sigma(Q) \text{ is } k\text{-}solo\}$.

The following lemma relates the above systems:

Lemma 1. *For all finite sets of processes P, $n \geq 1$, and $k \geq 1$:*

1. $\Sigma_{P,n}^1 = \Sigma_{P,1}^k = \Sigma_P$.
2. $\Sigma_{P,n}^{k+1} \subseteq \Sigma_{P,n}^k$. *Moreover, if $n \geq 2$ and $|P| \geq n$ then $\Sigma_{P,n}^{k+1} \subset \Sigma_{P,n}^k$.*
3. $\Sigma_{P,n+1}^k \subseteq \Sigma_{P,n}^k$. *Moreover, if $k \geq 2$ and $|P| \geq n+1$ then $\Sigma_{P,n+1}^k \subset \Sigma_{P,n}^k$.*

Some Notation. In the proof of this lemma (which is given below) and throughout this paper we use the following notation to describe some schedules of P. For two processes p and q in P, $(pq)^k$ is the schedule pq repeated k times; for example $(pq)^3$ is $pqpqpq$. Similarly, $\{p, q\}^k$ is the set all finite schedules of p and q that contains *at least* k steps of p and *at least* k steps of q, in any order. For example, $\{p, q\}^3$ includes the schedules $pqpqpq$, $pppqqqq$ and $ppqqpq$, but it does *not* include the schedule $pppqqq$ (because it contains a step of process r) or $ppppqq$ (because it contains fewer than 3 steps of q). The notation $(pq)^k$ and $\{p, q\}^k$ for p and q generalizes to any finite set of processes in the obvious way. We use the operator $\prod_{i=1}^{\infty}$ to repeat a schedule or a schedule pattern infinitely many times. More precisely, for a schedule σ, $\prod_{i=1}^{\infty} \sigma$ is the schedule $\sigma \cdot \sigma \cdot \sigma \cdot \ldots$, and for a set of schedules Σ, $\prod_{i=1}^{\infty} \Sigma$ is the set of schedules $\{\sigma \mid \sigma = \sigma_1 \cdot \sigma_2 \cdot \sigma_3 \cdot \ldots \text{ such that for all } i \geq 1, \sigma_i \in \Sigma\}$. For example, $\prod_{i=1}^{\infty}(p^kq^k)$ is the schedule $(p^kq^k)(p^kq^k)\ldots$, and $\prod_{i=1}^{\infty}\{p, q\}^k$ is the set of all schedules $\sigma_1 \cdot \sigma_2 \cdot \sigma_3 \cdot \ldots$ such that every $\sigma_i \in \{p, q\}^k$. If Σ is a set of schedules, e.g., $\{p, q\}^k$ or $\prod_{i=1}^{\infty}\{p, q\}^k$, say that σ is a schedule of the form Σ if $\sigma \in \Sigma$.

Proof. (Lemma 1) Let P be any finite set of processes, $n \geq 1$, and $k \geq 1$.

1. It is clear that $\Sigma_{P,n}^1 \subseteq \Sigma_P$ and $\Sigma_{P,1}^k \subseteq \Sigma_P$. We now show that $\Sigma_P \subseteq \Sigma_{P,n}^1$ and $\Sigma_P \subseteq \Sigma_{P,1}^k$. Consider any schedule $\sigma \in \Sigma_P$. For every $Q \subseteq P$ (of any size n), the schedule $\sigma(Q)$ is trivially 1-*solo*; so $\sigma \in \Sigma_{P,n}^1$. Thus, $\Sigma_P \subseteq \Sigma_{P,n}^1$.

For every $Q \subseteq P$ such that $|Q| \leq n = 1$, the schedule $\sigma(Q)$ is trivially k-*solo*, for every k; so $\sigma \in \Sigma_{P,1}^k$. Thus, $\Sigma_P \subseteq \Sigma_{P,1}^k$.

2. To prove that $\Sigma_{P,n}^{k+1} \subseteq \Sigma_{P,n}^k$, consider any schedule $\sigma \in \Sigma_{P,n}^{k+1}$. By definition of $\Sigma_{P,n}^{k+1}$, for every $Q \subseteq P$ of size n or less, the schedule $\sigma(Q)$ is $(k + 1)$-*solo*; thus $\sigma(Q)$ is also k-*solo*, and therefore $\sigma \in \Sigma_{P,n}^k$.

   Now let $n \geq 2$ and $P = \{p_1, p_2, \ldots, \}$. To prove that $\Sigma_{P,n}^{k+1} \subset \Sigma_{P,n}^k$, we give a schedule σ that is in $\Sigma_{P,n}^k$ but not in $\Sigma_{P,n}^{k+1}$. Consider the schedule $\sigma = \prod_{i=1}^{\infty}(p_1^k p_2^k) = (p_1^k p_2^k)(p_1^k p_2^k) \ldots$ of P. It is easy to see that for every $Q \subseteq P$ such that $|Q| \leq n$, the schedule $\sigma(Q)$ is k-*solo*, and so $\sigma \in \Sigma_{P,n}^k$. But for $Q = \{p_1, p_2\} \subseteq P$ of size $2 \leq n$, $\sigma(Q) = \sigma$ is *not* $(k + 1)$-*solo* for any $k \geq 1$ (because in σ no process ever takes $k+1$ consecutive steps alone). Thus, $\sigma \notin \Sigma_{P,n}^{k+1}$.

3. To prove that $\Sigma_{P,n+1}^k \subseteq \Sigma_{P,n}^k$, consider any schedule $\sigma \in \Sigma_{P,n+1}^k$. By definition of $\Sigma_{P,n+1}^k$, for every $Q \subseteq P$ such that $|Q| \leq n + 1$, the schedule $\sigma(Q)$ is k-*solo*. So for every $Q \subseteq P$ such that $|Q| \leq n$, $\sigma(Q)$ is k-*solo*; thus $\sigma \in \Sigma_{P,n}^k$.

   Now let $k \geq 2$ and $P = \{p_1, p_2, \ldots, p_{n+1}, \ldots\}$. To prove that $\Sigma_{P,n+1}^k \subset \Sigma_{P,n}^k$, we give a schedule σ that is in $\Sigma_{P,n}^k$ but not in $\Sigma_{P,n+1}^k$. Let $\sigma = \prod_{i=1}^{\infty}[(p_1 p_2)^k (p_1 p_3)^k (p_1 p_4)^k \ldots (p_1 p_{n+1})^k]$.

   We claim that for every $Q \subseteq P$ such that $|Q| \leq n$, the schedule $\sigma(Q)$ is k-*solo* (and so $\sigma \in \Sigma_{P,n}^k$). To see why this claim holds, suppose that $Q \cap \{p_1, p_2, \ldots, p_{n+1}\} \neq \emptyset$.[3] The are two possible cases: either (a) $p_1 \notin Q$, and so some $p_j \in Q$, or (b) $p_1 \in Q$, and, since $|Q| \leq n$, some $p_j \notin Q$. Since the subsequence $(p_1 p_j)^k$ appears infinitely often in σ, in case (a) the subsequence p_j^k appears infinitely often in $\sigma(Q)$, and in case (b) the subsequence p_1^k appears infinitely often in $\sigma(Q)$. Thus in all cases, $\sigma(Q)$ is k-*solo*. So $\sigma \in \Sigma_{P,n}^k$.

   For the subset $Q = \{p_1, p_2, \ldots, p_{n+1}\}$ of P of size $|Q| = n + 1$, however, the schedule $\sigma(Q) = \sigma$ is *not* k-*solo* for any $k \geq 2$ (because in σ no process ever takes two consecutive steps alone). Thus, $\sigma \notin \Sigma_{P,n+1}^k$ for any $k \geq 2$.

□ Lemma 1

2.3   Examples of Schedules in $\Sigma_{P,n}^k$

To provide some intuition about the systems $\Sigma_{P,n}^k$ that we defined, we now give a few simple examples of schedules that are in $\Sigma_{P,n}^k$ and of schedules are not in $\Sigma_{P,n}^k$.[4] Let $P = \{p_1, p_2, \ldots, p_{n+1}, \ldots\}$ be a finite set of at least $n + 1$ processes, where $n \geq 3$. We start with examples of schedules $\sigma \in \Sigma_{P,n}^k$:

---

[3] If $Q \cap \{p_1, p_2, \ldots, p_{n+1}, \ldots\} = \emptyset$ then $\sigma(Q)$ is the empty sequence, and so it is trivially k-*solo*.

[4] The examples of schedules that we give here are very simple (they have a simple repetitive pattern). It should be clear, however, that the set of schedules of $\Sigma_{P,n}^k$ is very "rich": it contains schedules that are much more varied and complex than the few simplistic ones given for illustration here.

(a) σ is any schedule of the form $\prod_{i=1}^{\infty}[\{p_1,p_2\}^k\{p_1,p_3\}^k\{p_1,p_4\}^k\ldots\{p_1,p_{n+1}\}^k]$. Note that for every $Q \subseteq P$ such that $|Q| \le n$, the schedule $\sigma(Q)$ is k-solo (the proof is similar to one that we gave for the claim in the proof of Lemma 1 part (3)). Thus, $\sigma \in \Sigma_{P,n}^k$.

(b) $\sigma = \prod_{i=1}^{\infty}[(p_1p_2)^k(p_2p_3)^k(p_3p_4)^k\ldots(p_{n-1}p_n)^k(p_np_{n+1})^k(p_{n+1}p_1)^k]$. It is easy to see that for every $Q \subseteq P$ such that $|Q| \le n$, the schedule $\sigma(Q)$ is k-solo, so $\sigma \in \Sigma_{P,n}^k$.

The following schedule is not in system $\Sigma_{P,n}^k$, i.e., $\sigma \notin \Sigma_{P,n}^k$, for any $k \ge 2$:

(a) $\sigma = \prod_{i=1}^{\infty}[(p_1p_2)^k(p_3p_4)^k\ldots(p_{n-2}\ p_{n-1})^k(p_n\ p_{n+1})^k]$. To see that $\sigma \notin \Sigma_{P,n}^k$, note that for the subset $Q = \{p_1,p_2\}$ of P of size $2 \le n$, the schedule $\sigma(Q) = \prod_{i=1}^{\infty}(p_1p_2)^k = p_1p_2p_1p_2p_1p_2\ldots$ is not k-solo for any $k \ge 2$.

3   Consensus

In the well-known consensus problem, each process has an initial value and must decide a value such that the following three properties hold:

- *Agreement:* If correct processes p and q decide v and v', respectively, then $v = v'$;
- *Integrity:* If a correct process decides v, then v is the initial value of some process.
- *Termination:* Every correct process eventually decides some value.

The initial value and decision value of a process are also called the *input value* and *output value* of this process.

In a stronger variant of consensus, called *uniform consensus*, agreement and integrity also apply to faulty processes. More precisely, uniform consensus requires: (a) *uniform agreement:* If *any* two processes p and q decide v and v', respectively, then $v = v'$; (b) *uniform integrity:* If *any* process decides v, then v is the initial value of some process; and (c) *termination:* Every correct process eventually decides some value.

Solving Consensus. We now explain what it means for an algorithm to solve consensus for a set of processes in a given system. The definitions that we give here are rather informal, but sufficient for understanding the statements and proofs of the paper's results. In the following, Q is a finite set of processes and $\mathcal{A}$ is an algorithm for Q.

A *run R of $\mathcal{A}$ by Q* is an execution of the algorithm $\mathcal{A}$ by the processes in Q: the run R specifies the initial state of each process in Q and the sequence of algorithm steps that the processes in Q take during their execution of $\mathcal{A}$. The *schedule σ_R of a run R of $\mathcal{A}$* is the sequence of processes that take steps of $\mathcal{A}$ in R, in the order in which these steps occur: e.g., if in run R process p_4 takes the first step of $\mathcal{A}$, then p_1 takes the next two steps of $\mathcal{A}$, and then p_3 takes the last step of $\mathcal{A}$, $\sigma_R = p_4p_1p_1p_3$. Note that if R is a run of $\mathcal{A}$ by Q, then σ_R is a schedule of Q. A process is *correct* in a run R of $\mathcal{A}$ if it is correct in the schedule σ_R of R, and it is *faulty* otherwise.

$\mathcal{A}$ *solves [uniform] consensus for Q in a run R*, if R is a run of $\mathcal{A}$ by Q that satisfies the three properties of [uniform] consensus for all the processes in Q. Let P be a finite set of processes, Q be a subset of P, and Σ be a system of P (i.e., Σ is a set of schedules of P). A run R of $\mathcal{A}$ by Q is *in system* Σ if the schedule σ_R of R is such that $\sigma_R = \sigma(Q)$ for some $\sigma \in \Sigma$. A *solves [uniform] consensus for Q in system Σ* if $\mathcal{A}$ solves [uniform] consensus for Q in every run R of $\mathcal{A}$ by Q in system Σ.

In general, an algorithm that solves consensus in a given system may not solve uniform consensus in that system (in fact, there are algorithms that solve consensus but not uniform consensus in synchronous systems). But for the systems $\Sigma_{P,n}^k$ that we use in our results, solving consensus and uniform consensus is equivalent. More precisely:

Lemma 2. *Let P be any finite set of processes and Q be any subset of P. For all $n \geq 1$, and all $k \geq 1$, if an algorithm $\mathcal{A}$ solves consensus for Q in system $\Sigma_{P,n}^k$, then it also solves* uniform *consensus for Q in system $\Sigma_{P,n}^k$.*

The proof of the above lemma uses standard arguments, and so it is omitted here.

4 Main Results

Let P be a set of processes. We now show that for all $n \geq 1$, there is a system Σ_n of P such that: (a) consensus *can* be solved for *every* set of n or fewer processes of P, and (b) consensus *cannot* be solved for *any* set of $n + 1$ or more processes of P. To prove this, we show that for all $n \geq 1$: (a) for some $k \geq 1$, consensus can be solved for every set of n or fewer processes in system $\Sigma_{P,n}^k$, and (b) for all $k \geq 1$, consensus cannot be solved for any set of $n+1$ or more processes in system $\Sigma_{P,n}^k$. Part (a) and (b) are shown in Sections 5 and 6, respectively.

5 Possibility of n-consensus in $\Sigma_{P,n}^k$

Theorem 1. *Let P be any finite set of processes. For all $n \geq 1$, there is a $k \geq 1$ such that: for all $Q \subseteq P$ such that $|Q| \leq n$, consensus can be solved for Q in system $\Sigma_{P,n}^k$.*

Proof (Proof Sketch). The proof of Theorem 1 follows from the existence of *bounded obstruction-free* consensus algorithms in shared-memory systems with SWMR registers [6] and so it is only sketched here.

From results in [6], we know that for every integer m, there is a constant k_m such that for every set Q of processes of size m, there is an algorithm $\mathcal{B}_Q$ that solves consensus for Q in every run where some process executes k_m steps solo; more precisely, in every run R of $\mathcal{B}_Q$ where the schedule of R is a k_m-*solo* schedule of Q, the validity, agreement and termination properties of consensus are satisfied for the processes in Q.

Let P be any finite set of processes, let $n \geq 1$, and let $k = \max_{1 \leq l \leq n} k_l$. Consider any subset $Q \subseteq P$ such that $|Q| = m \leq n$. We claim that consensus can be solved for Q in the system $\Sigma_{P,n}^k$ of P. To see this, suppose the m processes in Q execute the algorithm

$\mathcal{B}_Q$ among themselves in the system $\Sigma_{P,n}^k$ of P. Let σ be the schedule of P in this execution, so $\sigma \in \Sigma_{P,n}^k$. For the m processes in Q, this execution is indistinguishable from a run R of $\mathcal{B}_Q$ where they are the only processes that take steps; more precisely, for the m processes in Q, this execution is indistinguishable from a run R of $\mathcal{B}_Q$ where the schedule σ_R of R is $\sigma(Q)$. Since $\sigma \in \Sigma_{P,n}^k$ and $|Q| \leq n$, the schedule $\sigma(Q)$ is k-solo. Since $m \leq n$ and $k = \max_{1 \leq l \leq n} k_l$, $k \geq k_m$, and so the schedule $\sigma(Q)$ is also k_m-solo. We conclude that the execution of the algorithm $\mathcal{B}_Q$ by the m processes in Q in system $\Sigma_{P,n}^k$ is indistinguishable from a run R of $\mathcal{B}_Q$ where only the processes in Q take steps, and the schedule $\sigma(Q)$ of R is k_m-solo. Thus $\mathcal{B}_Q$ solves consensus for Q in this run.   □ Theorem 1

6   Impossibility of $(n+1)$-consensus in $\Sigma_{P,n}^k$

Theorem 2. *Let P be any finite set of processes. For all $n \geq 1$, and all $k \geq 1$: for all $Q \subseteq P$ such that $|Q| \geq n+1$, consensus cannot be solved for Q in system $\Sigma_{P,n}^k$.*

Proof. Consider any finite set of processes P and let $n \geq 1$. We now prove that for all $k \geq 1$, and all subsets of processes $Q \subseteq P$ such that $|Q| \geq n+1$, consensus cannot be solved for Q in system $\Sigma_{P,n}^k$.

For $n = 1$, the proof is straightforward. There are two possible cases. If $|P| \leq 1$, then there is no $Q \subseteq P$ such that $|Q| \geq n+1 = 2$, so the theorem trivially holds in this case. Now assume that $|P| \geq 2$. By Lemma 1, we have $\Sigma_{P,1}^k = \Sigma_P$, i.e., $\Sigma_{P,1}^k$ is the *asynchronous system* of the processes in P. From the results in [15,10], it is known that in the asynchronous system of P (which contains at least two processes), consensus cannot be solved for any set of processes $Q \subseteq P$ such that $|Q| \geq n+1 = 2$.

Now let $n \geq 2$. Suppose, for contradiction, that there is a $k \geq 1$, a subset of processes $Q_{m+1} \subseteq P$ such that $|Q_{m+1}| = m + 1 \geq n + 1$, and an algorithm $\mathcal{C}_{m+1}$ that solves consensus for Q_{m+1} in system $\Sigma_{P,n}^k$. Let $Q_{m+1} = \{q_1, q_2, \ldots, q_{m+1}\}$. We will show that two processes p_1 and p_2 can use $\mathcal{C}_{m+1}$ to solve consensus in the asynchronous system $\Sigma_{\{p_1,p_2\}}$ — contradicting the well-known impossibility result in [15,10].

To solve consensus among themselves, processes p_1 and p_2 *simulate* the execution of the consensus algorithm $\mathcal{C}_{m+1}$ by the $m + 1$ processes of Q_{m+1} such that: (a) the simulated executions of algorithm $\mathcal{C}_{m+1}$ by the processes in Q_{m+1} are in system $\Sigma_{P,n}^k$, and (b) if p_1 or p_2 do not crash, then at least one of the $m + 1$ processes in Q_{m+1} that they simulate does not crash. By property (a), the simulated runs of $\mathcal{C}_{m+1}$ solve consensus for Q_{m+1}, and by property (b), p_1 and p_2 can wait till one of the processes in Q_{m+1} decides a value (and then adopt this value as its own decision value).

To show how the above simulation works, we first define a set of schedules of the processes Q_{m+1} in system $\Sigma_{P,n}^k$ (we will later show that p_1 and p_2 can simulate these schedules in the asynchronous system $\Sigma_{\{p_1,p_2\}}$). Intuitively, this set consist of: (1) all

Shared variables:

/* Program Counters of simulated processes $q_1, q_2, \ldots, q_{m+1}$ */

$PC[1..m + 1]$: array of SWSR registers, initialized to [0..0]

CODE FOR PROCESS p_1:   /* process p_1 simulates process q_1 executing algorithm $\mathcal{C}_{m+1}$*/

1   input value of process q_1 in $\mathcal{C}_{m+1}$:= input value of process p_1
2   **forever do**
3        $PC[1] := PC[1] + 1$
4   execute one step of process q_1 running algorithm $\mathcal{C}_{m+1}$
5   **if** process q_1 decides some value v in $\mathcal{C}_{m+1}$ **then** decide v

CODE FOR PROCESS p_2:   /* process p_2 simulates processes $q_2, q_3, \ldots, q_{m+1}$
                                    executing algorithm $\mathcal{C}_{m+1}$*/

Local variables:

$pc[1..m + 1]$: array of integers
j: integer

6   **for** $j = 2$ **to** m **do**
7   input value of process q_j in $\mathcal{C}_{m+1}$:= input value of process p_2
8   **for** $i = 1, 2, \ldots$ **do**   /* simulation of $\{q_1, q_j\}^k$ steps of processes q_1 and q_j */
9            $j := 2 + (i - 1) \bmod m$ /* with $j = 2, 3, \ldots, m + 1, 2, \ldots$ in round-robin order */
10            $pc[1] := PC[1]$
11            $pc[j] := PC[j]$
12   **while** $(PC[1] \leq pc[1] + k)$ **or** $(PC[j] < pc[j] + k)$ **do**
13                $PC[j] := PC[j] + 1$
14   execute one step of process q_j running algorithm $\mathcal{C}_{m+1}$
15   **if** process q_j decides a value v in $\mathcal{C}_{m+1}$ **and** p_2 has not yet decided **then** decide v

Fig. 1. Processes p_1 and p_2 simulate the execution of $\mathcal{C}_{m+1}$ by processes $q_1, q_2, \ldots, q_{m+1}$ in system $\Sigma_{P,n}^k$.

schedules of the form $\prod_{i=1}^{\infty} [\{q_1, q_2\}^k \{q_1, q_3\}^k \ldots \{q_1, q_{m+1}\}^k]$,[5] (2) all the finite prefixes of such schedules, and (3) all the finite prefixes of such schedules followed by q^{∞} (i.e.. an infinite sequence of steps of q) for some process $q \in Q_{m+1}$. More precisely:

Lemma 3. *Let Σ be the set of schedules of the form $\prod_{i=1}^{\infty} [\{q_1, q_2\}^k \{q_1, q_3\}^k \ldots \{q_1, q_{m+1}\}^k]$. Let σ be any schedule such that: (1) σ is in Σ, or (2) σ is a finite prefix of a schedule in Σ, or (3) $\sigma = \sigma' q^{\infty}$ where σ' is a finite prefix of a schedule in Σ, and q is a process in Q_{m+1}. Then σ is a schedule of system $\Sigma_{P,n}^k$.*

Proof. Let σ be any schedule as defined above. To prove that σ is in $\Sigma_{P,n}^k$, we must show that for all $Q \subseteq P$ such that $|Q| \leq n$, the schedule $\sigma(Q)$ is k-solo. There are three possible cases:

---

[5] Recall that $\{q, q'\}^i$ is any sequence of steps of q and q' that contains *at least* i steps of q and *at least* i steps of q', in any order.

1. σ is schedule of the form $\prod_{i=1}^{\infty}[\{q_1, q_2\}^k \{q_1, q_3\}^k \cdots \{q_1, q_{m+1}\}^k]$.

   First note that if $Q \cap Q_{m+1} = \emptyset$, then $\sigma(Q)$ is the empty schedule, and so it is trivially k-*solo*. Now assume that $Q \cap Q_{m+1} \neq \emptyset$.

   Suppose $q_1 \in Q$. Since $|Q| \leq n$ and $|Q_{m+1}| = m + 1 \geq n + 1$, there is a process $q_j \in Q_{m+1} \setminus Q$. Note that subsequences of the form $\{q_1, q_j\}^k$ appears infinitely often in σ. Thus, since $q_1 \in Q$ and $q_j \notin Q$, the subsequence q_1^k appears infinitely often in $\sigma(Q)$. In other words, process q_1 runs solo for k steps infinitely often in $\sigma(Q)$. So $\sigma(Q)$ is k-*solo*.

   Suppose $q_1 \notin Q$. Since $Q \cap Q_{m+1} \neq \emptyset$, there is a process $q_j \in Q \cap Q_{m+1}$. Note that subsequences of the form $\{q_1, q_j\}^k$ appears infinitely often in σ. Thus, since $q_1 \notin Q$ and $q_j \in Q$, the subsequence q_j^k appears infinitely often in $\sigma(Q)$. So $\sigma(Q)$ is k-*solo*.

2. σ is finite. Then $\sigma(Q)$ is also finite, and it is trivially k-*solo*.

3. $\sigma = \sigma' q^{\infty}$ for some finite σ' and a process q. If $q \in Q$, then q^{∞} is a suffix of $\sigma(Q)$, so $\sigma(Q)$ is k-*solo*. If $q \notin Q$, then $\sigma(Q)$ is finite, and so it is trivially k-*solo*.

So in all possible cases, $\sigma(Q)$ is k-*solo*.   □ Lemma 3

We now show that when processes p_1 and p_2 execute the algorithm in Figure 1 in the asynchronous system $\Sigma_{\{p_1, p_2\}}$, they simulate the $m + 1$ processes of Q_{m+1} executing the algorithm $\mathcal{C}_{m+1}$ in system $\Sigma_{P,n}^k$.

Lemma 4. *When processes p_1 and p_2 execute the algorithm in Figure 1 in the asynchronous system $\Sigma_{\{p_1, p_2\}}$, they simulate runs of $\mathcal{C}_{m+1}$ by the processes Q_{m+1} in system $\Sigma_{P,n}^k$, i.e., the schedules of these simulated runs are schedules of $\Sigma_{P,n}^k$.*

Proof. First note that each time process p_1 executes an iteration of its forever loop (lines 2-5), it increments $PC[1]$ and does one step of process q_1 executing algorithm $\mathcal{C}_{m+1}$. Similarly, each time process p_2 executes an iteration of its while loop (lines 12-15) for a process $q_j \in \{q_2, \ldots, q_{m+1}\}$, it increments $PC[j]$ and does one step of process q_j executing algorithm $\mathcal{C}_{m+1}$. Thus, it is clear that p_1 and p_2 simulate runs of $\mathcal{C}_{m+1}$ by the processes in Q_{m+1}. It remains to show that the schedules of these simulated runs are schedules of the system $\Sigma_{P,n}^k$. In the following, we prove that they are either: (1) schedules of the form $\prod_{i=1}^{\infty}[\{q_1, q_2\}^k \{q_1, q_3\}^k \cdots \{q_1, q_{m+1}\}^k]$, or (2) finite prefixes of such schedules, or (3) finite prefixes of such schedules followed by q^{∞} for some process $q \in Q_{m+1}$. By Lemma 3, all these schedules are indeed schedules of system $\Sigma_{P,n}^k$.

From the code and the termination condition of process p_2's while loop of line 12, it is clear that p_2 completes each execution of the while loop that it starts, unless it crashes or process p_1 crashes (and stops incrementing $PC[1]$). Thus, unless p_1 or p_2 crash, process p_2 executes an infinite number of iterations of the for-loop of line 8. Note that during its i-th iteration of this for-loop, process p_2 simulates the steps of process q_j for $j = 2 + (i - 1) \bmod m$. So in the successive iterations of this for-loop, process p_2 simulates the steps of the processes $q_2, q_3, \ldots, q_{m+1}$ in round-robin order.

Let t_i be the time when process p_2 starts its i-th iteration of the for-loop of line 8; t_i is undefined if p_2 never starts this iteration. From the above, we have the following:

Observation 3. *If, for some $\ell \geq 1$, t_ℓ is undefined then process p_1 or p_2 (or both) crash.*

To show that p_1 and p_2 simulate schedules of $\Sigma_{P,n}^k$, we consider the steps of the processes in Q_{m+1} that p_1 and p_2 simulate from time 0 (when p_1 or p_2 start executing the simulation algorithm) to time t_1, from time t_1 to time t_2,..., from time t_j to time t_{j+1},... until we reach a time t_k that is undefined if such a time exists.

Note first that if t_1 is not defined, then p_2 crashes before executing its first for-loop of line 8, so p_2 never simulates any step. Since process p_1 simulates only the steps of process q_1, the resulting simulated schedule of Q_{m+1} is simply q_1^∞ or some finite prefix of q_1^∞ (if p_1 crashes). By Lemma 3 this is a schedule of $\Sigma_{P,n}^k$.

Henceforth assume that t_1 is defined. During the interval $[0, t_1]$ process p_2 does not simulate any step, and process p_1 simulates only steps of process q_1. So during interval $[0, t_1]$ the simulated schedule is some finite prefix of $q_1 \infty$.

Now suppose that, for some $i \geq 1$, t_i is defined. Let $j = 2 + (i - 1) \bmod m$. As we noted before, q_j is the (only) process of Q_{m+1} that p_2 simulates during its i-th iteration of the for-loop of line 8 that starts at time t_i.

There are two possible cases:

(1) t_{i+1} is defined. In this case, we show that during the interval of time $[t_i, t_{i+1}]$, processes p_1 and p_2 simulate a sequence of steps of the form $\{q_1, q_j\}^k$.

Claim. In the interval $[t_i, t_{i+1}]$ processes p_1 and p_2 simulate only the steps of processes q_1 and q_j, and they simulate at least k steps of q_1 and at least k steps of q_j.

Proof. First note that during interval $[t_i, t_{i+1}]$, process p_1 simulates only steps of process q_1, and process p_2 simulates only steps of process q_j.
Process p_2 stores the value of $PC[1]$ in $pc[1]$ at some time τ_1, and p_2 stores the value of $PC[j]$ in $pc[j]$ at some time τ_2, such that $t_i \leq \tau_1 \leq \tau_2 < t_{i+1}$. Furthermore, the while loop that p_2 executes during the interval $[t_i, t_{i+1}]$ ends at some time $\tau_3 \leq t_{i+1}$, when p_2 finds that $(PC[1] > pc[1] + k)$ and $(PC[j] \geq pc[j] + k)$ holds. Since $PC[1] = pc[1]$ at time τ_1 and $PC[1] > pc[1] + k$ at time τ_3, then at least k steps of process q_1 are simulated during the interval $[\tau_1, \tau_3]$. Similarly, since $PC[j] = pc[j]$ at time τ_2 and $PC[j] \geq pc[j] + k$ at time τ_3, then at least k steps of process q_j are simulated during the interval $[\tau_2, \tau_3]$. We conclude that during interval $[t_i, t_{i+1}]$, only steps of processes q_1 and q_j are simulated, and at least k steps of q_1 and at least k steps of q_j are simulated. $\square$ claim

(2) t_{i+1} is undefined.

Claim. After time t_i, only the steps of processes q_1 and q_j are simulated. Furthermore, there is a time $\tau \geq t_i$ after which only steps of process q_1 are simulated, or only steps of process q_j are simulated, or no steps are simulated.

Proof. Since t_{i+1} is undefined, process p_2 never starts its $(i + 1)$-th iteration of the for-loop of line 8. Thus, after time t_i process p_2 can simulate only the steps of process q_j. Since p_1 simulates only the steps of process q_1, after time t_i only the steps of q_1 and q_j can be simulated. Furthermore, since t_{i+1} is undefined, by Observation 3 process p_1 or process p_2 (or both) crash. If p_2 crashes then after this crash occurs no steps of q_j are simulated. If p_1 crashes then after this crash occurs no steps of q_1 are simulated. So there is a time $\tau \geq t_i$ after which only steps of process q_1 are simulated, or only steps of process q_j are simulated, or no steps are simulated. □ claim

From the above, it is clear that when p_1 and p_2 execute the algorithm in Figure 1 in the asynchronous system $\Sigma_{\{p_1,p_2\}}$, they simulate a run R of the consensus algorithm C_{m+1} by the $m+1$ processes of Q_{m+1} such that the schedule σ of R is of the form $\prod_{i=1}^{\infty}[\{q_1,q_2\}^k\{q_1,q_3\}^k \ldots \{q_1,q_{m+1}\}^k]$, or it is a finite prefix of such a schedule, or it is a finite prefix of such a schedule followed by q^{∞} for some process $q \in Q_{m+1}$. By Lemma 3, this schedule σ is in $\Sigma_{P,n}^k$. Thus, when p_1 and p_2 execute the algorithm in Figure 1 in the asynchronous system $\Sigma_{\{p_1,p_2\}}$, they simulate a run of algorithm C_{m+1} by the $m+1$ processes of Q_{m+1} in system $\Sigma_{P,n}^k$. □ Lemma 4

We now show that when p_1 and p_2 execute the algorithm in Figure 1 in the asynchronous system $\Sigma_{\{p_1,p_2\}}$, they solves consensus among themselves.

Consider an execution of the algorithm in Figure 1 where p_1 and p_2 have input value v_1 and v_2, respectively. In this execution, process p_1 simulates the steps of process q_1 executing algorithm C_{m+1} with input v_1 (see line 1); if process q_1 decides a value v in C_{m+1}, then p_1 also decides v. Similarly, process p_2 simulates the steps of processes $q_2, q_3, \ldots, q_{m+1}$ executing algorithm C_{m+1} with input v_2 (see line 7). If any process in $q_2, q_3, \ldots, q_{m+1}$ decides a value in C_{m+1}, then p_2 also decides this value. By Lemma 4, this execution simulates a run of the consensus algorithm C_{m+1} by the $m+1$ processes of Q_{m+1} in system $\Sigma_{P,n}^k$. We now show that p_1 and p_2 reach consensus.

- *(Uniform) Agreement:* If p_1 and p_2 decide, then p_1 decides the value that process q_1 decides, and p_2 decides the value that some process $q_j \in \{q_2, q_3, \ldots, q_{m+1}\}$ decides, in the simulated run of C_{m+1} by Q_{m+1} in system $\Sigma_{P,n}^k$. Since the algorithm C_{m+1} solves consensus for Q_{m+1} in system $\Sigma_{P,n}^k$, by Lemma 2, it also solves *uniform* consensus for Q_{m+1} in this system. So by the uniform agreement property, q_1 and q_j decide the same value. Thus, p_1 and p_2 also decide the same value.
- *Termination:* If process p_1 is correct, then the process q_1 that it simulates is also correct (i.e., q_1 takes an infinite number of steps) in the simulated run of C_{m+1} by Q_{m+1} in system $\Sigma_{P,n}^k$. Since C_{m+1} solves uniform consensus for Q_{m+1} in system $\Sigma_{P,n}^k$, by the termination property, correct process q_1 decides a value in this simulated run of C_{m+1}. So p_1 also decides a value.
  If process p_2 is correct, then at least one process $q_j \in \{q_2, q_3, \ldots, q_{m+1}\}$ that p_2 simulates is also correct in the simulated run of C_{m+1} by Q_{m+1} in system $\Sigma_{P,n}^k$. By the termination property, correct process q_j decides a value in this simulated run of C_{m+1}. So p_2 also decides a value.
- *(Uniform) Integrity:* If p_1 or p_2 decides a value v, then some process $p \in Q_{m+1}$ decides v in the simulated run of C_{m+1} by Q_{m+1} in system $\Sigma_{P,n}^k$. Since C_{m+1} solves uniform consensus for Q_{m+1} in system $\Sigma_{P,n}^k$, by the uniform integrity property, v

must be the input value of some process $q \in Q_{m+1}$ in this run of C_{m+1}. Note that the input value of q in this simulated run of C_{m+1} is the input value of p_1 or p_2 (algorithm lines 1 and 7). So v is the input value of p_1 or p_2.

Therefore the algorithm in Figure 1 solves consensus for p_1 and p_2 in the asynchronous system $\Sigma_{\{p_1,p_2\}}$ — contradicting the results in [15,10]. $\square$ Theorem 2

From Theorems 1 and 2, we have the following result:

Theorem 4. *Let P be any finite set of processes. For all $n \geq 1$, there is a system of P such that:*

(a) consensus can be solved for every subset of P with at most n processes, and
(b) consensus cannot be solved for any subset of P with at least $n + 1$ processes.

By setting $P = \{p_1, p_2, \ldots, p_{n+1}\}$ in the above theorem, we have the following:

Corollary 1. *For all $n \geq 1$, there is a system of $P = \{p_1, p_2, \ldots, p_{n+1}\}$ such that:*

(a) consensus can be solved for every proper subset of P, and
(b) consensus cannot be solved for P.

7 Concluding Remarks

The synchrony, asynchrony, and partial synchrony of systems can be defined in a simple and natural way by the set of process schedules that these systems allow. In this paper, we consider such schedule-based systems in the context of read-write shared-memory, and solve the following basic question: are there read-write shared-memory systems that can solve consensus for every subset of n processes but not for $n + 1$ processes? Since the answer is "yes", this work provides a step towards a hierarchy of systems defined in terms of sets of schedules, akin Herlihy's hierarchy for wait-free objects, where a shared object is at level n if it can solve consensus among any set of n processes but cannot solve consensus for $n + 1$ processes. In this sense, this work may also provide a step towards a possible unification of the "separate worlds" of partial synchrony and shared objects.

References

1. Aguilera, M.K., Delporte-Gallet, C., Fauconnier, H., Toueg, S.: Partial synchrony based on set timeliness. Distributed Computing 25(3), 249–260 (2012)
2. Aguilera, M.K., Toueg, S.: Adaptive progress: a gracefully-degrading liveness property. Distributed Computing 22(5-6), 303–334 (2010)
3. Biely, M., Robinson, P., Schmid, U.: The generalized loneliness detector and weak system models for k-set agreement. IEEE Trans. Parallel Distrib. Syst. 25(4), 1078–1088 (2014)
4. Borowsky, E., Gafni, E.: Generalized FLP impossibility result for t-resilient asynchronous computations. In: Proceedings of the 25th Annual ACM Symposium on Theory of Computing (STOC), pp. 91–100 (1993)

5. Chandra, T.D., Toueg, S.: Unreliable failure detectors for reliable distributed systems. Journal of the ACM 43(2), 225–267 (1996)
6. Delporte-Gallet, C., Fauconnier, H., Gafni, E., Rajsbaum, S.: Black art: Obstruction-free k-set agreement with $|MWMR\ registers| < |proccesses|$. In: Gramoli, V., Guerraoui, R. (eds.) NETYS 2013. LNCS, vol. 7853, pp. 28–41. Springer, Heidelberg (2013)
7. Delporte-Gallet, C., Fauconnier, H., Guerraoui, R.: Tight failure detection bounds on atomic object implementations. Journal of the ACM 57(4), April 2010
8. Delporte-Gallet, C., Fauconnier, H., Guerraoui, R., Tielmann, A.: The disagreement power of an adversary. Distributed Computing 24(3-4), 137–147 (2011)
9. Dolev, D., Dwork, C., Stockmeyer, L.J.: On the minimal synchronism needed for distributed consensus. Journal of the ACM 34(1), 77–97 (1987)
10. Fischer, M.J., Lynch, N.A., Paterson, M.: Impossibility of distributed consensus with one faulty process. Journal of the ACM 32(2), 374–382 (1985)
11. Gafni, E., Kuznetsov, P.: The weakest failure detector for solving k-set agreement. In: Proceedings of the 28th ACM Symposium on Principles of Distributed Computing (PODC), pp. 83–91 (2009)
12. Herlihy, M.: Wait-free synchronization. ACM Trans. Program. Lang. Syst. 13(1), 124–149 (1991)
13. Herlihy, M., Luchangco, V., Moir, M.: Obstruction-free synchronization: Double-ended queues as an example. In: ICDCS 2003: Proceedings of the 23rd International Conference on Distributed Computing Systems, pp. 522–529. IEEE Computer Society, May 2003
14. Herlihy, M., Shavit, N.: The topological structure of asynchronous computability. Journal of the ACM 46(6), 858–923 (1999)
15. Loui, M., Abu-Amara, H.: Memory requirements for agreement among unreliable asynchronous processes. Advances in Computing Research 4(31), 163–183 (1987)
16. Pike, S.M., Sastry, S., Welch, J.L.: Failure detectors encapsulate fairness. In: Lu, C., Masuzawa, T., Mosbah, M. (eds.) OPODIS 2010. LNCS, vol. 6490, pp. 173–188. Springer, Heidelberg (2010)
17. Rajsbaum, S., Raynal, M., Travers, C.: The iterated restricted immediate snapshot model. In: Hu, X., Wang, J. (eds.) COCOON 2008. LNCS, vol. 5092, pp. 487–497. Springer, Heidelberg (2008)
18. Saks, M., Zaharoglou, F.: Wait-free k-set agreement is impossible: The topology of public knowledge. SIAM J. Comput. 29(5), 1449–1483 (2000)
19. Zielinski, P.: Anti-Ω: the weakest failure detector for set agreement. Distributed Computing 22(5-6), 335–348 (2010)

Under the Hood of the Bakery Algorithm: Mutual Exclusion as a Matter of Priority*

Yoram Moses** and Katia Patkin

Technion - Israel Institute of Technology, Haifa 32000, Israel

Abstract. A new approach to the study and analysis of Mutual Exclusion (ME) algorithms is presented, based on identifying the priority relation that the ME algorithm constructs. It is argued that by analyzing how a process detects that it has priority over all other processes, ME algorithms can be better understood and improved. The approach is illustrated by applying it to Lamport's celebrated Bakery algorithm in the safe register SWMR model. By analyzing how Bakery established and detects priority, cases in which the Bakery algorithm causes processes to block unnecessarily are identified. Namely, a process that already knows that it has priority over another process is made to perform reads and wait on registers of the other process. An optimized version of the Bakery algorithm, called Boulangerie, is proposed, and is shown to be free of any unnecessary blocking. A second contribution of the approach is obtaining a clear explanation for how the Bakery algorithm uses reads from safe registers to detect that a process has priority. Our analysis provides more insight into the workings of the Bakery algorithm than is obtained by other proofs of its correctness.

Keywords: mutual exclusion, Bakery algorithm, safe registers, Boulangerie algorithm.

1 Introduction

Mutual Exclusion (ME) is a fundamental problem in distributed system. Indeed, many consider the introduction of ME by Dijkstra in [5] as the starting point of the field of distributed computing. Intuitively, at the heart of every mutual exclusion algorithm lies a priority relation among processes. To be in the critical section, a process must have priority over all other processes, who are denied access. We suggest that explicitly identifying the priority relation underlying a given ME algorithm is helpful for gaining a better understanding of the workings of the algorithm. One of the benefits of such an understanding can be identifying inefficiencies in ME algorithms and improving them. This papers applies this approach to Lamport's Bakery algorithm in the safe register SWMR model, providing new insights into the algorithm, and explaining how it obtains its goals despite the use of safe registers. This allows us to identify inefficiencies and offer optimizations that strictly improve the Bakery algorithm.

---

* This work was supported in part by ISF grant 1520/11.
** Yoram Moses is the Israel Pollak academic chair at the Technion.

C. Scheideler (Ed.): SIROCCO 2015, LNCS 9439, pp. 399–413, 2015.
DOI: 10.1007/978-3-319-25258-2_28

Lamport's Bakery algorithm for mutual exclusion from 1974 [9] is a very influential early solution to mutual exclusion.[1] Moreover, he was surprised to discover that the algorithm is correct under the weak memory assumption that registers are *safe*. For safe registers, a read operation on a register that overlaps a write to the same register can return an arbitrary value. As a result, the value obtained by a read is not necessarily a value that was ever written to it.

There are several proofs of correctness of other variants of the Bakery algorithm [1,4,18,19], as well as mechanical proofs of Lamport's original Bakery [8,15]. Typically, such a proof is based on an inductive invariant, involving a conjunction of several claims, that are all shown to be maintained by every step of the algorithm. While the invariants capture an essential aspect of the algorithm, they do not necessarily provide a clear explanation of the algorithm's rationale. Our goal is to provide a new analysis that explains the algorithm in a more transparent fashion.

In a mutual exclusion algorithm, if some process enters the critical section (CS), then, until it leaves the CS, no other process can enter. In this sense, a process in the CS has priority over all others. Identifying the priority relation underlying an ME algorithm and analyzing how processes detect that they have priority over others allows insight into the workings of the algorithm. The current paper studies the priority relation underlying the Bakery algorithm. This provides insight into the role of its different components. Moreover, it points to unnecessary blocking and waiting in the Bakery algorithm, and allows a strict improvement to be obtained. Indeed, we prove that improved version, which we call *Boulangerie*, does not suffer from unnecessary blocking.

The Bakery algorithm gets its name from the scheme used in some bakeries or shops, whereby a customer obtains a number upon entry, and the one with the smallest number has priority over the others. In the algorithm, for two processes that are both far enough along in pursuing the critical section (having "entered the bakery"), the one with the smaller number has priority over the other, with ties broken according to the process IDs. For processes that are already in the bakery, priority induces a total ordering. But the Bakery algorithm allows a process to enter the CS even if some or all of the others are not in the bakery. In this case, the asynchrony and concurrency of process operations complicate the picture. Interestingly, the general priority relation implemented by the Bakery algorithm is not even a partial order on processes. Rather, it is an antisymmetric binary relation, which suffices for mutual exclusion.

By studying the Bakery algorithm's priority relation, it is shown that the algorithm sometimes unnecessarily blocks a process from entering the CS and requires it to wait unnecessarily for reads of other processes even after the process can deduce that it has priority over them. Two improvements to the algorithm are shown, one of which takes advantage of the fact that a safe register will not show inconsistent readings for a register that is not being written to, and another that removes unnecessary reads and unnecessary blocking when contention for the CS is low. It is shown that the resulting algorithm, which we call the Boulangerie algorithm, does not suffer from unnecessary waiting or unnecessary blocking. Deployment of the Boulangerie algorithm can be incremental, in the following sense. Even if an arbitrary subset of the processes follow Boulangerie,

---

[1] On Lamport's web page on his writings, he mentions that "...*the bakery algorithm marked the beginning of my study of distributed algorithms.*"

while the others follow the original Bakery algorithm, the result is a correct ME algorithm. The Boulangerie users may gain efficiency, but the Bakery users do not suffer any inefficiency or degradation from the fact that others are following Boulangerie. While our analysis provides a rigorous mathematical argument for the correctness of the Boulangerie algorithm, Lamport has reported that he has successfully completed a mechanical proof of Boulangerie, based on his earlier mechanical proof of the Bakery algorithm [14].

The paper is organized as follows. The next section presents the model of computation and some preliminary definitions. Section 3 reviews the Bakery algorithm and discusses two ways of breaking it into blocks. It identifies the priority relation underlying the Bakery algorithm, and uses priority to reason about the algorithm and identifies the role of central steps of the algorithm. Section 4 identifies superfluous blocking in the Bakery algorithm, describes the Boulangerie algorithm, which is an optimization of the Bakery algorithm, and shows that it does not contain superfluous blocking. Finally, section 5 provides some concluding remarks.

2 Preliminary Definitions

This paper studies the Bakery algorithm in the asynchronous shared-memory model with safe single-writer, multi-reader (SWMR) registers. Access to these registers is obtained only via **read**($local\_var \leftarrow$ shared_reg) operations, which read a shared register shared_reg into a local variable $local\_var$, and by **write**(shared_reg $\leftarrow x$) operations, which write a (local variable or constant) value x to shared_reg. Other than that, computation makes use of local variables only. In the SWMR case, every shared register has an "owner" who is the process that can write to it. The others can only perform read operations on the variable. In contrast to local commands such as assignments to local variables, reads and writes are not executed instantaneously. Every **read** or **write** operation is associated with a *starting* time, which is when the process initiates the operation, and a *completion* time, after which the process proceeds to the next line of code in its program. In between, the process is suspended.

We assume that shared registers are "safe" in the sense of [12]. Such a register is considered to have a *stable* value at time t if its owner is not in the midst of a **write** operation to the register at that time. In this case, we define its value at time t to be the last value written to it by its owner (or at initialization). When a register's value is unstable, we find it convenient to define its value to be '?'. While safe register are sometimes modelled as having a different value at every moment [8,15], we find the direct approach of modelling it as "arbitrary", denoted by '?', to be more natural.

We consider an asynchronous but fair model of computation, in which every process is activated infinitely often, and every **read** and **write** operation that is initiated completes in finite time. At any given time, each process is at a well-defined control state captured by its program counter. At any point in time, the scheduler chooses an arbitrary subset of the processes and "activates" them, causing each of them to take a single step of computation according to its program. A process that is about to perform a **read** or **write** will initiate this operation and move to an "*i/o-suspended*" state. Activating a process in an i/o-suspended state advances it to the next command. Moreover, for a **read**(*local_var* ← shared_reg) it also assigns a value to *local_var*. If the register's value is '?', then the scheduler may assign an arbitrary value to *local_var*,[2] and otherwise *local_var* will be assigned the current value of shared_reg.

A **run** *r* of a given program is identified with an infinite sequence of *configurations*, and we refer to the configuration at time *t* in *r* by (r, t). Each configuration determines the values of all local variables, shared registers, as well as the program counters for all processes. (Recall that shared registers may have a value of '?' as described above.)

Reasoning with propositions. It will be convenient to reason about what is true or false in a given configuration. We shall write $(r, t) \models \varphi$ to state that a formula φ is true, or **holds**, at (r, t). Our formulas are boolean combinations[3] of basic propositions, where the basic propositions are either statements regarding the values of variables or registers, or propositions of the form $\text{in}_i(\ell)$, where *i* is a process and ℓ is a line in the program. We define $(r, t) \models \text{in}_i(\ell)$ to hold if *i* is either about to execute line ℓ at (r, t), or if the line contains a **read** or **write**, and *i* is suspended at (r, t) in the middle of the operation at line ℓ. Given a region *L* consisting of a set of lines of *i*'s program, it is natural to use $\text{in}_i(L)$ as shorthand for $\bigvee_{\ell \in L} \text{in}_i(\ell)$. (In our analysis, *L* will be a region of the Bakery algorithm, such as the doorway or the bakery.)

3   The Bakery Algorithm

We consider Dijkstra's mutual exclusion problem [5] for $N > 1$ processes in an asynchronous shared-memory setting with safe SWMR registers. Lamport's Bakery algorithm for this model [9] is a protocol $P = (P_1, \ldots, P_N)$, where for each $i = 1, \ldots, N$ the protocol P_i for process *i* is given by:

The Bakery protocol for process *i*

---

0 **Initialize:** number[*i*] = 0; choosing[*i*] = **false**;
1 **while true do**
2   non-critical section;
3   choosing[*i*] := **true**;
4   number[*i*] := 1 + max{number[1], ..., number[*N*]};
5   choosing[*i*] := **false**;
6   **forall the** $j \leq N$ **s.t.** $j \neq i$ **do**
7   **await** choosing[*j*] = **false** ;
8   **await** number[*j*] = 0 $\lor$ ⟨number[*i*], *i*⟩ $<_L$ ⟨number[*j*], *j*⟩ ;
9   critical section;
10   number[*i*] := 0;

Intuitively, every process begins operation in the non-critical section. It proceeds to choose a number in lines **3-5**. At this stage, the process checks that it is secure w.r.t. each

---

[2] E.g., if the value of shared_reg was '0', and '1' is currently being written to it, then a **read** can see an arbitrary value, such as 7456. It is required that the value assigned conform to the register's type, however.

[3] We will freely use boolean operators '¬' (NOT), '∧' (AND) and '⇒' (IMPLIES), with their standard interpretation.

of the other processes on **6-8**, after which it enters the critical section. Upon exiting the critical section, it resets its number register to 0. Traditionally, four main regions are distinguished in the Bakery algorithm: line **2** is called the *non-critical*, lines **3-5** are the *doorway*, lines **6-9** are the *bakery* and line **10** is the *exit*. Within the bakery region, lines **6-8** are called the *testing* region, and line **9** is the *critical section* (CS). Our analysis will use a slightly different partition of the algorithm into phases, defined in section 3.2, in which the doorway and bakery regions are modified. For thorough discussions of the Bakery algorithm, see [2,16,3,17]. On line **8**, we use $<_L$ to denote the lexicographical ordering on pairs. Namely, $\langle \text{number}[i], i \rangle <_L \langle \text{number}[j], j \rangle$ will hold if either $\text{number}[i] < \text{number}[j]$ or $\text{number}[i] = \text{number}[j]$ and $i < j$.

3.1 The Bakery Algorithm in the SWMR Model

The exposition of the Bakery Algorithm above is very close to the original version from [9]. However, our SWMR shared-memory model restricts access to shared registers to consist only of **read** and **write** operations. Clearly, lines **3**, **5** and **10** correspond to simple **write** operations. Lines **4** and **6-8** are shorthand for longer bits of code.[4]

Progress Assumptions. The convention imposed by Dijkstra's definition of the mutual exclusion problem is that a process that has no interest in entering the critical section, which in the bakery setting means that it is in the non-critical section (i.e., line 2), does not need to participate in the algorithm. Thus, we consider the non-critical as shorthand for code from which the process may choose, but is not required to, resume operation in the Bakery code. We assume that the model is asynchronous but fair, in the sense that processes not in the non-critical will be scheduled to move infinitely often. Observe that the only part of the algorithm in which a process may be blocked waiting for another process to move is in the testing region of lines **6-8**. Everywhere else, processes progress in a wait-free fashion.

3.2 Analysis of the Bakery Algorithm

A New Partition into Regions. Recall that in a contiguous region of i's code that contains no writes to a particular safe register, this register has a fixed value. Any **read** performed to it by other processes while process i is in that region will correctly return this value. Therefore, to facilitate the analysis of the algorithm, we modify the traditional partition slightly. Our purpose is to ensure that the doorway involves no writes to choosing$[i]$ and the bakery region has no writes to number$[i]$. To this end, we shrink the doorway region to consist only of line **4**, and shift the bakery region up by one line. We combine the writes on lines **3** and **10** and the non-critical section of line **2** to form a new region that we call the *outside* region. While we abuse the language slightly and maintain the old names for the bakery and doorway regions, we will add a dot on top of the letters used to denote each of the regions, to signal that our regions are slightly modified. The regions under the new partition are as follows: The doorway region consists of line **4** and denoted by $\dot{D}_i$, while the outside region consists of lines **1-3**, **10** and is denoted by O_i. Finally, the bakery region consists of lines **5-9** and denoted by $\dot{B}_i$.

---

[4] See, e.g., lines **4.1–4.3** in the related algorithm in the Appendix.

Observe that the computation of each process i cycles through the sequence of regions $(O_i; \dot{D}_i; \dot{B}_i;)^*$. Starting on the outside, it can move to the doorway and proceed to the bakery. There, after setting choosing$[i]$ to **false** on line **5**, it proceeds to the testing region, and can only exit the bakery if it passes through the critical section. Upon leaving the critical section, a process is on the outside.

Recall that at any configuration of the algorithm, the program counters of the processes are recorded. Thus, each process is associated with a unique line ℓ of the program, that the process is either in the middle of executing, or is about to execute, when in that configuration. In this case we say that the process is *on* line ℓ. We use the notation for the regions $\dot{B}_i$, $\dot{D}_i$ and O_i as shorthand for specific propositions of the form $\text{in}_i(L)$ for the lines contained in each respective region. At any given configuration (r, t) in a run r of the Bakery algorithm, exactly one of $\dot{B}_i$, $\dot{D}_i$ or O_i will hold.

According to the new partition of the Bakery algorithm, no writes are performed on the choosing$[i]$ register in the (modified) doorway region $\dot{D}_i$ and no writes are performed on number$[i]$ in the (modified) bakery region $\dot{B}_i$. This implies the following invariants:

Lemma 1. *Throughout the Bakery algorithm,*

(a) $\dot{D}_i \Rightarrow$ choosing$[i] = $ **true**, *and*
(b) $\dot{B}_i \Rightarrow 0 <$ number$[i]$, $\dot{B}_i \Rightarrow$ number$[i] \neq$ '?', *and* number$[i]$ *is stable while* $\dot{B}_i$
    holds.

Both (a) and (b) are immediate from the new partition and the sequential nature of the algorithm

3.3   A Priority Relation for the Bakery Algorithm

Intuitively, we'd like to think of process i as having priority over j at a given point, if it is guaranteed there that j cannot enter the CS before i has entered (an exited) the CS. It is natural to consider a process i that writes a value to number$[i]$ in line **4** as obtaining a *"ticket"* for entering the CS, consisting of the pair $\langle$number$[i], i\rangle$. Tickets are ordered by the lexicographical ordering '$<_L$' on ordered pairs of numbers. Recall that, by lemma 1(b), the value of number$[i]$ (and hence also the ticket $\langle$number$[i], i\rangle$) remains unchanged when i is in the bakery region. Since number$[h] \neq$ '?' for all processes h in the bakery region, the lexicographical ordering induces a total order on the tickets of all of these processes. Suppose that i and j that are both in the bakery region, and that $\langle$number$[i], i\rangle <_L \langle$number$[j], j\rangle$. Then i should be able to successfully test against j on line 8, while j will not be able to do so against i, before i enters the CS. Roughly speaking then, i should be viewed as having priority over j. It is not sufficient to define priority when among processes that are both in the bakery region, because a process should be able to enter the CS when other processes are in the non-critical section, for example. To this end, the Bakery algorithm is designed in such a way that if process i enters the bakery before j enters the doorway, then $\langle$number$[i], i\rangle <_L \langle$number$[j], j\rangle$ will hold if and when j might later join i in the bakery region. The algorithm thus guarantees that j cannot enter the CS before i once $\dot{B}_i \wedge O_j$ holds. Indeed, the same is true even if j then advances into the doorway. Based on these observations, we proceed as follows.

Our formulation of a priority relation will make use of the "Since" operator 'S' in temporal logic, whose definition is: $(r,t) \models \varphi S \psi$ if for some time $t' \leq t$ both (a) $(r,t') \models \varphi \wedge \psi$ and (b) $(r,m) \models \varphi$ for all m in the range $t' \leq m \leq t$. In particular, if $(r,t) \models \varphi S \psi$ then $(r,t) \models \varphi$. The main "Since" property that we will be interested in is $\dot{\mathsf{B}}_i \, S O_j$, which in words states that i is now in its bakery region, and it has been in the bakery region ever since a point in time at which j was in the outside region (and so $\dot{\mathsf{B}}_i \wedge O_j$ was true). The definition of S immediately implies:

Lemma 2. *If $\dot{\mathsf{B}}_i \, S O_j$ holds, then $\dot{\mathsf{B}}_i \, S O_j$ continues to hold as long as $\dot{\mathsf{B}}_i$ holds (i.e., while i remains in the bakery).*

We are now ready to define a binary priority relation ' $\lhd$ ' among processes for the Bakery algorithm:

Definition 1 (PRIORITY). *We say that i **has priority over** j at (r,t), which we denote by $(r,t) \models i \lhd j$, if either*

(i) $(r,t) \models \dot{\mathsf{B}}_i \wedge \dot{\mathsf{B}}_j \wedge \langle \mathsf{number}[i], i \rangle <_{\mathsf{L}} \langle \mathsf{number}[j], j \rangle$, or

(ii) $(r,t) \models \dot{\mathsf{B}}_i \wedge \neg \dot{\mathsf{B}}_j \wedge \dot{\mathsf{B}}_i \, S O_j$.

Notice that $i \lhd j$ can hold only when process i is in the bakery (i.e., when $\dot{\mathsf{B}}_i$ holds). When $i \lhd j$ is obtained by definition 1(i), both processes are in the bakery. Thus, both $\mathsf{number}[i] \neq$ '?' and $\mathsf{number}[j] \neq$ '?', by lemma 1(b). The lexicographical ordering is well-defined in this case. Formally speaking, the priority relation $i \lhd j$ is simply shorthand for a simple temporal formula, obtained by taking the OR of the formulas in parts (i) and (ii) of definition 1.

We note that Lamport has argued in [11] that it is hard to formally specify what one means by a priority relation in a distributed setting. We do not attempt to specify priority. Rather, we show that ' $\lhd$ ' satisfies three properties that, in our opinion, justify our use of the term *priority* for this relation. First, we will show that the relation is antisymmetric, so that if $i \lhd j$ holds then $j \lhd i$ does not hold.[5]

Then, we will show that once $i \lhd j$ holds it will remain true until process i (enters and) exits the critical section. Finally, we will show that a process can enter or be in the critical section only if it has priority over all other processes. We start with the first property:

Lemma 3. *The priority relation ' $\lhd$ ' is antisymmetric.*

Proof. We need to show that $(r,t) \not\models (i \lhd j) \wedge (j \lhd i)$ for all points (r,t) that arise in the Bakery algorithm. Assume, by way of contradiction, that $(r,t) \models (i \lhd j) \wedge (j \lhd i)$. Then, by definition, we have that $(r,t) \models \dot{\mathsf{B}}_i \wedge \dot{\mathsf{B}}_j$. By definition 1, both $\langle \mathsf{number}[j], j \rangle <_{\mathsf{L}} \langle \mathsf{number}[i], i \rangle$ and $\langle \mathsf{number}[i], i \rangle <_{\mathsf{L}} \langle \mathsf{number}[j], j \rangle$ hold at (r,t). This is a contradiction, since '$<_{\mathsf{L}}$' is an ordering relation. □

---

[5] The relation ' $\lhd$ ' is not a partial order, because it is not transitive. While the transitive closure of ' $\lhd$ ' is a partial order, the Bakery algorithm detects priority only using the basic clauses in the definition of $\lhd$.

Our next goal is to show that once the priority relation $i \lhd j$ holds, it persists for as long as i is in the bakery. We first show that if $i \lhd j$ holds by definition 1(ii), i.e., when j is not in the bakery, then $i \lhd j$ will continue to hold even if j enters the bakery $\left(\text{at which time } i \lhd j \text{ will hold by clause (i)}\right)$, for as long as i remains in the bakery.

Lemma 4. *The formula* $\dot{B}_i SO_j \Rightarrow i \lhd j$ *is true throughout the Bakery algorithm.*

Proof. By definition 1(ii), $\left(\neg \dot{B}_j \wedge \dot{B}_i SO_j\right) \Rightarrow i \lhd j$ is valid. We will show that $\left(\dot{B}_j \wedge \dot{B}_i SO_j\right) \Rightarrow i \lhd j$ is valid as well. Let r be a run of the bakery algorithm, and assume that $(r,t) \models \dot{B}_j \wedge \dot{B}_i SO_j$. It follows that $(r,t) \models \dot{B}_i \wedge \dot{B}_j$ and for some time $t_1 < t$ both (a) $(r,t_1) \models \dot{B}_i \wedge O_j$, and (b) for all times m in the range $t_1 \leq m \leq t$ we have that $(r,m) \models \dot{B}_i$. Without loss of generality assume that t_1 is the latest time with this property. By lemma 1(b) we have that number$[i] \neq$ '?' and it remains unchanged throughout the interval $[t_1, t]$. Between time t_1 at which j is in O_j and time t at which j is in the bakery, process j executes line **4**, after which number$[j] >$ number$[i]$ holds as long as both processes remain in the bakery region. It follows that $(r,t) \models i \lhd j$ by definition 1(i). □

Lemma 4 can be used to show the following:

Corollary 1. *If* $i \lhd j$ *holds, then it remains true as long as* $\dot{B}_i$ *holds. Formally:* $\dot{B}_i S(i \lhd j) \Rightarrow i \lhd j$ *is valid.*

Proof. Suppose that $(r,t) \models \dot{B}_i S(i \lhd j)$, and we will show that $(r,t) \models i \lhd j$. By definition of S we have that $(r,t) \models \dot{B}_i$, and that there exists a time $t' \leq t$ such that both $(r,t') \models \dot{B}_i \wedge i \lhd j$ and $(r,m) \models \dot{B}_i$ holds for all times m in the range $t' \leq m \leq t$. We prove the claim by induction on $k = t - t'$. The claim is immediate if $t - t' = 0$ since then $t = t'$ and $(r,t') \models \dot{B}_i \wedge i \lhd j$. Let $t - t' = k > 0$, and assume inductively that the claim is true for $k - 1$. Since $(t-1) - t' = k - 1 \geq 0$, we have by the inductive assumption that $(r,t-1) \models i \lhd j$. By definition of the '$\lhd$' relation, we consider two cases:

(a) $(r,t-1) \models \dot{B}_j \wedge \langle$number$[i],i\rangle <_L \langle$number$[j],j\rangle$. In this case, if $(r,t) \models \dot{B}_j$ then the values of number$[i]$ and of number$[j]$ are unchanged from $(r,t-1)$, so that $(r,t) \models \dot{B}_i \wedge \dot{B}_j \wedge \langle$number$[i],i\rangle <_L \langle$number$[j],j\rangle$ and $(r,t) \models i \lhd j$ holds by definition 1(i). If, however, $(r,t) \not\models \dot{B}_j$ then j moved out of the bakery between time $t-1$ and t, and so $(r,t) \models O_j$. It follows that $(r,t) \models \dot{B}_i \wedge \neg \dot{B}_j \wedge \dot{B}_i SO_j$, and so $(r,t) \models i \lhd j$ holds by definition 1(ii).

(b) $(r,t-1) \models \dot{B}_i SO_j$. Since $(r,t) \models \dot{B}_i$ we have that $(r,t) \models \dot{B}_i SO_j$, and so $(r,t) \models i \lhd j$ follows by lemma 4.
□

3.4 Proving Mutual Exclusion

Intuitively, the iteration of the testing region by process i (lines **6** to **8**) performed with parameter j is intended to establish and detect i's priority over j. For ease of exposition, we will denote by 7_j and 8_j the instances of lines 7 and 8 performed by i in this iteration. We now turn to seeing how this is achieved. First, we consider the precise role that the wait for choosing$[j] =$ **false** on line 7_j serves. We show that if this wait by i succeeds, it is guaranteed that j has been out of the doorway region at some point during the wait. Formally, we state this as:

Lemma 5. $\dot{\mathsf{B}}_i \, \mathcal{S} \neg \dot{\mathsf{D}}_j$ holds whenever process i leaves line 7_j.

Proof. Recall that choosing$[j] =$ **true** when process j enters $\dot{\mathsf{D}}_j$. Moreover, in the $\dot{\mathsf{D}}_j$ region, process j does not perform writes to choosing$[j]$. Thus, a read of choosing$[j]$ that completely overlaps $\dot{\mathsf{D}}_j$ will necessarily return **true**. Suppose that process i leaves line 7_j at (r, t), having successfully completed the wait on line 7_j. Its last r/w operation on line 7_j is a <u>read</u> of the register choosing$[j]$, which returns **false**. Since choosing$[j] =$ **true** $\neq$ '?' whenever j is in the doorway, it follows that at some time $t' < t$ this <u>read</u> operation did not overlap $\dot{\mathsf{D}}_j$. Moreover, at all times between t' and t in r, process i is in the bakery. It follows that $(r, t) \models \dot{\mathsf{B}}_i \, \mathcal{S} \neg \dot{\mathsf{D}}_j$, as claimed. $\qquad\square$

Lemma 5 implies the following very useful fact:

Corollary 2. *If i completes 7_j at (r, t) and $(r, t) \models i \not\vartriangleleft j$, then $(r, t) \models \dot{\mathsf{B}}_i \wedge \dot{\mathsf{B}}_j \wedge \langle$number$[j], j\rangle <_L \langle$number$[i], i\rangle$.*

Proof. Suppose that i completes 7_j at (r, t). By lemma 5 we have that $(r, t) \models \dot{\mathsf{B}}_i \, \mathcal{S} \neg \dot{\mathsf{D}}_j$. Thus, there is an earlier time $t' < t$ such that $\dot{\mathsf{B}}_i$ holds continuously between t' and t, and $(r, t') \models \neg \dot{\mathsf{D}}_j$. If $(r, t') \models O_j$, then $(r, t) \models \dot{\mathsf{B}}_i \, \mathcal{S} O_j$ and so $(r, t) \models i \vartriangleleft j$ holds by lemma 4. The assumption that $(r, t) \models i \not\vartriangleleft j$ implies that $(r, t) \models \neg(\dot{\mathsf{B}}_i \, \mathcal{S} O_j)$. In particular, $(r, t') \models \neg O_j$, and since $(r, t') \models \neg \dot{\mathsf{D}}_j$ we have that $(r, t') \models \dot{\mathsf{B}}_j$. Since (a) $\dot{\mathsf{B}}_i$ holds continuously between times t' and t, (b) $(r, t) \models \neg(\dot{\mathsf{B}}_i \, \mathcal{S} O_j)$, and (c) $(r, t') \models \dot{\mathsf{B}}_j$, it follows that $\dot{\mathsf{B}}_i \wedge \dot{\mathsf{B}}_j$ also holds continuously between times t' and t in r. In particular, $(r, t) \models \dot{\mathsf{B}}_i \wedge \dot{\mathsf{B}}_j$. Since $(r, t) \models i \not\vartriangleleft j$ holds by assumption, we thus obtain that $(r, t) \models \dot{\mathsf{B}}_i \wedge \dot{\mathsf{B}}_j \wedge \langle$number$[j], j\rangle <_L \langle$number$[i], i\rangle$, as claimed. $\qquad\square$

When $\dot{\mathsf{B}}_i \wedge \dot{\mathsf{B}}_j \wedge \langle$number$[j], j\rangle <_L \langle$number$[i], i\rangle$ holds, we have in particular that number$[j]$ is stable for as long as $\dot{\mathsf{B}}_j$ holds. If i later reads a larger value for number$[j]$ (so that test for $\langle$number$[i], i\rangle <_L \langle$number$[j], j\rangle$ succeeds), this will indicate that j has left the bakery region. But then $\dot{\mathsf{B}}_i \, \mathcal{S} O_j$ is true—i was in the bakery since j was outside. By lemma 4, we have $i \vartriangleleft j$ at that point. Line 8_j waits precisely until such a value is read by i. We thus have

Lemma 6. $i \vartriangleleft j$ holds whenever process i leaves line 8_j.

Proof. Suppose that process i leaves line 8_j at (r, t), having successfully completed the wait. Let $t' < t$ be the most recent time at which i completed 7_j. If $(r, t') \models i \vartriangleleft j$ then $(r, t) \models i \vartriangleleft j$ holds by corollary 1, since i does not leave the bakery between time t' and time t. Otherwise, $(r, t') \models \dot{\mathsf{B}}_i \wedge \dot{\mathsf{B}}_j \wedge \langle$number$[j], j\rangle <_L \langle$number$[i], i\rangle$ holds, by corollary 2. But line 8_j is completed after reading a value k for number$[j]$ that must be larger than the one for which $\langle \mathsf{number}[j], j \rangle <_{\mathrm{L}} \langle \mathsf{number}[i], i \rangle$ held at time t'. It follows that j must have left the bakery region at some time t'' in the range $t' < t'' \leq t$. Thus, $(r, t'') \models \dot{\mathsf{B}}_i \mathcal{SO}_j$, implying that $(r, t) \models \dot{\mathsf{B}}_i \mathcal{SO}_j$ as well, and so $(r, t) \models i \vartriangleleft j$, as claimed. $\square$

Theorem 1. *A process i can enter or be in the CS (i.e., be in line **9**) only if $i \vartriangleleft j$ holds for all $j \neq i$.*

Sketch of Proof.   Recall that the testing region of the algorithm on lines **6** to **8** is wholly contained in the bakery region. Hence, by corollary 1, if $i \vartriangleleft j$ holds in that region, it remains true as long as $\dot{\mathsf{B}}_i$ holds. By the time process i reaches line **9** it has completed line **8**$_j$ for all $j \neq i$. For each $j \neq i$, lemma 6 implies that $i \vartriangleleft j$ holds at some point when i is in the testing region. The claim follows. $\square$

By the antisymmetry of the ' $\vartriangleleft$ ' relation (lemma 3), at most one process can have priority over all others, and so theorem 1 immediately yields:

Corollary 3. *The Bakery algorithm guarantees mutual exclusion: At most one process is in the CS at any time.*

3.5   Liveness and Fairness of the Bakery Algorithm

In addition to the mutual exclusion property, the priority relation and our modified partition facilitate reasoning about other properties of the Bakery algorithm. For example, the algorithm is known to satisfy a form of FCFS (first-come first-serve) fairness. In [2,16,18], for example, the FCFS property shown is that if i enters (our) bakery region before j enters the doorway, which in our terminology means that $\dot{\mathsf{B}}_i \mathcal{SO}_j$ holds, then i will enter the critical section before j does. This follows immediately from lemma 4, corollary 1, and theorem 1. We can now state and prove slightly finer fairness properties of the Bakery algorithm:

Theorem 2 (Fairness). *In all runs r of the Bakery algorithm and times t:*

(a) *If $(r, t) \models \dot{\mathsf{B}}_i$ then no process j can enter the CS twice after time t in r before i enters the CS at least once, and*

(b) *If $(r, t) \models \dot{\mathsf{D}}_i$ then no process j can enter the CS three times after time t in r before i enters the CS at least once.*

Recall from the progress assumptions of section 3.1 that any process that leaves the non-critical section (line **3**) will reach the bakery in a finite amount of time, in a wait-free fashion. For completeness, we use this fact to state and prove a natural liveness condition for the Bakery algorithm as follows (similar proofs appear elsewhere; see, e.g., [3]):

Theorem 3. *If at least one process reaches the bakery region, then at least one process will enter the CS.*

4 *Boulangerie:* A Better Bakery Algorithm

In the testing region (lines **6-8**) of the Bakery algorithm process i detects that $i \lhd j$ holds, for each of the $j \neq i$. As long as i is unable to establish that $i \lhd j$ based on the checks in lines 7_j and 8_j, process i will block waiting for j to make progress. Our analysis showed how succeeding in the tests on both lines guarantees that $i \lhd j$ holds. We now wish to consider whether the blocking imposed by 7_j and 8_j is always justified.

Taking Advantage of Inconsistent Reads. Let us first consider the blocking imposed by line 8_j. As our analysis shows (in corollary 2), when 7_j is completed either $i \lhd j$ is already true, or both i and j are in the bakery, and j has a better ticket. Very roughly speaking, process i blocks on 8_j until it reads a value that contradicts the fact that j has a better ticket. Suppose, for example, that number$[i] = 10$ and that i reads a value of 5 for number$[j]$. It blocks correctly, since 5 could be the stable value that j has. Observe that number$[j]$ is a safe register, and so **read** operations on it may return arbitrary, and inconsistent, values when number$[j] = $ '?'. So now suppose that i performs another **read** on number$[j]$, and obtains a value, say 4 or 6, that is different from 5. This still corresponds to a better ticket than $\langle 10, i \rangle$ for j. But as long as j is in the bakery it performs no writes on number$[j]$. Thus, number$[j]$ is stable and all reads to it must return the same value. If i reads two different values for number$[j]$, it has proof that j was on the outside at some point. Thus, $i \lhd j$ is true, and i can stop blocking on j and move on to test the next process. We can avoid this case of unnecessary blocking as follows. When number$[j]$ is read for the second consecutive time or later on line 8_j, let previous[j] denote the previous value read by i from number$[j]$, or undefined if no such read has yet occurred. (See lines 8_ja to 8_je in the detailed version in the Appendix.) Then we can replace line 8_j in the Bakery algorithm by the following:

8_j **await** number$[j] = 0 \lor \langle$number$[i], i\rangle <_{\text{L}} \langle$number$[j], j\rangle \lor$
       number$[j] \neq$ previous$[j]$;

Optimizing for Low Contention. We now consider the blocking imposed by line 7_j. Roughly speaking, the justification for this blocking is that j may be in the doorway region, and it might come out of the region with a small number, and thus a winning ticket (at least one with priority over i). However, suppose that the testing process i has obtained number$[i] = 1$. Then the only processes j that can *ever* have a better ticket are ones whose ID is smaller than i (so that $\langle 1, j\rangle <_{\text{L}} \langle 1, i\rangle$). It follows that when number$[i] = 1$, there is no need to perform 7_j and 8_j for values $j > i$. To avoid this form of unnecessary blocking, we can replace line **6** of the Bakery algorithm by the following two lines:

6a   **if** number$[i] = 1$ **then** Limit $:= i - 1$ **else** Limit $:= N$;
6b   **forall the** $j \leq$ Limit **s.t.** $j \neq i$ **do**

While the values of number$[i]$ can grow without bound in the Bakery algorithm, we claim that the case of number$[i] = 1$ is not always a boundary case. Observe that such unbounded growth requires continuous contention for the critical section. Critical sections come in many flavors, and in many cases they do not experience continuous contention. Indeed, it is generally believed that contention for a critical section is rare in a well-designed system (see [13]). Notice that whenever all processes are in the non-critical region at once, even for a brief instant, their number values are all 0. The next process to leave the doorway will do so with a number value of 1 (others may attain the same number too). Therefore, when mutual exclusion is applied to a critical section that repeatedly experiences low contention, this optimization will repeatedly result in a reduction in the amount of blocking.[6]

We call the optimized variant of the Bakery algorithm that incorporates both changes the ***Boulangerie*** algorithm. Its full detailed description is given in the Appendix.

The optimization for the case of $\mathsf{number}[i] = 1$, which is beneficial under low contention, utilizes an aspect of priority that the Bakery algorithm admits, but does not try to detect. Namely, if $\mathsf{number}[i] = 1$ and $i < j$, then j will not be able to beat i to the CS. Since Boulangerie makes explicit use of this fact, we need to modify ' ⊲ ' slightly in order to capture the notion of priority that corresponds to the Boulangerie algorithm:

Definition 2. *The priority relation* i ◄ j *is said to hold exactly if* $(i \lhd j) \vee (\dot{\mathsf{B}}_i \wedge \neg \dot{\mathsf{B}}_j \wedge \mathsf{number}[i] = 1 \wedge i < j)$.

By definition 2, the new relation ' ◄ ' is strictly stronger than ' ⊲ '. When both processes are in the bakery region, the two relations coincide and reduce to lexicographic ordering. A very similar analysis as that used to establish lemma 3 and corollary 1 can show:

Lemma 7. *In the Boulangerie algorithm, both*

(a) *The priority relation* ' ◄ ' *is antisymmetric, and*
(b) *If* i ◄ j *holds, then it remains true as long as* $\dot{\mathsf{B}}_i$ *holds. Formally:* $\dot{\mathsf{B}}_i \, \mathcal{S} \, (i$ ◄ $j) \Rightarrow i$ ◄ j *is valid.*

In the full paper we show that the safety proof for our optimized algorithm, using the new priority relation ' ◄ ' is analogous to the proof for the Bakery algorithm presented in section 3.3 and section 3.4. We make some modifications to account for the new algorithm and the new relation.

4.1   Boulangerie has No Unnecessary Blocking

In a precise sense, Boulangerie strictly improves on the Bakery algorithm. Moreover, it achieves savings without needing to modify any of the `write`'s performed by the Bakery algorithm. Following the observation that the Bakery protocol causes processes to wait unnecessarily in some cases, we proposed Boulangerie as an optimization in which some of this waiting is avoided. It is natural, then, to ask whether the new protocol has unnecessary waiting. As we now show, the answer is No: Boulangerie does not suffer from unnecessary waiting.

Theorem 4. *Let* (r, t) *be a point of the Boulangerie algorithm in which process* i *is in the testing region, and assume that* i *has not completed line* $\mathbf{8}_j$ *since entering the*

---

[6] A particular case in which the savings with this optimization can be striking is when there are $N = 2$ processes. In this case, process 1 will not need to perform the testing region when $\mathsf{number}[1] = 1$.

bakery. If number$[i] > 1$ *or* $j < i$, *then based on the sequence of events it has seen thus far, process* i **does not know** *that* j *is out of the critical section at* (r, t).

Theorem 4 is stated in terms of what process i knows. This is a formal claim using the theory of knowledge in distributed systems (see [6,7]). Here we consider knowledge based on the agent's complete local history (the sequence of events that it has observed), to show that the protocol cannot be improved. The proof of theorem 4 is the main technical contribution of this paper, and is omitted from the proceedings version due to lack of space. It consists of an alternative-scenario argument showing that if the conditions of the theorem hold at a point (r, t), where r is a run of Boulangerie, then there is another point (r', t') satisfying that (a) r' is a run of Boulangerie, (b) process i has the same local history in r' up to the point at (r', t') as it does in r up to (r, t), and (c) $(r', t') \models CS_j$, i.e., process j is in the critical section at (r', t'). This implies that if process i enters the CS without completing line 8_j, then there is a run (the run r') in which it would do so when process j is in the CS, and violate the mutual exclusion property.

The scenario argument is shown by assuming that a point (r, t) satisfies the conditions of the theorem, and constructing a point (r', t') satisfying (a)-(c) above. The run r' coincides with r up to and including the latest time $t_0 < t$ at which i is in the non-critical section in r. From time t_0 on, r' is constructed so that process i obtains the same local history up to time t' as it does between time t_0 and t in r, while process j is in the CS at time t'. The detailed construction is made by way of a case analysis, depending on i's local history between (r, t_0) and (r, t). We leave the complete details to the full paper.

5 Conclusions

This paper offers a behind-the-scenes look at the workings of Lamport's Bakery algorithm in the somewhat challenging case of safe registers. We identified the priority relation that the algorithm implements and detects, and used it to formally capture the role of the main components of the algorithm. Based on the analysis, we were able to find two ways in which the Bakery algorithm admits unnecessary and potentially costly blocking. An improved version, called Boulangerie, fixes these flaws and is shown to contain no unnecessary blocking.

Reading from safe registers provides limited information about the state of the system, since a value can be read from such a register without ever being written to it. Indeed, when a test whether number$[j] = 0$ succeeds, this does *not* mean that number$[j] = 0$ was in fact true at some point during the **read** operation. As a result, it is tricky to interpret the tests performed on lines 7_j and 8_j of the Bakery algorithm.

As lemma 5 illustrates, a value read from a safe register can provide information about the register and about the state of its writer **by way of elimination**. Namely, if v is read, then no other value $w \neq v$ was stably written to the register throughout the time of the read operation. In particular, if a large value v is read, then it is not the case that a smaller value was stably written throughout the read. With the proper definition of a priority relation among the processes, these insights allow a process to conclude that it has priority over other processes by performing the tests on lines 7_j and 8_j (lemma 6).

Our analysis of the Bakery algorithm is based on the observation that every mutual exclusion algorithm breaks symmetry among the processes by implementing a priority relation of some sort. We saw that Lamport's original Bakery algorithm is based on the relation $i \vartriangleleft j$, which combines a lexicographic relation over the register values with a temporal condition. In particular, i has priority over j if i is in the bakery ever since j was on the "outside", before j ever entered the doorway. A slightly different relation, ' $\blacktriangleleft$ ', corresponds to the Boulangerie algorithm and to the Bakery algorithm for regular registers of [10].

This paper proposes that ME algorithms should be studied by considering the priority relations that they construct and detect. This essential aspect of ME, which has been in the background all along, can be brought to the fore. As our analysis proves in the case of the Bakery algorithm, even very familiar solutions can be seen in a new light, resulting in a better understanding as well as genuine improvements in the algorithm. Lamport reports that he has been able to machine-verify the Boulangerie algorithm by fairly simple modifications of his machine proof of the Bakery algorithm. We see this as both a testament to the quality of his verification tools and to the fact that the logic underlying Boulangerie is a refinement and direct improvement of the Bakery algorithm. We believe that the view of Mutual Exclusion as a matter of creating and detecting priority proposed in this paper promises to provide new insight to other existing ME algorithms, and lead to new ones.

References

1. Abraham, U.: Logical classification of distributed algorithms (Bakery Algorithms as an example). Theor. Comput. Sci. 412(25), 2724–2745 (2011)
2. Anderson, J.H.: Lamport on mutual exclusion: 27 years of planting seeds. In: Proceedings of the 20th ACM PODC Conference, pp. 3–12 (2001)
3. Attiya, H., Welch, J.: Distributed computing: fundamentals, simulations, and advanced topics, vol. 19 (2004)
4. Chaudhuri, K., Doligez, D., Lamport, L., Merz, S.: Verifying safety properties with the TLA+ proof system. In: Giesl, J., Hähnle, R. (eds.) IJCAR 2010. LNCS, vol. 6173, pp. 142–148. Springer, Heidelberg (2010)
5. Dijkstra, E.W.: Solution of a problem in concurrent programming control. Commun. ACM 8(9), 569 (1965)
6. Fagin, R., Halpern, J.Y., Moses, Y., Vardi, M.Y.: Reasoning about Knowledge (2003)
7. Halpern, J.Y., Moses, Y.: Knowledge and common knowledge in a distributed environment. Journal of the ACM 37(3), 549–587 (1990)
8. Hesselink, W.H.: Mechanical verification of Lamport's Bakery Algorithm. Science of Computer Programming 78(9), 1622 (2013)
9. Lamport, L.: A new solution of Dijkstra's concurrent programming problem. Commun. ACM 17(8), 453–455 (1974)
10. Lamport, L.: A new approach to proving the correctness of multiprocess programs. ACM Trans. Program. Lang. Syst. 1(1), 84–97 (1979)
11. Lamport, L.: What it means for a concurrent program to satisfy a specification: Why no one has specified priority. In: Proceedings of the 12th ACM POPL, pp. 78–83 (1985)
12. Lamport, L.: On Interprocess Communication. Part I: Basic Formalism. Distributed Computing 1(2), 77–85 (1986)

13. Lamport, L.: A fast mutual exclusion algorithm. ACM Trans. Comput. Syst. 5(1), 1–11 (1987)
14. Lamport, L.: A TLA+ mechanical proof of the Boulangerie Algorithm (2015). http://research.microsoft.com/en-us/um/people/lamport/tla/boulangerie.html
15. Lamport, L.: The TLA+ Hyperbook (2015). http://research.microsoft.com en-us/um/people/lamport/tla/hyperbook.html
16. Lynch, N.A.: Distributed Algorithms (1996)
17. Raynal, M., Beeson, D.: Algorithms for Mutual Exclusion (1986)
18. Rosenzweig, D., Börger, E., Gurevich, Y.: The bakery algorithm: yet another specification and verification. In: Börger, E. (ed.) Specification and Validation Methods, pp. 231–243 (1995)
19. Sedletsky, E., Pnueli, A., Ben-Ari, M.: Formal verification of the ricart-agrawala algorithm. In: Kapoor, S., Prasad, S. (eds.) FST TCS 2000. LNCS, vol. 1974, pp. 325–335. Springer, Heidelberg (2000)

Appendix

Detailed Boulangerie Algoritm for process i

---

0 **Initialize:** $num[i] = \text{number}[i] = 0$; choosing$[i] = tmp\_c = \textbf{false}$; Limit $= N$; $prev\_n = tmp\_n = \bot$;

1 **while true do**

2   non-critical section;

3   <u>**write**</u>(choosing$[i] \leftarrow$ **true**) ;

4.1   **forall the** $j \leq N$; $j \neq i$ **do** <u>**read**</u>($num[j] \leftarrow$ number$[j]$) ;

4.2 $num[i] \leftarrow 1 + \max\{num[1], \ldots, num[N]\}$;

4.3   <u>**write**</u>(number$[i] \leftarrow num[i]$) ;

5   <u>**write**</u>(choosing$[i] \leftarrow$ **false**) ;

6a   **if** number$[i] = 1$ **then** Limit $:= i - 1$ **else** Limit $:= N$;

6b   **forall the** $j \leq$ Limit **s.t.** $j \neq i$ **do**

7$_j$ **repeat** <u>**read**</u>($tmp\_c \leftarrow$ choosing$[j]$) **until** ($tmp\_c =$ **false**);

8$_j$a $tmp\_n \leftarrow \bot$;

8$_j$b   **repeat**

8$_j$c $prev\_n \leftarrow tmp\_n$;

8$_j$d   <u>**read**</u>($tmp\_n \leftarrow$ number$[j]$)

8$_j$e   **until** ($tmp\_n = 0 \vee \langle num[i], i \rangle <_{\text{L}} \langle tmp\_n, j \rangle \vee (tmp\_n \neq prev\_n \wedge prev\_n \neq \bot)$);

9   critical section;

10.1 $num[i] \leftarrow 0$;

10.2   <u>**write**</u>(number$[i] \leftarrow num[i]$) ;

The Computability of Relaxed Data Structures: Queues and Stacks as Examples

Nir Shavit[1] and Gadi Taubenfeld[2]

[1] MIT and Tel-Aviv University
[2] The Interdisciplinary Center, P.O. Box 167, Herzliya 46150, Israel
shanir@csail.mit.edu, tgadi@idc.ac.il

Abstract. Most concurrent data structures being designed today are versions of known sequential data structures. However, in various cases it makes sense to relax the semantics of traditional concurrent data structures in order to get simpler and possibly more efficient and scalable implementations. For example, when solving the classical producer-consumer problem by implementing a concurrent queue, it might be enough to allow the *dequeue* operation (by a consumer) to return and remove one of the two oldest values in the queue, and not necessarily the oldest one. We define infinitely many possible relaxations of several traditional data structures: queues, stacks and multisets, and examine their relative computational power.

Keywords: Relaxed data structure, consensus number, synchronization, wait-freedom, queue, stack, multiset.

1 Introduction

1.1 Motivation

Early in our computer science education, we learn how to implement sequential data structures. In the context of sequential data structures, implementing a queue in which it is fine for a dequeue operation to return one of the two oldest items in the queue, instead of always returning the oldest item, does not help in making the problem of efficiently implementing a queue easier to solve. Maybe for that reason, we sometimes tend to overlook the fact that in the context of concurrent programming, such relaxations might help a lot.

Assume that you need to solve the classical producer-consumer synchronization problem by implementing a concurrent queue. In some cases, it might be fine to allow the consumer to return and remove one of the two oldest items in the queue, and not necessarily the oldest one as is usually required. More generally, in some cases it makes senses to relax the semantics of traditional concurrent data structures in order to get more efficient and scalable concurrent implementations.

There is a trade-off between synchronization and the ability of an implementation to scale performance with the number of processors. Amdahl's law, implies that even a small fraction of inherently sequential code limits scaling. Using semantically weaker data structures may help in reducing the synchronization requirements and hence improves scalability for many-core systems. As a result, there is a recent trend towards

© Springer International Publishing Switzerland 2015
C. Scheideler (Ed.): SIROCCO 2015, LNCS 9439, pp. 414–428, 2015.
DOI: 10.1007/978-3-319-25258-2_29

implementing semantically weaker data structures for achieving better performance and scalability [18].

Important research has already been done on implementing semantically weaker data structure (see for example, [2,3,15,18,19]).While these implementations address complexity issues, less research has been done on the computability of relaxed data structures. In this paper we investigate the computability of (wait-free) relaxed data structures, by considering infinitely many possible relaxations of several traditional data structures: queues, stacks and multisets (i.e., bags), and examine their relative computational power. Our results demonstrate, for example, that for a concurrent queue small changes in its semantics dramatically effects its computational power, and that similar results do not apply for a concurrent stack.

1.2 Data Structures with Relaxed Specifications

We will assume that processes can try to access a shared object at the same time, however, although operations of concurrent processes may overlap, each operation should appear to take effect instantaneously. In particular, operations that do not overlap should take effect in their "real-time" order. This type of correctness requirement for shared objects is called *linearizability* [13].

A concurrent queue is a linearizable data structure that supports *enqueue*, *dequeue* and *peek* operations, by several processes, with the usual queue semantics.[1] Below we generalize this traditional notion of a concurrent queue.

A concurrent queue w.r.t. the numbers a, b and c, denoted $queue[a, b, c]$, is a linearizable data structure that supports the $enq.a(v)$, $deq.b()$ and $peek.c()$ operations, by several processes, with the following semantics: The $enq.a(v)$ operation inserts the value v at one of the a positions at the end of the queue;[2] the $deq.b()$ operation returns and removes one of the values at the b positions at the front of the queue; the $peek.c()$ operation returns one of the values at the c positions at the front of the queue without removing it. If the queue is empty the $deq.b()$ and the $peek.c()$ operations return a special symbol. We emphasize that the queue $queue[a, b, c]$, is implemented w.r.t. some fixed numbers a, b and c; these number are defined a priori and are *not* parameters that are passed at run time.

When defining the queue $queue[a, b, c]$, the numbers a, b and c can take the values of any positive integer, and the two special values 0 and $*$. When a, b or c equals 0, it means that the corresponding operation is not supported; when it equals $*$, it means that the corresponding operation can insert, remove or return (depending on the type of operation) a value at a random position (i.e., a position chosen by an adversary).

Thus, $queue[1, 1, 1]$ is the traditional FIFO queue (which is sometimes called augmented queue), where the values are dequeued in the order in which they were

---

[1] The enqueue operation inserts a value to the queue and the dequeue operation returns and removes the oldest value in the queue. That is, the values are dequeued in the order in which they were enqueued. The *peek* operation reads the oldest value in the queue without removing it. If the queue is empty the dequeue and the peek operations return a special symbol.

[2] A position of an item in a queue or in a stack is simply the number of items which precede it plus one.

enqueued, and where the *peek* operation reads the oldest value in the queue without removing it; $queue[1, 1, 0]$ is a queue which supports the standard *enqueue* and *dequeue* operations but does not support a *peek* operation; $queue[1, 1, *]$ is a queue where the *peek* operation returns a random value that is currently in the queue; finally $queue[*, *, 0]$ is exactly a linearizable *multiset* object that supports *insert* and *remove* operations, by several processes, with the usual multiset semantics.

Relaxed versions of other traditional data structures are defined similarly. A relaxed concurrent stack, denoted $stack[a, b, c]$, is a linearizable data structure that supports the $push.a(v)$, $pop.b()$ and $top.c()$ operations, by several processes, which the obvious semantics. The object $stack[*, *, 0]$ is equivalent to the object $queue[*, *, 0]$ and corresponds to a *multiset* object. The object $stack[1, 0, 1]$ is exactly an *atomic read/write register*, where the push and top operations correspond to the write and read operations, respectively. For $k \geq 1$, the object $stack[1, 0, k]$ is exactly a k-atomic register as defined in [21].

1.3   Consensus Numbers

The (binary) consensus problem is to design an algorithm in which all non-faulty processes reach a common decision based on their initial opinions. The problem is defined as follows: There are n processes $p_1, p_2, \ldots, p_n$. Each process p_i has an input value $x_i \in \{0, 1\}$. The requirements of the consensus problem are that there exists a *decision value* v such that: (1) each non-faulty process eventually decides on v, and (2) $v \in \{x_1, x_2, \ldots, x_n\}$. In particular, if all input values are the same, then that value must be the decision value.

The notion of a consensus number is central to our investigation and is formally defined below. A *wait-free* implementation of an object guarantees that any process can complete any operation in a finite number of steps, regardless of the speed of the other processes. A *register* is an object that supports read and write operations. With an *atomic* register, it is assumed that operations on the register (i.e, on the same memory location) occur in some definite order. That is, reading or writing an atomic register is an indivisible action.

The *consensus number* of an object of type o, denoted $CN(o)$, is the largest n for which it is possible to solve consensus for n processes in a wait-free manner using any number of objects of type o and any number of atomic registers. If no largest n exists, the consensus number of o is infinite (denoted ∞). Classifying objects by their consensus numbers is a powerful technique for understanding the relative computational power of shared objects.

The *consensus hierarchy* is an infinite hierarchy of objects such that the objects at level i of the hierarchy are exactly those objects with consensus number i. It is known that, in the consensus hierarchy, for any positive i, in a system with i processes: (1) no object at level less than i together with atomic registers can implement any object at level i; and (2) each object at level i together with atomic registers can implement any object at level i or at a lower level [10].

1.4   Contributions

New Definitions. The definitions of concurrent queues and stacks with relaxed specifications together with the following technical results provide a deeper understanding of the computability issues which are involved in the development of relaxed data structures.

Relaxing the Enqueue Operation. First we show that, while $CN(queue[1,1,1]) = \infty$, the consensus number drops to *two* when the *enqueue* operation is allowed to insert an item at any position at random, regardless whether the *peek* and *dequeue* operations are relaxed or not. That is,

$$CN(queue[*,1,1]) = 2. \tag{R1}$$

It follows from R1 and the known result that $CN(queue[*,*,0]) = 2$ (i.e., that the consensus number of a *multiset* object is 2), that: for every $b \in Z^+ \cup \{*\}, c \in Z^+ \cup \{0,*\} : CN(queue[*,b,c]) = 2$. ($Z^+$ is the set of all positive integers.) Next, we show that the consensus number of all the queues in which the *peek* operation is not relaxed (i.e., peek always returns the element at the front of the queue) is *infinity*, even when the *enqueue* operation is allowed to insert an item at any one of the last k positions for any fixed k. That is,

$$\text{For every } a \in Z^+ : CN(queue[a,0,1]) = \infty \tag{R2}$$

In contrast with R2, the consensus numbers of *all* possible relaxations of a concurrent *stack* are at most 2. In particular, $CN(stack[1,1,1]) = 2$ and $CN(stack[1,0,1]) = 1$ [7,10,16] (as already mentioned, the object $stack[1,0,1]$ is exactly an atomic read/write register).

Relaxing the Peek Operation. Next, we show that the consensus number of all the queues in which the *peek* operation is relaxed (i.e., peek is not required to always return the oldest value in the queue), is exactly two, regardless of how far the *enqueue* and *dequeue* operations are relaxed, as long as these operations are supported. That is,

$$CN(queue[1,1,2]) = 2. \tag{R3}$$

It follows from R3 and the known result that the consensus number of a *multiset* object is 2 [14], that: for every $a \in Z^+ \cup \{*\}, b \in Z^+ \cup \{*\}, c \neq 1 : CN(queue[a,b,c]) = 2$.

Not Supporting the Dequeue Operation. The situation changes dramatically when dequeue is not supported. The consensus number of all the queues where the dequeue operation is not supported *and* the peek operation is slightly relaxed, is just 1. That is,

$$CN(queue[1,0,2]) = 1. \tag{R4}$$

Thus, while $CN(queue[1,0,1]) = \infty$ and $CN(queue[1,1,2]) = 2$, by removing the dequeue operation from the object $queue[1,1,2]$, we get an object with consensus number one. It follows from R4 that: for every $a \in Z^+ \cup \{0,*\} : CN(queue[a,0,2]) = 1$.

Atomic Registers vs. Relaxed Queues. It is known that $CN(atomic\ register) = 1$ [16]. It is easy to see that a $queue[*, 0, 2]$ has a trivial wait-free implementation from a single atomic register. While, for every $a \in Z^+$, atomic registers and $queue[a, 0, 2]$ both have consensus number 1, we observe that,

> A $queue[a, 0, c]$ has no wait-free implementation from atomic registers,
> for every two positive integers a and c.   (R5)

The above results hold for both an initialized queue and an uninitialized queue. These two cases differ, for example, when the enqueue operation is not supported. For an initialized queue, $CN(queue[0, 1, 0]) = 2$ [10], while for an uninitialized queue, it is obvious that $CN(queue[0, 1, 1]) = 1$.

1.5 Related Work

The design of concurrent data structures has been extensively studied [9,20]. However, there are limitations in achieving high scalability in their design [4,6]. Two progress conditions that have been proposed for data structures which avoid locking are wait-freedom [10] (defined earlier), and obstruction-freedom [11]. Obstruction-freedom, guarantees that an active process will be able to complete its pending operations in a finite number of its own steps, if all the other processes "hold still" long enough.

It is shown in [4] that the worst-case operation time complexity of obstruction-free implementations is high, even in the absence of step contention. In [6], an $\Omega(n)$ lower bound is proven on the time to perform a single instance of an operation in any implementation of a large class of data structures shared by n processes, such as counters, stacks, and queues. It is suggested in [6] that "it might be beneficial to replace linearizable implementations of strongly ordered data structures, such as stacks and queues, with more relaxed data structures, such as pools and bags".

In [18], it is pointed out that concurrent data structures will have to go through a substantial "relaxation process" in order to support scalability: "The data structures of our childhood – stacks, queues, and heaps – will soon disappear, replaced by looser *unordered* concurrent constructs based on distribution and randomization". A few examples are given in [18] showing how relaxing a stack's LIFO ordering guarantees can result in higher performance and greater scalability.

Another approach to weaken the requirement of traditional data structures is not to change at all the definition of the data structures, but rather to relax the traditional correctness requirements. A tutorial which describes many issues related to memory consistency models can be found in [1]. In the context of relaxing the consistency condition linearizability [13], two relaxations of a queue were presented in [3]. In [15], a k-FIFO queue was implemented, which may dequeue elements out of FIFO order up to a constant $k \geq 0$. There are various implementations of relaxed data structures where insertion-order is of no importance, such as pools and bags, see for example [2,19]. In [8], a systematic and formal framework is presented for obtaining new data structures by quantitatively relaxing existing ones.

The impossibility result that there is no consensus algorithm that can tolerate even a single crash failure was first proved for the asynchronous message-passing model in

[7], and later has been extended for the shared memory model with atomic registers in [16]. The impossibility result that, for $1 \leq k \leq n - 1$ there is no k-resilient k-set-consensus algorithm for n processes using atomic registers, is from [5,12,17] (set-consensus is defined in section 6). It is shown in [10] that traditional data types, such as sets which support insert and remove operations, queues which support enqueue and dequeue operations (i.e., $queue[1, 1, 0]$), stacks which supports push and pop operations (i.e., $stack[1, 1, 0]$), all have consensus number exactly two. In the proofs of [10], it is assumed that the data structures are initialized. The same results also hold for the case of uninitialized data structures [14,20]. It is trivial to show that $CN(queue[1, 1, 1]) = \infty$.

For $k \geq 1$, the object $stack[1, 0, k]$ is exactly a k-atomic register [21]. In [21], it is shown that, for every $k \geq 1$, an atomic register can be implemented from k-safe bits. This result implies that, for every $k \geq 1$, a $stack[1, 0, 1]$ object can be implemented from $stack[1, 0, k]$ objects.

2 Preliminaries

2.1 Model of Computation

Our model of computation consists of an asynchronous collection of $n \geq 2$ processes that communicate via shared objects. We use P to denote the set of all processes. An *event* corresponds to an atomic step performed by a process. For example, the events which correspond to accessing registers are classified into two types: read events which may not change the state of the register, and write events which update the state of a register but do not return a value. We use the notation e_p to denote an instance of an arbitrary event at a process p.

A *run* is a pair (f, R) where f is a function that assigns initial states (values) to the objects and R is a finite or infinite sequence of events. An implementation of an object from a set of other objects, consists of a non-empty set C of runs, a set P of processes, and a set of shared objects O. For any event e_p at a process p in any run in C, the object accessed in e_p must be in O. Let $x = (f, R)$ and $x' = (f', R')$ be runs. Run x' is a *prefix* of x (and x is an *extension* of x'), denoted $x' \leq x$, if R' is a prefix of R and $f = f'$. When $x' \leq x$, $(x - x')$ denotes the suffix of R obtained by removing R' from R. Let $R; T$ be the sequence obtained by concatenating the finite sequence R and the sequence T. Then $x; T$ is an abbreviation for $(f, R; T)$.

Process p is *enabled* at the end of run x if there exists an event e_p such that $x; e_p$ is a run. For simplicity, whenever we say that p is enabled at x we mean that p is enabled at *the end* of x. Also, we write xp to denote either $x; e_p$ when p is enabled in x, or x when p is not enabled in x. Register r is a *local* register of p if only p can access r. For any sequence R, let R_p be the subsequence of R containing all events in R which involve p. Runs (f, R) and (f', R') are *indistinguishable* for p, denoted by $(f, R)[p](f', R')$, iff $R_p = R'_p$ and $f(r) = f'(r)$ for every local register r of p.

The runs of an asynchronous implementation of an object must satisfy several properties. For example, if a *write* event which involves p is enabled at run x, then the same event is enabled at any finite run that is indistinguishable to p from x. In the following proofs, we will implicitly make use of few such straightforward properties.

2.2 Three Simple Observations

The following lemmas are easy consequences of the above properties and definitions.

Lemma 1. *Let w, x and y be runs of an algorithm and p be a process such that (1) $w \leq x$ and $w[p]y$, and (2) the states of all the objects (local and shared) that p can access are the same in w and y, and $(x - w)$ contains only events of p. Then, $z = y$; $(x - w)$ is a run of the algorithm and $x[p]z$.*

Proof. By induction on the length of $(x - w)$.   □

Next, we state two simple lemmas regarding relaxed queues. The first states that in any component, going from $a \in Z^+$ to $a + 1$ or to $*$ does not increase the power of the object since it just gives the adversary more choices of what to return. The second lemma states that going from $a \in Z^+ \cup \{0, *\}$ to 0 in any component does not increase the power of the object, since it just eliminates a possible operation.

Lemma 2. *For every $a_1, b_1, c_1, a_2, b_2, c_2$ in $Z^+ \cup \{0, *\}$,*
if $((a_2 = * \wedge a_1 \neq 0) \vee 0 < a_1 \leq a_2 \vee a_2 = 0) \wedge ((b_2 = * \wedge b_1 \neq 0) \vee 0 < b_1 \leq b_2 \vee b_2 = 0) \wedge ((c_2 = * \wedge c_1 \neq 0) \vee 0 < c_1 \leq c_2 \vee c_2 = 0)$
then $CN(queue[a_1, b_1, c_1]) \geq CN(queue[a_2, b_2, c_2])$.

Proof. The proof of the lemma follows immediately from the definitions.   □

Lemma 3. *For every a, b, c in $Z^+ \cup \{0, *\}$,*

1. $CN(queue[0, b, c]) \leq CN(queue[a, b, c])$, *and*
2. $CN(queue[a, 0, c]) \leq CN(queue[a, b, c])$, *and*
3. $CN(queue[a, b, 0]) \leq CN(queue[a, b, c])$.

Proof. The proof of the lemma follows immediately from the definitions.   □

2.3 Known Results

Lemma 4. *(a) $CN(queue[*, *, 0]) = 2$, (b) $CN(queue[1, 1, 0]) = 2$, and (c) $CN(stack[1, 1, 1]) = 2$.*

The proofs that $CN(queue[*, *, 0]) = 2$, $CN(queue[1, 1, 0]) = 2$, and $CN(stack[1, 1, 0]) = 2$ (with and without initialization) are from [10,14]. The wait-free consensus algorithm which uses a single queue and registers from [14], is also correct when the queue is replaced with a stack or with a multiset. Proving that $CN(stack[1, 1, 1]) = 2$, can be establish by modifying the existing proof from [10], that $CN(queue[1, 1, 0]) = 2$.[3]

---

[3] To our surprise, we could not find any publication in which it is claimed that $CN(stack[1, 1, 1]) = 2$. Nevertheless, we consider it as a known result.

3 Basic Properties of Wait-Free Consensus Algorithms

The first four lemmas below are known and have appeared (using different notations) or follow from known impossibility proofs for wait-free consensus. The definitions below refer to runs of a given consensus algorithm. A (finite) run x is v-*valent* if in all extensions of x where a decision is made, the decision value is v ($v \in \{0,1\}$). A run is *univalent* if it is either 0-valent or 1-valent, otherwise it is *bivalent*. We say that two univalent runs are *compatible* if they have the same valency, that is, either both runs are 0-valent or both are 1-valent. A run is *critical* if: (1) it is bivalent, and (2) any extension of the run is univalent. A run (f, R) is an *empty* run if the length of R is 0 (that is, no process has taken a step yet). Recall that $n \geq 2$.

Lemma 5. *In every wait-free consensus algorithm, if two univalent runs are indistinguishable for some process p, and the states of all the objects that p can access are the same at these runs, then these (univalent) runs must be compatible.*

Proof. Let w and y be univalent runs, such that $w[p]y$ and the states of all the objects (local and shared) that p can access are the same at w and y. By the wait-free property, w has an extension x such $x - w$ contains only events of process p, and p has decided in x. Let w be v-valent, for $v \in \{0,1\}$. Then p decide v in x. (The event in which p decides on v, may be implemented by p writing v into a special single-writer output register.) By Lemma 1, $z = y$; $(x - w)$ is a run of the algorithm such that $z[p]x$. Since p decides on v (i.e., p writes v to its output register) in z, z is v-valent. Hence, since $y \leq z$, y must also be v-valent.   □

Lemma 6. *Every wait-free consensus algorithm has a bivalent empty run.*

Proof. We show that a bivalent empty run must exist. Assume to the contrary that every empty run is univalent. The empty run with all 0 inputs must be 0-valent, and similarly the empty run with all 1 inputs must be 1-valent. Thus, by Lemma 5, all the empty runs with all but one 0 inputs are 0-valent, and similarly all the empty runs with all but one 1 inputs are 1-valent. By repeatedly applying this argument i times we get that, all the empty runs with all but i 0 inputs are 0-valent, and similarly all the empty runs with all but i 1 inputs are 1-valent. Thus, when i is half the number of processes, we get that there are two empty runs x_0 and x_1 that differ only at the value of a single input, for process p, such that x_0 is 0-valent and x_1 is 1-valent. However, this contradicts Lemma 5. Hence, an empty bivalent run exists.   □

Lemma 7. *Every wait-free consensus algorithm has a critical run.*

Proof. Let $Cons$ be an arbitrary wait-free consensus algorithm. By Lemma 6, $Cons$ has an empty bivalent run x_0. We begin with x_0 and pursue the following round-robin *bivalence-preserving scheduling* discipline (Recall that P denotes a set of processes, x and y denote runs and yp is an extension of the run y by one event of process p):

```
1   x := x_0; P := ∅; i := 0                    /* initialization */
2   repeat
3           if x has a bivalent extension yp_i    /* which involves p_i */
```

```
4              then x := yp_i              /* bivalent extension of x */
5              else P := P ∪ {p_i}         /* no such bivalent extension */
6              i := i + 1(mod n)                        /* round-robin */
7  until |P| = n.
```

If the above procedure does not terminate, then there is an infinite run with only bivalent finite prefixes. However, the existence of such a run contradicts the definition of a wait-free consensus algorithm. Hence, the procedure will terminate with some critical run x.   □

Lemma 8. *Let x be a critical run of a wait-free consensus algorithm and let p and q be two different processes such that the runs xp and xq are not compatible. Then, in their next events from x, p and q are accessing the same object, and this object is not a register.*

Proof. We consider the following three possible cases, and show that each one of them leads to a contradiction. We will assume that in the last event in xp process p is accessing some object, say o, and in the last event in xq process q is accessing some object, say o'.

Case 1. $o \neq o'$. Since the next events from x of p and q are independent, $xpq[p]xqp$, and the values of all objects are the same in both xpq and xqp. Hence, by Lemma 5, xpq and xqp are compatible; since xpq is an extension of xp and xqp is an extension of xq, it must be that xp and xq are also compatible. A contradiction.

Case 2. $o = o'$ is a register and in xp the last event is a *write* event by p to o. Since p writes to o in its next operation from x, the value of o must be the same in xp and xqp. (Here we use the fact that the write by p overwrites the possible changes of o made by q.) Hence, $xp[p]xqp$ and the values of all the objects, which are not local to q, are the same in xp and xqp. By Lemma 5, xp and xqp are compatible. Since xqp is an extension of xq, it must be that xp and xq are also compatible. A contradiction.

Case 3. $o = o'$ is a register and in xp the last event is a *read* event by p. Thus, $xpq[q]xq$, and the values of all the objects, which are not local to p, are the same in both xpq and xq. Hence by Lemma 5, xpq and xq are compatible. Since xpq is an extension of xp, it must be that xp and xq are also compatible. A contradiction.

Thus, it must be the case that $o = o'$ and o is not a register.   □

Lemma 9. *Let x be a critical run of a wait-free consensus algorithm, and assume that the next event of p from x is a relaxed peek event which may return one of the two oldest items in a queue. Let xp^1 (resp. xp^2) denotes an extension of x by a peek event by p that has returned the oldest (resp. second oldest) item in a queue. Then, xp^1 and xp^2 are compatible.*

Proof. Let p and q be two different processes. Because the value the peek operation by p returns (i.e., the first or second) does not affect the state of the queue object visible to q, it follows that $xp^1[q]xp^2$ and and the states of all the objects that q can access are the same at these runs. Thus, by Lemma 5, xp^1 and xp^2 are compatible.   □

4   Relaxing the Enqueue Operation

It is obvious that $CN(queue[1, 0, 1]) = \infty$. Each process inserts its input value into the queue using an enqueue operation, and then uses a peek operation to find out what is the value at the front of the queue and decides on it. Also, it is obvious that, for an uninitialized queue, $CN(queue[0, 1, 1]) = 1$.[4] That is, a relaxed uninitialized queue where the enqueue operation is not supported is useless. Assume a queue object where only the enqueue operation may be relaxed. We show that only when the enqueue operation can insert a value at a random position, the consensus number drops to two; otherwise, in all other possible relaxations in which the enqueue operation is supported, the consensus number is not effected (i.e., it is ∞).

Theorem 1. $CN(queue[*, 1, 1]) = 2$.

Proof. It follows immediately from Lemma 2, Lemma 3, and Lemma 4(a) that $CN(queue[*, 1, 1]) \geq 2$. We prove that $CN(queue[*, 1, 1]) \leq 2$. A possible correct behavior of a $queue[*, 1, 1]$ object, is that every enqueue operation always inserts a data item at the head of the queue. In such a case, the $queue[*, 1, 1]$ object, behaves like a $stack[1, 1, 1]$ object. This implies that $CN(queue[*, 1, 1]) \leq CN(stack[1, 1, 1])$. Thus, by Lemma 4(c), $CN(queue[*, 1, 1]) \leq 2$. □

Corollary 1. *For every* $b \in Z^+ \cup \{*\}$, $c \in Z^+ \cup \{0, *\} : CN(queue[*, b, c]) = 2$.

Proof. The corollary follows from Lemma 2, Lemma 3, Lemma 4(a) and Theorem 1. □

Next we show that when the enqueue operation is relaxed but *can not* insert a value at a random position, the consensus number is infinity.

Theorem 2. *For every* $a \in Z^+ : CN(queue[a, 0, 1]) = \infty$.

Proof. For any given number $a \in Z^+$, we present a simple consensus algorithm for any number of processes using a singe $queue[a, 0, 1]$ object. Each process first enqueues its input value $a + 1$ times. Then, the process uses a peek operation to find out the value of the first item in the queue, and decides on that value. Clearly, once some process finishes to enqueue its input value $a + 1$ times, the value of the item at the head of the queue never changes. The result follows. □

Corollary 2. *For every* $a \in Z^+$, $b \in Z^+ \cup \{0, *\} : CN(queue[a, b, 1]) = \infty$.

Proof. The corollary follows immediately from Lemma 2, Lemma 3 and Theorem 2. □

5   Relaxing the Peek Operation

Assume a queue object where only the *peek* operation may be relaxed. We show that in *all* possible relaxations of the peek operation the consensus number drops (from infinity) to two.

---

[4] This is false, if the queue initially contains one element. In such a case, two processes can solve consensus, by deciding on the input of the process that successfully dequeues the element.

Theorem 3. $CN(queue[1, 1, 2]) = 2$.

Proof. It follows from Lemma 3 and Lemma 4(b) that $CN(queue[1, 1, 2]) \geq 2$. Below we prove that $CN(queue[1, 1, 2]) \leq 2$. By contradiction, assume that we have a wait-free consensus algorithm for *three* processes p, q and g using only $queue[1, 1, 2]$ objects and registers. By Lemma 7, the algorithm has a critical run x. By definition of a critical run, for two of the processes, say p and q, a run resulting by an extension of x by a single event of p and a run resulting by an extension of x by a single event of q are not compatible. Thus, by Lemma 8, in their next events from x, p and q are accessing the same object, which must be a $queue[1, 1, 2]$ object. By Lemma 9, if the next event of p (resp. q) from x is a relaxed peek event which may return one of the two oldest items in a queue, xp^1 and xp^2 (resp. xq^1 and xq^2) are compatible. Below, when the next event of p from x is a peek event, xp refers to xp^1 and xp^2.

Without loss of generality, we can assume the xp is 0-valent and xq is 1-valent. Since xp is 0-valent also xpq is 0-valent. Since xq is 1-valent also xqp is 1-valent. Thus, xpq and xqp are not compatible. Next, we consider all the possible cases, regarding the next two events of p and q from x and show that each one of these cases leads to a contradiction.

Case 1. Both events are *peek* events. Because a peek operation does not have any effect on the state of a $queue[1, 1, 2]$ object, it follows that $xpq[g]xqp$ and the states of all the objects that g can access are the same at these runs. Thus, by Lemma 5, xpq and xqp must be compatible, a contradiction. Notice that we do not really care what value a peek operation returns (i.e., the oldest or second oldest), since this will not affect the state of the object visible to g.

Case 2. Exactly one of the two events is a *peek* event. Because the peek operation does not have any effect on the state of a $queue[1, 1, 2]$ object and the other operation has the same effect in both xpq and xqp, it follows that $xpq[g]xqp$ and the states of all the objects that g can access are the same at these runs. Thus, by Lemma 5, xpq and xqp must be compatible, a contradiction. Notice that again we do not really care what value the peek operation returns.

Case 3. Both events are *dequeue* events. In the last two events in xpq and xqp the same two items were removed from the queue, thus, $xpq[g]xqp$ and the states of all the objects that g can access are the same at these runs. Thus, by Lemma 5, xpq and xqp must be compatible, a contradiction.

Case 4. One event is a *enqueue* and the other is a *dequeue*. Assume w.l.o.g. that the enqueue event is by p and the dequeue event is be q. If the queue is nonempty, the two events commute since each operates on a different end of the queue. Thus, xpq and xqp are indistinguishable for all the processes and the states of all the objects is the same in xpq and xqp, and thus by Lemma 5 the contradiction is immediate. If the queue is empty, $xp[g]xqp$ and and the states of all the objects that g can access are the same at these runs. Thus, by Lemma 5, xp and xqp must be compatible, a contradiction.

Case 5. Both events are *enqueue* events. Assume that p enqueues the value v_p and q enqueues the value v_q. Consider the runs xpq and xqp. The valency of each one of these two runs is determined by the process that has taken the first step from x. If p or q runs uninterrupted starting from either xpq or xqp, the only way for each one of them to observe the queue's state is via a *dequeue* or a *peek* operation. However, since the peek operation can return one of the first two items at the head of the queue, a *peek* can not be used to determine which process enqueue operation was first. That is, once the values v_p and v_q are at the head of queue, a peek operation by p can always return v_p, and a peek operation by q can always return v_q. Thus, the only way for a process to determine which process went first is via dequeue operations. Next we consider the following two extensions of xpq and xqp.

- Let y be an extension of xpq that results from the following execution: Starting from x let p enqueue v_p and then let q enqueue v_q. Run p uninterrupted until it dequeues v_p (as explained above this is the only way for p to observe which process went first). Then, run q uninterrupted until it dequeues v_q.
- Let y' be an extension of xqp that results from the following execution: Starting from x let q enqueue v_q and then let p enqueue v_p. Run p uninterrupted until it dequeues v_q Then, run q uninterrupted until it dequeues v_p.

Since y is and extension of xpq, y is 0-valent, and since y' is and extension of xqp, y' is 1-valent. Clearly, $y[g]y'$ and the states of all the objects that g can access are the same at these runs. Thus, by Lemma 5, y and y' must be compatible, a contradiction.   □

Corollary 3. *For every* $a \in Z^+ \cup \{*\}, b \in Z^+ \cup \{*\}, c \neq 1 : CN(queue[a, b, c]) = 2.$

Proof. The corollary follows from Lemma 2 and Lemma 4(a) and Theorem 3.   □

6 Not Supporting the Dequeue Operation

The consensus number of all the queues where the dequeue operation is not supported *and* the peek operation is relaxed, is just 1. Put another way, while $CN(queue[1, 0, 1])$ $= \infty$ and $CN(queue[1, 1, 2]) = 2$, by removing the dequeue operation from the object $queue[1, 1, 2]$, we get an object with consensus number one. That is,

Theorem 4. $CN(queue[1, 0, 2]) = 1.$

Proof. By contradiction, assume that we have a wait-free consensus algorithm for *two* processes p and q using only $queue[1, 0, 2]$ objects and registers. By Lemma 7, the algorithm has a critical run x. By definition of a critical run, a run resulting by an extension of x by a single event of p and a run resulting by an extension of x by a single event of q are not compatible. By Lemma 8, in their next events from x, p and q are accessing the same object, which must be a $queue[1, 0, 2]$ object. By Lemma 9, if the next event of p (resp. q) from x is a relaxed peek event which may return one of the two oldest items in a queue, xp^1 and xp^2 (resp. xq^1 and xq^2) are compatible. Below, when the next event of p from x is a peek event, xp refers to xp^1 and xp^2.

Without loss of generality, we assume the xp is 0-valent and xq is 1-valent. Since xp is 0-valent also xpq is 0-valent. Since xq is 1-valent also xqp is 1-valent. Thus, xpq

and xqp are not compatible. Next, we consider all the possible cases, regarding the next events of p and q from x and show that each one of these cases leads to a contradiction.

Case 1. Both events are *peek* events. Because a peek operation does not have any effect on the states of the $queue[1, 1, 2]$ object, it follows that $xp^1[p]xqp^1$ and the states of all the objects that p can access are the same at these runs. Thus, by Lemma 5, xp^1 and xqp^1 must be compatible, a contradiction. We do not really care what value a peek operation by q returns since this will not affect the state of the object visible to p.[5]

Case 2. Exactly one of the two events is a *peek* event. Assume w.l.o.g. that the peek event is by process q. Because the peek operation does not have any effect on the states of the $queue[1, 0, 2]$ object and the operation by p has the same effect in both xp and xqp, it follows that $xp[p]xqp$ and the states of all the objects that p can access are the same at these runs. Thus, by Lemma 5, xp and xqp must be compatible, a contradiction. Notice that again we do not really care what value the peek operation returns.

Case 3. Both events are *enqueue* events. Assume that p enqueues the value v_p and q enqueues the value v_q. Consider the 0-valent run xpq and the 1-valent run xqp. The valency of each one of these two runs is determined by the process that has taken the first step from x. If p or q runs uninterrupted starting from either xpq or xqp, the only way for each one of them to observe the queue's state is via a *peek* operation. Since the peek operation can return one of the first two items at the head of the queue, a *peek* can not be used to determine which process enqueue operation was first.

More precisely:

1. If the queue is *not* empty at x then after the two enqueue events by p and q, the adversary can force every peek event to always return the item at the head of the queue, and thus it is not possible for p or q to decide which process enqueue event was first.

2. If the queue is empty at x then after the two enqueue events by p and q, the values v_p and v_q are at the head of queue (in some order). Now the adversary can force every peek operation by p to always return v_p, and every peek operation by q can always return v_q. Thus, again, it is not possible to decide which process enqueue event was first.

Thus, it can not be that both events are enqueue events, a contradiction. □

Corollary 4. *For every* $a \in Z^+ \cup \{0, *\}$, $c \geq 2 : CN(queue[a, 0, c]) = 1$.

Proof. The corollary follows from Lemma 2, Lemma 3 and Theorem 4. □

7   Atomic Registers vs. Relaxed Queues

It is known that $CN(atomic\ register) = 1$ [16]. It is easy to see that a $queue[*, 0, 2]$ has a trivial wait-free implementation from a single atomic register, which raises the question whether also $queue[1, 0, 2]$ has a wait-free implementation from atomic registers. The answer to this question is negative. We prove the following general result:

---

[5] Notice that we reach a contradiction, by assuming that p's peek operation returns the first element in both passes. Since this implies that it cannot be the case that both events by p and q are *peek* events, there is no need to consider the sub-case where p's peek operation returns the first element in one path and the second element is the other path.

Theorem 5. *A queue$[a, 0, c]$ has no wait-free implementation from atomic registers, for every two integers $a \geq 1$ and $c \geq 1$.*

Proof. The (n, k)-*set consensus problem* is to find a solution for n processes, where each process starts with an input value from some domain, and must choose some participating process' input as its output. All n processes together may choose no more than k distinct output values. An (n, k)-set consensus object (or algorithm) is an object which solves the (n, k)-set consensus problem. One of the most celebrated impossibility results in distributed computing is that, for any $1 \leq k < n$, a wait-free (n, k)-set consensus object can not be implemented using any number of wait-free $(n, k + 1)$-set consensus objects and atomic registers [5,12,17].

We observe that, for any $1 \leq k < n$, a wait-free (n, k)-set consensus object has a simple wait-free implementation using a single (initially empty) queue$[a, 0, c]$ object where a and c are positive integers and $k = a+c-1$, as follows. Each process p_i inserts its input value v_i into the queue using an $enq.a(v_i)$ operation, and then uses a $peek.c()$ operation to find a value in one of the c positions at the front of the queue, and decides on it. During the execution any one of the $a + c - 1$ values that are inserted first into the queue can occupy (at some point in time) one of the c positions at the front of the queue. Any value that is inserted later will never occupy one of the c positions at the front of the queue. Thus, processes together will never choose more than $k = a+c-1$ distinct output values. Since, for every two positive integers a and c, it is possible to solve in a wait-free manner the $(a + c, a + c - 1)$-set consensus problem using a queue$[a, 0, c]$ object, but it is not possible to solve it in a wait-free manner using atomic registers, the result follows. $\square$

8 Discussion

Synchronization inherently limits parallelism. As a result, there is a recent trend towards implementing semantically weaker data structures which reduce the need for synchronization and thus achieve better performance and scalability. We have considered infinitely many possible relaxations of queues and stacks, and examined their relative computational power by determining their consensus numbers.

Our results demonstrate, somewhat surprisingly, that each one of the infinitely many relaxed objects considered has one of the following three consensus numbers: 1, 2 or ∞. Another conclusion is that a queue is more sensitive than a stack to changes in its semantics. It would be interesting to extend our results to other data structures.

It would be interesting to find out the internal structure among relaxed objects in the same level of the consensus hierarchy. In particular, for $i \in Z^+$, is it possible to implement a queue$[1, 1, i+1]$ object using queue$[1, 1, i+2]$ objects and registers? Is it possible to implement a queue$[1, 1, 2]$ object using queue$[1, 1, 0]$ objects and registers?

Acknowledgements. Support is gratefully acknowledged from the National Science Foundation under grants CCF-1217921, CCF-1301926, and IIS-1447786, and the Department of Energy under grant ER26116/DE-SC0008923.

References

1. Adve, S.V., Gharachorloo, K.: Shared memory consistency models: A tutorial. IEEE Computer 29(12), 66–76 (1996)
2. Afek, Y., Korland, G., Natanzon, M., Shavit, N.: Scalable producer-consumer pools based on elimination-diffraction trees. In: D'Ambra, P., Guarracino, M., Talia, D. (eds.) Euro-Par 2010, Part II. LNCS, vol. 6272, pp. 151–162. Springer, Heidelberg (2010)
3. Afek, Y., Korland, G., Yanovsky, E.: Quasi-linearizability: Relaxed consistency for improved concurrency. In: Lu, C., Masuzawa, T., Mosbah, M. (eds.) OPODIS 2010. LNCS, vol. 6490, pp. 395–410. Springer, Heidelberg (2010)
4. Attiya, H., Guerraoui, R., Hendler, D., Kuznetsov, P.: The complexity of obstruction-free implementations. J. ACM 56(4), 1–33 (2009)
5. Borowsky, E., Gafni, E.: Generalizecl FLP impossibility result for t-resilient asynchronous computations. In: Proc. 25th ACM Symp. on Theory of Computing, pp. 91–100 (1993)
6. Ellen, F., Hendler, D., Shavit, N.: On the inherent sequentiality of concurrent objects. SIAM Journal on Computing 41(3), 519–536 (2012)
7. Fischer, M.J., Lynch, N.A., Paterson, M.S.: Impossibility of distributed consensus with one faulty process. Journal of the ACM 32(2), 374–382 (1985)
8. Henzinger, T., Kirsch, C., Payer, H., Sezgin, A., Sokolova, A.: Quantitative relaxation of concurrent data structures. SIGPLAN Not 48(1), 317–328 (2013)
9. Herlihy, M., Shavit, N.: The Art of Multiprocessor Programming, p. 508. Morgan Kaufmann Publishers (2008)
10. Herlihy, M.P.: Wait-free synchronization. ACM Trans. on Programming Languages and Systems 13(1), 124–149 (1991)
11. Herlihy, M.P., Luchangco, V., Moir, M.: Obstruction-free synchronization: Double-ended queues as an example. In: Proc. of the 23rd International Conference on Distributed Computing Systems, p. 522 (2003)
12. Herlihy, M.P., Shavit, N.: The topological structure of asynchronous computability. Journal of the ACM 46(6), 858–923 (1999)
13. Herlihy, M.P., Wing, J.M.: Linearizability: a correctness condition for concurrent objects. TOPLAS 12(3), 463–492 (1990)
14. Jayanti, P., Toueg, S.: Some results on the impossibility, universality, and decidability of consensus. In: Rozenberg, G. (ed.) APN 1993. LNCS, vol. 674, pp. 69–84. Springer, Heidelberg (1993)
15. Kirsch, C.M., Payer, H., Röck, H., Sokolova, A.: Performance, scalability, and semantics of concurrent FIFO queues. In: Xiang, Y., Stojmenovic, I., Apduhan, B.O., Wang, G., Nakano, K., Zomaya, A. (eds.) ICA3PP 2012, Part I. LNCS, vol. 7439, pp. 273–287. Springer, Heidelberg (2012)
16. Loui, M.C., Abu-Amara, H.: Memory requirements for agreement among unreliable asynchronous processes. Advances in Computing Research 4, 163–183 (1987)
17. Saks, M., Zaharoglou, F.: Wait-free k-set agreement is impossible: The topology of public knowledge. SIAM Journal on Computing 29 (2000)
18. Shavit, N.: Data structures in the multicore age. Communications of the ACM 54(3), 76–84 (2011)
19. Sundell, H., Gidenstam, A., Papatriantafilou, M., Tsigas, P.: A lock-free algorithm for concurrent bags. In: Proc. of the Twenty-Third Annual ACM Symposium on Parallelism in Algorithms and Architectures, SPAA 2011, pp. 335–344 (2011)
20. Taubenfeld, G.: Synchronization Algorithms and Concurrent Programming, 423 p. Pearson / Prentice-Hall (2006). ISBN 0-131-97259-6
21. Taubenfeld, G.: Weak read/write registers. In: Frey, D., Raynal, M., Sarkar, S., Shyamasundar, R.K., Sinha, P. (eds.) ICDCN 2013. LNCS, vol. 7730, pp. 423–427. Springer, Heidelberg (2013)

Comparison-Based Interactive Collaborative Filtering

Yuval Carmel and Boaz Patt-Shamir*

School of Electrical Engineering, Tel Aviv University, Tel Aviv, Israel

Abstract. We study the interactive model of comparison-based collaborative filtering. Each *player* prefers one *object* from each pair of objects. However, revealing what is a player preference between two objects can be done only by asking the player specifically about that pair, an action called *probing*. The goal is to (approximately) reconstruct the players' preferences with the smallest possible number of probes per player. The per-player number of probes can be reduced if there are many players who share a similar taste, but a priori, players do not know who to collaborate with. In this paper, we present the model of comparison-based interactive collaborative filtering, analyze a few possible taste models and present distributed algorithms whose output is close to the best possible approximation to the players' taste.

Is there anyone so wise as to learn by the experience of others?　　　*Voltaire*

1 Introduction

Recommendation systems have become a significant part of our lives in the past few years. Most people encounter recommendation systems on a daily basis, while buying a book, choosing which movie to watch, buying groceries in the supermarket, or even finding a life mate. Collaborative Filtering is one of the prevalent approaches to recommendation systems, especially large scale systems (such as Netflix [7]). The idea in collaborative filtering is to take advantage of the existence of many players with similar preferences which can collaborate by sharing the load of trying the objects and identifying objects they perceive as good.

Following Drineas et al. [10], we distinguish between *interactive* and *non-interactive* recommendation systems, which differ in assumption and usage. In non-interactive recommendation systems, the algorithm is fed all known preferences as collected from the users in the past, and the goal is to output (possibly few) unknown preferences. This model is very popular, and conceptually easy to implement, but it does not take into account the dynamics of the system after the output is made.

---

* Supported in part by the Israel Science Foundation (grant No. 1444/14).

© Springer International Publishing Switzerland 2015
C. Scheideler (Ed.): SIROCCO 2015, LNCS 9439, pp. 429–443, 2015.
DOI: 10.1007/978-3-319-25258-2_30

In interactive recommendation systems (called "competitive" in [10]), on the other hand, while the goal remains to predict preferences, it is assumed that no preference is known a priori, and the focus is on the evolution of the system. Specifically, an interactive algorithm proceeds by asking players to reveal specific preferences in an action called *probe*. The results of probes not only determine the predictions the algorithm makes, but also determine the identity of future objects to be probed. The goal of interactive algorithms is to predict players' preferences while minimizing the number of probes, since probing is assumed to be costly.

The interactive model is as follows (cf. Section 2). There are n players and m objects; each player has his[1] preferences over the objects, represented by his *preference vector* or *taste*. An entry in the vector of a player can be revealed only by asking that player to preform the appropriate probe. Probe results are assumed to be posted on a shared "billboard" (modeling eBay feedback records, IMDb reviews, etc.), so that each player can run his algorithm to find which probe to do next, as well as compute preference predictions.

The existence of a billboard does not solve the problem, since players still need to decide whose results to adopt. We assume that some tastes are popular, namely many players have them. Concretely, given a *popularity factor* $0 < \alpha \leq 1$, and a *distance parameter* $D \geq 0$, we say that a preference vector v_j is (α, D)-*popular* if there are at least αn players whose preference vectors differ from v_j by at most D entries. Note that in order to reconstruct any taste (preference vector), $\Omega(m)$ probes are required just to cover all objects, and hence, to reconstruct his preferences to within $O(D)$ errors, the average number of probes per player with an (α, D)-popular taste cannot be less than $\Omega(\frac{m-D}{\alpha n})$.

In most previous work, preferences are simply an absolute grade for each object. The grades are typically binary, with the interpretation of "the user likes/dislikes object a." In this paper we introduce a *comparison-based* interactive collaborative filtering model, where preferences are expressed only over pairs of objects, with the interpretation of "the user prefers object a over object b." In the interactive model, a comparison-based probe means that the user is presented with two objects, and responds with his preference between them (which may also include "equal" or "incomparable"). We note that it is well known that comparison-based preferences are more intuitive, consistent and accurate than absolute grading (we elaborate below). However, it is not quite clear what can be assumed about the structure of comparison-based preferences. In this paper we study a few simple user models. The simplest model is that the user preference between a pair of objects is independent of his preferences over other pairs, and possibly the most structured model is when the pairwise preferences are induced by an underlying total order over all objects. In between, one can consider pairwise preference induced by partial order.

Our Contribution. The main technical contribution of this paper is a comparison-based algorithm for reconstructing preferences induced by an underlying total order. First we present Algorithm $\mathcal{DP}$ (Section 3) for instances with distance

---

[1] Or her. For uniformity, we have arbitrarily chosen to refer to players as males.

parameter $D = 0$. With high probability, Algorithm $\mathcal{DP}$ reconstructs $(\alpha, 0)$-popular preference vectors exactly, incurring at most $O(\frac{1}{\alpha} \log n (\log \log n + \log \frac{1}{\alpha}))$ probes per player, assuming $m = n$.[2]

Our main result is Algorithm $\mathcal{DPD}$ (Section 4), that uses Algorithm $\mathcal{DP}$ as a subroutine, and solves problem instances with distance parameter $D > 0$ w.h.p. Algorithm $\mathcal{DPD}$ reconstructs (α, D)-popular preference vectors with at most $O(D)$ errors and using at most $O(\frac{D^2}{\alpha} \log^2 n (\log \log n + \log \frac{1}{\alpha}))$ probes per player. We also consider the case where each user perceives the object set as a disjoint union of a few *object categories*, such that objects within a category are totally ordered, but objects in different classes are incomparable. This model is appropriate in the case that the object set is eclectic, e.g., cars and restaurants. We show how to reconstruct the taste in this case without prior knowledge on the categorizations, and even when different users have different categorizations.

Related Work. Collaborative filtering is studied quite intensively, but mostly from the non-interactive perspective (see, e.g., [7, 11, 16]). The interactive model was introduced by Drineas et al. [10] (referred to there as "competitive recommendation systems"). In the absolute grade model, where a preference vector specifies a grade for each object, it was shown by Awerbuch et al. [5], that all $(\alpha, 0)$-popular preferences can be reconstructed exactly, using $O(\frac{1}{\alpha} \log n)$ user probes (assuming $m = \Theta(n)$ for simplicity). Alon et al. [4] extend this result with instances in which the distance parameter is $D > 0$ and provide a reconstruction for all (α, D)-popular users preferences using $O(\frac{1}{\alpha} D^{3/2} \log^2 n)$ user probes with at most $O(D)$ errors. In addition, [4] provides an algorithm which reconstructs all (α, D)-popular tastes with $O(D/\alpha)$ errors using $O(\log^{3.5} n / \alpha^2))$ user probes. Nisgav and Patt-Shamir [15] improve these results, presenting algorithms for reconstructing (α, D)-popular users preferences with $O(D)$ errors and probe complexity either $O(\frac{D}{\alpha} \log^2 n)$ or $O(\frac{1}{\alpha} \log^3 n)$.

Comparison-based recommendation systems are considered superior to absolute grading systems in stability and more natural for human interaction based ratings. For example, in an experiment held by Jones et al. [12], user preferences expressed by comparisons were measured as 20% more stable over time than preferences expressed by 5-star grading scale. Since a 10% increase in accuracy is considered significant (that was the goal of the million-dollar Netflix challenge [7]), a 20% reduction in inconsistency is clearly meaningful in this context. Moreover, users tend to prefer comparison-based grading, finding it more intuitive than 5-star rating-based grades [8, 12].

It is therefore not surprising that there are some proposals for comparison-based model recommendation systems. Loepp et al. [14] provide an interactive collaborative filtering algorithm which uses a priori knowledge about the objects to optimize the estimations of user preferences. Desarkar et al. [9] provide a comparison-based non-interactive algorithm to predict preferences using matrix factorization. Both [9] and [14] present experimental results, which show that comparison-based preferences perform well against other user correlation

---

[2] In general, the probe complexity results should be multiplied by $\lceil m/n \rceil$. For simplicity, we usually omit this factor and assume $m = n$.

methods in collaborative filtering algorithms. From the theoretical-algorithmic viewpoint, Ailon [1] gives an active learning algorithm to produce approximate ranking from pairwise preferences, which reconstructs a single vector (equivalent to a single taste) using $O(n \cdot \text{polylog}(n, \epsilon^{-1}))$ queries, where $0 < \epsilon < 1$ is the error tolerance parameter. However, to the best of our knowledge, the current paper is the first to present an interactive comparison-based recommendation system with worst-case guarantees.

Our Techniques. Our algorithms use ideas from the interactive non comparison-based algorithms of [3, 5]. We also use the parallel merge sort algorithm of Valiant [17] to speed up computation, and the approximation algorithm for minimum feedback arc set in tournaments from [2, 13] to obtain approximate total order.

Organization. The remainder of this paper is organized as follows. In Section 2 we define the comparison-based model and on the different possible user models. Then, focusing on the total order case, in Section 3 we give our algorithm $\mathcal{DP}$ for problem instances with distance parameter $D = 0$, and in Section 4 we give our second algorithm $\mathcal{DPD}$, which solves problem instances with distance parameter $D > 0$. In Section 5 we consider the case where objects can be classified into disjoint categories.

2   Preliminaries

Instances. An instance of the comparison-based reconstruction problem is defined as follows. There is a set P of n *players* (a.k.a. *users*), and a set O of m *objects*. For every two objects $i, i' \in O$ and player $j \in P$, there is a *comparison value* $c_j(i, i') \in \{-1, 0, 1, \perp\}$, called the *personal preference* of player j over objects i, i', interpreted as follows. If $c_j(i, i') = 1$ then player j prefers i over i', and if it is -1 then player j prefers i'; if $c_j(i, i') = 0$ then player j likes i and i' the same, and if $c_j(i, i') = \perp$ then player j cannot compare objects i and i'. We define the vector $v_j := (c_j(1, 2), c_j(1, 3), \ldots, c_j(m, m-1))$ of length $\binom{m}{2}$ to be the *preference vector* or *taste* of player j.

Distances and Popularity. Fix an instance of the problem, and let $O' \subseteq O$ be an object set. The *distance* between two players j, j' w.r.t O' is defined by

$$\text{dist}_{O'}(v_j, v'_j) = |\{(i, i') : i, i' \in O' \wedge c_j(i, i') \neq c'_j(i, i')\}| .$$

Namely, $\text{dist}_{O'}(v_j, v'_j)$ is the number of object pairs from O' on which j and j' disagree. A *taste* is a vector v containing a comparison value $c(i, i')$ for each pair $(i, i') \in O$. Given $0 < \alpha \leq 1$ and $D \geq 0$, a taste v is (α, D)-*popular* if there are at least αn players, whose taste is at distance at most D from v. A player is (α, D)-*popular* if his taste is (α, D)-popular. Note that for any taste v and $0 \leq \alpha \leq 1$ there is a $0 \leq D \leq \binom{m}{2}$ such that v is (α, D)-popular, and similarly, for any given $0 \leq D \leq \binom{m}{2}$ there exists an $0 \leq \alpha \leq 1$ such that v is (α, D)-popular.

Algorithms. An Algorithm proceeds in parallel rounds, where in each round the algorithm may present to each player j a pair of objects (i, i') and obtain

$c_j(i, i')$. This action is called a *probe* by player j. Players communicate through a shared billboard, i.e., all probes results are immediately visible to all players. An algorithm makes output and chooses which objects to probe based on the results of all past probes (including probes taken by other players), and possibly coin tosses. The algorithms we consider in this paper are required to output preference vectors that approximate the pairwise preferences of the players, up to some specified number of errors. The maximal number of probes a player is asked to perform in an algorithm execution is the *probe complexity* of the algorithm. Our algorithms are randomized and all guarantees are *with high probability*, i.e., holds with probability $1 - n^{-\Omega(1)}$, where probability is taken over the coin tosses of the algorithm.

We assume random partitioning and sampling is always done uniformly. Splitting a set in two, for instance, is done by choosing, for each element, its part independently with probability $1/2$.

2.1   User Models

We consider the following variants of the comparison-based preferences model.

General Model. In the general model, the $\binom{m}{2}$ comparison values of a player are completely independent. In particular, no transitivity of preferences is assumed. This model is equivalent to a model of $\binom{m}{2}$ virtual objects where each virtual object represents a pair of two physical objects, with four possible grades. For an upper bound, one can apply Algorithm 1 of Azar et al. [6] for recommendations with discrete grades, which reduces the problem to the binary grade model. Plugging in that algorithm the algorithm of [15] for binary recommendations, we obtain the following result.

Theorem 1. *In the comparison-based model, any (α, D)-popular taste can be reconstructed to within $O(D)$ errors in probe complexity $O(\left\lceil \frac{m^2}{n} \right\rceil \frac{1}{\alpha} \log^3(m+n))$.*

The general model also admits easy lower bounds. In particular, the total probe complexity (the sum of individual probe complexity over all players) in the general model is $\Omega(\left\lceil \frac{m^2 - D}{n} \right\rceil)$. Intuitively, even in the case where all players of some (α, D)-popular taste are perfectly coordinated, each individual object (with the exception of at most D objects) must be probed at least once. Therefore, $\Omega(\binom{m}{2} - D)$ total probes must be taken by these players just to cover all pairs. Next, consider the case where all players know all the possible tastes in advance (they are common knowledge). Still, to find his own taste, each user must make at least one probe, giving rise to total probe complexity of at least $\Omega(n)$. Hence the average probe complexity is $\Omega(\frac{\binom{m}{2} - D + n}{n}) = \Omega(\left\lceil \frac{m^2 - D}{n} \right\rceil)$.

Total Order Model. In this model we assume the player taste is induced by a total order over the objects. Comparisons may not have "⊥" as a value: $c_j(i, i') \in \{-1, 0, 1\}$ for all players j and objects i, i'. In this model, a preference vector can be represented by a permutation of the objects. We note that in this case distances, as defined above, are just the number of *transpositions* of one permutation with respect to the other.

We have the following straightforward lower bound on the number of probes in this model.

Theorem 2. *Consider an (α, D)-popular taste which is a total order. The number of probes for reconstructing it is $\Omega(\frac{(m-2D)\log(m-2D)}{\alpha n})$ probes on average.*

Proof: [sketch] We use a variant of the sphere packing bound. The number of permutations that are at distance at most $2D$ from a given permutation is at most $\sum_{i=0}^{2D} \binom{m}{i} i! \le (2D+1)m^{2D}$. Therefore, if we cannot specify at least $\frac{m!}{(2D+1)m^{2D}}$ distinct inputs, there will be instances containing two tastes at distance greater than $2D$, and both will have the same output, so necessarily one of them is at distance more than D from its output. Since specifying $\frac{m!}{(2D+1)m^{2D}}$ distinct outcomes requires $\Omega((m-2D)\log(m-2D))$ bits, and since each probe provides only $O(1)$ bits, the result follows. ∎

Disjoint Categories Model. In this model we assume that the object set can be broken into disjoint *classes*, or *categories*, where only objects in the same category are comparable. The categorization of objects may be different for different tastes.

Specifically, we assume that for each taste, the object set can be partitioned into k disjoint sets $O = \bigcup_{i=1}^{k} O_i$, and that there is a total order σ_i defined over each O_i. The preference between two objects is either $\perp$ if they belong to different subsets, or given by σ_i if they both belong to O_i for some i. This model is appropriate in cases where the objects may be of incompatible nature, e.g., O_1 consists of films and O_2 is a set of pets: it doesn't make sense to compare a film with a pet. Moreover, different tastes may have different categorizations, which allows our algorithm to be applicable even when the categories are unanimously acceptable.

3   Exact Total Order: Algorithm $\mathcal{DP}$

In this section we consider the case where the algorithm receives as input a popularity factor α, and the goal is to reconstruct the preferences precisely for all $(\alpha, 0)$-popular players, i.e., players whose *exact* taste is shared by at least αn players. Most importantly, we assume that these $(\alpha, 0)$-popular tastes are induced by total orders over the objects. Our solution to this case is a randomized distributed algorithm called $\mathcal{DP}$, based on Algorithm ZERO_RADIUS of [4] for the binary grade model. While $D = 0$ is an interesting case in its own right, we shall use Algorithm $\mathcal{DP}$ as a subroutine in the algorithm for $D > 0$, presented in Section 4.

The algorithm works as follows (see pseudo code in Algorithm 3.1). The input consists of a set of players P and a set of objects O. If either P or O is small enough, each player in the current player set P sorts all objects in the current object set O, using a straight-forward adaptation of any efficient comparison-based sorting algorithm, such as Quicksort, Heapsort, Mergesort, etc. Otherwise, the algorithm randomly splits the object set and the player set into O', O'' and P', P'' respectively, and calls itself recursively: P' with O' and P'' with O''. When the recursive call returns, each player in P' knows his complete preferences (ordering in our case) over O' and each player in P'' knows his ordering over O''. Consider w.l.o.g. a player $j \in P'$: j needs to find the ordering over O'' it agrees with, and then to merge the order of O' with the order of O'', to obtain a sorted order on O.

This is done as follows. Looking at the billboard, a player in P' first identifies the preference vectors over O'' with popularity larger than $\alpha/2$ (line 8) as "candidate tastes." From these candidates, he selects the vector compatible with his taste by probing *controversial pairs*, i.e., pairs on which some candidate vectors differ. Formally, for two vectors v_1 and v_2, there exist a pair (i, i'), for which, $i <_{v_1} i'$ and $i >_{v_2} i'$, namely i is preferred over i' in v_2 but not in v_1 (lines 11–12). Each such probe eliminates at least one candidate. Eventually, the player is left with a single vector. This vector, which is an ordering of O'', is then merged with the vector the player has already computed for O'. This is done using the parallel merging algorithm of Valiant [17].

Algorithm 3.1. $\mathcal{DP}(P, O)$ Pseudo-code executed by player $j \in P$

1: **if** $\min(|P|, |O|) < \frac{16c \ln n}{\alpha}$ **then**   ▷ *base case*
2:     Sort O using any reasonable sorting algorithm and **return** the sorted vector.
3: **end if**
4: Randomly partition $P = P' \cup P''$, and $O = O' \cup O''$.
5: **if** $j \in P'$ **then** call $\mathcal{DP}(P', O')$
6: **else** call $\mathcal{DP}(P'', O'')$
7: **end if**
    Assume w.l.o.g. that $j \in P'$. Upon returning, j has v' his complete order over O', and sees the order selected by each player $j' \in P''$
8: Let V be a set of vectors of O'' chosen by at least $\alpha/2$ players from P''.
9: **while** $|V| > 1$ **do**
10:     Let $C = \{(i, i') \in O'' \times O'' \mid \exists v_1, v_2 \in V \text{ s.t. } i <_{v_1} i' \text{ and } i >_{v_2} i'\}$
11:     Choose an arbitrary pair $(i, i') \in C$.
12:     Let $c = \text{probe}(i, i')$
13:     Remove from V all vectors whose value on (i, i') is not c.
14: **end while**
15: Suppose $V = \{v''\}$.   ▷ $|V| = 1$ *w.h.p. if j is in an α-popular tatse*
16: Let $P_j = \{j' \in P' : \text{player } j' \text{ chose vector } v''\}$.
17: **return** $\text{Merge}(P_j, v', v'')$.

Analysis. Consider an α-popular taste. The algorithm's success critically depends on having more than $|P|\alpha/2$ players of that taste in any invocation with player set P. The following lemma ensures this w.h.p.

Lemma 1. *Fix an α popular taste v, and let P_v be the set of players of taste v. Then, with probability $1 - n^{-\Omega(1)}$, in all invocations of $\mathcal{DP}(P, O)$ at least $|P|\alpha/2$ of the players in P have taste v, i.e., $|P_v \cap P| \geq \alpha |P|/2$.*

Proof: Observe that the set of players P in any invocation is a random sample of the set of all players. By the Chernoff bound we have that the number of players in any random sample P is

$$\Pr\left[\text{\# players of taste } v < \alpha |P|/2\right] \leq e^{-\frac{\alpha |P|}{8}} \leq e^{\frac{2c \ln n}{\alpha}} = n^{-2c},$$

because by the code, $|P| \geq \frac{16c}{\alpha} \ln n$. Since the total number of invocations is bounded by n, the result follows from the union bound. ∎

Next, we state the correctness of Algorithm Merge(P, v_1, v_2) used in line 17.

Lemma 2. *Let v_1, v_2 be two sorted vectors with $|v_1| = \Theta(|v_2|)$ and let P be the set of players agreeing on the joint order of the objects in v_1 and v_2. Then the merged sorted vector can be computed using probe complexity $O(\frac{|v_1|}{|P|} \log \log |v_1|)$ per player.*

The algorithm is based on [17] in a straightforward way. Details omitted.

Theorem 3. *Under Algorithm $\mathcal{DP}$, with probability $1 - n^{-\Omega(1)}$, all outputs by α-popular players are correct. The number of probes per player is bounded by $O(Y(\log Y + \log \log m))$, where $Y = \frac{\log n}{\alpha} \lceil \frac{m}{n} \rceil$.*

Proof: Fix an α-popular taste v. By Lemma 1, w.h.p., all invocations have at least $\alpha/2$-fraction of players whose taste is v. Hence the set V selected in line 8 contains the projection of v onto O'', and it will be the only vector not eliminated by any player from P' whose taste is v. Since this is true symmetrically also for P'', it follows that with high probability, all outputs are correct.

Regarding complexity, note that probing is done by the algorithm only in lines 2, 12, and 17. The probing of line 2 is done according to the sorting algorithm, at the cost of $O(|O| \log |O|)$ probes. Let P and O be the set of players and objects respectively, when the sorting algorithm is executed. Since the base condition is $\min(|P|, |O|) < \frac{16c \ln n}{\alpha}$ and since both the player and the object sets are approximately halved in every recursive call, we have that, with high probability, $|O| = O(Y)$, and hence the number of probes due to line 2 is $O(Y \log Y)$. Note that line 2 is executed once throughout the algorithm by each player.

Next, consider line 8. Since each vector in V represents at least an $\frac{\alpha}{2}$ fraction of the players of P'', $|V| \leq \frac{2}{\alpha}$. Since in each iteration of the while loop, at least one vector is removed from V in line 13, we have that each player makes at most $2/\alpha$ probes in line 12 in an invocation of Algorithm $\mathcal{DP}$. Finally, the probing of line 17 is done using Valiant's merging algorithm [17]. By Lemma 2, we have that each player makes $O(\frac{m}{n\alpha} \log \log m)$ probes in line 17. It follows that the total number of probes due to lines 8 and 17 in a single invocation of the algorithm is $O(\frac{1}{\alpha} \log \log m)$. Since the number of recursive levels is $O(\log n)$, the result follows. ∎

Corollary 1. *If $m = \Theta(n)$ and $\alpha = (\log n)^{-\Omega(1)}$, the probe complexity of Algorithm $\mathcal{DP}$ is $O\left(\frac{1}{\alpha} \log n \log \log n\right)$.*

We note that the probe complexity of Algorithm $\mathcal{DP}$ is larger than the lower bound of Theorem 2 only by a factor of $O(\log \log n)$.

4   Approximate Total Order: Algorithm $\mathcal{DPD}$

In this section we present our main result. As in Section 3, we assume that the taste of each player is induced by a total order on the objects, but whereas previously we assumed that there are αn players whose taste is *exactly* the same, here we require only that the taste is (α, D)-popular, namely there are αn players such that any two players in the set may disagree on the outcomes of at most D comparisons. We use Algorithm $\mathcal{DP}$ from Section 3 as a subroutine, but we can still reconstruct the taste to within $O(D)$ errors with polylogarithmic overhead. Our solution is a distributed algorithm we call $\mathcal{DPD}$, which extends Algorithm $\mathcal{S}$ from [15] to the comparison-based model.

Algorithm Description. The algorithm consists of three conceptual steps as follows (see pseudocode in Algorithm 4.1). In the first step, we split the object set uniformly at random into $S = cD$ disjoint subsets, for some constant $c > 8$. We apply Algorithm $\mathcal{DP}$ to each subset by all players. This splitting and application of $\mathcal{DP}$ is repeated $K = \Theta(D \log m)$ times, thus making sure (w.h.p.) that each pair of objects appears in the same subset $\Theta(\log m)$ times. After this step, we have, for each pair of objects, $\Theta(\log m)$ estimates for each player. Next, using these estimates, each player j builds a directed graph $G_j = (O, E_j)$ as follows. There is a directed edge between every pair of objects, whose direction is determined by the majority of outcomes computed in the first step. Note that G_j is a tournament, but it may be inconsistent, i.e., contain cycles. These are eliminated in the last step by applying the algorithm of Ailon et al. [2] to G_j, which finds a 3-approximation to *MFAST*: MFAST is the problem of deleting the minimum number of edges from a tournament so that the result is acyclic.[3] No probing is required for this step. The result is the output of player j.

Algorithm Analysis. We first define some notation. Let v_j denote the taste of player j, and $v_j(i, i')$ denote the preference of player j on the pair (i, i'). For each player j, define $P(j) = \{j' \in P : \text{dist}(v_j, v_{j'}) \leq D\}$, namely $P(j)$ is the set of players whose preference vectors differ from j on at most D object pairs.

Now, define, for each player j, the set of object pairs $O(j)$ to be the pairs on which player j agrees with the majority of the players in $P(j)$, i.e.,

$$O(j) = \left\{(i, i') \in O \times O : |\{j' : v_{j'}(i, i') = v_j(i, i')\}| > \frac{|P(j)|}{2}\right\}$$

---

[3] MFAST stands for Minimum Feedback Arc Set in Tournaments (an NP-hard problem, according to Alon [3]). The approximation algorithm of Ailon et al. [2] has running time $O(|O| \log |O|)$. Kenyon-Mathieu and Schudy [13] give a PTAS for MFAST with running time $O(|O|^6)$).

Algorithm 4.1. $\mathcal{DPD}$ reconstruct approximate taste

Require: P, O, α, D

1: **for all** $k \in 1, .., K$ **do** $\triangleright K = \Theta(D \log m)$
2:     Partition O into $S = cD$ disjoint subsets $O = \bigcup_{s \in 1..S} O_s$. $\triangleright c > 8$ *is a constant*
3:     **for all** $s \in 1, .., S$ **do**
4:         Invoke $\mathcal{DP}(P, O_s, \alpha/4)$
5:     **end for**
6:     Let $C_j^k(i, i')$ be the reconstructed output of pair (i, i') for player j on round k.
7: **end for**
8: For all object pairs (i, i'), let $L(i, i')$ denote the set of iterations in which the objects i and i' are in the same subset.
9: For all $i, i' \in O$, let $C_j(i, i')$ be the majority of $\{C_j^k(i, i') : k \in L(i, i')\}$
10: Let $G_j = (O, E_j)$ be a directed graph, where, $E_j = \{(i, i') : C_j(i, i') \geq 0\}$
11: Invoke *MFAST-Approx*(G_j) for each player $j \in P$
12: Output the resulted ranking for each player j.

The following lemma helps to get a lower bound on $|O(j)|$.

Lemma 3. *For all $0 < \delta \leq 1$ it holds that any player j disagrees with at most a δ-fraction of the players in $P(j)$ on at most D/δ object pairs.*

Proof: By definition, each player in $P(j)$ disagrees on at most D pairs with player j. Hence the total number of disagreements between j and all players in $P(j)$ is less than $|P(j)|D$. Therefore the number of players in $P(j)$ with which j disagrees on more than D/δ pairs is at most $\delta|P(j)|$. ∎

Next, we show that for any $(i, i') \in O(j)$, in each iteration $k \in L(i, i')$ of algorithm $\mathcal{DPD}$ (cf. line 8), Algorithm $\mathcal{DP}$ computes a correct estimate of $v_j(i, i')$ in line 4 with "good" probability.

Lemma 4. *For all $j \in P$, $(i, i') \in O(j)$ and iteration $k \in L(i, i')$, we have that $\Pr[C_j^k(i, i') = v_j(i, i')] \geq 1 - \frac{4}{c}$.*

Proof: Fix player j and an object pair (i, i'). Consider a random subset O_s for which $i, i' \in O_s$. Let $P_s(j)$ be the set of players that agree with player j on all objects in O_s, i.e., $P_s(j) = \{j' : \mathrm{dist}_{O_s}(j, j') = 0\}$. If $|P_s(j)| \geq \frac{\alpha|P|}{4}$, then $\mathcal{DP}$ will return correct results. It therefore suffices to show that $\Pr[|P_s(j)| \geq \frac{\alpha|P|}{4}] \geq 1 - \frac{4}{c}$, where the probability here is over the choice of O_s.

To do that, we first observe that for any $0 < \beta < 1$, we have that if $\sum_{j' \in P(j)} \mathrm{dist}_{O_s}(j, j') \leq \beta \cdot |P(j)|$, then $|P_s(j)| \geq (1 - \beta)|P(j)|$. It follows that it suffices to bound the probability that the sum of distances for players in $P(j)$ is at most $\frac{3}{4}|P(j)|$, i.e., the probability that $\sum_{j' \in P(j)} \mathrm{dist}_{O_s}(j, j') < \frac{3}{4}|P(j)|$.

Let $O_s^* = O_s \setminus \{(i, i')\}$. Consider the random variable $\sum_{j' \in P(j)} \mathrm{dist}_{O_s^*}(j, j')$, namely, the sum of distances between j and all $P(j)$ players, ignoring the pair (i, i'). As $(i, i') \in O(j)$, we have that $\sum_{j' \in P(j)} |v_j(i, i') - v_{j'}(i, i')| < \frac{|P(j)|}{2}$.

Clearly,

$$\Pr\left[\sum_{j'\in P(j)} \mathrm{dist}_{O_s}(j,j') \le \frac{3}{4}|P(i)|\right] \ge \Pr\left[\sum_{j'\in P(j)} \mathrm{dist}_{O_s^*}(j,j') \le \frac{|P(j)|}{4}\right].$$

Since distance is measured by object pairs and from the definition of $P(j)$, we have

$$D|P(j)| \ge \sum_{j'\in P(j)} \mathrm{dist}_O(j,j') \ge \sum_{\ell=1}^{S}\sum_{j'\in P(j)} \mathrm{dist}_{O_\ell}(j,j'),$$

which implies that $\sum_{j'\in P(j)} \mathrm{dist}_{O_\ell}(j,j') > \frac{|P(j)|}{4}$ for at most $4D$ subsets O_ℓ. From that we have

$$\Pr\left[\sum_{j'\in P(j)} \mathrm{dist}_{O_s^*}(j,j') \le \frac{|P(j)|}{4}\right] \ge \frac{S-4D}{S} = 1 - \frac{4}{c}.$$

Since we consider (α, D)-popular players, the definition of $P(j)$ implies that $|P(j)| \ge \alpha n$. Therefore, $|P_s(j)| \ge \frac{|P(j)|}{4} \ge \frac{\alpha n}{4}$ with probability $\ge 1 - \frac{4}{c}$. The result follows. ∎

Lemma 5. *W.h.p., each pair (i, i') occurs together in $\Omega(\log n)$ invocations of $\mathcal{DP}$.*

Proof: Consider object i. The probability that i' is chosen to the same subset as i in a given iteration is $\frac{D}{S} = \frac{1}{c}$. It follows that the expected number of subsets they occur together is $\frac{K}{c} = \Omega(\log m)$. Since choices in different iterations are independent, the result follows from the Chernoff bound. ∎

By Lemma 4, the probability that each invocation of $\mathcal{DP}$ is successful is at least $1-4/c > 1/2$ because $c > 8$. By Lemma 5, each pair occurs together in $\Omega(\log m)$ invocations of $\mathcal{DP}$. Therefore, by the Chernoff bound and the union bound, we have that w.h.p, the majority value is correct for all pairs for all players.

Lemma 6. $\Pr\left[C_j(i, i') \ne v_j(i, i') \text{ for some } j \text{ and } (i, i')\in O(j)\right] \le n^{-\Omega(1)}$.

We can now summarize the properties of Algorithm $\mathcal{DPD}$ as follows.

Theorem 4. *With probability $1 - n^{-\Omega(1)}$, algorithm $\mathcal{DPD}$ predicts for each player $j \in P$ its preference vector with at most $O(D)$ errors. The probe complexity for each player is $O(D^2 \log m \cdot T_{\mathcal{DP}}(n, m/cD, \alpha/4))$, where $T_{\mathcal{DP}}(n, m, \alpha)$ is the probe complexity of Algorithm $\mathcal{DP}$ with n users, m objects and popularity factor α.*

Corollary 2. *For $m = \Omega(nD)$ and $\alpha = (\log n)^{-O(1)}$, the probe complexity of Algorithm $\mathcal{DPD}$ for reconstructing an (α, D)-popular taste is*

$$O\left(\frac{m}{n}\frac{D}{\alpha} \log^2 m \log\log m\right).$$

Proof: [of Theorem 4] Fix a player j. Forevery object pair $(i, i') \in O(j)$, we have that $C_j(i, i') = v_j(i, i')$ w.h.p. Since this is guaranteed only for object pairs in $O(j)$, and since $|O(j)| \geq m - 2D$ (by Lemma 3), we have $\text{dist}_O(v_j, C_j) \leq 2D$, i.e., the number of object pairs on which the results of the majorities differ from the true preferences of j (which is a consistent total order) is at most $2D$. Since the Algorithm *MFAST-Approx* of [2] finds a 3-approximation to the optimal number of edges that need to be reversed, *MFAST-Approx* finds a permutation whose distance from v_j is at most $6D$ errors. The query complexity follows from Theorem 3, along with the fact that $\mathcal{DP}$ is invoked $KS = KcD$ times, with $|P| = n$, $|O_s| = \lceil \frac{m}{D} \rceil$ and popularity factor $\alpha/4$. The probability of correctness follows from Lemma 4 and Lemma 6.   ∎

5   The Case of Disjoint Categories

In many cases, the object set is eclectic in the sense that it consists of a few kinds of objects, e.g., cats and cars. Typically, in these cases one may have preference over pairs of objects of the same kind, but there is no sense in comparing objects of different kinds. The problem becomes more complicated when there may be ambiguity about object classification: For example, some users may classify a jaguar a as a cat, while others may classify it as a car.

In this section we model this situation and present algorithms to reconstruct preferences.

Preference Model. A user j is said to have a *categorized taste* if the object set can be partitioned into k disjoint subsets $O = \bigcup_{\ell=1}^{k} O_\ell^j$ called the *categories* of user j, and if j has a total order σ_ℓ^j over each category O_ℓ^j. Formally, if user j has a categorized taste then

$$c_j(i, i') = \begin{cases} 0 & \text{if } i = i' \\ 1 & \text{if } i, i' \in O_\ell^j \text{ for some } \ell \text{ and } \sigma_\ell^j(i) > \sigma_\ell^j(i') \\ -1 & \text{if } i, i' \in O_\ell^j \text{ for some } \ell \text{ and } \sigma_\ell^j(i) < \sigma_\ell^j(i') \\ \bot & \text{otherwise.} \end{cases}$$

For simplicity, we say that users have the same taste if their preferences are identical. However, in the case of categorized tastes, we also define the notion of *crude taste*: two users j and j' are said to have the same crude taste if they have the same object categorization (but not necessarily the same ordering within each category), i.e., $\left\{ O_1^j, \ldots, O_k^j \right\} = \left\{ O_1^{j'}, \ldots, O_k^{j'} \right\}$.

Algorithm. Clearly, if the user categorizations are given, one can run algorithm $\mathcal{DP}$ on each user category with all users that have this category; if a taste is α-popular. Formally, we have the following result.

Lemma 7. *Suppose that in a given instance, all tastes are categorized. If the categories of all users are known, then preferences can be reconstructed w.h.p. with probe complexity $O((\frac{m}{\alpha n} + k) \frac{\log n}{\alpha} \log(\lceil \frac{m}{n} \rceil \frac{\log m}{\alpha}))$, where k is the number of categories.*

Proof: Consider an arbitrary user, say j. Let $m_\ell = |O_\ell^j|$, i.e., m_ℓ is the size of category ℓ of user j. Let n_ℓ be the number of users with category O_ℓ^j. Note that $n_\ell \geq \alpha n$ because at least all users with the taste of j have category O_ℓ^j, and possibly others too. User j takes part in k invocations of $\mathcal{DP}$, where invocation ℓ has m_ℓ objects and $n_\ell \geq \alpha n$ players participating. By Theorem 3, the total probe complexity for user j is therefore

$$\sum_{\ell=1}^{k} T_{\mathcal{DP}}(n, m_\ell, n_\ell/n) = \sum_{\ell=1}^{k} \left\lceil \frac{m_\ell}{n_\ell} \right\rceil O\left(\frac{\log n}{\alpha} \log \left(\left\lceil \frac{m_\ell}{n_\ell} \right\rceil \frac{\log(m+n)}{\alpha} \right) \right)$$

$$\leq O\left(\left(\frac{m}{\alpha n} + k \right) \frac{\log n}{\alpha} \log \left(\left\lceil \frac{m}{n} \right\rceil \frac{\log(m+n)}{\alpha} \right) \right).$$

∎

Note that by using algorithm $\mathcal{DPD}$, Lemma 7 can be extended to handle (α, D)-popular tastes, so long as each category is shared by at least αn users.

In view of Lemma 7, we now consider the question of reconstructing the *crude* taste of the users. This can be done by adapting algorithm $\mathcal{DP}$ to *crude* probes. A crude probe, or c-probe for short, is just a regular comparison probe, except that its return value is either 1 if the two object probed are comparable (no matter what is the comparison result), or 0 is the pair if incomparable (i.e., the comparison probe returns $\perp$). Algorithm $\mathcal{DPC}$ (see Algorithm 5.1 for pseudocode) for reconstructing crude tastes has the same structure as Algorithm $\mathcal{DP}$, but with different base procedure (called Classify, line 2) and different merging procedure (called Combine, line 17). We explain these two now.

Procedure Classify works by picking an unclassified object i_0 and applying c-probe of i_0 against all other unclassified objects. If the result of c-probe$(i_0, i) = 1$ then i is in the same category as i_0, and otherwise i remains unclassified. Clearly the probe complexity of a set O is $O(k|O|)$.

Procedure Combine gets as input two classifications v', v'', each with at most k categories. It applies c-probe for each pair of categories: one from v' and the other from v''. These probes are done using representatives from the categories, i.e., $O(k^2)$ probes need to be executed in total. Moreover, these probes can be split among all players in P_j, resulting in individual probe complexity of $O(k^2/|P_j|)$.

We can therefore summarize the properties of Algorithm $\mathcal{DPC}$ as follows.

Theorem 5. *W.h.p., Algorithm $\mathcal{DPC}$ outputs the α-popular crude taste for each user with probe complexity $O(k(\lceil \frac{m}{n} \rceil \log n + k) + \frac{\log n}{\alpha})$.*

Proof: The probing of the base case (line 2) costs $O(\lceil m/n \rceil k \log n)$ because in the base case, the number of objects is $O(\lceil m/n \rceil \log n)$. The elimination step (line 12) costs $O(1/\alpha)$ probes in each iteration as before. regarding the probing due to Combine (line 17), let n_t denote the number of users in P_j in iteration t. Then the total cost of Combine for user j over all iterations is

$$\sum_{t=1}^{\log(\alpha n)} \left\lceil \frac{k^2}{n_t} \right\rceil = O(k^2 + \log(\alpha n)) \sum_{t=1}^{\log(\alpha n)} \frac{1}{\alpha n 2^{-t}} \leq O(k^2 + \log n),$$

Algorithm 5.1. $\mathcal{DPC}(P, O)$ reconstruct crude taste   Pseudo-code executed by player $j \in P$

1: **if** $\min(|P|, |O|) < \frac{16c \ln n}{\alpha}$ **then**   ▷ *base case*
2:     **return** Classify(O).   ▷ *see text*
3: **end if**
4: Randomly partition $P = P' \cup P''$, and $O = O' \cup O''$.
5: **if** $j \in P'$ **then** call $\mathcal{DPC}(P', O')$
6: **else** call $\mathcal{DPC}(P'', O'')$
7: **end if**
    Assume w.l.o.g. that $j \in P'$. Upon returning, j has v' as his classification of O', and sees the classifications of each player $j' \in P''$.
8: Let V be a set of classifications of O'' chosen by at least $\alpha/2$ players from P''.
9: **while** $|V| > 1$ **do**
10:     Let C be the set of object pairs $(i, i') \in O'' \times O''$ for which there are classifications $v_1, v_2 \in V$ such that v_1 classifies i, i' together and v_2 classifies them in different categories.
11:     Choose an arbitrary pair $(i, i') \in C$.
12:     Let $c = $ c-probe(i, i')
13:     Remove from V all classifications whose value on (i, i') does not agree with c.
14: **end while**
15: Suppose $V = \{v''\}$.   ▷ $|V| = 1$ *w.h.p. if j is in an α-popular taste*
16: Let $P_j = \{j' \in P' : $ player j' chose classification $v''\}$.
17: **return** Combine(P_j, v', v'').   ▷ *see text*

---

because w.h.p., $n_t = O(\alpha n / 2^t)$. The result follows.   ∎

Corollary 3. *For $m = \Theta(n)$, the probe complexity of Algorithm $\mathcal{DPC}$ for reconstructing an α-popular crude taste is $O((\frac{1}{\alpha} + k) \log n + k^2)$.*

6   Conclusions and Open Problems

In this paper we showed that preferences can be reconstructed in a comparison-based model if tastes are derived from a total order. We also showed how to deal with tastes which can be decomposed into unrelated categories, assuming that within each category objects are totally ordered. Our results are tight up to a polylogarithmic factor, except for Algorithm $\mathcal{DPD}$, whose complexity has an extra factor of $O(D^2)$ when the number of objects is roughly the same as the number of users.

While we know that if the comparison results are arbitrary the probe complexity of taste reconstruction is $\Omega(\lceil m^2/n \rceil)$ in the worst case, we leave open the question of reconstructing tastes derived from a general partial order (i.e., assuming transitivity) in the comparison-based model.

References

[1] Ailon, N.: Active learning ranking from pairwise preferences with almost optimal query complexity. In: Proc. NIPS, pp. 810–818 (2011)

[2] Ailon, N., Charikar, M., Newman, A.: Aggregating inconsistent information: ranking and clustering. J. ACM 55(5), 23 (2008)

[3] Alon, N.: Ranking tournaments. SIAM Journal on Discrete Mathematics 20(1), 137–142 (2006)

[4] Alon, N., Awerbuch, B., Azar, Y., Patt-Shamir, B.: Tell me who I am: an interactive recommendation system. Theory of Computing Systems 45(2), 261–279 (2009)

[5] Awerbuch, B., Azar, Y., Lotker, Z., Patt-Shamir, B., Tuttle, M.R.: Collaborate with strangers to find own preferences. Theory of Computing Systems 42(1), 27–41 (2008)

[6] Azar, Y., Nisgav, A., Patt-Shamir, B.: Recommender systems with non-binary grades. In: Proc. 23rd SPAA, pp. 245–252. ACM (2011)

[7] Bell, R.M., Koren, Y.: Lessons from the netflix prize challenge. SIGKDD Explorations 9(2), 75–79 (2007)

[8] Carterette, B., Bennett, P.N., Chickering, D.M., Dumais, S.T.: Here or there. In: Proc. 30th European Conf. on Advances in Information Retrieval, pp. 16–27 (2008)

[9] Desarkar, M.S., Saxena, R., Sarkar, S.: Preference relation based matrix factorization for recommender systems. In: Masthoff, J., Mobasher, B., Desmarais, M.C., Nkambou, R. (eds.) UMAP 2012. LNCS, vol. 7379, pp. 63–75. Springer, Heidelberg (2012)

[10] Drineas, P., Kerenidis, I., Raghavan, P.: Competitive recommendation systems. In: Proc. 34th Ann. ACM Symp. on Theory of Computing, pp. 82–90. ACM (2002)

[11] Goldberg, K., Roeder, T., Gupta, D., Perkins, C.: Eigentaste: A constant time collaborative filtering algorithm. Information Retrieval 4(2), 133–151 (2001)

[12] Jones, N., Brun, A., Boyer, A.: Comparisons instead of ratings: Towards more stable preferences. In: Proc. Int. Conf. on Web Intelligence and Intelligent Agent Technology, pp. 451–456. IEEE Computer Society (2011)

[13] Kenyon-Mathieu, C., Schudy, W.: How to rank with few errors. In: Proc. 39th Ann. ACM Symp. on Theory of Computing, pp. 95–103. ACM (2007)

[14] Loepp, B., Hussein, T., Ziegler, J.: Choice-based preference elicitation for collaborative filtering recommender systems. In: Proc. 32nd Ann. ACM Conf. on Human Factors in Computing Systems, pp. 3085–3094 (2014)

[15] Nisgav, A., Patt-Shamir, B.: Improved collaborative filtering. In: Asano, T., Nakano, S.-i., Okamoto, Y., Watanabe, O. (eds.) ISAAC 2011. LNCS, vol. 7074, pp. 425–434. Springer, Heidelberg (2011)

[16] Sarwar, B., Karypis, G., Konstan, J., Riedl, J.: Analysis of recommendation algorithms for e-commerce. In: Proc. 2nd ACM Conf. on Electronic Commerce, pp. 158–167. ACM (2000)

[17] Valiant, L.G.: Parallelism in comparison problems. SIAM J. on Computing 4(3), 348–355 (1975)

Coalescing Walks on Rotor-Router Systems[*]

Colin Cooper[1], Tomasz Radzik[1], Nicolás Rivera[1], and Takeharu Shiraga[2]

[1] Department of Informatics, King's College London, United Kingdom
{colin.cooper,tomasz.radzik,nicolas.rivera}@kcl.ac.uk
[2] Theoretical Computer Science Group, Department of Informatics,
Kyushu University, Fukuoka, Japan
shiraga@tcslab.csce.kyushu-u.ac.jp

Abstract. We consider the rotor-router mechanism for distributing particles in an undirected graph. If the last particle passing through a vertex v took an edge (v, u), then the next time a particle is at v, it will leave v along the next edge (v, w) according to a fixed cyclic order of edges adjacent to v. The system works in synchronized steps and when two or more particles meet at the same vertex, they coalesce into one particle. A k-particle configuration of such a system is *stable*, if it does not lead to any coalescing. For $2 \leq k \leq n$, we give the full characterization of stable k-particle configurations for cycles. We also show sufficient conditions for regular graphs with n vertices to admit n-particle stable configurations.

1 Introduction

We consider an undirected connected graph $G = (V, E)$ and the *rotor-router* mechanism which keeps moving simple entities along the edges of G in synchronized steps. We call these entities *particles*, but terms like agents, tokens or chips may be used by others. Each edge $\{v, u\}$ is viewed as a pair of opposite arcs (v, u) and (u, v), and for each vertex, the arcs outgoing from this vertex are kept in a fixed cyclic order. In each step, each particle moves from its current vertex to an adjacent vertex. For each vertex v, if (v, u) is the most recently traversed arc outgoing from v, then the next particle leaving v will traverse the next arc (v, w). This is implemented by maintaining at each vertex v the *vertex pointer* π_v which indicates which arc outgoing from v should be taken next. While particles keep passing through v, the pointer π_v is the "rotor" moving around the cyclic order of arcs outgoing from v. This model was introduced by Priezzhev *et al.* [17], was further studied and popularised by James Propp, and hence also referred to as the *Propp machine*.

The rotor-router mechanism can be viewed as a model of graph exploration by simple mobile entities and the efficiency of such exploration has been extensively studied. While the earlier works refer mostly to single-particle rotor-router exploration, there are now also a few recent results concerning the multi-particle

[*] This work was supported in part by EPSRC grant EP/M005038/1, "Randomized algorithms for computer networks".

C. Scheideler (Ed.): SIROCCO 2015, LNCS 9439, pp. 444–458, 2015.
DOI: 10.1007/978-3-319-25258-2_31

case. In both the single-particle and the multi-particle cases, some similarities with graph exploration by random walks have been observed. This further motivates investigations of the rotor-router mechanism as a possible deterministic alternative to random walks. For example, one random walk on a cycle of length n covers (visits) all vertices in expected $\Theta(n^2)$ time and a single-particle rotor-router does this in deterministic $\Theta(n^2)$ worst-case time. Wagner *et al.* [18, 19] showed that for an arbitrary connected n-vertex m-edge graph and an arbitrary initial configuration of the single-particle rotor-router system (arbitrary cyclic orders of arcs, arbitrary initial setting of the vertex pointers and an arbitrary starting vertex for the particle) the particle visits all vertices of this graph in $O(nm)$ steps. Subsequently a number of more detailed analyses of single-particle rotor-router systems in various types of graphs have been published [4–6,20], but it can be shown that $\Theta(nm)$ is the worst-case bound for the general graphs. The expected cover time by a random walk has the same general bound $O(nm)$. More recently the cover time by (parallel) k random walks has been analyzed and speedups over a single random walk between $\Theta(\log k)$ and $\Theta(k)$ have been shown for various classes of graphs and various initial settings [3,9,11,12]. The similar range of speedups for k-particle rotor router (over the single-particle system) have been demonstrated in [10, 14].

In this paper we look at another aspect of multi-particle systems: *coalescence of particles*. Whenever two or more particles meet at the same step in the same vertex of the graph, then they coalesce (merge) into one particle. This particle continues moving through the graph, following the underlying protocol (for example, the random-walk protocol or the rotor-router mechanism). Coalescing random walks is a long established topic, attracting research interest partly due to its close relation to the randomized *pull-voting* process [1]. The main case considered is when initially each vertex has one particle and the question is to provide good bounds on the expected time (number of steps) until full coalescing into one particle. Aldous [2] conjectured that this expected full-coalescence time is at most of the order of the maximum hitting time of a single random walk. This conjecture has not been fully settled yet, but considerable progress has been made [8,16].

Systems with coalescing particles may find applications in parallel computing. For example, Israeli and Jalfon [13] proposed coalescing random walks as the basis of a self-stabilizing mutual exclusion algorithm. In a network of interconnected processing units (PUs) competing for access to some resource, each PU creates a token and sends it for a random walk through the network. Tokens coalesce whenever they meet at the same PU, eventually only one token remains in the network (provided the network is connected and non-bipartite) and the PU with the token gets exclusive access to some resource.

In this paper we want to initiate investigation of rotor-router systems with coalescing particles. While the coalescing random walks will always eventually merge into a single walk (or two walks in the case of bipartite graphs), and good bounds for expected coalescence time are known, it is not difficult to come up with examples of rotor-router configurations with multiple particles which do not lead to any coalescence. Thus a reasonable first question is to characterize such stable multiple-particle configurations. Other interesting questions are to bound the probability that a random initial configuration (defined, for example, by random initial settings of the vertex pointers) leads to full coalescing, and to analyze the coalescence time. We give some answers to the first question, leaving the other two as directions for further research. In particular, we give a full characterization of stable configurations in cycles. This characterization implies that if the length n of the cycle is prime, then any initial configuration with $k < n$ particles leads to full coalescing. We also show that all n-vertex even-degree regular graphs admit n-particle stable configurations and give a sufficient condition for odd-degree regular graphs to admit such configurations. For graphs which have vertices of degree greater than 2 and for $k < n$ particles, the full coalescence may depend more on the structure of the graph than on the primality of the number of edges or the number of vertices. As an example of this, we show a graph with m edges and constant maximum degree, which admits k-particle stable configuration, for any sufficiently large m and any $2 \le k \le \sqrt[3]{m/6}$.

The important property of non-coalescing rotor-router systems which we use in our work is the long run behaviour of such systems. For single-particle rotor-router systems, Bhatt $et\ al.$ [6] showed that within $O(nm)$ steps, the particle $enters\ (establishes)\ an\ Eulerian\ cycle$. More precisely, after the initial $stabilisation\ period$ of $O(nm)$ steps, the particle keeps repeating the same Eulerian cycle of the whole set $\overrightarrow{E}$ of directed arcs. The long run behaviour of multiple-particle rotor-routers was open for a long time, but has been recently settled by Chalopin et al. [7]. They showed that in polynomial number of steps the system reaches a stable configuration S, that is a configuration which will be repeated after some (potentially exponential) number of steps. Most importantly, they provide a strong characterization of the way the particles will be moving around the graph starting from a stable configuration. The set $\overrightarrow{E}$ can be partitioned into arc-disjoint Eulerian circuits and the particles can be assigned to these circuits such that each particle will be perpetually following the circuit it is assigned to. The circuits are arc disjoint, but may share vertices, and two or more particles can be assigned to the same circuit. Our analysis of coalescing rotor-router systems is directly based this characterization of the stable configurations of non-coalescing rotor-router systems.

2   Preliminaries

We consider an undirected, simple (no loops or multiple edges), connected graph $G = (V, E)$ with $n \ge 3$ vertices and m edges. We define $\overrightarrow{E} = \{(v, w), (w, v) : \{v, w\} \in E\}$ as the set of (directed) arcs in G. For each $v \in V$, the arcs outgoing from v are arranged in a fixed cyclic order. The vertex pointer π_v indicates the arc outgoing from v which will be taken by the next particle leaving v. When a particle leaves v along the arc indicated by the pointer π_v, the pointer advances to the next arc outgoing from v. The system works in synchronised steps and each particle moves in each step (that is, a particle never waits in the same vertex). In a coalescing rotor-router system, if two or more particles arrive at the same vertex v at the same step, they coalesce into one particle, which moves out of v in the next step in the direction indicated by the vertex pointer. In a non-coalescing system, if $q \geq 2$ particles arrive at the same vertex v at the same time, they all leave v in the next step taking the q consecutive arcs outgoing from v, starting from the arc indicated by the vertex pointer and wrapping-around, if q is greater than the degree of v. The vertex pointer changes to the arc next after the last arc taken by a particle. We do not distinguish among particles, so the order in which the particles leave vertex v in the same step is not important.

A k-particle *configuration* S is defined by the values of the vertex pointers and the position of the particles. For a configuration S, we denote by $\sigma(S)$ the set of all configurations visited starting from S (we assume the system works perpetually). We say that a configuration S is *stable*, if after starting from S we eventually return to S. Clearly, a configuration S is stable, if and only if, for each $S' \in \sigma(S)$, $\sigma(S') = \sigma(S)$. A set of configurations is stable, if it is equal to $\sigma(S)$ for a stable configuration S. Two configurations of a rotor-router system are isomorphic, if there is a one-to-one mapping on V which preserves the cyclic orders of arcs, the vertex pointers and the particle counts at vertices. We say that a graph G admits a k-particle stable configuration, if there exist cyclic orders of arcs at the vertices of G, the initial positions of vertex pointers and the initial locations of k particles, which define a k-particle stable configuration of the coalescing rotor-router system.

By definition, the rotor-router system is *locally fair*, sending, in the long run, the same number of particles into each of the arcs outgoing from the same vertex. More precisely, if S is a stable configuration and it takes T steps to return to S, then during these T steps each of the arcs outgoing from the same vertex is traversed the same number of times (otherwise the vertex pointer does not return to its initial position). This local fairness implies the *global fairness*.

Lemma 1. *If S is a stable configuration and it takes T steps to return to S, then during these T steps each arc has been traversed the same number of times.*

Proof. For each vertex v, during these T steps each arc outgoing from v has been traversed the same number of times. Let $\alpha(v)$ denote this number, let $\alpha_{\min} = \min\{\alpha(v) : v \in V\}$ and assume, by contradiction, that $U = \{v \in V : \alpha(v) = \alpha_{\min}\} \neq V$. Each arc from U to $V \setminus U$ has been traversed $\alpha_{\min}$ times but each arc from $V \setminus U$ to U has been traversed more than $\alpha_{\min}$ times. This contradicts the assumption that after T steps we are back in the initial configuration S, so the number of traversals from U to $V \setminus U$ must have been the same as the number of traversals from $V \setminus U$ to U.

Chalopin et al. [7] proved the following strong characterization of stable configurations of non-coalescing rotor-router systems, which is valid also for the coalescing systems.

Theorem 1. *[7] A configuration S is stable, if and only if, there exists a decomposition of $\overrightarrow{E}$ into arc-disjoint Eulerian circuits and an assignment of particles* *to circuits (possibly with more than one particle assigned to the same circuit) such that starting from S, each particle follows perpetually the circuit to which it is assigned.*

This theorem says that while a rotor-router system keeps changing from configuration to configuration within a stable set, each particle keeps tracing the same Eulerian circuit. The circuits are arc disjoint and cover the whole set of arcs $\overrightarrow{E}$. Two or more particles can trace the same circuit and each circuit must be traced by at least one particle (see Lemma 1). Note that opposite arcs (v, w) and (w, v) may belong to the same circuit or to two different circuits.

Corollary 1. *Let $C_0, C_1, \ldots, C_{q-1}$ be a circuit decomposition associated with a stable set σ and let $k_i \geq 1$ denote the number of particles which follow the circuit C_i. Then the ratio $|C_i|/k_i$ is the same for each circuit.*

For a stable configuration of a coalescing rotor-router system, the Eulerian circuit decomposition of Theorem 1 must be unique. Note also that while each stable configuration has an associated Eulerian circuit decomposition, the converse is not true. A decomposition of $\overrightarrow{E}$ into arc-disjoint Eulerian circuits might not correspond to any stable configuration, because it might not be possible to set up the vertex pointers and the initial positions of particles to make the particles follow the circuits.

3   Stable Configurations in a Cycle

We consider the coalescing rotor-router system based on the n-vertex cycle C_n. We first show various types of stable k-particle configurations and then prove that these are the only possible stable configurations. Throughout this section we assume that $n \geq 3$ and $k \geq 2$. Let $C_n = (v_0, \ldots, v_{n-1}, v_0)$, so $\overrightarrow{E} = \{(v_i, v_{i+1}), (v_{i+1}, v_i) : i = 0, 1, \ldots, n-1\}$, assuming $v_n \equiv v_0$. Each decompositions of $\overrightarrow{E}$ into arc-disjoint Eulerian circuits is either of the *Cycle type* (the C type) or the *Path type* (the P type), with the latter split further into two categories P1 and P2.

C: Two Eulerian circuits $(v_0, \ldots, v_{n-1}, v_0)$ and $(v_0, v_{n-1}, v_{n-2}, \ldots, v_0)$.

P: The Eulerian circuits are defined by a partitioning of the edges of the cycle C_n into edge-disjoint paths $P_0, \ldots, P_{q-1}$, $q \geq 1$. The last vertex of path P_i is the first vertex of path P_{i+1}, for $i = 0, 1, \ldots, q-1$, with $P_q \equiv P_0$. Each of these paths $P = (w_0, w_1, \ldots, w_j)$ defines the Eulerian circuit $(w_0, w_1, \cdots, w_{j-1}, w_j, w_{j-1}, \cdots, w_1, w_0)$.

P1: There is only one path, which covers the whole cycle, that is, there is only one Eulerian circuit, which covers all arcs in $\overrightarrow{E}$. Each such circuit is isomorphic to the circuit $(v_0, v_1, \cdots, v_{n-1}, v_0, v_{n-1}, \cdots, v_1, v_0)$.

P2: There are at least two paths, so there are at least two circuits in the decomposition.

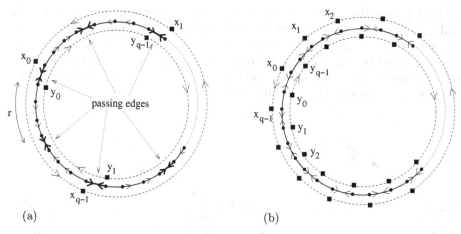

Fig. 1. Stable k-particle configurations of type C, with $k = 2q$ and $n = k(r + 1)$, for integers $q \geq 1$, $r \geq 0$. The particles on one circuit are passing the particles on the other circuit when traversing the "passing edges." There are r edges between two consecutive passing edges. (a) The case $r \geq 1$. (b) The case $r = 0$: each edge is a passing edge.

We say that a stable configuration, or a stable set, is of type X, if the associated circuit decomposition is of type X. Figure 1 shows configurations representing stable sets of type C. There are $k = 2q$ particles, for an integer $q \geq 1$, which are marked on the diagrams with small black squares. Particles $x_0, x_1, \ldots, x_{q-1}$ are assigned to the anti-clockwise circuit and the particles $y_0, y_1, \ldots, y_{q-1}$ are assigned to the clockwise circuit. The particles are evenly spaced along both circuits, with $2(r+1)$ edges between each two consecutive particles on one circuit, for an integer $r \geq 0$ (in Figure 1(a), $r = 3$). Thus the length of the cycle is $n = k(r + 1)$. Consider the relative positions of particles x and y as shown in Figure 1, when the particles x_0 and y_0 are about to traverse the same edge in opposite directions. The arrow at a vertex shows the direction where the next particle will leave this vertex. In this configuration, each pair of particles x_i and y_{q-i}, for $i = 0, 1, \ldots, q - 1$ (with $y_q \equiv y_0$), is about to traverse the same edge in opposite directions. Such traversing of the same edge in opposite directions is repeated every $r + 1$ steps, and each edge which is traversed at some step by a particle x_i in one direction and a particle y_j in the other direction is called a *passing edge*. There are $n/(r + 1) = k$ passing edges, evenly spaced along the cycle. In the case $r = 0$ (shown in Figure 1(b)), there are n particles in total, one on each vertex, and each edge is a passing edge.

We now show two different stable sets of type P2, one with one particle assigned to each circuit and one with two particles assigned to each circuit. We refer to these two types of stable sets as types P2.1 and P2.2, respectively. Stable sets of type P2.1 are illustrated in Figure 2. The cycle has $n = kr$ vertices, for an integer $r \geq 1$, the circuit decomposition is defined by k paths of equal length r, and each circuit has one particle assigned to it. Consider one configuration, which is defined by the positions of the particles in their circuits. To avoid collisions, these positions are restricted by the following condition. Let C_i and C_{i+1} be adjacent circuits which share vertex u_{i+1}, and let x_i and x_{i+1} be the particles assigned to these circuits. The distance from x_i to u_{i+1} along the circuit C_i must be different than the distance from x_{i+1} to u_{i+1} along the circuit C_{i+1}. Figure 2 shows two stable configurations of type P2.1, which belong to two different (non-isomorphic) stable sets.

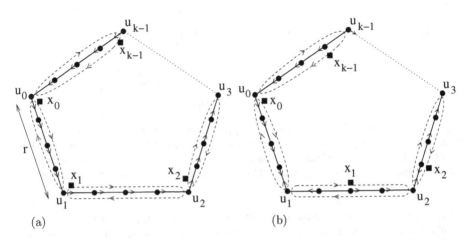

Fig. 2. Stable k-particle configurations of type P2.1; $n = kr$ and $r \geq 1$. Configurations (a) and (b) belong to two different (non-isomorphic) stable sets. In (a), each particle x_i is in the same position within its Eulerian circuits, that is, within the same distance from vertex u_i, moving in the same direction along the cycle (all clock-wise or all anti-clockwise). In (b), the particles are in different positions within their Eulerian circuits.

Figure 3 illustrates stable sets of type P2.2: the Path type, at least 2 circuits and exactly 2 particles in each circuit. The cycle has $n = q(2r + 1)$ vertices, for integers $q \geq 2$ and $r \geq 1$, and the circuit decomposition is defined by q paths of equal length $2r + 1$. Each circuit has two particles assigned to it, so $k = 2q \geq 4$. The two particles x_i and y_i assigned to the same circuit C_i are exactly half-way around the circuit from each other. They will pass each other every $2r + 1$ steps, traversing in opposite directions the middle edge of the path which defines this circuit. To avoid collisions, we have a condition restricting the relative positions of particles on the adjacent circuits, which is analogous as in the stable configurations of type P2.1 described above.

The following theorem, proven in Sections 3.1 and 3.2, gives a full characterization of the stable sets in a cycle.

Theorem 2. *Assume $n \geq 3$ and $2 \leq k \leq n$, and consider the coalescing rotor-router system based on the cycle C_n.*

   (i) *If k is odd and n is a multiple of k, then there exist k-particle stable sets and they all are of type P2.1 shown in Figure 2.*

(ii) If k is even and n is a multiple of k, then there exist k-particle stable sets and each stable set is either of type C shown in Figure 1, or of type P2.1 shown in Figure 2.

(iii) If $k \geq 4$ is even and n is an odd multiple of $k/2$, then there exist k-particle stable sets and they all are of type P2.2 shown in Figure 3.

(iv) For any other combination of n and k, each k-particle configuration leads to at least one coalescing.

Corollary 2. *Consider the coalescing rotor-router system based on the cycle C_n, where $n \geq 3$ is prime. If $2 \leq k \leq n - 1$, then each k-particle configuration leads to full coalescing (into one particle). For $k = n$, there is only one unique (up to isomorphism) stable configuration, which is shown in Figure 2(a) with $r = 1$.*

Proof. The first part follows by repeatedly applying the case (iv) of Theorem 2, while $k > 1$. The second part follows from the case (i) of Theorem 2.

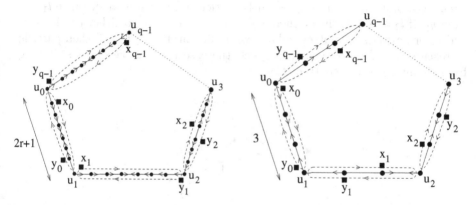

Fig. 3. Stable k-particle configurations of type P2.2, with $k = 2q$ and $n = q(2r + 1)$, for integers $q \geq 2$ and $r \geq 1$. (a) The general case $r \geq 1$. (b) The case $r = 1$, where each circuit is in one of three states, illustrated by the three circuits with particles x_0 and y_0, x_1 and y_1, and x_2 and y_2.

3.1   Stable Configurations of the Cycle Type

Lemma 2. *If n is odd or k is odd, then there is no stable set with Eulerian decomposition of type C.*

Proof. Let σ be a k-particle stable set for C_n with an Eulerian decomposition of type C. Corollary 1 implies that the same number of particles must be assigned to each of the two circuits, so k must be even. Let C_1 and C_2 denote the two circuits and let x be a particle assigned to C_1 and y a particle assigned to C_2. Consider a configuration $S \in \sigma$ such that x and y face each other along an edge $\{v, u\}$: x is at v and will move to u in the next step, while y is at u and will move to v in the next step. If n were odd, then after $(n + 1)/2$ steps particles x and y would collide on the opposite side of the cycle.

Lemma 3. *For n and k both even, if there is a stable set with Eulerian decomposition of type C, then each of the two circuits has the same number of particles assigned to it, and the particles assigned to the same circuit must be evenly spaced along this circuit.*

Proof. The condition that each of the two circuits has the same number of particles assigned to it follows from Lemma 1. To prove the second part of the lemma, assume by contradiction that particles on one of the circuits (or on both of them) are not evenly spaced. This implies that there are two consecutive particles x_1 and x_2 on C_1 (x_1 next after x_2 in the direction of C_1) and two consecutive particles y_1 and y_2 on C_2 (y_1 next after y_2 in the direction of C_2) such that x_1 is ahead of x_2 by l_1 arcs and y_1 is ahead of y_2 by l_2 arcs, for some $l_1 \neq l_2$. Assume by symmetry that $l_1 < l_2$ and consider the step when particles x_1 and y_1 have just passed each other, as shown in Figure 4. Particle x_1 is at a vertex v and will be moving towards particle y_2, which is at distance $l_2 - 1$ from x_1. Particle y_1 is at the vertex u next after to vertex v in the direction of circuit C_2, and will be moving towards particle x_2, which is at distance $l_1 - 1$ from y_1. If $l_2 = l_1 + 1$, then either $l_2 - 1$ or $l_1 - 1$ is even, so either particles x_1 and y_2 or particles y_1 and x_2 collide: contradiction. If $l_2 \geq l_1 + 2$, then particle x_2 reaches vertex v before particle y_2 gets there, so x_2 turns back at v, switching from circuit C_1 to C_2: contradiction.

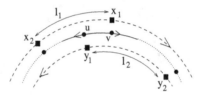

Fig. 4. For the proof of Lemma 3: Eulerian decomposition of type C and particles are not evenly spaced along the circuits.

Lemma 4. *For n and k both even and $k \nmid n$, there is no stable set with Eulerian decomposition of type C.*

Proof. Let n and k be both even and $k \nmid n$, that is, $k = 2q$ and $n = pk + 2r$, for some positive integers q, p, r such that $r < q$. Observe that we must have $q \geq 2$. Assume, by contradiction, that there is a stable set σ with an Eulerian decomposition $\{C_1, C_2\}$ of type C, and start in a configuration $S \in \sigma$. Each of the two circuits has q particles assigned to it and they are evenly spaced along the circuit (Lemma 3). This implies that $q \mid n$, so we must have $q = 2r$. The distance between two consecutive particles on the same circuit is equal to $n/q = 2p + 1$, so it is odd. Consider a particle x on circuit C_1 and two consecutive particles y_1 and y_2 on the other circuit. Since the distance between y_1 and y_2 is odd, then either the distance along circuit C_1 from x to y_1 is even or the distance along C_1 from x to y_2 is even. Thus x will collide with y_1 or y_2: contradiction.

Lemma 5. *For n and k both even and k | n, there exists a unique (up to isomorphism) stable set with an Eulerian decomposition of type C. This unique stable set is illustrated in Figure 1.*

Proof. Let $k = 2q$ and $n = k(r + 1)$ for an integer $r \geq 0$. From Lemma 3, a stable set with an Eulerian decomposition of type C has q particles assigned to each circuit and the distance between any two consecutive particles on the same circuit is equal to $2(r + 1)$. This is the stable set illustrated in Figure 1.

3.2   Stable Configurations of the P Type

Lemma 6. *Let σ be a stable set of type P and let C be one of the circuits in the Eulerian decomposition associated with σ. Then no more than two particles are assigned to C.*

Proof. Recall that C spans a path P in the cycle and the particles assigned to C keep walking along the path from one end to another. For any two particles x and y assigned to C, the parity of the distance along P between these particles remains constant. This distance must be odd, because if it were even, then particles x and y would eventually meet. If there were three particles assigned to C, then not all three pairwise distances between these particles could be odd, so two of the three particle would have to meet.

Lemma 7. *Let σ be a stable set of type P and let C be one of the circuits of the Eulerian decomposition associated with σ, and assume that two particles are assigned to C. Then (i) the two particles are at the same time at the opposite ends of the path P which is spanned by C, (ii) path P has an odd number of edges and (iii) the two particles on C always pass each other when traversing (in the opposite directions) the middle edge of the path.*

Proof. Let x and y be the particles assigned to C and let u and v be the end vertices of the path P. To show (i), assume by contradiction that at some step particle x is at vertex v but particle y is not at the other end u. Assume that y is moving towards vertex v. (If y is moving towards u, wait until y reaches u to get an analogous arrangement: y at u and x not at the other end of the path and moving towards u.) Particles x and y will now be moving towards each other, eventually overpassing at step t along some edge $\{w, r\}$: particle x traverses this edge from r to w while particle y traverses from w to r. When now x leaves vertex w to go towards u, the pointer at w is changed to arc (w, r). The distance between w and u is at least the distance between w and v, so the next time a particle comes to w, it will be particle y and it will go back from w to r, contradicting the movement of both particles along the circuit C.

We have shown that at some step particles x and y are the opposite ends of path P. If P had an even $2q$ number of edges, then the particles would meet after the next q steps. The particles must be passing each other when traversing (in the opposite directions) the middle edge of the path, or otherwise they would not be at the end vertices of the path at the same time.

Corollary 3. *There is no stable set of type P1, that is, for each stable set of type P, its Eulerian circuit decomposition must have at least two circuits.*

Lemma 8. *Suppose σ is a stable set of type P2. Then all circuits of the Eulerian decomposition associated with σ have the same length and the same number of particles. That is, each stable set of type P2 is either of type P2.1 shown in Figure 2, or of type P2.2 shown in Figure 3.*

Proof. Assume that the Eulerian decomposition contains a circuit C_1 with one particle and a circuit C_2 with two particles. Then Corollary 1 implies that $|C_1| = |C_2|/2$, but $|C_1|$ is even while Lemma 7 implies that $|C_2|/2$ is odd; contradiction. Thus all circuits in the Eulerian decompositions must have the same number of particles, either one or two, and Corollary 1 further implies that they all must have the same length.

Lemma 9.
   (i) *If n is a multiple of k, then each stable set of type P2 is of type P2.1.*
   (ii) *If k is even and n is an odd multiple of $k/2$, then each stable set of type the P2 is of type P2.2.*

Proof. Lemma 8 says that each stable set of type P2 must be either P2.1 or P2.2. Part *(i)* holds because if we have a stable set of type P2.2, then k is even and Lemma 7 implies that $n = (k/2)(2r+1)$, for some integer $r \geq 0$, so n is not a multiple of k. Part *(ii)* holds because if we have a stable set of type P2.1, then Lemma 8 implies that n is a multiple of k.

4   General Graphs

For general graphs, we first look at the case $k = n$. We saw that in cycles there can be only two different n-particle stable sets. One requires an even n and is shown in Figure 1(b), while the other applies to an arbitrary $n \geq 3$ and is shown in Figure 2(a) with $r = 1$. The stable set in Figure 1(b) can be viewed in the following way. The set of edges E of an even-length cycle is partitioned into two perfect matchings M_1 and M_2. For any configuration in this stable set, either all vertex pointers are set onto the edges in M_1 or all of them are set onto the edges in M_2. We describe now a generalization of such stable sets to graphs with higher vertex degrees, considering perfect matchings as well as 2-factors (collections of vertex-disjoint cycles covering all vertices). We show first that n-particle stable sets can exist only in regular graphs.

Theorem 3. *If a connected n-vertex graph has an n-particle stable set, then the graph must be regular.*

Proof. Consider a connected n-vertex graph $G = (V, E)$ which has an n-particle stable set. Let C be a circuit in the Eulerian decomposition for this stable set. In each step one particle leaves each vertex, so in each step each vertex pointer advances to the next arc. Therefore, if a particle x assigned to C passes in the current step through an arc (v, w), then the next particle on C will pass through (v, w) in exactly $\deg(v)$ steps, where $\deg(v)$ is the degree of v in G. This means that the particles assigned to C must be equally spaced around C, with distance $\deg(v)$ between the consecutive particles. Since v is an arbitrary vertex on C, all vertices on C must have the same degree. Thus if two circuits share a vertex, then all vertices on these two circuits have the same degree. Since the graph is connected, all vertices must have the same degree.

Lemma 10. *A connected d-regular n-vertex graph has an n-particle stable configuration if and only if the set of arcs $\overrightarrow{E}$ can be partitioned into sets H_1, $H_2, \ldots, H_d$, such that each H_i is a collection of vertex-disjoint simple arc-cycles.*

Proof. If there is an n-particle stable configuration S in a d-regular graph, then denote by S_i, for $i = 1, 2, \ldots, d$, the configuration at the beginning of step i, starting form configuraton $S = S_1$. Let H_i be the set of pointer arcs in configuration S_i. Each H_i is a collection of vertex-disjoint simple arc-cycles (the movement of the n particles in a given step defines a one-to-one mapping on V) and each arc belongs to exactly one H_i (from the rotor-router property).

Conversely, if H_1, $H_2, \ldots, H_d$ are collections of vertex-disjoint simple arc-cycles and these collections partition $\overrightarrow{E}$, then the d arcs outgoing from any vertex belong to different collections. For each vertex v, set the order of the arcs outgoing from v so that the i-th arc is the arc belonging to H_i, and initialize the vertex pointer to the first arc. These orders of arcs, the vertex pointers, and the assignment of one particle to each vertex define an n-vertex stable configuration.

Lemma 11. *If the edges of a connected regular graph can be partition into 2-factors and perfect matchings, then this graph admits an n-particle stable configuration.*

Proof. Each perfect matching defines one vertex-disjoint collection of simple arc-cycles covering V: each edge of the matching defines one two-arc cycle. Each 2-factor gives two vertex-disjoint collections of arc-cycles: for each cycle in the 2-factor, one orientation of this cycle is included in one collection and the other orientation in the other collection.

Corollary 4. *Let d be a positive integer. Each n-vertex, (2d)-regular graph admits an n-particle stable configuration. Each n-vertex, (2d + 1)-regular graph which has a perfect matching admits an n-particle stable configuration. These graphs include all (2d + 1)-regular (2d)-connected graphs.*

Proof. Petersen's 2-factor theorem says that every regular graph of even degree has a 2-factor, so (by iterating this theorem) it can be partitioned into 2-factors (see [15]). If a $(2d + 1)$-regular graph has a perfect matching, then the edges which are not in this perfect matching form a $(2d)$-regular graph.

Petersen's matching theorem says that every 3-regular, 2-connected graph has a perfect matching. Babler's generalization of this theorem says that every $(2d + 1)$-regular, $(2d)$-connected graph has a perfect matching (see [15]).

We now consider the case when $k < n$. In cycles, k-particle stable configurations exist only if n is a multiple of k, or a multiple of $k/2$ for an even $k \geq 4$. Thus in a cycle of prime length any initial configuration leads to full coalescing. There are no similar strong conditions for general graphs. Actually, if we allow vertices of degree 3 or higher, then the coalescence seems to depend more on the structure of the graph than on the primality of n or m. As an example, we show that for each sufficiently large m (which can be prime) and each $2 \leq k \leq \sqrt[3]{m/6}$, there is a connected graph with m edges which admits a k-particle stable configuration. Both the number of edges and the number of nodes in this example can be co-prime with the number of particles.

Our example is illustrated in Figure 5. The set of edges E is partitioned into $k + 2$ components: a tree T with $k + 1$ leaves $r_0, r_1, \ldots, r_k$, and at most $k - 1$ internal vertices, each of degree at least 3, and vertex-disjoint connected sub-graphs $H_0, H_1, \ldots, H_k$. Sub-graph H_i shares vertex r_i with T and has either $\lfloor h \rfloor$ or $\lceil h \rceil$ edges, where $h = (m - |T|)/(k + 1)$ and $|T| \leq 2k - 1$ is the number of edges in T. We now show a k-particle stable configuration in this graph. Fix an Eulerian cycle C of the whole set $\overrightarrow{E}$. The arcs of H_i form one segment of C, which is an Eulerian circuit C_i of the arcs of H_i. If we remove all circuits C_i from C, then the remaining arcs form an Eulerian tour of T. The numbering $r_0, r_1, \ldots, r_k$ of leaves of T is consistent with the reverse order of this Eulerian tour of T. For each component H_i, we set the cyclic orders of arcs and the vertex pointers in such a way that one particle starting from vertex r_i would first follow the whole circuit C_i before entering tree T. The positions of the pointers are not final yet; they will be adjusted. The (cyclic) order $(v, w_1), \ldots, (v, w_{\deg(v)})$ of the arcs outgoing from an internal vertex v in T is consistent with the numbering of the leaves of T: if arcs $(v, w_1), \ldots, (v, w_j)$ lead to leaves $r_{i_1}, \ldots, r_{i_j}$, respectively, then $r_{i_1} < r_{i_2} < \cdots < r_{i_j}$ (the anti-clockwise order in Figure 5). The pointers at the internal vertices in T are set in the direction of vertex r_k.

The final stage of our construction of a stable configuration is the placement of the k particles $x_0, x_1, \ldots, x_{k-1}$ and the adjustment of the vertex pointers. All particles will be following the Eulerian circuit C and we show their initial positions in relation to this circuit. We place particle x_0 at vertex r_0 and change the vertex pointer at r_0 to the arc (r_0, p) of the tree T. This will be the next arc on C taken by x_0. We place particle x_1 on C at distance either $\lfloor g \rfloor$ or $\lceil g \rceil$ arcs *behind* particle x_0, where $g = 2m/k$. Generally, we place particles $x_1, x_2, \ldots, x_k$ so that each distance from x_i to x_{i-1} (including from x_0 to x_{k-1}) is either $\lfloor g \rfloor$ or $\lceil g \rceil$. Thus the distance from x_i to x_0 is between $i \lfloor g \rfloor$ and $i \lceil g \rceil$. The values h and g and the assumption that $k \leq \sqrt[3]{m/6}$ imply that for each $i = 1, 2, \ldots, k - 1$, $i \lfloor g \rfloor \geq 2(|H_0| + |H_1| + \cdots + |H_{i-1}| + |T|)$ and $i \lceil g \rceil \leq 2(|H_0| + |H_1| + \cdots + |H_i|)$. This means that particle x_i is in H_i and H_k is empty (does not have any particle). Finally, for each $i = 1, 2, \ldots, k - 1$, we adjust the vertex pointers in H_i by simulating the rotor-router movement of a "ghost" particle from r_i to the position of particle x_i. With this adjustment of vertex pointers in H_i, particle x_i will complete traversing circuit C_i (the traversing started by the ghost particle) and then will enter tree T (assuming no interference from other particles). This completes the construction of a stable configuration.

Starting from the constructed configuration, the particles will move according to the following pattern. First particle x_0 moves to r_k along the $r_0 - r_k$ path in T in $O(k)$ steps, while the other particles move inside their initial H sub-graphs. Then particle x_1 completes the traversing of H_1, arrives at vertex r_1 and is ready to enter tree T. This completes the first phase and at this point, we have a configuration similar to the initial configuration, but now H_0 is empty. In the next phase, first particle x_1 moves to r_0 along the $r_1 - r_0$ path in T, while the other particles move inside their current H sub-graphs and x_2 reaches r_2. In the subsequent phases, particle x_2 moves from H_2 to H_1, then particle x_3 moves from H_3 to H_2, and so on. After $k(k+1)$ phases, the system is back in the initial configuration. It can be shown that no two particles will be at the same time in T or in the same H component, so no two particles ever collide.

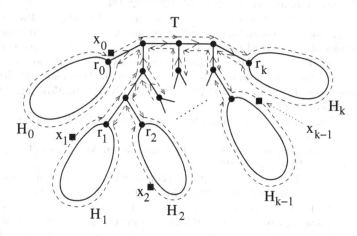

Fig. 5. A graph with m edges and a stable k-particle configuration, where $k = \Theta(m^{1/3})$.

References

1. Aldous, D., Fill, J.A.: Reversible markov chains and random walks on graphs 2002. Unfinished monograph, recompiled (2014).
   http://www.stat.berkeley.edu/~aldous/RWG/book.html
2. Aldous, D.J.: Meeting times for independent markov chains. Stochastic Processes and their Applications 38(2), 185–193 (1991)
3. Alon, N., Avin, C., Koucky, M., Kozma, G., Lotker, Z., Tuttle, M.R.: Many random walks are faster than one. In: Proc. 20th Annual Symposium on Parallelism in Algorithms and Architectures, SPAA 2008, pp. 119–128. ACM (2008)
4. Bampas, E., Gąsieniec, L., Hanusse, N., Ilcinkas, D., Klasing, R., Kosowski, A.: Euler tour lock-in problem in the rotor-router model. In: Keidar, I. (ed.) DISC 2009. LNCS, vol. 5805, pp. 423–435. Springer, Heidelberg (2009)
5. Bampas, E., Gasieniec, L., Klasing, R., Kosowski, A., Radzik, T.: Robustness of the rotor-router mechanism. In: Abdelzaher, T., Raynal, M., Santoro, N. (eds.) OPODIS 2009. LNCS, vol. 5923, pp. 345–358. Springer, Heidelberg (2009)

6. Bhatt, S.N., Even, S., Greenberg, D.S., Tayar, R.: Traversing directed eulerian mazes. J. Graph Algorithms Appl. 6(2), 157–173 (2002)
7. Chalopin, J., Das, S., Gawrychowski, P., Kosowski, A., Labourel, A., Uznanski, P.: Lock-in problem for parallel rotor-router walks. CoRR, abs/1407.3200 (2014)
8. Cooper, C., Elsässer, R., Ono, H., Radzik, T.: Coalescing random walks and voting on connected graphs. SIAM J. Discrete Math. 27(4), 1748–1758 (2013)
9. Cooper, C., Frieze, A.M., Radzik, T.: Multiple random walks in random regular graphs. SIAM J. Discrete Math. 23(4), 1738–1761 (2009)
10. Dereniowski, D., Kosowski, A., Pajak, D., Uznanski, P.: Bounds on the cover time of parallel rotor walks. In: 31st International Symposium on Theoretical Aspects of Computer Science, STACS 2014, pp. 263–275 (2014)
11. Efremenko, K., Reingold, O.: How well do random walks parallelize? In: Dinur, I., Jansen, K., Naor, J., Rolim, J. (eds.) APPROX and RANDOM 2009. LNCS, vol. 5687, pp. 476–489. Springer, Heidelberg (2009)
12. Elsässer, R., Sauerwald, T.: Tight bounds for the cover time of multiple random walks. Theor. Comput. Sci. 412(24), 2623–2641 (2011)
13. Israeli, A., Jalfon, M.: Token management schemes and random walks yield self-stabilizing mutual exclusion. In: Proceedings of the Ninth Annual ACM Symposium on Principles of Distributed Computing, PODC 1990, pp. 119–131. ACM (1990)
14. Kosowski, A., Pajak, D.: Does adding more agents make a difference? A case study of cover time for the rotor-router. In: Esparza, J., Fraigniaud, P., Husfeldt, T., Koutsoupias, E. (eds.) ICALP 2014, Part II. LNCS, vol. 8573, pp. 544–555. Springer, Heidelberg (2014)
15. Lovász, L., Plummer, D.: Matching Theory. AMS Chelsea Publishing Series. American Mathematical Soc. (2009)
16. Oliveira, R.: On the coalescence time of reversible random walks. Trans. Amer. Math. Soc. 364, 2109–2128 (2012)
17. Priezzhev, V.B., Dhar, D., Dhar, A., Krishnamurthy, S.: Eulerian walkers as a model of self-organized criticality. Phys. Rev. Lett. 77, 5079–5082 (1996)
18. Wagner, I.A., Lindenbaum, M., Bruckstein, A.M.: Smell as a computational resource - A lesson we can learn from the ant. In: ISTCS, pp. 219–230 (1996)
19. Wagner, I.A., Lindenbaum, M., Bruckstein, A.M.: Distributed covering by ant-robots using evaporating traces. IEEE T. Robotics and Automation 15(5), 918–933 (1999)
20. Yanovski, V., Wagner, I.A., Bruckstein, A.M.: A distributed ant algorithm for efficiently patrolling a network. Algorithmica 37(3), 165–186 (2003)

Secure Multi-party Shuffling*

Mahnush Movahedi, Jared Saia, and Mahdi Zamani

Department of Computer Science, University of New Mexico
{movahedi,saia,zamani}@cs.unm.edu

Abstract. In secure multi-party shuffling, multiple parties, each holding an input, want to agree on a random permutation of their inputs while keeping the permutation secret. This problem is important as a primitive in many privacy-preserving applications such as anonymous communication, location-based services, and electronic voting. Known techniques for solving this problem suffer from poor scalability, load-balancing issues, trusted party assumptions, and/or weak security guarantees.

In this paper, we propose an unconditionally-secure protocol for multi-party shuffling that scales well with the number of parties and is load-balanced. In particular, we require each party to send only a polylogarithmic number of bits and perform a polylogarithmic number of operations while incurring only a logarithmic round complexity. We show security under universal composability against up to about $n/3$ fully-malicious parties. We also provide simulation results showing that our protocol improves significantly over previous work. For example, for one million parties, when compared to the state of the art, our protocol reduces the communication and computation costs by at least three orders of magnitude and slightly decreases the number of communication rounds.

1 Introduction

Shuffling a sequence of values is a fundamental tool for randomized algorithms; applications include fault-tolerant algorithms, cryptography, and coding theory. In *secure multi-party shuffling (MPS)* problem, a group of parties each holding an input value want to randomly permute their inputs while ensuring no party can map any of the outputs to any of the input holders better than can be done with a uniform random guess. An MPS protocol is a useful primitive for achieving privacy and robustness in many applications such as anonymous communication [12], location-based services [23], electronic voting [28], secure auctions [18], and general multi-party computation [9].

Despite many applications of MPS, we are not aware of any technique that can be used to achieve a scalable and secure MPS protocol. We believe this is of increasing importance with the growth of modern networks. Moreover, most protocols lack load-balancing – a crucial requirement for protocols running in large networks. With the rise of sophisticated cyber-attacks, it is now essential to provide provable guarantees against strong adversaries. Also, relying on trusted parties has become a major security issue in today's world.

---

* An extended version of this paper is available at
  http://cs.unm.edu/~zamani/papers/sirocco15

© Springer International Publishing Switzerland 2015
C. Scheideler (Ed.): SIROCCO 2015, LNCS 9439, pp. 459–473, 2015.
DOI: 10.1007/978-3-319-25258-2_32

In this paper, we address these concerns by proposing a scalable and load-balanced protocol for MPS that is unconditionally-secure against malicious attacks and does not rely on trusted parties.

Our Contribution. We first propose a formal definition of security for MPS. Our definition is different from the standard definition of security for *multi-party computation (MPC)* [7], where a group of parties each holding a *private input* want to compute a known function over their inputs, without revealing any more information about their inputs than what is revealed by the output of the function. Instead of focusing on inputs privacy, we base our definition on the secrecy of the output permutation.

Next, we propose an unconditionally-secure MPS protocol that scales polylogarithmically with the number of parties, tolerates malicious faults, and is load-balanced. Simulations of our protocol suggest that it compares favorably with the current state of the art in terms of communication cost, computation cost, and the number of communication rounds.

In our protocol, we achieve sublinear per-party communication complexity by requiring each party to only communicate with polylogarithmic-size groups of parties rather than with *all* parties. This approach, however, introduces important technical challenges to our model; the most important one is to guarantee the adversary cannot break the security of our protocol via coalitions of corrupted parties in more than one group, when we share the same secret information with the parties in these groups. Some prior work solve this by relaxing the load-balancing requirement [9], the resiliency bound [33], or practical efficiency [17]. We propose a novel technique called *share renewal* without relaxing any of these requirements.

When a protocol is concurrently executed alongside other protocols, one requires to ensure this composition preserves the security of the protocol. Since our goal is to design *modular* MPS protocols that can be flexibly used with other protocols, we show security of our protocol under the universal composability framework as described by Canetti [11].

Our Model. Consider n parties $P_1, ..., P_n$ in a fully-connected synchronous network with private and authenticated channels. We assume the parties have no access to any trusted party and/or to any reliable broadcast channel. We consider a *malicious adversary* who corrupts at most $t < n$ of the parties and can see (and analyze) the entire traffic in the network, but cannot see the content of messages transmitted between uncorrupted parties since we assume private links. The corrupted parties not only can gossip their information with other corrupted parties but also can deviate from the protocol in any arbitrary manner, *e.g.*, by sending invalid messages or remaining silent. We finally assume that the adversary is *static* meaning that it has to select the set of corrupted parties at the start of the protocol.

Problem Statement. Let $\mathbb{F}$ be a finite field, and $\pi : \{1, ..., n\} \to \{1, ..., n\}$ denote a *permutation*; a one-to-one and onto function that maps a sequence of n elements $(x_1, ..., x_n) \in \mathbb{F}^n$ to another sequence $(x_{\pi(1)}, ..., x_{\pi(n)}) \in \mathbb{F}^n$. For $i \in \{1, ..., n\}$, every party P_i holds an input $x_i \in \mathbb{F}$. A *multi-party shuffling (MPS)* protocol allows these parties to agree on a permutation π of the sequence $(x_1, ..., x_n)$. We consider two variants of this problem. In the first variant, which we call *single-output MPS*, each party P_i is required to receive only one of the shuffled inputs $x_{\pi(i)}$. In the second variant, which we call *all-output MPS*, each party receives the entire output sequence $(x_{\pi(1)}, ..., x_{\pi(n)})$. We now define our notion of security.

Definition 1. *An MPS protocol is said to be t-secure if and only if, in the presence of a malicious adversary corrupting up to $t < n$ of the parties, the protocol ensures*

- Unlinkability: *the adversary can guess π correctly with probability at most $\frac{1}{(n-t)!}$. We refer to the set of possible permutations from which the adversary tries to guess the secret permutation π as the* unlinkability set.

- Correctness: *each party is guaranteed that the output it receives is one of the inputs (for single-output MPS) or contains all (and only all) the inputs (for all-output shuffle).*

- Output delivery: *corrupted parties cannot prevent honest parties from receiving their output.*

In this paper, we consider a relaxed version of Definition 1. This allows us to achieve the highest level of efficiency in our protocol in exchange of a very small increase in the success probability of the adversary.

Definition 2. *We say an MPS protocol is* almost *t-secure if and only if in the presence of a malicious adversary corrupting up to $t < n$ of the parties, the protocol guarantees correctness and output delivery, and that the adversary can guess π correctly with probability at most $\frac{1}{(n-t)!}(1 + \delta)$, where $\delta = o(1)$ is called the* deviation factor.

1.1 Our Results

We prove the following main theorem in [1].

Theorem 1. *There exists a universally-composable MPS protocol such that with probability $1 - O(n^{-3})$, it guarantees the following properties:*

- *The protocol is almost t-secure against a computationally-unbounded malicious adversary with static corruptions, where $t < (1/3 - \epsilon)n$, for some positive constant ϵ.*

- *The deviation factor is $O(2^{-2^{k\sqrt{\log n}}})$, for some constant $k > 1$.*

- *Each party sends $\tilde{O}(1)$ bits and computes $\tilde{O}(1)$ operations.*[1]

- *The protocol terminates after $O(\log n)$ rounds of communication.*

In [1], we also construct a computationally-secure variant of Theorem 1 to observe (via simulations) how much cryptographic techniques can influence practical efficiency of our protocol. This protocol provides the same guarantees as

---

[1] The symbol $\tilde{O}$ is used as a variant of the big-O notation that ignores the logarithmic factors. Thus, $f(n) = \tilde{O}(g(n))$ means $f(n) = O(g(n) \log^k g(n))$, for some k.

Theorem 1 except for a polynomially time-bounded adversary. We provide our simulation results in Section 4.

1.2 Related Work

Shuffling in the multi-party setting has already been studied, primarily in the context of *mix-nets*. As first defined by Chaum [12], a mix-net consists of a chain of servers (called *mixes*) that randomly reorder a sequence of messages in a way that the correlation with the corresponding input messages remains hidden. To ensure honest behavior in the malicious setting, a *verifiable shuffling* [2,28] technique is often used, where each mix is asked to prove correctness of its shuffles without leaking how the shuffle was performed.

Unfortunately, Mix-nets and verifiable shuffling techniques rely on cryptographic assumptions. Moreover, mix-nets require semi-trusted servers and are known to be vulnerable to *traffic-analysis* attacks [29]. Protocols such as [8,30] attempt to solve this with provable guarantees. However, they are either complicated and scale linearly with the number of parties [30], or are not secure against malicious attacks and an adversary monitoring all communication channels [8].

Chaum [13] uses MPC to introduce the *dining cryptographers network (DC-net)* for achieving unlinkability between inputs and outputs; a crucial requirement for both anonymous communication and MPS. The DC-net eliminates the two limitations of Mix-nets: cryptographic assumptions and traffic-analysis vulnerability.

Although the original DC-net allows only one participant to broadcast at a time, there are variants such as [31] that implement all-to-all anonymous broadcast and thereby enable multi-party shuffling of the inputs. Unfortunately, DC-nets suffer from collision and jamming attacks. Although several work address these issues [14,21,31], they either do not scale well with the number of parties [21,31] or require a few highly-available servers [14].

MPS is closely related to *data-oblivious* protocols [20]. A protocol is data-oblivious if its control flow is independent of input data. Such a protocol can be used to anonymize access patterns or prevent an adversary from taking over a certain fraction of protocol inputs. Customized shuffling techniques are designed in the context of oblivious RAMs [20], oblivious database manipulation [26], oblivious sorting [22,24,34], and evaluation of sublinear functions [9].A multi-party sorting protocol such as that of [24,34] can be used to perform MPS. Although these protocols scale well with n, they scale poorly with the number of parties.

Rackoff and Simon [30] show that if all parties send at each time step, then the traffic-analysis problem can be solved using MPC. This means that a general MPC scheme such as [7] that can securely compute any functionality (including shuffling), can be used to design an MPS protocol with traffic-analysis resistance. Although much theoretical progress has been made in the MPC literature to achieve polylogarithmic overhead [9,17], there is a lack of practical solutions, especially for large number of parties. Moreover, most of these techniques cannot be easily implemented due to a lack of detailed protocol specifications.

In Table 1, we compare our main protocol with several other ones that can be used to solve the MPS problem. To make a fair comparison with the MPC proto-

Table 1. Comparison of MPS techniques

Protocol	Adversarial Power	Malicious Adversary?	Fraction of Parties Controlled	MPS Security	Latency	Bandwidth	Easy to Implement?
Chaum [12]	Computational	No	$O(1)^\dagger$	See note§	polylog(n)	polylog(n)	Yes
Rackoff et al. [30]	Computational	No	$O(1)^\dagger$	Statistical£	polylog(n)	$\tilde{O}(n)$	No
Berman et al. [8]	Computational	No	$O(1)^\dagger$	Statistical£	polylog(n)	polylog(n)	Yes
Boyle et al. [9]	Computational	Yes	$1/3 - \epsilon$	Almost	polylog(n)	$\tilde{O}(n)$	No
Dani et al. [17]	Unconditional	Yes	$1/3 - \epsilon$	Almost	$O(\log n)$	$\tilde{O}(\sqrt{n})$	No
This paper	Unconditional	Yes	$1/3 - \epsilon$	Almost	$O(\log n)$	$\tilde{O}(1)$	Yes

† This protocol assumes the rest of parties are trusted.
‡ [29] shows traffic-analysis attacks on this protocol if all links are monitored by the adversary.
§ Originally supposed to generate perfect shuffles but known attacks reduce shuffle security.
£ Measures the statistical distance between the distribution generated by the system and the uniform distribution [30].

cols of [9,17], we use their techniques to compute our own shuffling functionality described in Section 3. In this table, by bandwidth we mean the communication complexity per shuffled message delivered.

2 Preliminaries

We now define our notation and describe the tools used throughout this paper.

Notation. For prime p, let $\mathbb{F}_p$ denote a finite field with p elements. We say an event occurs *with high probability*, if it occurs with probability $1 - 1/n^c$, for some positive constant c.

Verifiable Secret Sharing. A *secret sharing* protocol allows a party (called *the dealer*) to share a secret among n parties such that any set of t or less parties cannot gain any information about the secret, but any set of at least $t + 1$ parties can reconstruct it. A *verifiable secret sharing (VSS)* protocol is a secret sharing protocol with the additional property that after the sharing phase, a corrupted dealer is either disqualified or the honest parties can reconstruct the secret, even if shares sent by corrupted parties are spurious. In our protocol, we use the VSS scheme of Ben-Or et al. [7]. We refer to the sharing protocol of this scheme as VSS-Share and to its reconstruction protocol as VSS-Reconst.

Theorem 2 ([7]). *There exists a synchronous linear VSS scheme for $t < n/3$ that is unconditionally-secure against a static malicious adversary.*

Quorum Building. King et al. [25] give a protocol that can be used to bring all parties to agreement on a collection of n quorums. A *quorum* is a set of $N = O(\log n)$ parties, where it is guaranteed that at most a fixed fraction of the parties in the set are corrupted. In general, one can use any BA algorithm (such as [10]) to build a set of quorums in the way described in [25].

Theorem 3 ([10,25]). *There exists an unconditionally-secure protocol that brings all honest parties to agreement on n quorums with probability $1 - O(n^{-3})$. The protocol has $\tilde{O}(n)$ amortized communication and computation complexity over the number of parties, and it can tolerate up to $(1/3 - \epsilon)n$ corrupted parties, for any positive ϵ. Each quorum is guaranteed to have $T < N/3$ corrupted parties.*

We refer to this protocol as Build-Quorums. Several recent MPC schemes [9,33] make use of quorums to achieve scalability. We are particularly inspired by Dani *et al.* [17].

Sorting Networks. A *sorting network* is a network of *comparators*. Each comparator is a gate with two input wires and two output wires. When two values enter a comparator, it outputs the lower value on the top output wire, and the higher value on the bottom output wire. Ajtai *et al.* [4] describe an asymptotically-optimal $O(\log n)$ depth sorting network. However, this network is not practical due to large constants hidden in the depth complexity. Leighton and Plaxton [27] propose a *probabilistic sorting circuit* with depth $7.44 \log n$ that sorts a randomly chosen input permutation with very high probability meaning that it sorts all but $\sigma \cdot n!$ of the $n!$ possible input permutations, where $\sigma = 1/2^{2^{\kappa\sqrt{\log n}}}$, for some constant $\kappa > 0$.[2]

Secure Comparison. Given two linearly secret-shared values a, b, Damgård *et al.* [16] propose an efficient protocol for computing a new secret-shared value $\rho = (a \leq b)$ meaning that ρ is 1 if $a \leq b$ and 0 otherwise. Their protocol is unconditionally secure, has $O(1)$ rounds, and requires $O(\ell)$ invocations of a secure multiplication protocol, where ℓ is the bit-length of elements to be compared. We denote this protocol by Compare. For multiplication of secret-shared values, we use the protocol of Ben-Or *et al.* [7] with the simplifications of Gennaro *et al.* [19]. By plugging the VSS of Theorem 2 into the protocol of [19], we achieve an unconditionally-secure multiplication protocol, which we denote by Multiply.

3   Our Protocol

We now describe our MPS protocol. Consider two finite fields $\mathbb{F}_p$ and $\mathbb{F}_q$ of prime orders p and q respectively. The high-level idea is as follows: for each party P_i holding an input $x_i \in \mathbb{F}_p$, a uniform and independent random value $r_i \in \mathbb{F}_q$ is chosen to form an input pair (r_i, x_i), where $i \in [n]$. Then, the sequence of pairs $((r_1, x_1), ..., (r_n, x_n))$ is sorted according to the first elements of the pairs. We show that, for sufficiently large prime q, this algorithm randomly shuffles the sequence of inputs $(x_1, ..., x_n)$ with high probability.

To compute this functionality securely, we construct the circuit shown in Figure 1, which we denote by $\mathcal{M}$. This circuit consists of the probabilistic sorting circuit of [27] augmented by n input gates; the functionality of each gate is computed by a quorum. $\mathcal{M}$ is created jointly by all parties before the protocol starts during an input-independent setup phase. Then, it is jointly evaluated by all parties possibly many times to shuffle many input sequences[3].

---

[2] This gives a Monte Carlo guarantee: for $(1 - \sigma)n!$ of input permutations, the circuit sorts correctly, but for the rest $\sigma n!$ permutations, it simply fails and gives no guarantees.

[3] This setup phase is information-theoretically secure and does not rely on one-time pads. Thus, the same $\mathcal{M}$ can be used *any* number of times for shuffling many input sequences.

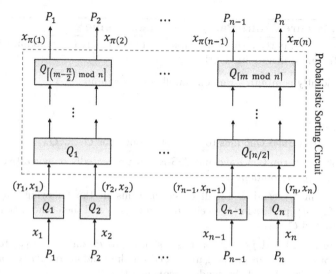

Fig. 1. MPS circuit

The circuit $\mathcal{M}$ is constructed in the following way. First, we create n quorums $Q_1, ..., Q_n$ each with $N = O(\log n)$ parties. We assign each party P_i to Q_i, for all $i \in [n]$. This quorum is responsible for receiving P_i's input x_i and choosing a random value r_i on behalf of P_i. Now, let $\mathcal{C}$ denote the probabilistic sorting network of [27] and $m = \Theta(n \log n)$ be the number of gates in $\mathcal{C}$.

For all $j \in [m]$, we assign the j-th gate of $\mathcal{C}$ to $Q_{(j \bmod n)}$. This quorum is later used for secure evaluation of the gate's functionality. For simplicity of notation, we assume the quorums associated with the output gates of $\mathcal{C}$ are $Q_1, ..., Q_{\lceil n/2 \rceil}$.[4] When used to receive inputs of $\mathcal{M}$, we refer to $Q_1, ..., Q_n$ as *input quorums*. When used to send outputs of $\mathcal{M}$ to all parties, we refer to $Q_1, ..., Q_{\lceil n/2 \rceil}$ as *output quorums*.

Creating the probabilistic sorting circuit $\mathcal{C}$ requires $O(\log^2 n)$ random bits known to all parties. We generate these bits by asking one of the quorums to agree on a sequence R of $O(\log^2 n)$ random bits, and then send R to all parties via a binary tree of quorums. This randomness is then used by the parties to agree on the structure of $\mathcal{C}$ using the random butterfly tournament procedure described in [27].

To ensure privacy, every quorum in $\mathcal{M}$ receives and maintains its inputs in a secret-shared format, *i.e.*, each party receives only a share of each input rather than the actual input. Moreover, all computations in these quorums are performed over secret-shared values. When we say a party *VSS-shares* (or *secret-shares*) a value s in a quorum Q (or among a set of parties), we mean the party participates as the dealer with input s in the protocol VSS-Share with all parties in Q (or in the set of parties). As a result, the parties agree on a random polynomial $f(x)$ such that $f(0) = s$, and the i-th party receives $f(i)$ as his verified share of s.

---

[4] Note that a quorum can be *re-used* any number of times for local computations as long as its inputs for each use are secret-shared independently from other uses.

Protocol 1. Secure Multi-Party Shuffling Scheme

Inputs. For all $i \in [n]$, party P_i holds an input x_i. Let $\mathcal{C}$ denote the probabilistic sorting network of [27] and d denote its depth.

Goal. Parties jointly compute a random shuffle of their inputs.

The protocol:

1. **Setup.**
   (a) Parties run Build-Quorums to agree on n quorums $Q_1, ..., Q_n$.
   (b) Parties in Q_1 run Gen-Rand and VSS-Reconst repeatedly to generate a sequence R of $\Theta(\log^2 n)$ random bits.
   (c) Parties in Q_1 send R to all other quorums in the following way. For all $i \in \{2, .., n\}$, parties in Q_i receive R from $Q_{\lfloor i/2 \rfloor}$, and then send it to all parties in Q_{2i} and Q_{2i+1}.
   (d) For all $i \in [n]$ and $j \in [m]$, parties assign Q_i to P_i and $Q_{(j \bmod n)}$ to the j-th gate of $\mathcal{C}$, and connect the gates based on the random butterfly tournament described in [27] and the random sequence R.

2. **Input Sharing.** Party P_i VSS-shares his input x_i with Q_i.

3. **Random Generation.** Parties in input quorum Q_i perform the following steps:
   (a) Run Gen-Rand to generate a random secret-shared value $r_i \in \mathbb{F}_q$, where $q > \frac{3}{2} kn^2 \log n$ for any $k > 0$.
   (b) Run Renew-Shares to send the secret-shared pair (r_i, x_i) to $Q_{\lceil i/2 \rceil}$.

4. **Sorting.** $\mathcal{C}$ is evaluated level-by-level starting from the input gates. For each gate G in $\mathcal{C}$ and quorum Q assigned to G, parties in Q perform the following steps:
   (a) **Comparison.** Let (r, x) and (r', x') be the secret-shared inputs of G. The parties run Compare to securely compare the secret-shared values r, r'. Let $\rho = (r \leq r')$ be the resulting secret-shared value. The parties compute the output secret-shared pairs (s, y) and (s', y') from

$$s = \rho \cdot r + (1 - \rho) \cdot r', \qquad y = \rho \cdot x + (1 - \rho) \cdot x'$$
$$s' = \rho \cdot r' + (1 - \rho) \cdot r, \qquad y' = \rho \cdot x' + (1 - \rho) \cdot x$$

   For every addition of secret-shared values a, b, parties locally compute $a + b$. For every multiplication, they run Multiply.
   (b) **Output Resharing.** Parties run Renew-Shares to send secret-shared values s, y, s', y' to the parent quorum.

5. **Output Delivery.** For all $i \in [n-1]$, let (s_i, y_i) and (s_{i+1}, y_{i+1}) be the pairs of secret-shared values the output quorum $Q_{\lceil i/2 \rceil}$ computes in the previous step.
   (a) Each party in this quorum sends his share of y_i to party P_i and his share of y_{i+1} to party P_{i+1}.
   (b) Parties P_i and P_{i+1} run VSS-Reconst to reconstruct y_i and y_{i+1} respectively.

Protocol 1 shows our main protocol, where $\mathcal{M}$ is evaluated level-by-level until the final outputs are generated by the output quorums. For all $i \in [n]$, parties in the output quorum Q_i send their shares directly to P_i who reconstructs the corresponding secret $x_{\pi(i)}$, where π denotes the permutation generated by the circuit.

It is left to implement two subprotocols used in Protocol 1: Renew-Shares and Ran-Gen. In Section 3.1, we describe Renew-Shares as a protocol that allows parties of a quorum to securely send a secret-shared value to parties of another quorum. In [1], we describe Ran-Gen as a protocol that allows a group of parties to agree on a uniformly random value. We prove the security of Protocol 1 (and Theorem 1) in [1]. In particular, we show that for sufficiently large $k > 0$ and $q = \Omega(kn^2 \log n)$, this protocol provides almost t-secure MPS with high probability.

3.1 Share Renewal

Once the computation of each gate is finished, parties in the quorum associated with that gate send the secret-shared result to any quorums associated with gates that need this result as input. Let Q denote a quorum at which the computation of a gate has finished, and let Q' denote a quorum that requires the output of that computation. In order to secret-share the result to Q' without revealing any information to any individual party (or to any coalition of corrupted parties in both Q and Q'), a *fresh sharing* of the result must be distributed in Q'. If s is secret-shared using a polynomial $f(x)$ of degree t, then a fresh sharing of s is a new secret sharing of s defined using another polynomial $g(x)$ of degree t chosen uniformly and independently at random. We refer to the problem of generating a fresh sharing of a secret-shared value among a new set of parties as *share renewal*.

Handling the share renewal problem efficiently and robustly is challenging. Dani *et al.* [17] solve it by masking the result in Q using a fresh random value and unmasking it in Q'. Although their approach is secure against up to $T < N/3$ corrupted parties in each quorum, they do not provide an explicit construction and simple constructions seem very expensive in terms of both communication and computation costs.[5]

Boyle *et al.* [9] overcome this problem by sending encrypted inputs to only one quorum which does all of the computation using fully-homomorphic encryption. This is not load-balanced, as it incurs a large computation and communication overhead to parties in that quorum. Zamani *et al.* [33] propose a simple technique for this problem that is, unfortunately, secure only against up to $T < N/6$ corrupted parties in each quorum.

We now describe a novel technique for share renewal that is secure against up to $T < N/3$ corrupted parties in each quorum. Let s denote the output of Q that is secret-shared among parties in Q using a random polynomial $f(x)$ of degree t. Our technique is based on the observation that if every share of s is *reshared* using a fresh random polynomial, then a specific linear combination of the new shares defines a new random polynomial $g(x)$ such that $g(0) = s$.

---

[5] Their approach relies on the existence of an unmasking circuit securely evaluated by parties in Q'. Such a circuit must implement an error-correcting technique which requires many multiplication gates.

Protocol 2. Renew-Shares

Inputs. A set of parties $P_1, ..., P_N$ jointly hold a secret-shared value s, *i.e.*, a polynomial $f(x)$ of degree $T < N/3$ is defined such that $f(0) = s$, and for all $i \in [N]$, P_i holds $f(i)$.

Goal. Generate a fresh sharing of s among another group of parties $P'_1, ..., P'_N$. This means that the protocol must calculate a polynomial $g(x)$ of degree T uniformly and independently at random such that $g(0) = s$, and for all $j \in [N]$, P'_j holds $g(j)$.

The protocol:

1. Each party P_i runs **Reshare** to VSS-share $f(i)$ among $P'_1, ..., P'_N$ using a random polynomial $h_i(x)$ of degree T such that $h_i(0) = f(i)$.

2. Each party P'_j locally computes its share of s from $g(j) = \sum_{i=1}^{N} \lambda_i h_i(j)$.

This was first observed by Gennaro *et al.* [19] as a simple method for polynomial randomization and degree-reduction in the multiplication protocol of [7].

Let $g(x) = s + a_1 x + ... + a_T x^T$. Our goal is to calculate the coefficients $a_1, ..., a_T$. Following [19], we write

$$\begin{bmatrix} 1 & 1 & \cdots & 1 \\ 1 & 2 & \cdots & 2^{N-1} \\ \vdots & & & \\ 1 & N & \cdots & N^{N-1} \end{bmatrix} \begin{bmatrix} s \\ a_1 \\ \vdots \\ a_N \end{bmatrix} = \begin{bmatrix} f(1) \\ f(2) \\ \vdots \\ f(N) \end{bmatrix},$$

where $a_{T+1}, \cdots, a_N = 0$. The matrix above is an N-by-N Vandermonde matrix that is non-singular and hence is invertible. Let $\begin{bmatrix} \lambda_1 & \lambda_2 & \cdots & \lambda_N \end{bmatrix}$ be the first row of the inverse matrix. Thus, $s = \lambda_1 f(1) + ... + \lambda_N f(N)$. For all $i \in [N]$, consider a fresh polynomial $h_i(x)$ of degree T, where $h_i(0) = f(i)$. We define $g(x) = \sum_{i=1}^{N} \lambda_i h_i(x)$. Since $g(0) = \lambda_1 f(1) + ... + \lambda_N f(N) = s$, the polynomial $g(x)$ defines a fresh sharing of s. Using this, we define our share renewal protocol **Renew-Shares** in Protocol 2.

In the first step of **Renew-Shares**, we ask each party to reshare its share $f(i)$ by running a protocol called **Reshare**. This protocol ensures that every corrupted party shares its correct share $f(i)$ instead of some random or maliciously-chosen value. Asharov and Lindell [5] implement a protocol (called *subshare*) that ensures this resharing process is done robustly. We refer to this protocol as **Reshare**. In [1], we prove **Renew-Shares** is UC-secure against at most $T < N/3$ corrupted parties in each quorum.

3.2   Remarks

In the following, we discuss alternative approaches that could be used to design different MPS protocols from Protocol 1.

All-Output MPS. Protocol 1 describes a single-output MPS construction, where each party receives only one element of the output sequence. Although this is useful in many applications such as data-oblivious protocols that often use MPS as an intermediate step, an all-output MPS protocol can be used in some applications such as anonymous broadcast. To achieve all-output MPS, the output delivery step of Protocol 1 becomes as follows. For all $i \in [n-1]$, parties in the output quorum $Q_{\lceil i/2 \rceil}$ run VSS-Reconst to reconstruct y_i and y_{i+1} and then send (y_i, y_{i+1}) to all n parties. Each party receiving a set of N pairs from each output quorum, chooses one pair via majority filtering and considers it as the output of that quorum.

Remark on Deterministic Sorting Networks. While the probabilistic sorting network of [27] is sufficient for us to achieve an almost t-secure MPS with logarithmic latency (Theorem 1), one can instead use a deterministic sorting network such as those of [3,6] to achieve t-secure MPS (*i.e.*, uniform shuffling) at the expense of increased latency, communication, and computation costs. We are not aware of a sorting network that can result in better asymptotic and practical costs than the sorting network of [27] in terms of latency, communication, and computation costs.

Remark on Permutation Networks. One approach for solving MPS is to securely evaluate a *permutation network* instead of obliviously sorting random values. A permutation network is a network of *swappers*, where each swapper is a gate with two inputs and two outputs; it permutes the inputs randomly with probability $1/2$. A permutation network with n input wires is typically used to generate a random permutation of n values. A network consisting entirely of switches with swapping probability of $1/2$ cannot generate uniform permutations of n values, because for a network with m swappers, there are 2^m different outcomes. Since $n!$ is not a power of 2, some of the possible $n!$ permutations are generated with higher probability than others.

Waksman [32] suggests an $O(n \log n)$ time and memory algorithm for generating unbiased permutations. The idea is to first choose a permutation uniformly at random and then compute a proper setting of swappers that represents the permutation. Unfortunately, it is not clear how this algorithm can be implemented efficiently in a load-balanced multi-party setting. Czumaj *et al.* [15] propose a permutation network with $O(1/n^2)$ statistical distance from the uniform distribution. To the best of our knowledge, this network provides the smallest distance among known networks with $\mathsf{polylog}(n)$ depth. Still, this result cannot be used to achieve an almost t-secure MPS (as in Definition 2) because in worst case, the adversary can guess the correct permutation with probability $1/n! + O(1/n^2)$ that is $\omega(1/n!)$.

3.3 Security Proofs Outline

The error probability in Theorem 1 comes entirely from the following steps of Protocol 1 failing to output correct results with some probability:

- *Setup:* Protocol Build-Quorums may fail to create good quorums. Theorem 3 shows this failure happens with probability $o(1)$.

- *Random Generation:* It is possible that two or more input quorums choose exactly the same random elements from $\mathbb{F}_q$. Such collisions increase the probability that the adversary can correctly guess the secret permutation generated by the protocol. In [1], we prove that, for sufficiently large q, failure due to collisions happens with probability $o(1)$.

– *Sorting:* The circuit of [27] may fail to sort correctly with probability $o(1)$.

All other components of our protocol are deterministic and thus have no error probability. For simplicity, we assume the three steps above return without failure. However, even assuming the sorting step of Protocol 1 returns without failure, the adversary can still take advantage of the *a priori* knowledge that a σ fraction of the input permutations are never sorted by the circuit, to reduce the set of possible input permutations; thus increasing his chance of correctly guessing the secret permutation. In [1], we show this *a priori* knowledge increases the chance of the adversary in correctly guessing the secret permutation by only a small (*i.e.*, $o(1)$) amount. Hence, Protocol 1 achieves an almost t-secure MPS.

4   Simulation Results

To study the feasibility of our scheme and compare it to previous work, we simulated a proof-of-concept prototype of our protocol (and the cryptographic variant described in [1]) along with two others that are based on a similar model to ours. These protocols are due to Dani *et al.* [17] and Boyle *et al.* [9]. To the best of our knowledge, these protocols are the most efficient in terms of communication cost, computation cost, and the number of rounds for large networks. Since the protocols of [17] and [9] are general MPC algorithms, we use them for computing our (single-output) shuffling functionality described in Section 3. We are interested in evaluating our protocols for large networks; thus, our choice of protocols for this section is based on their scalability for large values of n.

We set the parameters of our protocols in such a way that we ensure the probability of error for the quorum building algorithm of [10] is smaller than 10^{-5}. For the sorting circuit, we set $k = 2$ to get $\sigma < 10^{-8}$ for all values of n in the experiment. Clearly, for larger values of n, the error becomes superpolynomially smaller, *e.g.*, for $n = 2^{25}$, we get $\sigma < 10^{-300}$. For all protocols evaluated in this section, we assume cheating (by corrupted parties) happens in every round of the protocols. This is done by having $t = \lfloor n/3 \rfloor$ of the parties send random message in every round of the protocols.

Figure 2 illustrates the simulation results obtained for various network sizes between 2^5 and 2^{30} (*i.e.*, between 32 and about 1 billion). To get a system-independent estimation of the computation costs, we implemented a wrapper that counts the number of processor instructions evaluated during the execution of each protocol. We repeat each experiment five times and report the average for each network size. To better compare the protocols, the vertical and horizontal axis of all plots are scaled logarithmically.

In Figure 2, we report results from three different versions of our protocols. The first plot (marked with triangles) refers to our unconditionally-secure protocol (Protocol 1). The second plot (marked with circles) represents the cryptographic variant of Protocol 1 described in [1]. The third plot (marked with diamonds) shows the cost of our unconditionally-secure protocol with amortized (averaged) setup cost. To obtain this plot, we run the setup phase of Protocol 1

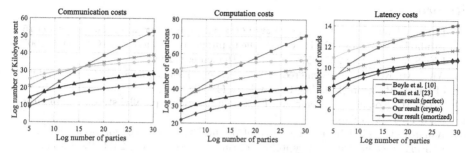

Fig. 2. Simulation results

once and then use the setup data to run the online protocol 100 times. The total number of bits sent was then divided by 100 to get the average cost.

We observe that our protocol performs significantly better than the prior work. For example, for $n = 2^{15}$ (about 33 thousand parties), our amortized protocol requires each party to send about 128MB of data, while the protocols of [9] and [17] each send more than one terabyte of data per party. For the computation cost, our amortized protocol requires each party to perform about one billion operations, while the other protocols require each party to perform more than 10^{13} operations. Finally, our amortized protocol requires about 500 rounds of communication, while the protocols of [9] and [17] require about 1500 and 4100 rounds of communication respectively.

5   Conclusion

We described a multi-party shuffling protocol that is fully decentralized and tolerates up to $t < (1/3 - \epsilon)n$ malicious faults. Moreover, our protocol is load-balanced and can tolerate traffic-analysis attacks. The amount of information sent and the number of computations performed by each party scales polylogarithmically with the number of parties. Scalability is achieved by performing local communications and computations in groups of logarithmic size.

Several open problems remain. First, can we decrease the number of rounds of our protocol using a smaller-depth sorting circuit? For example, since our protocol sorts uniform random numbers, it seems possible to use a smaller depth non-comparison-based sorting circuit like bucket sort. Second, can we improve performance even further by detecting and blacklisting parties that exhibit adversarial behavior? Finally, can we adopt our results to the asynchronous model of communication? We believe that this is possible for a suitably chosen upper bound on the fraction of faulty parties.

Acknowledgment. The authors would like to acknowledge supports from NSF under grants CCF-1320994, CCR-0313160, and CAREER Award 644058. We are also grateful for valuable comments from Ran Canetti (Boston University), Bryan Ford (Yale University), Shafi Goldwasser (MIT), Aniket Kate (Purdue University), Yehuda Lindell (Bar-Ilan University), and anonymous reviewers of this paper.

References

1. Extended version of this paper. http://cs.unm.edu/~zamani/papers/sirocco15
2. Adida, B., Wikström, D.: How to shuffle in public. In: Vadhan, S.P. (ed.) TCC 2007. LNCS, vol. 4392, pp. 555–574. Springer, Heidelberg (2007)
3. Ajtai, M., Komlós, J., Szemerédi, E.: An $0(n \log n)$ sorting network. In: Proceedings of STOC 1983, pp. 1–9. ACM, New York (1983)
4. Ajtai, M., Komlós, J., Szemerédi, E.: Sorting in $c \log n$ parallel steps. Combinatorica 3(1), 1–19 (1983)
5. Asharov, G., Lindell, Y.: A full proof of the BGW protocol for perfectly-secure multiparty computation. Cryptology ePrint Archive, Report 2011/136 (2011)
6. Batcher, K.E.: Sorting networks and their applications. In: Proceedings of the April 30–May 2, 1968, Spring Joint Computer Conference. AFIPS '68 (Spring), pp. 307–314. ACM, New York (1968)
7. Ben-Or, M., Goldwasser, S., Wigderson, A.: Completeness theorems for noncryptographic fault-tolerant distributed computing. In: Proceedings of the Twentieth ACM Symposium on the Theory of Computing (STOC), pp. 1–10 (1988)
8. Berman, R., Fiat, A., Ta-Shma, A.: Provable unlinkability against traffic analysis. In: Juels, A. (ed.) FC 2004. LNCS, vol. 3110, pp. 266–280. Springer, Heidelberg (2004)
9. Boyle, E., Goldwasser, S., Tessaro, S.: Communication locality in secure multiparty computation. In: Sahai, A. (ed.) TCC 2013. LNCS, vol. 7785, pp. 356–376. Springer, Heidelberg (2013)
10. Braud-Santoni, N., Guerraoui, R., Huc, F.: Fast Byzantine agreement. In: Proceedings of the 2013 ACM Symposium on Principles of Distributed Computing, PODC 2013, pp. 57–64. ACM, New York (2013)
11. Canetti, R.: Universally composable security: a new paradigm for cryptographic protocols. In: Proceedings of the 42nd Annual Symposium on Foundations of Computer Science, FOCS 2001, pp. 136–145, October 2001
12. Chaum, D.: Untraceable electronic mail, return addresses, and digital pseudonyms. Commun. ACM 24(2), 84–90 (1981)
13. Chaum, D.: The dining cryptographers problem: Unconditional sender and recipient untraceability. Journal of Cryptology 1, 65–75 (1988)
14. Corrigan-Gibbs, H., Wolinsky, D.I., Ford, B.: Proactively accountable anonymous messaging in verdict. In: Proceedings of the 22nd USENIX Security Symposium, Berkeley, CA, USA, pp. 147–162 (2013)
15. Czumaj, A., Kanarek, P., Lorys, K., Kutylowski, M.: Switching networks for generating random permutations (2001)
16. Damgård, I.B., Fitzi, M., Kiltz, E., Nielsen, J.B., Toft, T.: Unconditionally secure constant-rounds multi-party computation for equality, comparison, bits and exponentiation. In: Halevi, S., Rabin, T. (eds.) TCC 2006. LNCS, vol. 3876, pp. 285–304. Springer, Heidelberg (2006)
17. Dani, V., King, V., Movahedi, M., Saia, J.: Brief announcement: breaking the $o(nm)$ bit barrier, secure multiparty computation with a static adversary. In: Proceedings of the 2012 ACM Symposium on Principles of Distributed Computing, PODC 2012, pp. 227–228. ACM, New York (2012)
18. Frank, S., Anderson, R.: The cocaine auction protocol: On the power of anonymous broadcast. In: Pfitzmann, A. (ed.) IH 1999. LNCS, vol. 1768, pp. 434–447. Springer, Heidelberg (2000)

19. Gennaro, R., Rabin, M.O., Rabin, T.: Simplified VSS and fast-track multiparty computations with applications to threshold cryptography. In: Proceedings of the Seventeenth Annual ACM Symposium on Principles of Distributed Computing, PODC 1998, pp. 101–111. ACM, New York (1998)

20. Goldreich, O., Ostrovsky, R.: Software protection and simulation on oblivious RAMs. J. ACM 43(3), 431–473 (1996)

21. Golle, P., Juels, A.: Dining cryptographers revisited. In: Cachin, C., Camenisch, J.L. (eds.) EUROCRYPT 2004. LNCS, vol. 3027, pp. 456–473. Springer, Heidelberg (2004)

22. Goodrich, M.T.: Randomized shellsort: A simple data-oblivious sorting algorithm. J. ACM 58(6), 27:1–27:26 (2011)

23. Gruteser, M., Grunwald, D.: Anonymous usage of location-based services through spatial and temporal cloaking. In: Proceedings of the 1st International Conference on Mobile Systems, Applications and Services, MobiSys 2003, pp. 31–42. ACM, New York (2003)

24. Hamada, K., Kikuchi, R., Ikarashi, D., Chida, K., Takahashi, K.: Practically efficient multi-party sorting protocols from comparison sort algorithms. In: Kwon, T., Lee, M.-K., Kwon, D. (eds.) ICISC 2012. LNCS, vol. 7839, pp. 202–216. Springer, Heidelberg (2013)

25. King, V., Lonargan, S., Saia, J., Trehan, A.: Load balanced scalable byzantine agreement through quorum building, with full information. In: Aguilera, M.K., Yu, H., Vaidya, N.H., Srinivasan, V., Choudhury, R.R. (eds.) ICDCN 2011. LNCS, vol. 6522, pp. 203–214. Springer, Heidelberg (2011)

26. Laur, S., Willemson, J., Zhang, B.: Round-efficient oblivious database manipulation. In: Lai, X., Zhou, J., Li, H. (eds.) ISC 2011. LNCS, vol. 7001, pp. 262–277. Springer, Heidelberg (2011)

27. Leighton, T., Plaxton, C.G.: A (fairly) simple circuit that (usually) sorts. In: Proceedings of the 31st Annual Symposium on Foundations of Computer Science, FOCS 1990, pp. 264–274, October 1990

28. Neff, C.A.: A verifiable secret shuffle and its application to e-voting. In: Proceedings of the 8th ACM Conference on Computer and Communications Security, CCS 2001, pp. 116–125. ACM, New York (2001)

29. Pfitzmann, A., Waidner, M.: Networks without user observability – design options. In: Pichler, F. (ed.) EUROCRYPT 1985. LNCS, vol. 219, pp. 245–253. Springer, Heidelberg (1986)

30. Rackoff, C., Simon, D.R.: Cryptographic defense against traffic analysis. In: Proceedings of the Twenty-Fifth Annual ACM Symposium on Theory of Computing, STOC 1993, pp. 672–681. ACM, New York (1993)

31. von Ahn, L., Bortz, A., Hopper, N.J.: k-anonymous message transmission. In: Proceedings of the 10th ACM Conference on Computer and Communications Security, CCS 2003, pp. 122–130. ACM, New York (2003)

32. Waksman, A.: A permutation network. J. ACM 15(1), 159–163 (1968)

33. Zamani, M., Movahedi, M., Saia, J.: Millions of millionaires: Multiparty computation in large networks. Cryptology ePrint Archive, Report 2014/149 (2014)

34. Zhang, B.: Generic constant-round oblivious sorting algorithm for MPC. In: Boyen, X., Chen, X. (eds.) ProvSec 2011. LNCS, vol. 6980, pp. 240–256. Springer, Heidelberg (2011)

Author Index

Printed in the United States
By Bookmasters